Size- and Age-Related Changes in Tree Structure and Function

Tree Physiology

Volume 4

Series Editors: Frederick C. Meinzer and Ülo Niinemets

For further volumes:
http://www.springer.com/series/6644

Frederick C. Meinzer • Barbara Lachenbruch
Todd E. Dawson
Editors

Size- and Age-Related Changes in Tree Structure and Function

Springer

Editors
Dr. Frederick C. Meinzer
USDA Forest Service
Pacific Northwest Research Station
Corvallis, OR 97331
3200 SW Jefferson Way
USA
fmeinzer@fs.fed.us

Dr. Todd E. Dawson
Department of Integrative Biology
University of California, Berkeley
4006 Valley Life Sciences Bldg.
Berkeley, California
USA
tdawson@berkeley.edu

Dr. Barbara Lachenbruch
Department of Wood Science
and Engineering
Oregon State University
Richardson Hall 119
Corvallis, OR 97331
USA
barb.lachenbruch@oregonstate.edu

ISSN 1568-2544
ISBN 978-94-007-1241-6 e-ISBN 978-94-007-1242-3
DOI 10.1007/978-94-007-1242-3
Springer Dordrecht Heidelberg London New York

Library of Congress Control Number: 2011930866

Printed on acid-free paper

Springer is part of Springer Science+Business Media (www.springer.com)

Preface

Trees are globally important because of their roles in providing fuel, food, construction materials and fiber, and their more recently recognized role in providing diverse ecosystem services such as habitat for other organisms and the oxygen we all breathe. They typically increase in size and biomass by several orders of magnitude from the seedling to adult stage. Across this growth trajectory they undergo an enormous number of changes both within their own structure, and in their interactions with the environment. Some of these changes are well-known, and others we are just beginning to understand. These changes can be in features as diverse as foliar physiology, anatomy, and chemistry; wood structure and mechanical properties; overall architecture and allometry; reproductive ability; and interactions with herbivores. Many of these changes are presumed to allow them to acclimate to the environment and endure for millennia; indeed, the ability of certain tree species to attain spectacular heights has captivated society's imagination for centuries and has driven an increasing amount of research on the biophysical factors that limit maximum tree height. The ability to acclimate and endure, however, can also restrict future response options because of a legacy of structural adjustments that constrain physiological function. Despite the importance of these changes, until recently relatively little was known about their specific causes and the extent to which they are related strictly to tree size versus age. This book presents a synthesis of the current state of knowledge about the causes and consequences of ontogenetic changes in key features of structure and function at both the individual tree and the stand level.

The book explains why it is important to distinguish among attributes intrinsically related to size versus age and why we know so little about age- versus size-specific traits. Chapters highlight some of the implications of these size- and age-related changes for commercial forestry plantations with shortening rotational ages, and for predicting how current and future forests will respond to climate and other environmental changes.

The organization of this book reflects a philosophy that universal principles are likely to govern key processes and relationships between structure and function regardless of a species' evolutionary history and where it normally grows. Thus, angiosperms and gymnosperms as well as temperate and tropical species are often discussed in the same chapter. We have grouped the chapters into six sections beginning with a chapter that provides an historical perspective on prior research and relevant technological developments as well as a synthesis of the major topics and themes covered in the remaining chapters. This is followed by five sections organized according to processes and phenomena: reproduction and regeneration, integration of structure and function across ontogeny, photosynthetic gas exchange and carbon balance, water relations and transport processes, and scaling from individual traits to stands. Several chapters describe technological tools and innovative methodological approaches that can be used to address specific topics. Although much progress has been made in unravelling structure-function relationships in trees and forest stands during their ontogeny, important gaps in our knowledge need to be filled in order to understand the constraints on sustaining trees as viable components of ecosystems under current and future climate regimes.

We are grateful to the contributors for their willingness to participate in this project and to internal and external reviewers for their valuable comments on initial drafts of the chapters.

Frederick C. Meinzer
Barbara Lachenbruch
Todd E. Dawson

Contents

Contributors

Niels P.R. Anten Ecology and Biodiversity Group, Institute of Environmental Biology, Utrecht University, P.O. Box 800.84, 3508 TB, Utrecht, The Netherlands, N.P.R.Anten@uu.nl

Kasey E. Barton Botany Department, University of Hawai'i at Mānoa, 3190 Maile Way, St. John 101 Honolulu, HI, 96822, USA, kbarton@hawaii.edu

Karina Boege Instituto de Ecología, Universidad Nacional Autónoma de México, Apartado Postal 70–275, Ciudad Universitaria, UNAM, 04510 Mexico D.F., Mexico, kboege@servidor.unam.mx

Barbara J. Bond Department of Forest Ecosystems and Society, Oregon State University, 330 Richardson Hall, Corvallis, OR 97331, USA, barbara.bond@oregonstate.edu

Ben Bond-Lamberty Joint Global Change Research Institute, 8400 Baltimore Avenue, Suite 201, College Park, MD 20740, USA, bondlamberty@pnnl.gov

Sandra J. Bucci Consejo Ncional de Investigaciones Cientificas y Tecnicas (CONICET), Argentina

Departamento de Biología, Facultad de Ciencias Naturales, Universidad Nacional de la Patagonia San Juan Bosco, Comodoro Rivadavia, Argentina, sj_bucci@yahoo.com

Jan Čermák Faculty of Forestry and Wood Technology, Mendel University of Agriculture and Forestry Brno, Brno 61300, Czech Republic, Cermak@mendelu.cz

Todd E. Dawson Departments of Integrative Biology & Environmental Science, Policy and Management, University of California – Berkeley, Berkeley, CA 94720, USA, tdawson@berkeley.edu

Michael E. Day School of Forest Resources, University of Maine, Nutting Hall, Orono, ME 04469–5755, USA, daym@maine.edu

Lee T. Dickman Earth and Environmental Sciences Division, Los Alamos National Lab, MS-J495, Los Alamos, NM 87544, USA

Rodolfo Dirzo Department of Biology, Stanford University, 371 Serra Mall, Stanford CA 94305, USA, rdirzo@stanford.edu

Robert Evans CSIRO Materials Science and Engineering, Clayton, VIC, Australia

Brent E. Ewers Department of Botany, Program in Ecology, University of Wyoming, 1000 E. University Ave., 3165, Laramie, WY 82071, USA, beewers@uwyo.edu

Willa Fouts Division of Biological Sciences, The University of Montana, Missoula, MT 59812, USA

Guillermo Goldstein Consejo Ncional de Investigaciones Cientificas y Tecnicas (CONICET), Argentina

Laboratorio de Ecología Funcional, Departamento de Ecología, Genética y Evolución, Facultad de Ciencias Exactas y Naturales, Universidad de Buenos Aires, Buenos Aires, Argentina

Department of Biology, University of Miami, P.O. Box 249118, Coral Gables, FL 33124, USA, gold@bio.miami.edu

Michael S. Greenwood School of Forest Resources, University of Maine, Nutting Hall, Orono, ME 04469–5755, USA, greenwd@umenfa@maine.edu

Thomas M. Hinckley School of Forest Resources, University of Washington, Seattle, WA 98195, USA, hinckley@u.washington.edu

Günter Hoch Institute of Botany, University of Basel, Schönbeinstrasse 6, CH-4056 Basel, Switzerland, guenter.hoch@unibas.ch

Teemu Hölttä Department of Forest Ecology, University of Helsinki, Helsinki, Finland

Hiroaki Ishii Department of Plant Science, Graduate School of Agricultural Science, Kobe University, Kobe 657-8501, Japan, hishii@swallow.kobe-u.ac.jp

Daniel M. Johnson Department of Environmental and Plant Biology, Ohio University, Athens, OH 45701, USA, johnsod6@ohio.edu

David A. King Biological and Ecological Engineering, Oregon State University, Corvallis, OR 97331, USA, dkingaz@yahoo.com

Barbara Lachenbruch Department of Wood Science and Engineering, Oregon State University, Corvallis, OR 97331, USA, Barb.Lachenbruch@oregonstate.edu

D. Scott Mackay Department of Geography, State University of New York at Buffalo, 105 Wilkeson Quadrangle, Buffalo, NY 14261, USA

Jordi Martinez-Vilalta Centre for Ecology Research and Forestry Applications (CREAF), Universitat Autonoma de Barcelona, Bellaterra, Cerdanyola, Barcelona, Spain

Katherine A. McCulloh Department of Wood Science and Engineering, Oregon State University, Corvallis, OR 97331, USA, kate.mcculloh@oregonstate.edu

Nate G. McDowell Earth and Environmental Sciences Division, Los Alamos National Lab, MS-J495, Los Alamos, NM 87544, USA, mcdowell@lanl.gov

Frederick C. Meinzer USDA Forest Service, Pacific Northwest Research Station, 3200 SW Jefferson Way, Corvallis, OR 97331, USA, fmeinzer@fs.fed.us

Maurizio Mencuccini School of Geosciences, University of Edinburgh, Crew Building, West Mains Road, Edinburgh, EH9 3JN, UK

Centre for Ecology Research and Forestry Applications (CREAF), Universitat Autonoma de Barcelona, Bellaterra, Cerdanyola, Barcelona, Spain, m.mencuccini@ed.ac.uk

John R. Moore Forest Products Research Institute, Edinburgh Napier University, Edinburgh EH10 5DT, Scotland, UK

Nadezhda Nadezhdina Faculty of Forestry and Wood Technology, Mendel University of Agriculture and Forestry Brno, Brno 61300, Czech Republic

Ülo Niinemets Institute of Agricultural and Environmental Sciences, Estonian University of Life Sciences, Kreutzwaldi 1, Tartu 51014, Estonia

Nathan G. Phillips Department of Geography and Environment, Boston University, 675 Commonwealth Avenue, Boston, MA 02215, USA

Keith Reinhardt Department of Biological Sciences, Idaho State University, Pocatello, ID 83209, USA

Michael G. Ryan Laboratory for Studies of the Forest Carbon Cycle, USDA Forest Service, Rocky Mountain Research Station, 240 West Prospect Road, Fort Collins, CO 80526, USA

Colorado State University, Fort Collins, CO 80523, USA, mgryan@fs.fed.us

Anna Sala Division of Biological Sciences, The University of Montana, Missoula, MT 59812, USA, sala@mso.umt.edu

Fabian G. Scholz Consejo Ncional de Investigaciones Cientificas y Tecnicas (CONICET), Argentina

Departamento de Biología, Facultad de Ciencias Naturales, Universidad Nacional de la Patagonia San Juan Bosco, Comodoro Rivadavia, Argentina, fgscholz@unpata.edu.ar; fgscholz@yahoo.com

N. Galia Selaya Ecology and Biodiversity Group, Institute of Environmental Biology, Utrecht University, P.O. Box 800.84, 3508 TB, Utrecht, The Netherlands

Programa Manejo de Bosques de la Amazonía Boliviana (PROMAB), Casilla 107, Riberalta, Bolivia

Kathy Steppe Laboratory of Plant Ecology, Faculty of Bioscience Engineering, Ghent University, Coupure links 653, B-9000 Ghent, Belgium, Kathy.Steppe@UGent.be

Robert O. Teskey Daniel B. Warnell School of Forestry and Natural Resources, University of Georgia, Athens, GA 30602, USA

Sean C. Thomas Faculty of Forestry, University of Toronto, 33 Willcocks St., Toronto, ON M5S 3B3, Canada, sc.thomas@utoronto.ca

Steven L. Voelker Department of Wood Science and Engineering, Oregon State University, Corvallis, OR 97331, USA, steve.voelker@oregonstate.edu

David Whitehead Landcare Research, P.O. Box 40, Lincoln 7640, New Zealand

David R. Woodruff USDA Forest Service, Pacific Northwest Research Station, 3200 SW Jefferson Way, Corvallis, OR 97331, USA, dwoodruff@fs.fed.us

Part I
The Whole-Lifespan Perspective on Integration of Structure and Function

Chapter 1
A Lifespan Perspective on Integrating Structure and Function in Trees

Thomas M. Hinckley, Barbara Lachenbruch, Frederick C. Meinzer, and Todd E. Dawson

Abstract Trees are spectacular organisms that can accumulate a large amount of biomass, live for millennia, grow in stressful environments, and have global importance to organisms and human society as a result of their roles in contributing to diverse ecosystem services. They survive in an enormous range of environments with disturbances of various frequencies and severities, and by their very nature of being long-lived and sessile, they must possess the ability to change. This chapter provides a context for the chapters that follow, by describing a history of the investigations and of several important technological innovations that have enabled research to progress, and some of the foundations of water relations which underpin many of the later chapters. The chapter then discusses seven recurrent themes brought out by the diverse chapter authors: microclimate and other abiotic forces that change with tree and stand age; the changing structure and function employed as trees grow; what we can learn from scaling —which scaling laws pertain, when, and what is the biology or physics embedded in these laws; the tradeoffs that occur as structures and functions change through growth and development; the causes and

T.M. Hinckley (✉)
School of Forest Resources, University of Washington, Seattle, WA 98195, USA
e-mail: hinckley@u.washington.edu

B. Lachenbruch
Department of Wood Science and Engineering, Oregon State University, Corvallis, OR 97331, USA

F.C. Meinzer
USDA Forest Service, Pacific Northwest Research Station, 3200 SW Jefferson Way, Corvallis, OR 97331, USA

T.E. Dawson
Departments of Integrative Biology & Environmental Science, Policy and Management, University of California – Berkeley, Berkeley, CA 94720, USA

F.C. Meinzer et al. (eds.), *Size- and Age-Related Changes in Tree Structure and Function*, Tree Physiology 4, DOI 10.1007/978-94-007-1242-3_1,

correlates of the decline in productivity of trees and stands with age; the extent to which changes are plastic; and the meaning of whether changes occur relative to age versus size. The chapter concludes with a discussion on the emerging issues and questions raised in the chapters of this book. The overall aim is to give insights into the whole tree, whole lifespan perspective of the structural and functional strategies trees employ to persist in the face of internal and external factors that change as trees grow and age. This knowledge can be used to develop management approaches to manipulate trees and stands to provide a wide range of ecosystem goods and services.

1 Introduction

Using 17 chapters authored by 42 individuals, "Size- and Age-related Changes in Tree Structure and Function" captures the current state of knowledge regarding how age and size impact tree structure and function. The current chapter lays the foundations for this book with a historical perspective on how scientific curiosity, forest management, wood science, and technological innovations have led to advances in our understanding of tree and stand growth. It then gives a brief explanation of the basic plant water relations theory essential for understanding aspects of hydraulic architecture, water transport, growth and growth allometry dealt with in a number of the chapters. Next, it presents the major themes that were touched upon in the individual contributions, and lastly, the emerging issues and researchable questions on size- and age-related changes in the structure and function of trees.

Trees are spectacular organisms: they can accumulate a very substantial amount of biomass, can live for millennia, can grow in extraordinarily stressful environments, are globally important as a result of their current and historical roles in providing fuel and fiber and their more recent and emerging roles in providing diverse ecosystem services including clean air and water, conserving and cycling nutrients, capturing and storing carbon, synthesizing an extensive array of chemicals including pharmaceuticals, having living and dead structures that provide habitat for a diverse array of organisms and serving special cultural and spiritual needs. As a tree develops from seed, it expands progressively from a newly germinated seedling weighing a few milligrams to a large, complex organism containing over 3,000 Mg of dry biomass (Fig. 1.1). These and other special features of trees as compared to other plant and animal organisms have earned them a unique status throughout the history of civilization.

Because of their diverse evolutionary history, trees extend from all but the coldest and driest habitats through all terrestrial systems; and because of their ability to acclimate, they can endure for millennia. Collectively, trees employ an enormous range of reproductive, regenerative, and survival strategies. By their very nature of being long-lived and sessile, they must possess the ability to change. This change encompasses the genetic potential to alter both structure and function in association with ontogeny, as well as with abiotic and biotic factors that change

Fig. 1.1 Foliage at various life-stages in Douglas-fir (*Pseudotsuga menziesii* var. *menziesii*) (*left* to *right* are cotyledons in a newly germinated seed, foliage of a young tree, and foliage from the top of an old-growth tree)

either independently or concomitantly with the aging tree and stand. The chapters in this book provide a context for understanding the features associated with "success," and give insights into the mechanisms and strategies that facilitate their persistence in the face of these changing internal and external factors. We can then use this knowledge to predict tree and stand responses to perturbations, and to devise management approaches to manipulate trees and stands to provide a wide range of ecosystem goods and services.

The whole tree can be defined as a system of relatively fine organs (foliage and fine roots) that acquire and allocate resources and that are linked via a coarser system of roots, stems, branches and the bole to create a particular and often complex form. Changes in architectural form often result in trade-offs between structure and function that may scale according to orderly rules that we may or may not understand. These rules often begin with the trade-offs between architectural constraints and architectural opportunities, hydraulic and biomechanical safety and efficiency, vegetative versus reproductive growth, and growth rate versus longevity (e.g., Zimmermann 1983; Kennedy et al. 2010). Therefore, the visible form that a tree takes is the result of combined architectural and process rules (i.e., the hardwired or genetic component) and their interplay over time with the combined abiotic and biotic environment, which in itself changes dramatically within a day, within a crown, within a year, and over the life of the tree. During these changes in their abiotic and biotic environments, trees are responding through growth of height, diameter, crown, and roots; through physiological transitions from juvenile to reproductive and mature life history stages, and through developmental and plastic responses to a wide range of stimuli. They are also accumulating the structural and physiological legacy that potentially will restrict or alter their future response options.

In this chapter, we have attempted to present our perspective on tree growth, structure, and function to allow readers to more readily integrate the chapters as they read them. The next three sections focus on the history of whole lifespan tree and stand ecophysiology (Sect. 2), the role of technological innovations (Sect. 3), and the foundations of plant water relations research (Sect. 4). They provide context for the findings, innovations, and connections made by authors in later chapters. Section 5 focuses on seven themes that recur in many of the chapters. We suggest this recurrence indicates the current importance of these thematic areas for understanding tree function. The next section (Sect. 6) compiles the emerging issues and questions raised in the chapters of this book. We conclude (Sect. 7) with a brief assessment of the potential utility and broader applications of the findings discussed in the remaining chapters.

2 History of Investigations into Effects of Size and Age on Trees

A starting point for the history of tree physiology or silvics might be traced back to Leonardo da Vinci who puzzled about the relationship between the ratio of the sum of twig cross-sectional areas to small branch cross-sectional areas to branch cross-sectional areas to stem cross-sectional area in an apple tree. This example of an allometric relationship has currency whether one is using the pipe theory (Shinozaki et al. 1964), the West-Brown-Enquist power relationships (West et al. 1997, 1999), or the emerging conduit packing density approach (McCulloh et al. 2010; Savage et al. 2010) when addressing conduction through a plant or through the entire soil-plant-atmosphere continuum (SPAC). Similar to da Vinci, authors interested in the allometry of hydraulics have tended to focus on plant segments and even the junctions between segments (e.g., Zimmermann 1978, 1983) rather than the whole plant. The work of Stephen Hales is often cited as the first example of the use of experiments with woody plants to test the applicability of a physical or a vitalistic theory of how water is transported within a woody plant (Hales 1727). Individuals following Hales tended to focus on finding a robust and parsimonious explanation for how water flowed through the SPAC, what controlled its flow, and how it was able through turgor to sustain cell enlargement and control stomatal opening. Early authors, such as Bruno Huber (1924) focused on merging structural – functional relationships of tree hydraulics. All these approaches have either directly or indirectly provided much of the foundation for the way scientists think today about trees from a water and carbon perspective.

Similar to those interested in the basics of tree function and structure, practicing foresters soon recognized that tree height was dependent upon age, species and the quality of the site on which the trees were growing. Site index curves, which now exist for many species (Fig. 1.2) clearly illustrate that there is a maximum height for any given species even when growing on the best of sites. In addition, certain species, such as Douglas-fir (*Pseudotsuga menziesii*), coastal redwood (*Sequoia sempervirens*) and mountain ash (*Eucalyptus regnans*) have long been noted for their ability to achieve spectacular heights and perhaps it is the existence of very tall trees

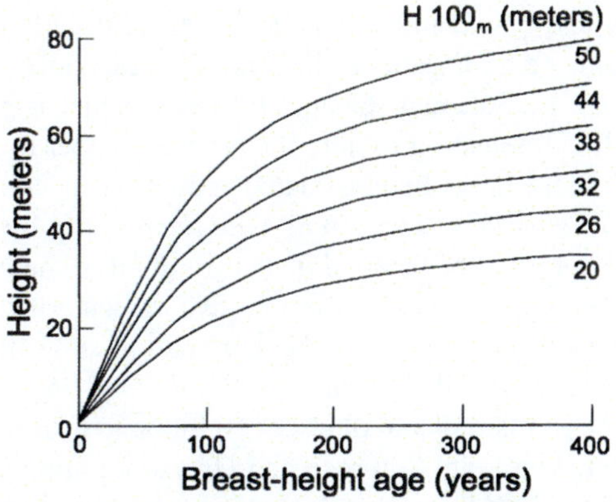

Fig. 1.2 Site index (or age versus height) curves for Douglas-fir (*Pseudotsuga menziesii* var. *menziesii*) growing in high elevation forests of the Cascades of Oregon and Washington (from DeMars and Herman 1987). Although each curve appears to be approaching an asymptote, curves have not flattened even at age 400 (see Chaps. 4, 6, 8 and 14 about height growth in Douglas-fir and other trees). It is important to note that the sample size was relatively small for trees 300 years old or older (21/240). Similar site index curves were noted for Douglas-fir growing at lower elevations on the eastside of the Cascades (Means and Helm 1985). In contrast, the range of heights at age 100 for Douglas-fir (*P. menziesii* var. *glauca*) from New Mexico was from 9 m on the poorest sites to 36.5 m on the best (Edminster and Jump 1976). Such curves have strong predictive powers, but very weak explanatory capabilities

that brings practitioners and experimenters together. Greenhill in 1881 published his still-cited paper that brings together diameter and taper to explain the form in which ship masts (tree trunks) can be found in nature, as well as how tall a tree can grow and still be stable. At about the same time, in 1895, Dixon and Joly described the Cohesion-Tension theory for water transport, which has been challenged multiple times (e.g., Zimmermann et al. 2004). Finally, after a series of ingenious experiments in the mid-1990s and a 2004 letter to the *New Phytologist* (Angeles et al. 2004), the cohesion-tension theory was accepted as the most robust and parsimonious explanation of how water moves to the top of trees, short or tall.

While the argument over the mechanism of water transport to the tops of tall trees was occurring, two elegant papers appearing in the early 1990s (Schoettle 1994; Yoder et al. 1994) demonstrated the effects of age (actually, height in these studies, which can be used as a proxy for age) on tree structure and function. Indeed, old (i.e., tall) trees were structurally different from younger trees and showed reduced photosynthesis. These observations were consistent with the declines in productivity with stand age that have been noted in the existing literature (e.g., Odum 1969; Grier et al. 1981; Gower et al. 1996; Ryan et al. 1997; see reviews by Binkley et al. 2002 and Ryan et al. 2010). Ryan and Yoder (1997) narrowed the explanation for the observations from these two 1994 papers (and others) to what is now call the hydraulic limitation hypothesis. Examinations of this hypothesis have increased in number, some supporting the original ideas and some not (Barnard and Ryan 2003; Bond 2000; Koch et al. 2004; Mencuccini et al. 2005; Burgess et al. 2006;

Burgess and Dawson 2007; Sala and Hoch 2009; Petit et al. 2011) including several chapters in this volume (e.g., Ewers et al. Chap. 18, Lachenbruch et al. Chap. 5, Mencuccini et al. Chap. 12, Sala et al. Chap. 11, Steppe et al. Chap. 9, Woodruff and Meinzer Chap. 14). It is interesting to note that a recent paper by Sillett and others (2010) suggests that there is no height-related reduction in volume increment in very large (and old) *Sequoia sempervirens* and *Eucalyptus regnans* trees. They reached this conclusion by a very detailed inventory of all of the growth increments that included all of the appendages in the massive crowns of these trees, which had not been considered previously. At the publication time of this book, these investigations continue, particularly with coast redwood (*S. sempervirens*) and further support the notion that tall and old trees continue to grow at the highest rates they ever have (Sillett, Van Pelt, Ambrose and Dawson, unpublished data). Many of the chapters in this book address the role that age may play in the observations of structure and function in tall trees. There is clearly a significant series of research questions that remain largely unanswered regarding (1) the biochemical and biophysical mechanisms that govern the interaction between age and height and how they individually and in combination result in the responses and morphologies observed and (2) a sufficient understanding of the interactions between the regulators of carbon pools, fluxes, and source – sink relationships as they play out over both small and large distances. Advancing knowledge in these areas and other will require new approaches and technology innovations as discussed next.

3 Technological Innovations

Early scientists were able to lay foundational elements that still stand today (e.g., Hales 1727; Dixon 1895; Huber 1924, etc.). However, a number of new techniques and technologies have given researchers improved access to trees and stands allowing characterization of all components in the soil-plant-atmosphere continuum including the soil (e.g., TDR, multi-sensor, frequency domain capacitance), the root system including mycorrhizal relationships (e.g., the high pressure flow meter developed by Tyree et al. (1995)), and the bole and crown. Access to tall trees was a particular problem; most often solved using mountaineering or arborist techniques. Triangular TV tower sections and scaffolding were also used, but were often relegated to canopies 45 m or less (e.g., Woodman 1971; Hinckley and Bruckerhoff 1975; Martin et al. 2001). With the advent (and use) of large construction cranes, finances became the main limitation to canopy access (e.g., Shaw et al. 2004). Many of the chapters in this book illustrate the use of a wide-range of canopy access technologies (e.g., ropes: Ishii Chap. 8; canopy construction crane: Chap. 14).

Sap flux and hydraulic conductivity can now be assessed in foliage (e.g. Brodribb and Holbrook 2003; Bucci et al. 2003) and in very small to large branches, roots, and stems (e.g., Čermák and Kucera 1981; Granier 1987; Salleo et al. 1992; Sperry et al. 1994; Burgess et al. 1998; Spicer and Gartner 1998; Čermák et al. 2007). We have methods to assess tissue water relations (e.g., Scholander—Hammel pressure chamber, Scholander et al. 1965) and stomatal conductance (e.g., the null balance porometer: Beardsell et al. 1972; Parkinson and Legg 1972 or the leaf or shoot cuvette: Ritchie 1969; Field et al. 1982). In addition, there has been a concomitant development of infrared techniques to measure the exchange of physiologically important gases such as CO_2 and water vapor at the tissue and canopy (e.g. eddy covariance, Baldocchi 2003) scales. Analysis of stable isotopes of carbon and oxygen in CO_2 and in the hydrogen and oxygen of liquid water and water vapor allow inferences concerning physiological controls on gas exchange, sources and sinks for carbon and water and transit and residence times of water in trees (Dawson et al. 2002; Meinzer et al. 2006; Dawson and Simonin 2011). These isotope techniques have allowed investigators to measure easily, rapidly and frequently the factors associated with the movement and use of carbon and water by trees. Two of the chapters in this book explore and document the value of technology and demonstrate how some unknowns may be cracked via an imaginative application of existing technologies (e.g., chapters by Čermák and Nadezhdina Chap. 15, McDowell et al. Chap. 10). Johnson et al. (Chap. 3) explore the other end of the size spectrum and discuss the need for adapting technology for measuring physiology of the seedlings.

Technology is a two edged sword – it presence and development provide powerful ways to acquire important information and to address previously difficult to impossible questions. On the other hand, science driven by technology alone is often handicapped in scope and imagination. Several chapters in this book illustrate how relatively simple measurement techniques, when applied to interesting questions, result in powerful studies.

Another area in which technological advances have had a tremendous impact has been in characterization of wood properties. Until the 1960s, measurements of within-plant density were done with tedious direct methods to capture mass and volume of variously-sized samples; precision depended largely on how small one could cut up a tree and how well one could measure volume. In the 1960s, however, Polge (1978) introduced x-ray densitometry in which slabs or cores cut to a precise thickness were x-rayed; the intensity of the x-ray was related to the density of the material. Refinements of this technique have made it possible to document radial trends in wood density of trees and to compile data for different heights in trees or to compare trees from different silvicultural treatments. With the proper software, these systems have allowed people to compile ring-by ring information on earlywood and latewood density as well as ring width and proportion of the ring that is latewood. Such work has been instrumental in tree breeding programs and wood quality studies (Cown et al. 2004), and maximum ring density has been used as an index of drought severity in dendrochronological studies (Cleaveland 1974; Yasue et al. 1997)

A further major advance in the study of xylem structure was the introduction of SilviScan and then SilviScan-2 (Evans 1999, 2006; Evans and Ilic 2001), a facility that scans precision-dimensioned microbeams with x-rays, light, and x-ray diffraction techniques at pre-determined step-sizes. In addition to giving the information from a typical x-ray densitometer, it produces estimates of fiber coarseness (related to cell wall thickness and lumen diameter) and orientation of the microfibrils in the thickest layer of the cell walls. With density and microfibril angle, a computer algorithm can then estimate the wood's stiffness (modulus of elasticity) for each measurement location, typically giving radial profiles of wood properties. These advances in wood properties have made it possible for people to relate mechanical and hydraulic function to structure at a much finer scale than was possible in the past (e.g., Lachenbruch et al. 2010). Other work implementing light- and electron microscopy techniques has allowed variation in structural features of xylem conduits to be linked to variation in functional traits such as hydraulic conductivity and resistance to embolism (e.g. Pittermann et al. 2005; Wheeler et al. 2005; Domec et al. 2006; Hacke et al. 2006; Choat et al. 2008). Application of cryo-scanning electron microscopy to the study of xylem water transport has yielded important new insights on the dynamics and potential mechanisms of embolism formation and reversal in both primary and secondary xylem (e.g. Canny 1997; McCully et al. 1998; Melcher et al. 2001; Johnson et al. 2009).

4 Foundations of Plant Water Relations

For this book, it is useful to revisit the key concepts of plant water relations because much of the discussion about the role of age/height depends upon understanding how water potential varies, the biotic and abiotic factors that influence it, and how it and the tree's ability to modify it might create limits to height and reduce productivity. Qualitatively, changes in plant water status may be considered proportional to the difference between water gained through the roots (and/or redistributed from other tissues, see Chap. 13) and that transpired by the foliage (Jarvis 1975) with the additional and important caveat that no water flow from roots or storage can occur until there is a water potential gradient. Thermodynamically defined, total water potential (Ψ) is a measure of the capacity of water at a particular point to do work as compared with free, pure water (Slatyer 1967). Water potential gradients, as indicators of differential energy states of water, provide the driving force for the movement of liquid water as well as for water vapor throughout the soil-plant-atmosphere continuum (SPAC). Mathematically, Ψ can be expressed as:

$$\Psi = \left(\mu_w - \mu_w{}^o\right) / V_w = \Psi_p + \Psi_\pi + \Psi_\tau \tag{1.1}$$

where μ_w and $\mu_w{}^o$ are the chemical potential of water in the system and that of free, pure water, respectively; V_w is the partial molal volume of water; Ψ_p is the pressure potential (or turgor pressure or xylem tension); Ψ_π is the osmotic potential (due to solutes); and Ψ_τ is the matric potential (due to surface forces, e.g., cell walls, soil particles).

As suggested in Eq. 1.1, Ψ is measured in terms of energy per unit volume which is equivalent to a force per unit area (commonly expressed in dynes cm^{-2}, bars or MPa). The pressure term (Ψ_p) can be positive while the others are always negative since these reduce the capacity of pure water to do work. Since the water potential of pure water at standard temperature and pressure is defined as zero, Ψ is usually equal to or less than zero, except during periods of plant exudation (or guttation) or periods of positive root pressure, which are of negligible importance to most woody plants (Kramer 1969). Thus, Eq. 1.1 provides a framework within which all the component potentials may be accounted for when the plant properties, which lead to internal adjustments of Ψ, are discussed (e.g., Chap. 3, Chap. 14). Although Eq. 1.1 shows the solution for an instantaneous measurement at a single point in space and time, temporal and spatial dynamics can be easily added when considering either a Höfler diagram or the results of pressure-volume curves (Tyree and Hammel 1972).

A much more useful equation for the discussion of temporal and spatial variation in water potential at any point "x" in a tree can be defined as:

$$\Psi_x = \Psi_{soil} - \rho gh - \Sigma_{soil}{}^{x} f_i r_i \tag{1.2}$$

where ρgh is the gravitational potential (a product of the density of water, the gravity constant, and the height above water in the root zone, respectively; equal to about –0.01 MPa m^{-1} height). The frictional potential, $\Sigma_{soil}{}^{x} f_i r_i$, represents the summation of the products of the partial fluxes (f_i) and the partial resistance (r_i) along the pathway(s) from the soil to point "x" in the tree (Richter 1972, 1973). The water potential of the bulk soil is indicated by Ψ_{soil}. It should be remembered that for a given point "x" and time period in the tree, the bulk soil and gravitation potential are fixed (the soil value can change with time and the gravitation with height above the ground). The frictional potential is the dynamic part of this equation. It is also important to note that trees can undergo adjustments in their hydraulic architecture to partly compensate for the increasing frictional component as they get taller, but this has no effect on the gravitational component other than to increase it as a result of the increase in height. Therefore, gravity is always felt even when water is not moving and thus is not a hydraulic constraint in the usual sense of the term. An interesting note is that Eq. 1.2 was proposed by Richter after reading (and translating) the original work of Bruno Huber (1924), indicating that we continue to rely on this conceptual framing of water potential in the crowns of tall trees.

Equation 1.2 was derived by Richter (1973) from a consideration of water flow through the SPAC using the Ohm's Law analogy where flow (q) is equated to the gradient divided by the pathway resistance (Huber 1924; van den Honert 1948). A popular, pragmatically useful, but mechanistically incorrect formulation of this analogy may be written (after van den Honert 1948);

$$q = \Delta\Psi_{soil\text{-}root} / R_{soil\text{-}root} = \Delta\Psi_{root\text{-}leaf} \cdot / R_{root\text{-}leaf} = \Delta\Psi_{leaf\text{-}atmos} / R_{leaf\text{-}atmosphere} \tag{1.3}$$

where R is the resistance along a particular pathway. Although the use of the Ohm's Law analogy, especially as offered by van den Honert (1948) where he viewed water flow through the SPAC as a catena (or chain), has aided in the understanding and modeling of water flow through the SPAC (e.g., see Elfving et al. 1972; Hinckley and Bruckerhoff 1975; Alarcon et al. 2003; Vasconcelos et al. 2010), several precautions must be taken to avoid erroneous applications of this concept. First, the ability of various parts of the tree to act as sources and/or sinks (i.e., capacitors) is not accounted for in this formulation. Second, an assumption of steady-state, non-elastic flow is made in which water movement through the stomata matches water uptake through the roots, hence the idea of a catena. Clearly, this cannot and does not happen. Third, stomatal gas and liquid phase resistances are not analogous, and phase changes at the leaf complicate $R_{leaf\text{-}atmos}$. Fourth, and probably most important, the Ohm's Law analogy only states that total water flow through all the roots, all the conduits of the stem, all the leaves is equal; it does not imply that flows through undefined sections of a single conduit running through the tree are numerically equal (Richter 1973). Much more sophisticated models, still initially based upon this simplified view of water flow, exist and can provide important insights into how different hydraulic conductances (1/resistance) can influence flow and vulnerability to embolism (Sperry et al. 1998). Models using very different approaches have also been successful (e.g., Aumann and Ford 2006; Lovisolo et al. 2010).

Several chapters build from these earlier and more simplistic foundations of plant water relations. It is often worth reverting to these earlier, perhaps simpler relationships in order to understand the relationships being used in these chapters. Finally, Hinckley (2004) provided a review of whole-tree physiology. The five topics there are worthy of review: (1) the nature and definition of the individual organism, (2) carbon, water, and nutrient acquisition, use, and distribution, (3) how trees and specifically large trees biophysically and biochemically "perceive" their internal and external environment and how then responses are mediated or signaled, (4) scaling and a critique of the use of branches as a surrogate for whole tree responses, and (5) the potential for "domestication" of trees.

5 Recurrent Themes in Book

These 17 chapters encompass many universal themes for the study of age- and size-related changes in tree structure and function and whether ontogenetic variation is size- or age-related. Recurrent themes include the microclimate experienced by trees of different stature and age, the extent to which morphological or functional traits change with age or size, the extent to which these changes follow simple scaling principles, tradeoffs of one trait or several traits for another trait or groups of traits, the decline of growth with tree or stand age, plastic vs. fixed control of traits with age or size, and the concept of age versus size relations for tree function.

5.1 Microclimate

The coverage of this book spans tree growth and development from day-old seedlings to towering giants. As a tree increases in size from a germinant to a large mature individual, its aboveground parts occupy different environments both due to growth into new microenvironments, and to changes it causes in its own environment such as self-shading (e.g., Woodman 1971; Lewis et al. 2000; Ishii et al. 2008, Chap. 10).

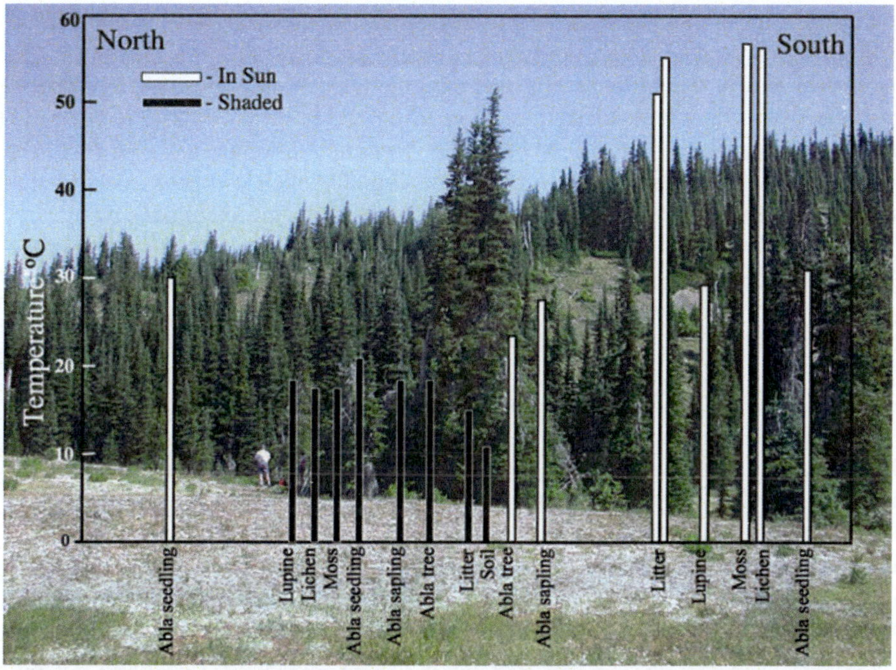

Fig. 1.3 Data are from a subalpine meadow in NE part of Olympic National Park, USA, July 28, 2005. The meadows have a 10% slope to the south and they are at 1,591 m on the SW side of a ridge that tends from the NW to the SE at Hurricane Ridge. Temperature measurements were taken with an infrared thermometer between 1,315 and 1,335 h (PDT); air temperature was between 24.1°C and 24.8°C. Measured temperatures ranged from 10.8°C to 5 cm beneath the soil surface under the sub-alpine clump to 56.3°C for dry, sunlit moss/lichen crusts on the south side of the clump. *White bars* represent things measured in direct sunlight, whereas *black bars* are for shaded measurements. Plant temperatures were largely based upon measurements of foliage temperatures

The distribution and microenvironment of belowground parts also change as touched on in Chap. 15. Chapter 3 details the extreme temperature, light and moisture regimes that new germinants may experience, as well as the patchiness of their environment in both space and time. It is not difficult to imagine the kinds of differences in micro- and mesoclimate that a tree might experience over its life, from day to night, and from summer to winter. Over the life of a bristlecone bristlecone pine tree (*Pinus aristata*), it may experience almost 5,000 years of climate, and in its mature stages, it may retain foliage that was produced 28 years before. Nonetheless, microclimatic differences among seedlings or plant parts may be as pronounced as the annual differences a tree might experience in 5 millennia. A newly germinated subalpine fir (*Abies lasiocarpa*) seedling located at 1,600 m in a subalpine meadow in the Olympic National Park, Washington (Waring and Franklin 1979) would experience surface temperatures over 50°C (Fig. 1.3, see also Seymour et al. 1983), temperatures that can kill the cambium (thus girdling the developing seedling) and that exceed the upper temperature threshold for positive carbon uptake leading to foliage death (Seymour et al. 1983; Teskey et al. 1984). Shaded plants and plants with sufficiently thick bark are not subject to such exposures. The impact of the harsh microclimate near the ground is illustrated by the fates of thousands of *Abies lasiocarpa* seeds that have been tracked for 17 growing season (Soll 1994, A. Woodward, pers. comm.) in subalpine meadows in the Olympic National Park (see Fig. 1.3 for one of the study meadows). There were initially 696±469 seeds per square meter of substrate from which 12.6 seeds proceeded to the germinant stage, 9.1 survived by the end of the first growth season, and 0.43 were alive 17 years later. At this last census date, the average seedling was only 12 cm-tall while the tallest is just over 30 cm. Germination and early survival were influenced by both soil aridity and temperature.

Mature trees are faced with large within canopy gradients of light and temperature. Evergreen trees must both modify (i.e., acclimate) existing foliage and produce somewhat different new foliage as the older, lower, or more interior foliage becomes increasingly shaded (Brooks et al. 1994, 1996). In addition to light, gradients in carbon dioxide, air and foliage temperature, and humidity exist affecting the overstory and understory vegetation as well as the epiphytes and other organisms occupying these different canopy zones. For example, Garrett and Cox (1978) observed differences as great as 100 ppm between early morning and mid-day in forest carbon dioxide levels collected between 2.8 and 18.3 m in a Missouri deciduous hardwood forest. Carbon dioxide levels between 700 and 450 ppm near ground level at sunrise are not uncommon in many forest types and these values can drop to below 280 ppm in the mid to lower canopy at mid-day. Similar, but even much more pronounced gradients were noted by K. Kavanagh (unpubl. data) in a coniferous forest in Northern Idaho and also in a Northern California redwood forest (Tu and Dawson, in review). When air temperature and relative humidity at different positions in the canopy are converted to vapor pressure deficit (VPD), vertical gradients in evaporative demand are observed. Depending upon the openness of the canopy and the atmospheric conditions, very shallow or very steep gradients in VPD are noted. Near the tops of coast redwoods and giant sequoia, VPDs of about 3–4 kPa, respectively, are observed; decreasing steeply to about 0.6–1.5 kPa within 50 m of the top (Fig. 1.4). Tropical lowland and montane forests also exhibit vertical gradients, but not as steep as the two California species (Fig. 1.4). The relatively shallow gradient observed in the old-growth Douglas-fir-western hemlock forest (Fig. 1.4) is likely the result of two factors: (1) measurements were taken from a tower versus within the crowns of the trees and (2) this forest has a very strong rumple factor increasing the likelihood of strong turbulent mixing and entraining of above canopy air.

In addition to the changes imposed on their own environment as trees increase in size, the resulting increase in mass must be supported physically. This support includes not only the requirements to avoid breakage from the static load of the tree, including its foliage, reproductive structures, and occasionally snow and ice loads, but also to avoid breakage in the wind. The forces of the shifting crown extend from the upper roots to the branch tip, and act on the wood from the center of the tree (which was formed when the tree was smaller) all the way out to the surface. The impact of increasing size on biomechanics is discussed in Chap. 5.

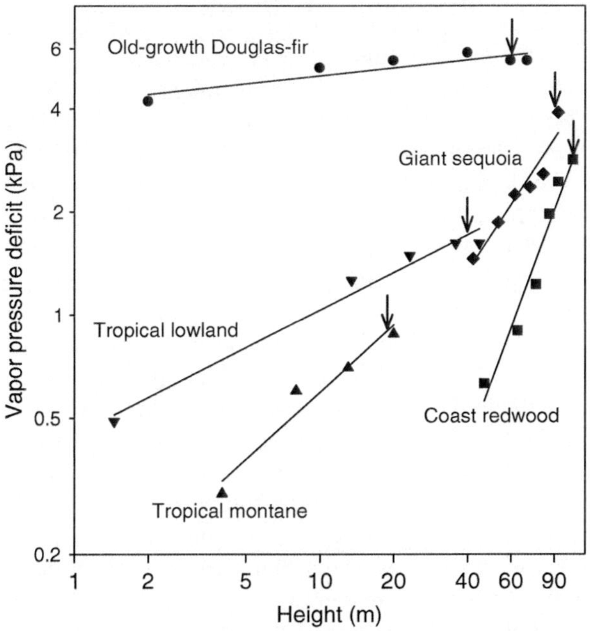

Fig. 1.4 Variation in vapor pressure deficit with height in five forest types differing in maximum canopy height: an old-growth Douglas-fir-western hemlock (*Pseudotsuga menziesii-Tsuga heterophylla*) forest in southwestern Washington State, USA, an old-growth giant sequoia (*Sequoiadendron giganteum*) forest in the Sierra Nevada Mountains, USA (Dawson, unpublished data), a coast redwood (*Sequoia sempervirens*) forest in California (Dawson, unpublished data), a tropical lowland forest in Brazil (Data from Roberts et al. 1990), and a tropical montane forest in Ecuador (Data from Motzer et al. 2005). *Arrows* indicate approximate maximum canopy height. Note that plot axes have log scales. Data were obtained during the dry season in each forest type

5.2 Changes in Structure, Function and Functional Relationships

Almost every chapter discusses the morphological and physiological adjustments that occur as trees get larger and/or older. Many changes are recorded in tree structure. Lachenbruch et al. (Chap. 5) propose using the radial variation in wood structure to tease out the hydraulic and biomechanical factors of importance to individuals through their growth trajectory. Similarly, Voelker (Chap. 17) uses the radial wood record present in tree rings together with regional climate to infer the factors of importance for radial growth across the lifespan of the trees. He showed, for example, that relative humidity explained half as much variation in radial growth for old as young trees, and that the age-dependency of growth varies by habitat and is much stronger at treeline than within the interior of a tree's geographic range. Another change is the enormous switch from the structure and physiology of cotyledons to that of the true leaves, which may be associated with a spike in mortality (Chap. 3). Ishii (Chap. 8) uses the record of the canopy's structure to explain that young trees with their hierarchical structures will have less buffering capacity to disturbances than will older, larger tree with their polyarchic structure, which is essentially a population of linked hierarchical structures. Similarly, Voelker (Chap. 17) discusses that the growth of old trees seems more buffered from large variations in climate than that of young trees. McDowell et al. (Chap. 10) used a record of isotopic composition rather than physical structure to infer past physiology. They found a universal decline in stable carbon isotope discrimination with increasing height for all foliar and wood datasets. This decline occurred at a rate different from what would be expected from the effects of gravity acting on the water column alone, giving sound support to the idea of changing physiology with height.

Other changes occur by virtue of size alone, such as the fact that large trees have more self-shading inside their crowns than small trees (Chap. 10). Another example is that small trees should have more synchronized within-crown induction response to herbivory than large trees because the transmission of the signaling compounds will occur over shorter distances (Boege et al. Chap. 7). If distal water-transport conduits tend to have smaller diameters than proximal conduits (called conduit tapering), and if distal conduit diameter at the leaf petiole is fixed at a minimum level by natural selection and biophysical forces, then the size of conduits at the plant's base will be determined by plant size (Mencuccini et al. Chap. 12). Mencuccini et al. also note the increased cost of construction of a long transport system as trees get taller. Another set of examples of changes that occur by virtue of size comes from the increase in tension in the water column as trees get taller that occurs both because of the greater path length, and also from the increased hydrostatic force (the effect of gravity on the taller water column). The increase in xylem tension high in the tree causes a decrease in the turgor that can be maintained during tissue expansion, causing permanent reductions in xylem conduit and other cell sizes leading to decreased hydraulic conductivity and increased leaf mass per area and its resultant reductions in growth efficiency, increased resistance to CO_2 diffusion to the carboxylation sites, and finally, a decrease in photosynthetic rate (Chap. 9; Chap. 14; Ambrose et al. 2009, 2010; Cavaleri et al. 2010; Oldham et al. 2010). The impact of height-related reductions in turgor and leaf expansion was originally and experimentally demonstrated by Richter (1974); by using overlapping saw cuts: he was able to transform a branch that produced shade morpho-type needles to one that produced sun types. Superimposed on the approximately linear gradient of decreasing water potential with increasing height is an exponential gradient of increasing light both of which affect vertical gradients in foliar morphology and physiology (Chap. 8). McDowell et al. (Chap. 10) use a meta-analysis of stable isotope data in tall trees to separate the hydrostatic from light effects on photosynthesis and leaf morphology.

Most of the other examples of changes in structure, function, or functional relationships are those that involve structural or physiological changes that mitigate some effect of the tree's increasing size. The apparency of a plant to herbivores will generally increase as a plant grows and develops (Chap. 7). It would be expected that trees have an increased tolerance to herbivory as they grow, and/or have dispersal and germination strategies to mitigate their increased apparency. Two such strategies are germination close to the parent, assuming the parent is more herbivore-resistant than the germinant, or germination at long distance from the parent, increasing its chances of moving away from the herbivores. The production of epicormic branches (from buds that had long been dormant) in some species appears to partially compensate for the increasing resistance for water flow in large old trees (Chap. 8) and possibly also compensates for the increasing ratio of respiration to photosynthesis in these large trees. Lastly, tall, old trees appear to have a higher hydraulic capacitance, permitting greater reliance on stored water to transiently replace transpirational losses and buffer fluctuations in xylem tension (Scholz et al. Chap. 13).

At the stand level, it is appropriate to ask which traits can best predict tree function with age. Ewers et al. (Chap. 18) showed that getting the correct estimate of stand stomatal conductance is key, and that either leaf area or sapwood area have more accurate predictive capability than sap flux per xylem area. Succession can be viewed as an age-related change at the stand level, and Anten and Selaya (Chap. 16) show that the light level is the most important environmental change through succession in tropical forests. Early successional species have adaptations that are useful in the highly resource competitive environment in which they live (thin leaves, strong shade avoidance response, thin wide-ranging roots). Late successional species have adaptations that are useful in the more variable light environment in which they grow but that also enable them to survive the stresses of the lower resource environment.

Some of the changes in structure or function that occur throughout the tree's lifetime are related to changes in its biotic environment. The clearest examples of this sort of change come from the relationship of ontogeny to plant defenses against herbivory (Chap. 7). Indirect defenses include traits that facilitate the visitation or colonization by mutualistic animals that defend plants against herbivores. The reliance on these indirect defenses can change during ontogeny because a tree may need to be sufficiently large to acquire mutualists. Similarly, seedlings may have facilitative or competitive interactions with other trees through a variety of factors including pathogens and mycorrhizae (Chap. 3). Another example of biotic factors that may shape the age- or size-related changes in trees is differential effect herbivory may have on reproductive success: female plants, bearing a higher reproductive effort, may be less tolerant of herbivory than male plants (Chap. 2). Lastly, height relative to surrounding vegetation is often more important than absolute height: consider the greater importance for males than females to be tall for pollen transfer (Chap. 2) and consider the effects of being an emergent tree on the risk of wind damage (Chap. 4) , and the effects of shading from its own and adjacent crowns. The preceding examples represent situations in which changes in functional traits and functional relationships during growth could not be studied in isolation from biotic factors of the environment.

5.3 Scaling of Structure and Function with Size

Another theme in many of the chapters is that some processes scale with tree size, whereas other processes occur at different sizes or ages. Mencuccini et al. (Chap. 12) gave the example of the pipe model, which scales relatively well with size (as leaf area/stem cross-sectional area) only when a stem contains no heartwood. They also summarize eight models for long-distance transport in xylem or phloem, telling explicitly whether the model is scalable by tree size, and explaining the manner in which functions are scaled. Day and Greenwood (Chap. 4) stated that there is no predictable ontogenetic trajectory for a number of physiological traits such as photosynthetic rate, water use efficiency, specific leaf mass, ratio of shoot to root mass and metabolic efficiency. Such lack of correlation means that age-related scaling models may need input from all ontogenetic stages. King (Chap. 6) showed that the aboveground mass allocation of trees appears to follow different scaling rules at different life stages.

With regard to scaling of resilience to disturbances, large trees and stands of large trees may have a greater capacity to buffer themselves than smaller trees and stands of smaller young trees. Therefore, it would be inappropriate to scale up the response of a young stand to a perturbation to that of an old stand of large trees. Not only do young and old trees and stands have different responses to disturbance, but Ewers et al. (Chap. 18) reported that stand age has a large effect on forest transpiration. For example, accurate modeling of stand transpiration requires substantially different old-growth and young stand values of the canopy average reference stomatal conductance and average sensitivity of canopy conductance to vapor pressure deficit.

5.4 Tradeoffs

The changes in tree structure and function that occur throughout growth (Sect. 5.2) are presumably those that give the highest fitness through promotion of survival and reproduction, regardless of whether they follow simple scaling rules (Sect. 5.3). Many of these changes are likely to involve tradeoffs that may be difficult to uncover. These tradeoffs are considered explicitly in some scientific formulations and are implicitly assumed without considering alternatives in others. In an explicit discussion of the optimal design of a transport system for tall trees Mencuccini et al. (Chap. 12) suggest that among the answers of what should be maximized are total conductance of the tree or organ-level efficiency; tree growth; total net carbon gain; and tree level efficiency of use of a particular resource. Such explicit consideration can move the field forward tremendously. We can also be misled by preconceptions that a trait is traded off for another when in reality entirely different factors may be of importance (Chaps. 2 and 5). Moreover, many features have more than one role in a tree, so it is a challenge to take multi-functionality into account as in the case of the cost of construction of a xylem conduit wall, which functions both in hydraulic reinforcement and mechanical stability of the tree (Chap. 12).

Adaptations and tradeoffs that serve an individual well at one life stage may be mal-adaptive to survival at another life-stage (Chap. 4). A similar comparison could be made involving "young" (early successional) species and "old" (late successional) species. The early successional species tend to have a high allocation to leaves, enabling them to have high assimilation rates for a period of time. The late successional species, with lower leaf allocation and more allocation to mechanical support have slower growth but then have a longer life span, for they can maintain higher assimilation rates in the later years of life than can the early successional species that are mal-adapted to the late-successional habitat (Chap. 16).

5.5 *Decline in Productivity of Individual Trees and Stands*

Many chapters discuss and evaluate hypothetical causes for the well-documented decline in productivity of individual trees and stands with age. These hypotheses include declining photosynthetic rates (Chap. 11) the age-related increasing cost of reproduction (Thomas Chap. 2) or of structural support (Chap. 6). The decreasing photosynthetic rate observed in taller trees could result from many factors, including a feedback inhibition of photosynthesis resulting from reduced sink demand for carbohydrates associated with the impact of reduced turgor on expansion of cells in leaves and shoot tips (Chap. 14). Evidence from studies that characterize pool sizes of non-structural carbohydrates suggests that carbon does not limit growth of old trees, and that other causes of size-related growth reductions must be sought (Chap. 11) such as difficulty in sustaining adequate rates of long-distance transport of nutrients and assimilates. Ishii (Chap. 8) points out that the study of these limiting factors is made more difficult by extremely complex architecture of the oldest trees: what is the appropriate functional architectural unit of study in old trees? As pointed out above (Sect. 2), Sillett et al. (2010) showed an increase in the growth of mature (>800 years old) redwood trees, a trend opposite to many previous reports. They emphasized that the reason they saw an opposite trend is that because estimates in most past studies relied on ground-based measurements and not on measurements obtained in the upper crown where most of the growing biomass is. Thus, even though shoot extension in tall trees may be limited by reduced turgor, the multiplication of meristematic regions (apical and lateral buds, cambium) in large trees may allow total volume growth to continue unabated. These findings raise yet another question: are we making our growth measurements in the appropriate places in large and old trees? Without access to time or the entire tree, how does one access total biomass increment by a tree or a stand, and how does one include the belowground portion?

5.6 *Plasticity*

Several chapters deal with whether the observed differences between young and old trees are fixed traits or plastic responses to the changing environment sensed by the plant. Day and Greenwood (Chap. 4) distinguish between traits regulated by factors that are intrinsic vs. extrinsic to the meristem. Those traits regulated by intrinsic factors are manifested by permanent ontogenetic changes in meristem behavior (i.e. "fixed" developmental patterns), whereas traits regulated by extrinsic factors, are plastic responses that can follow changes in the environment. They provide a list of the predominant intrinsically and extrinsically regulated traits in germinants, seedlings, saplings, young-mature, and old-growth trees. (Note that Johnson et al. Chap. 3 state that seedlings exhibit a high degree of phenotypic plasticity in traits associated with water transport and light, whereas Day and Greenwood's summary says seedlings exhibit very little phenotypic plasticity.) Similarly, the chapter on radial changes in wood characteristics (Chap. 5) shows examples of species in which certain wood traits appear to be intrinsically controlled and others in which the control is plastic. A good example is wood density: in many species, it increases by growth ring number until it reaches an asymptote. In many of the pines, the density value is more closely correlated with growth ring number than with the diameter of the cambium at the time the wood was produced. This suggests that the changes follow a fixed trajectory, by age. Other species have clear plastic responses. Many lianoid species, for example, show a radical increase in wood density when the self-supported young stem encounters support and begins lianoid growth.

If a tree experiences a defoliating event such as herbivory, its ability to replace the lost biomass is termed compensation. Compensation is a form of plastic reaction to the disturbance. Boege et al. (Chap. 7) state that in general saplings are able to compensate to a greater degree than are seedlings, presumably by virtue of sapling's larger photosynthetic area, reserves, and bud banks.

Several authors mentioned traits that appear to follow specific rules, but that have variants on the basic pattern that are related to environment. Examples include the effects of crowdedness or wind on whether trees grow with geometric similarity, (Chap. 6), and the offset between height-related declines in carbon isotope discrimination for sunlit and shaded foliage (Chap. 10) (Fig. 1.5).

5.7 *Age vs. Size*

Several chapters make distinctions between age- and size-related features. Thomas (Chap. 2) stated that monocarpic plants reproduce a given age rather than size, although Day and Greenwood (Chap. 4) state that precocious flowering is regulated by both size and maturation state. Tree size is more important than meristem age for the growth rate of buds from grafted meristems (Chaps. 8 and 4). Size (height) is more important than age for processes driven by turgor such as leaf expansion and shoot extension (Chaps. 9 and 14) and for processes driven by xylem tension such as vessel and tracheid features that confer resistance to embolism (Chap. 5). Size has also been linked to mechanical influences on radial variation in stems (Chap. 5).

Lachenbruch et al. (Chap. 5) list the most apparent similarities and differences between hypothetical trees of the different size but the same age, and different age but the same size. The list is long and includes a wide range of traits (Table 1, Chap. 5). Had there been only one trait that differed and many that were the same (i.e. "small trees have less stress on the stem periphery than do large trees; all other traits were

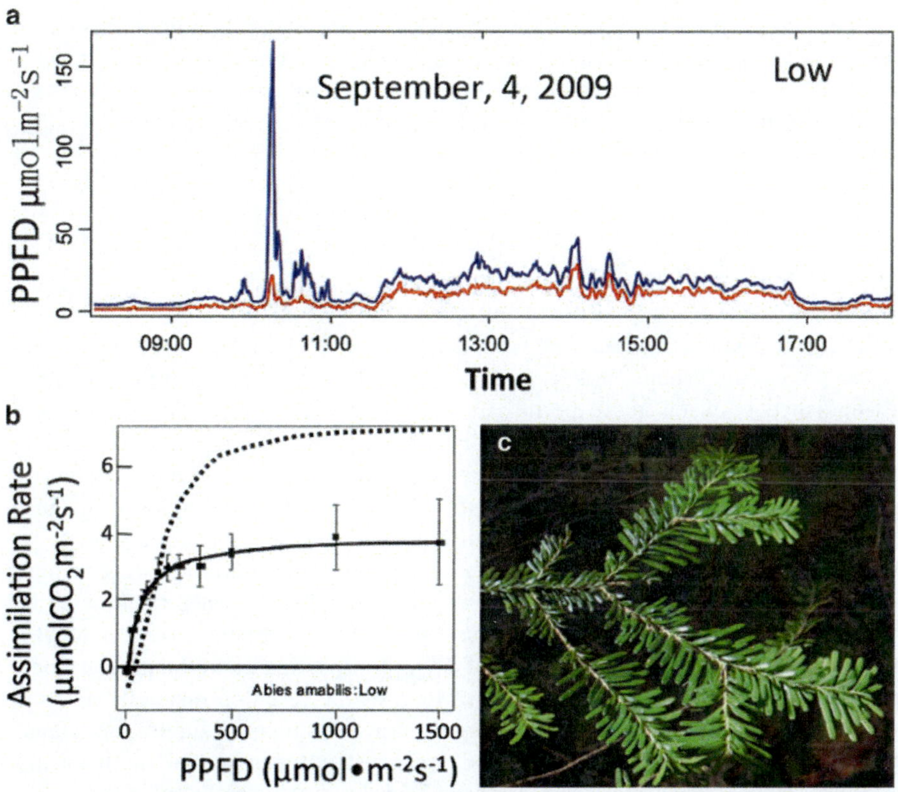

Fig. 1.5 (**a**) Daily pattern of direct beam radiation (*blue line*) and diffuse radiation (*red line*) in a dense western hemlock stand (*Tsuga heterophylla*) in the NW part of the Olympic Peninsula, Washington State. (**b**) Assimilation versus photosynthetic photon flux density curve (*solid line*) for shade foliage (see panel **c**) from *Abies amabilis* saplings growing under the stand in panel (**a**) The *dashed line* is for sun foliage for an *Abies amabilis* sapling growing in a thinned stand. (**c**) Photograph of typical shade foliage and branch from *Abies amabilis* (Data from Ma 2010)

the same between small and large trees") then logic would allow us to hypothesize that the size-related changes were driven by the stress on the stem periphery. The list demonstrates, however, that there is not any one obvious factor that is driving changes for either size- or age-related changes.

6 Emerging Issues and Open Questions

A number of gaps in our knowledge, mechanistic questions, and integrative issues remain and likely will continue to grow. Two key levels of integrative issues remain in our minds. The first involves the pronounced changes in structure and function that occur over the life cycle of a tree. The second issue involves the controls on integration over time and space of carbon, water and nutrient functioning, its changes in integration, and feedbacks at multiple scales within a tree. In addition to these themes, some authors have called for more standardization in techniques, terminologies, and the way in which different measures are reported (Chaps. 13 and 15) and warn of the importance of being explicit about the variables under discussion (Chap. 12).

6.1 Gaps in Knowledge

Authors have pointed out that more work is needed to clarify differences in patterns among taxonomic groups, biomes, and specific life stages. Most of the chapters distinguished between angiosperms and gymnosperms, noting that they often followed different developmental and physiological rules. Boege et al. (Chap. 7) state that to understand the macro-evolutionary patterns of the ontogeny of defense, a phylogenetic context is needed. Indeed, it is intriguing when angiosperms and gymnosperms have similar (rather than different) ranges of values for a trait, such as intrinsic capacitance, C, of xylem tissue (Chap. 13). This similarity suggests that the trait (or something strongly correlated with it) is of importance for fitness, as well as being governed by laws of physics. Other examples are the maximum height of trees, and the near-universal pattern of asymptotic increases in tracheid or fiber cell length with cambial age (Chap. 5). Much of the research discussed in this book comes from research on tropical angiosperms or temperate angiosperms and gymnosperms; more work is needed in other ecosystem types (boreal, dry tropical, and other special ecosystems) and in other woody growth forms (shrubs, prostrate and climbing plants and hemiepiphytes). This information will help us ask more informed questions, to ultimately better understand, predict, and manipulate species, stands, and ecosystems. For example, the knowledge that tropical angiosperms had higher carbon isotope discrimination than did either temperate gymnosperms or angiosperms (Chap. 10) gave insights into the strong effect of precipitation and the lack of freezing temperatures on this variable.

Most research has focused on saplings to mature trees, and young to mature stands of economic value, with relatively less research on the seedling to sapling stage, reproductive effects on tree growth, and senescence, or very young and very old forests. There is little research on the physiology of newly emerged seedlings, in spite of the importance of the regeneration niche for establishment of new individuals (Chap. 3). Likewise, in spite of a large body of theory on life history strategies and their use in predicting growth and reproductive dynamics of trees and stands, there is relatively less research that integrates reproductive costs into our understanding of age-related changes in tree ecophysiology (Chap. 2) and the challenge of how to measure "costs" remains as well.

Senescence is not discussed very explicitly in this book, perhaps because of the nature of the individual contributions, but also perhaps because of less research in this area. It was mentioned in the context of monocarpy, a reproductive strategy in which the plant dies soon after reproduction (Chap. 2), as maximum height attainable for biomechanical considerations (Chap. 6) and for early vs. late successional species (Chap. 16). Mortality was mentioned in the context of establishment success (Chap. 3), questions about vulnerability during specific ontogenetic transitions, such as when shade-tolerant conifer saplings transition into the upper canopy (Chap. 4), and the effects of herbivory at life stages that differ in plant architecture and metabolic stores (Chap. 7). Ishii (Chap. 8) discussed the buffering effects of large, complex canopies on the impacts of disturbance on individuals, but did not discuss tree death, and Anten and Selaya (Chap. 16) discussed the modeling results of leaf longevity vs. longevity of trees growing under the forest canopy and in gaps of different sizes. The physiological data on factors such as carbon isotope discrimination (Chap. 10) and tolerance to herbivory at the oldest ages will be particularly interesting in the light of trends toward increased non-structural carbon stores in mature trees (Chap. 11).

6.2 Questions About Integration and Feedbacks

Numerous questions remain surrounding the integration of woody plant growth over space and time, including the changes, controls and feedbacks associated with the usage of carbon, water and nutrients. The most basic question is how does the tree regulate the development of appropriate structures that work together, in an apparently coordinated manner? This question includes the mechanisms of coordination (like epigenetic changes at the meristem, or plastic responses), as well as the causal agent or signal to direct the appropriate development and physiological responses (Chaps. 4 and 7). The second question is what variables does a tree or a tree part sense in order to develop and function differently at different stages? Does it "know" its size or age, are there intrinsic directional changes in the meristem that tend to correlate with size or age, or does it sense its size or age more indirectly from the environment (such as by sensing bending stresses near the cambium)? Root to shoot distances and the number of divisions a given meristem has undergone all likely play some role. The phloem is largely overlooked in this book: Mencuccini et al. (Chap. 12) point out the importance of developing suitable empirical methods to estimate the physiological properties of phloem (fluxes of water and solute fluxes, turgor pressure, hydraulic conductivity) so that the integration and interactions between xylem and phloem can be studied.

Below are some of the specific questions related to integration or feedbacks that were raised in the chapters. To better understand the degree to which the age-related decline in productivity is related to the geometrical distribution of biomass, we need data on life-long trajectories of tree allometry that we can compare to the allometric patterns generated by models that embed our hypotheses (Chap. 6). An empirical approach could make use of individual trees, comparing mass distribution before and at periods after the removal of their tops.

The newly emerging picture of carbon economy suggests that trees may not be specifically carbon-limited in their growth. Many more cases will need to be studied to understand the range of species and conditions for which this is true, and also to integrate this knowledge with our current paradigms in which resource availability directs the organismal physiology and resulting growth or reproduction (Chap. 12). Sala et al. (Chap. 11) highlight the need to find the most sensitive and informative time and tissues in which the study of non-structural carbon (e.g. carbohydrates) concentrations best reflects the actual imbalance between sources and sinks. It is possible that some of this stored carbon becomes inaccessible at some point: the study of heartwood non-structural carbon stores would be enlightening in species with extremely large stores of starch in their sapwood, such as *Toxicodendron radicans*, (Rock 1972). A more general picture of the degree to which trees are carbon-limited throughout their lifetime would have immediate application to better understand resiliency to disturbance of individuals and stands of different ages.

The decline in assimilation and tree growth with age may be related to physical factors and feedbacks, but also to intrinsic changes in tissues. More research using rootstocks and scions of different sizes and ages will provide more insights (Chap. 9). Measuring growth at the points in a tree crown where most of it is happening also needs to be done more frequently because it is revealing that apparently "old trees" are not showing growth declines at all (Sillett et al. 2010). Such experimental methods and others may help us answer questions about whether tree growth at different developmental stages has the same sensitivity to biophysical drivers. This age-specific information will be of great value for models of tree productivity and net primary productivity under different climate projections (Chap. 17).

6.3 Questions About Changes in Structure and Function over the Tree's Life-Cycle

There are also many remaining questions about the changes in structure and function that occur during the tree's life cycle. Many of these questions are aimed at understanding the effect of a factor on a trait through time. The factors may relate to phylogeny, genetics, previous history, the physical environment (its stability, its change, or the tree's changed perception of the environment), biotic interactions that themselves may change as the tree grows, or a particular observable characteristic displayed by the organism such as its xylem anatomy. The affected traits are also varied, and can include functional traits (such as water transport) or a population-level characteristic (like survivorship). Many of the questions are not easy to address, and for this reason, experimental methods must be designed carefully. It is exceedingly easy to misinterpret tradeoffs (Chap. 12). Thomas (Chap. 2) discussed an example in which rainfall had a positive effect on growth and a negative effect on acorn production and could have been interpreted as a negative correlation of growth and acorn production (Knops et al. 2007), when in fact, the correlation did not represent causation. Another example is when a trait (e.g. embolism) affects more than one function (it may have provided stored water; it may decrease water transport) (Chap. 12).

7 Conclusions

This book brings together current information from diverse fields on the age- and size-related changes in structure and function in trees, successional groups of trees and stands, as well as providing frameworks for understanding, characterizing, and further researching age- and size-related changes. Many voids in our knowledge were exposed, and ideas for how to move forward are presented. We know that we have not covered all of the relevant topics or all of the tree species where relevant work has been done. Nonetheless, the findings from the types of research that are presented in the chapters that follow do highlight the state of our knowledge in many areas of tree research and its many applications. In all cases, the work presented is designed to lead to a better understanding of how trees work. For example, by understanding how xylem structure integrates the sometimes-conflicting requirements of water transport and biomechanics at multiple stages of life within an individual, we have created a working model for how these functions are integrated in other systems. In some cases, the information could potentially be used by breeders to increase biomass productivity, increase fruit production, or decrease losses to herbivory. In other cases, information could be used to better understand the magnitude of vulnerabilities to different biotic or abiotic factors at different life stages, which could lead to management strategies to encourage or discourage a population to flourish. These same insights could be very helpful in predicting and shaping performance and biogeography of species and assemblages of species at the individual and population levels under current and changing climate regimes. An example is the insights provided by Woodruff and Meinzer (Chap. 14), who state that "It is noteworthy that elevated carbon dioxide levels will not benefit trees if they are turgor limited as opposed to carbon limited. Furthermore, increased temperatures and/or droughts are likely to exacerbate turgor-related constraints on growth." It is with the goal of bringing together this information and inspiring others to continue contributing to the growing understanding and use of whole lifespan biology of trees that the editors and authors put forward this volume.

Acknowledgments A special thanks to Dr. Andrea Woodward, USGS Research Scientist, Seattle, Washington for providing unpublished data for us to use in this chapter. Funding provided by the David RM Scott Professorship.

References

Alarcon JJ, Domingo R, Green SR, Nicolas E, Torrecillas A (2003) Estimation of hydraulic conductance within field-grown apricot using sap flow measurements. Plant Soil 251:125–135

Ambrose AR, Sillett SC, Dawson TE (2009) Effect of tree height on branch hydraulics, leaf structure and gas exchange in California redwoods. Plant Cell Environ 32:743–757

Ambrose AR, Sillett SC, Koch GW, Van Pelt R, Antoine ME, Dawson TE (2010) Effects of height on treetop transpiration and stomatal conductance in coast redwood (*Sequoia sempervirens*). Tree Physiol 30:1260–1272

Angeles G, Bond B, Boyer JS, Brodribb T, Brooks JR, Burns MJ, Cavender-Bares J, Clearwater M, Cochard H, Comstock J, Davis SD, Domec J-C, LDonovan L, Ewers F, Gartner B, Hacke U, Hinckley T, Holbrook NM, Jones HG, Kavanagh K, Law B, Lopez-Portillo J, Lovisolo C, Martin T, Martinez-Vilalta J, Mayr S, Meinzer FC, Melcher P, Mencuccini M, Mulkey S, Nardini A, Neufeld HS, Passioura J, Pockman WT, Pratt RB, Rambal S, Richter H, Sack L, Salleo S, Schubert A, Schulte P, Sparks JP, Sperry J, Teskey R, Tyree M (2004) The cohesion-tension theory. New Phytol 163:451–452

Aumann CA, Ford ED (2006) Simulation of the effects of wood microstructure on water transport. Tree Physiol 26:285–301

Baldocchi DD (2003) Assessing the eddy covariance technique for evaluating carbon dioxide exchange rates of ecosystems: past, present and future. Glob Change Biol 9:479–492

Barnard HR, Ryan MG (2003) A test of the hydraulic limitation hypothesis in fast-growing *Eucalyptus saligna*. Plant Cell Environ 26:1235–1245

Beardsell MF, Jarvis PG, Davidson B (1972) A null-balance diffusion porometer suitable for use with leaves of many shapes. J Appl Ecol 9:677–690

Binkley D, Ryan MG, Stape JL, Barnard H, Fownes JH (2002) Age-related decline in forest ecosystem growth: an individual-tree, stand-structure hypothesis. Ecosystems 5:58–67

Bond BJ (2000) Age-related changes in photosynthesis of woody plants. Trends Plant Sci 5:349–353

Brodribb TJ, Holbrook NM (2003) Stomatal closure during leaf dehydration, correlation with other leaf physiological traits. Plant Physiol 132:2166–2173

Brooks JR, Hinckley TM, Sprugel DG (1994) Acclimation responses of mature *Abies amabilis* sun foliage to shading. Oecologia 100:316–324

Brooks JR, Hinckley TM, Sprugel DG (1996) The effects of light acclimation during and after foliage expansion on photosynthetic function of *Abies amabilis* foliage within the canopy. Oecologia 107:21–32

Bucci SJ, Scholz FG, Goldstein G, Meinzer FC, da Sternberg LSL (2003) Diurnal changes in hydraulic conductivity of petioles of two savanna tree species: factors and mechanisms contributing to the refilling of embolized vessels. Plant Cell Environ 26:1633–1645

Burgess SOO, Dawson TE (2007) Predicting the limits to tree height using statistical regressions of leaf traits. New Phytol 174:626–636

Burgess SSO, Adams MA, Turner NC, Ong CK (1998) The redistribution of soil water by tree root systems. Oecologia 115:306–311

Burgess SOO, Pitterman JA, Dawson TE (2006) Hydraulic efficiency and safety of branch xylem increases with height in *Sequoia sempervirens* (D. Don) crowns. Plant Cell Environ 29:229–239

Canny MJ (1997) Vessel contents during transpiration-embolism and refilling. Am J Bot 84:1223–1230

Cavaleri MA, Oberbauer SF, Clark DB, Clark DA, Ryan MG (2010) Height is more important than light in determining leaf morphology in a tropical forest. Ecology 91:1730–1739

Čermák J, Kucera J (1981) The compensation of natural temperature gradient in the measuring point during the sap flow rate determination in trees. Biol Plant 23:469–471

Čermák J, Kučera J, Bauerle WL, Phillips N, Hinckley TM (2007) Tree water storage and its diurnal dynamics related to sap flow and changes of stem volume in old-growth Douglas-fir trees. Tree Physiol 27:181–198

Choat B, Cobb AR, Jansen S (2008) Structure and function of bordered pits: new discoveries and impacts on whole-plant hydraulic function. New Phytol 177:608–626

Cleaveland MK (1974) Climatic response of densitometric properties in semi-arid site tree rings. Tree Ring Bull 46:3–29

Cown DJ, Ball RD, Riddell MJC (2004) Wood density and microfibril angle in 10 *Pinus radiata* clones: distribution and influence on product performance. New Zeal J For Sci 34:293–315

Dawson TE, Simonin KS (2011) The roles of stable isotopes in forest hydrology and biogeochemistry. In: Levia D, Carlyle-Moses D, Tanaka T (eds.) Forest hydrology and biogeochemistry: synthesis of research and future directions (in press), pp 75–93, *Ecological Studies* 216, Springer-Verlag, DOI 10.1007/978-94-007-1363-5_7

Dawson TE, Mambelli S, Plamboeck AH, Templer PH, Tu KP (2002) Stable isotopes in plant ecology. Annu Rev Ecol Syst 33:507–559

DeMars DJ, Herman FR (1987) Estimates of site index and height growth for Douglas-fir in high-elevation forests of the Oregon-Washington Cascade Range: curves and tables for field application. Res. Pap. PNW-RP-378, USDA Forest Service, Portland, 16 pp

Dixon HH, Joly J (1895) On the ascent of sap. Philos Trans R Soc Lond B 186:563–576

Domec J-C, Lachenbruch B, Meinzer FC (2006) Bordered pit structure and function determine spatial patterns of air-seeding thresholds in xylem of Douglas-fir (Pseudotsuga menziessi; Pinaceae) trees. Am J Bot 93:1588–1600

Edminster, CB, Jump LH (1976) Site index curves for Douglas-fir in New Mexico. USDA Forest Service Research Note RM – 326, USDA, Portland, 3 pp

Elfving DC, Hall AE, Kaufmann MR (1972) Interpreting leaf water potential measurements with a model of the soil-plant-atmosphere continuum. Physiol Plant 27:161–168

Evans R (1999) A variance approach to the x-ray diffractometric estimation of microfibril angle in wood. Appita J 52:283–289

Evans R (2006) Wood stiffness by x-ray diffractometry. In: Stokke DD, Groom LH (eds.) Charactrization of the cellulosic cell wall. Blackwell Publishing, Ames, pp 138–146

Evans R, Ilic J (2001) Rapid prediction of wood stiffness from microfibril angle and density. For Prod J 51:53–57

Field CB, Berry JA, Mooney HA (1982) A portable system for measuring carbon dioxide and water vapor exchanges of leaves. Plant Cell Environ 5:179–186

Garrett HE, Cox GS, Roberts JE (1978) Spatial and temporal variations in carbon-dioxide concentrations in an oak-hickory forest ravine. For Sci 24:180–190

Gower ST, McMurtrie RE, Murty D (1996) Aboveground net primary production decline with stand age: potential causes. Trends Ecol Evol 11:378–382

Granier A (1987) Evaluation of transpiration in a Douglas-fir stand by means of sap flow measurements. Tree Physiol 3:309–320

Greenhill AG (1881) Determination of the greatest height consistent with stability that a vertical pole or mast can be made, and of the greatest height to which a tree of given proportions can grow. Proc Cambr Philos Soc 4:65–73

Grier CC, Vogt KA, Keyes MR, Edmonds RL (1981) Biomass distribution and above- and below-ground production in young and mature *Abies amabilis* zone ecosystems of the Washington Cascades. Can J For Res 11:155–167

Hacke UG, Sperry JS, Wheeler JK, Castro L (2006) Scaling of angiosperm xylem structure with safety and efficiency. Tree Physiol 26:689–701

Hales S (1727) Vegetable staticks (Innys WJ, Woodward T compilers). Scientific Book Guild, London

Hinckley TM (2004) Tree physiology: a whole tree perspective. In: Jeffery Burley (ed.) Encyclopedia of forest sciences. Elsevier, Oxford, pp 1578–1583. doi:10.1016/B0-12-145160-7/00098-3. ISBN 978-0-12-145160-8

Hinckley TM, Bruckerhoff DN (1975) The effects of a major drought on xylem pressure potential, leaf resistance and stem circumference on a forest-grown white oak (*Quercus alba* L.) in mid-Missouri. Can J Bot 53:62–72

Huber B (1924) Die Beurteilung des Wasserhaushaltes der Pflanze. Ein Beitrag vergleichenden Physiol Jahrb Wiss Bot 64:1–120

Ishii HT, Jennings GM, Sillett SC, Koch GW (2008) Hydrostatic constraints on morphological exploitation of light in tall *Sequoia sempervirens* trees. Oecologia 156:751–763

Jarvis PG (1975) Water transfer in plants. In: deVries DA, van Allen NK (eds.) Heat and mass transfer in the environment of vegetation. Scripta Book Co., Washington, DC, pp 369–394

Johnson DM, Meinzer FC, Woodruff DR, McCulloh KA (2009) Leaf xylem embolism detected acoustically and by cryo-SEM, corresponds to decreases in leaf hydraulic conductance in four evergreen species. Plant Cell Environ 32:828–836

Kennedy M, Hinckley TM, Ford ED (2010) Defining how aging *Pseudotsuga* and *Abies* compensate for multiple stresses through multi-criteria assessment of a functional-structural model. Tree Physiol 30:3–22

Knops JMH, Koenig WE, Carmen WJ (2007) Negative correlation does not imply a tradeoff between growth and reproduction in California oaks. Proc Natl Acad Sci USA 104:16982–16985

Koch GW, Sillett SC, Jennings GM, Davis SD (2004) The limits to tree height. Nature 428:851–854
Kramer PJ (1969) Plant and soil water relationships: a modern synthesis. McGraw-Hill, New York, 482 p
Lachenbruch B, Johnson GR, Downes GM, Evans R (2010) Relationships of density, microfibril angle, and sound velocity with stiffness and strength in mature wood of Douglas-fir. Can J For Res 40:55–64
Lewis JD, McKane RB, Tingey DT, Beedlow PA (2000) Vertical gradients in photosynthetic light response within an old-growth Douglas-fir and western hemlock canopy. Tree Physiol 20:447–456
Lovisolo C, Perrone I, Carra A, Farrandino A, Flexas J, Medrano H, Schubert A (2010) Drought-induced changes in development and function of grapevine (*Vitis* spp.) organs and in their hydraulic and non-hydraulic interactions at the whole-plant level: a physiological and molecular update. Funct Plant Biol 57:98–116
Ma Z (2010) Understory light and its effect on photosynthesis in conifer saplings: study of *Abies amabilis* and *Tsuga heterophylla* in a western Olympic Peninsula mature forest. Unpublished MS thesis, University of Washington, Seattle
Martin TA, Brown KJ, Hinckley TM, Kucera J, Meinzer FC, Sprugel DG (2001) Control of transpiration in a 220-year-old *Abies amabilis* forest. For Ecol Manage 152:211–224
McCulloh KA, Sperry JS, Lachenbruch B, Meinzer FC, Reich PB, Voelker S (2010) Moving water well: comparing hydraulic efficiency in twigs and trunks of coniferous, ring-porous, and diffuse-porous saplings from temperate and tropical forests. New Phytol 186:439–450
McCully ME, Huang CX, Ling LE (1998) Daily embolism and refilling of xylem vessels in the roots of field-grown maize. New Phytol 138:327–342
Means JE, Helm ME (1985) Height growth and site index curves for Douglas-fir on dry sites in the Willamette National Forest. USDA Forest Service Research Paper PNW – 341, USDA, Portland, 17 pp
Meinzer FC, Brooks JR, Domec J-C, Gartner BL, Warren JM, Woodruff DR, Bible K, Shaw DC (2006) Dynamics of water transport and storage in conifers studied with deuterium and heat tracing techniques. Plant Cell Environ 29:105–114
Melcher PJ, Goldstein G, Meinzer FC, Yount DE, Jones TJ, Holbrook NM, Huang CX (2001) Water relations of coastal and estuarine *Rhizophora mangle*: xylem pressure potential and dynamics of embolism formation and repair. Oecologia 126:182–192
Mencuccini M, Martinez-Vilalta J, Vanderklein D, Hamid HA, Korakaki E, Lee S, Michiels B (2005) Size-mediated ageing reduces vigour in tall trees. Ecol Lett 8:1183–1190
Motzer T, Munz N, Küppers M, Schmitt D, Anhuf D (2005) Stomatal conductance, transpiration and sap flow of tropical montane rain forest trees in the southern Ecuadorian Andes. Tree Physiol 25:1283–1293
Odum EP (1969) The strategy of ecosystem development. Science 164:262–270
Oldham AR, Sillett SC, Tomescu AMF, Koch GW (2010) The hydrostatic gradient, not light availability, drives height-related variation in *Sequoia sempervirens* (Cupressaceae) leaf anatomy. Am J Bot 97:1087–1097
Parkinson KJ, Legg BJ (1972) A continuous flow porometer. J Appl Ecol 9:669–675
Petit G, Anfodillo T, Carraro V, Grani F, Carrer M (2011) Hydraulic constraints limit height growth in trees at high altitude. New Phytol 189:241–252
Pittermann J, Sperry JS, Hacke UG, Wheeler JK, Sikkema EH (2005) Torus-margo pits help conifers compete with angiosperms. Science 310:1924
Polge H (1978) Fifteen years of wood radiation densitometry. Wood Sci Technol 12:187–196
Richter H (1972) Wie entstehen Saugspannungsgradienten in Bäumen? Ber Deut Bot Ges 85:341–351
Richter H (1973) Frictional potential losses and total water potential in plants: a reevaluation. J Exp Bot 24:983–994
Richter H (1974) Erhöhte Saugspannungswerte und morphologische Veräinderungen durch transversale Einschnitte in einem *Taxus* Stamm. Flora 163:291–309
Ritchie GA (1969) Cuvette temperatures and transpiration rates. Ecology 50:667–670

Roberts J, Cabral OMR, De Aguiar LF (1990) Stomatal and boundary-layer conductances in an Amazonian terra firme rain forest. J Appl Ecol 27:336–353

Rock BN (1972) The wood and flora of the florida keys: "pinnatae". In: Smithsonian contributions to botany. Smithsonian Institution Press, Washington, DC, pp 9–10

Ryan MG, Yoder BJ (1997) Hydraulic limits to tree height and tree growth. Bioscience 47:235–242

Ryan MG, Binkley D, Fownes JH (1997) Age-related decline in forest productivity: pattern and process. Adv Ecol Res 27:213–262

Ryan MG, Stape JL, Binkley D, Fonseca S, Loos R, Takahashi EN, Silva CR, Silva HR, Ferreira JM, Lima AM, Gava JL, Leite FP, Silva G, Andrade H, Alves JM (2010) Factors controlling eucalyptus productivity: how water availability and stand structure alter production and carbon allocation. For Ecol Manage 259:1695–1703

Sala A, Hoch G (2009) Height-related growth declines in ponderosa pine are not due to carbon limitation. Plant Cell Environ 32:22–30

Salleo S, Hinckley TM, Kikuta SB, LoGullo MA, Weilgony P, Yoon T-M, Richter H (1992) A method for inducing xylem emboli *in situ*: experiments with a field-grown tree. Plant Cell Environ 15:491–497

Savage VM, Bentley LP, Enquist BJ, Sperry JS, Smith DD, Reich PB, von Allmen EI (2010) Hydraulic trade-offs and space filling enable better predictions of vascular structure and function in plants. Proc Natl Acad Sci USA. doi:10.1073/pnas.1012194108

Schoettle AW (1994) Influence of tree size on shoot structure and physiology of *Pinus-contorta* and *Pinus-aristata*. Tree Physiol 14:1055–1068

Scholander PR, Hammel HT, Bradstreet ED, Hemmingsen EA (1965) Sap pressure in vascular plants. Science 143:339–346

Seymour VA, Hinckley TM, Morikawa Y, Franklin JF (1983) Foliage damage in coniferous trees following volcanic ashfall from Mt. St Helens Oecologia 59:339–343

Shaw DC, Franklin JF, Bible K, Klopatek J, Freeman E, Greene S, Parker GG (2004) Ecological setting of the wind river old-growth forest. Ecosystems 7:427–439

Shinozaki K, Yoda K, Hozumi K, Kira T (1964) A quantitative analysis of plant from – the pipe model theory. I Basic analysis Jpn J Ecol 14:97–105

Sillett SC, Van Pelt R, Koch GW, Ambrose AR, Carroll AL, Antoine ME, Mifsud BM (2010) Increasing wood production through old age in tall trees. For Ecol Manage 259:976–994

Slatyer RO (1967) Plant-water relationships. Academic, New York, 366 pp

Soll JA (1994) Seed number, germination and first year survival of subalpine fir (*Abies lasiocarpa*) in subalpine meadows of the northeastern Olympic Mountains. Unpublished MS thesis, University of Washington, Seattle, 120 pp

Sperry JS, Nichols KL, Sullivan JEM, Eastlack SE (1994) Xylem embolism in ring-porous, diffuse-porous, and coniferous trees of Northern Utah and Interior Alaska. Ecology 75:1736–1752

Sperry JS, Adler FR, Campbell GS, Comstock JC (1998) Limitation of plant water use by rhizosphere and xylem conductances: results from a model. Plant Cell Environ 21:347–359

Spicer R, Gartner BL (1998) Hydraulic properties of Douglas-fir (Pseudotsuga menziesii) branches and branch halves with reference to compression wood. Tree Physiol 18:777–784

Teskey RO, Grier CC, Hinckley TM (1984) Change in photosynthesis and water relations with age and season in *Abies amabilis*. Can J For Res 14:77–84

Tyree MT, Hammel HT (1972) The measurement of the turgor pressure and water relations of plants by the pressure-bomb technique. J Exp Bot 23:267–282

Tyree MT, Patiño S, Bennink J, Alexander J (1995) Dynamic measurements of root hydraulic conductance using a high-pressure flowmeter in the laboratory and field. J Exp Bot 46:83–94

Van den Honert TH (1948) Water transport in plants as a catenary process. Discuss Faraday Soc 3:146–153

Vasconcelos LFL, Ribeiro RV, de Oliveira RF, Machado EC (2010) Variation of sap flow density and leaf water potential in the east and west side of the canopy of Valencia sweet orange plant. Rev Bras Frutic 32:35–46

Waring RH, Franklin JF (1979) Evergreen coniferous forests of the Pacific Northwest. Science 204:1380–1386

West GB, Brown JH, Enquist BJ (1997) A general model for the origin of allometric scaling laws in biology. Science 276:122–126

West GB, Brown JH, Enquist BJ (1999) A general model for the structure and allometry of plant vascular systems. Nature 400:664–667

Wheeler JK, Sperry JS, Hacke UG, Hoang N (2005) Inter-vessel pitting and cavitation in woody Rosaceae and other vesselled plants: a basis for a safety vs. efficiency trade-off in xylem transport. Plant Cell Environ 28:800–812

Woodman JN (1971) Variation of net photosynthesis with crown of a large forest-grown conifer. Photosynthetica 5:50

Yasue K, Funada R, Fukazawa K, Ohtani J (1997) Tree-ring width and maximum density of *Picea glehnii* as indicators of climatic changes in northern Hokkaido, Japan. Can J For Res 27:1962–1970

Yoder BJ, Ryan MG, Waring RH, Schoettle AW, Kaufmann MR (1994) Evidence of reduced photosynthetic rates in old trees. For Sci 40:513–527

Zimmermann MH (1978) Hydraulic architecture of some diffuse-porous trees. Can J Bot 56:2286–2295

Zimmermann MH (1983) Xylem structure and the ascent of sap. Springer, Berlin, 143 pp

Zimmermann U, Schneider H, Wegner LH, Haase A (2004) Water ascent in tall trees: does evolution of land plants rely on a highly metastable state? New Phytol 162:575–615

Part II
Reproduction and Regeneration

Chapter 2
Age-Related Changes in Tree Growth and Functional Biology: The Role of Reproduction

Sean C. Thomas

Abstract Population biologists have generally assumed that reproduction entails a reduction in growth, future reproduction, and/or increase in mortality: such "costs of reproduction" are central to life-history models that predict patterns of reproductive allocation, growth curves, and other aspects of plant ontogeny. However, the apparent importance of biophysical processes (such as hydraulic limitation effects) has led to a widespread view that physiological effects of reproduction play a minor role, if any, in driving age-related patterns of tree growth and functional traits. This chapter assesses available data to detect and measure the role of reproduction on tree functional biology, where possible by use of meta-analysis to integrate results across studies. Trees generally show a long period of pre-reproductive growth, and a pattern of increasing reproductive allocation with size, as predicted by life-history theory. The widespread existence of substantial physiological costs of reproduction in trees is indicated by several sources of comparative data. (1) Reproductive allocation generally increases with tree size, and recent studies suggest that >40% of fixed carbon may be allocated annually to reproductive structures in mature forests. (2) There are ~100 species of monocarpic trees in which rapid senescence is triggered by a single reproductive event. (3) Dendrochronological studies comparing stem growth increment to measures of fruit production across years, generally find negative correlations, even when statistically accounting for environmental effects such as growing season temperature and precipitation. These studies also document carry-over effects, in which intense reproductive events reduce tree performance over multiple years. (4) Shoots near reproductive structures generally show reduced shoot extension growth, leaf size, and leaf N content, among other changes. There is also some evidence that leaf traits related to carbon gain (such as leaf size, photosynthetic capacity and leaf N) increase with size among small pre-reproductive trees,

S.C. Thomas (✉)
Faculty of Forestry, University of Toronto, 33 Willcocks St., Toronto, ON M5S 3B3, Canada
e-mail: sc.thomas@utoronto.ca

F.C. Meinzer et al. (eds.), *Size- and Age-Related Changes in Tree Structure and Function*, Tree Physiology 4, DOI 10.1007/978-94-007-1242-3_2,

but decrease following reproduction. (5) Studies of sexual dimorphism in dioecious tree species generally find patterns consistent with a higher cost of reproductive in female (pistillate) individuals, which show reduced growth. Analyses of growth curves of some dioecious tree species also indicate sexual dimorphism in patterns of growth decline late in tree ontogeny, consistent with a strong role of reproduction in determining the form of growth declines. In conclusion, reproduction cannot be dismissed as of only minor importance as a determinant of age-related trends in tree functional biology. The physiological mechanisms for reproductive effects, their comparative biology, and their interactions with other processes that limit growth deserve much greater research attention.

1 Introduction

Trees are remarkable organisms in many respects, their size, longevity, and strong interactions with global biogeochemistry being some of the traits in which trees reign superlative among extant life forms. A less commonly noted but equally remarkable aspect of tree biology is their prolonged and often prolific reproductive output. With few exceptions, trees show an initial juvenile period in which no reproductive structures are produced, followed by an adult phase lasting decades to centuries characterized by periodic bouts of reproduction that continue throughout the tree's lifespan (Harper and White 1974). The reproductive output of trees is commonly a central driver of community processes in forest ecosystems. For example, mast fruiting patterns in the forests of Southeast Asia and parts of the temperate zone largely determine population dynamics of regenerating seedlings, of organisms that depend on seeds as resources, and through trophic cascades have large impacts on many organisms within the ecosystem (e.g., Ostfield et al. 1996; Jones et al. 1998; Curran and Leighton 2000). Historically, the capacity of forests to produce acorns and beech mast (used for pannage, the fattening of domesticated pigs in forests), was the main "ecosystem function" by which the forests of Europe were valued (Shaw 1940; Rackham 1980; Birrell 1987).

A pronounced dichotomy exists in the literature on age-related changes in tree growth and physiology as it relates to reproduction. On the one hand, there is a well-developed literature focused on life-history strategies as a basis for predicting patterns of reproductive and growth dynamics of trees and other perennial plants (e.g., Kozlowski and Uchmanski 1987; Iwasa and Cohen 1989; Clark 1991; Thomas 1996a; Iwasa 2000; Zhang 2006). A fundamental assumption in essentially all of this work is that there are "costs" of reproduction that result in tradeoffs between tree growth and reproductive output (Obeso 2002). This literature has generally not explicitly considered biophysical limitations to tree growth (but see Mäkelä 1985). In contrast, much of the recent ecophysiological literature examining "age-related" patterns of growth and physiology has tacitly assumed that biophysical processes such as water transport are the primary causal agents of growth and physiological patterns, and has ignored, downplayed or dismissed reproductive "costs" as playing any important role (e.g., Gower et al. 1996; Ryan et al. 1997; Bond 2000; Bond et al. 2007; Kutsch et al. 2009).

There are some strong arguments, both theoretical and empirical, that both reproductive "cost" effects and biophysical limitations may be of importance in driving age-related patterns of tree growth and physiology, but also that the relative importance of these factors varies among species and with environmental conditions. Weiner and Thomas (2001) compared efforts to identify "the" physiological basis for limitations of tree size to efforts in prior decades to identify the (sole) limit to population growth in populations of organisms. Through much of the early history of population ecology it was assumed that there must be some main factor that acted to limit population carrying capacity, with proponents of differing camps arguing, for example, that density-dependent competition, predation, or abiotic factors were most likely to be that single limiting factor (Andrewartha and Birch 1954). This historical debate in the literature was ultimately settled by the realization that different factors were clearly of central importance in different populations of organisms.

An important theoretical argument related to constraints on tree growth is the hypothesis that selection should act such that multiple limiting factors generally will occur (Bloom et al. 1985). In contrast to the assumption codified as "Liebig's Law of the Minimum", if plant growth is under positive selection pressure, then selection will act to favor phenotypes that are less limited by that particular resource. If another resource then becomes limiting, selection will similarly act to reduce limitations of that resource, with the ultimate result being that plants are equally co-limited by a set of resources. The original hypothesis of Bloom et al. has been critiqued from a variety of perspectives: in particular, access to or uptake of different resources may be interdependent (Gleeson and Tilman 1992), and the assumption that one resource may be "bought" by another resource in excess may not hold (Bazzaz et al. 1987). However, there is empirical evidence that co-limitation of plant growth by at least two resources (though often with one dominant limitation) is common (Chapin et al. 1987; Elser et al. 2007). Studies of mature trees and forests, mainly focused on nutrient limitation, have likewise commonly found co-limitation (e.g., Tanner et al. 1998; Gradowski and Thomas 2006, 2008; Paoli and Curran 2007; Vadeboncoeur 2010). Reproductive costs are similar to other limitations of plant growth in that selection should act to reduce their costs, thus potentially enhancing the importance of other limiting factors. However, the selection scenario articulated by Bloom et al. (1985) should ultimately act through reproductive fitness rather than growth, since growth in an evolutionary context is the means by which an organism can achieve reproductive success. Thus, evolutionary theory predicts that multiple resources should constrain tree growth early in ontogeny, but that effects of reproductive allocation should become increasingly important as trees age. A single causal agent for age-related declines in growth or for associated changes in physiological processes is very unlikely from this perspective.

The main objective of this review is to examine the empirical evidence for costs of reproduction in trees, where possible using meta-analyses of data from published studies. I start with a short overview of the salient aspects of life-history theory as applied to trees, and of empirical data on patterns of size-dependent reproduction in trees. I then review comparative studies on monocarpic trees, as these species constitute a special case where reproduction is apparently closely linked to rapid senescence. I then consider (1) dendrochronological studies in which tree ring series have been related to long-term data on reproductive output, (2) studies of canopy growth in which patterns of shoot extension and physiology have been related to production of reproductive structures on tree shoots or branches, and (3) studies of dioecious tree species that have compared growth patterns and functional biology of male (staminate) and female (pistillate) individuals. Finally, I summarize the ensemble of results of different approaches to evaluating reproductive effects on age-related patterns of tree growth and physiology, and suggest important avenues for future work.

2 Life History Theory and Tree Reproduction: An Overview

Life-history theory as applied to patterns of age-related growth and reproduction generally starts with an assumption of a "principle of allocation" (Levins 1968): specifically, that resources are finite, and that resources allocated to reproduction must have some commensurate negative consequence to growth processes. An important distinction has also generally been made between "reproductive allocation" and "reproductive effort". Reproductive allocation is the "proportion of the total resource supply devoted to reproductive structures" (Bazzaz et al. 2000). Reproductive effort is a more holistic concept of the total resources diverted from vegetative growth as a consequence of reproduction. From evolutionary first principles, there is an expectation that reproduction must entail some fitness cost (i.e., to future growth, survivorship, or reproduction): an organism that has "free" reproduction will be selected to increase reproduction to a point where there is at least some cost. There is also an expectation that reproductive effort generally will exceed reproductive allocation, since there are physiological costs in the development of any tissues above and beyond those represented by carbon, nutrients, and energy in the tissues themselves. These costs (sometimes labeled "somatic costs of reproduction": Bazzaz et al. 2000; Obeso 2002) may include energetic construction costs, costs associated with transport of metabolites, and "opportunity costs" of meristem allocation and storage prior to allocation.

A useful starting point for theoretical analyses to predict age-related patterns of growth and reproduction is annual semelparous plants (Cohen 1968, 1976). In cases where there is a predictable termination of the life cycle (e.g., the end of the growing season), the optimal strategy for reproductive investment is generally a "bang-bang" strategy: all resources are invested in somatic growth, and then there is a complete shift to complete resource allocation to reproduction. This strategy is an evolutionary stable strategy in a game-theoretical context: a population of phenotypes with a "graded" reproductive investment pattern will be invaded by a phenotype with a "bang-bang" strategy. However, in the case where there is uncertainty in the termination of the life cycle (e.g., a stochastic growing season), graded reproductive investment is favored (King and Roughgarden 1982). In all cases the optimal strategy involves a pattern of increasing reproductive allocation with age (and size).

The generalization of life-history models to consider long-lived perennials, such as trees, has been based on treating reproductive allocation as a series of allocation decisions made over sequential growing seasons (Kozlowski and Uchmanski 1987; Iwasa and Cohen 1989; Iwasa 2000). Within each growing season allocation strategies follow a similar logic to that of an annual plant: resources initially captured are initially allocated to somatic structures, and then increasingly to reproductive structures. This pattern is predicted to occur both within each growing season, and through ontogeny. There is strong empirical evidence that increases in reproductive effort with age are, with very few exceptions, the rule in animals (Roff 1992). Increases in the costs of reproduction though plant ontogeny have also been found in several herbaceous plants (Reekie and Bazzaz 1992; Worley and Harder 1996; Hemborg and Karlsson 1998), but no consistent size or developmental trends were found in several other studies (Obeso 2002; Cheplick 2005). The relevant empirical data for woody perennial plants appear not to have been synthesized.

3 Size at Onset of Maturity

One basic but important prediction of the life-history theories outlined above is that trees should show a prolonged juvenile period, particularly in environments where early allocation to reproduction would come at high cost to future growth and reproduction. While a long juvenile period of trees has been casually noted by generations of plant biologists (Harper and White 1974), quantitative analyses have been surprisingly limited. An empirical challenge has been the lack of long-term data on reproductive status, since the most direct method for assessing age- and size-dependent reproductive status would be to follow cohorts of trees through time. Although notes on age at first reproduction are commonly made of trees grown in arboreta or in breeding trials (and are the usual basis for age to reproduction given in silvics manuals and similar sources), quantitative analyses for natural tree populations are very scant.

If trees consistently reproduce (or can be recognized as reproductive) following reproductive onset, then a function describing the probability of reproduction as a function of size (or age) in a given reproductive episode will correspond to the cumulative distribution function of size (or age) at first reproduction (Thomas 1996b). This property allows the distribution of size at reproductive onset to be estimated using static data. Analyses using this approach indicate great variation among species and between forests in reproductive size thresholds (Thomas 1996b; Davies and Ashton 1999; Wright et al. 2005). The relative size at onset of maturity (RSOM) may be quantified as the ratio of height at onset of maturity to maximum (asymptotic) height (Thomas 1996b). RSOM also shows high variation among tree species, with reported values ranging from ~0.2 to 0.8. (Thomas 1996b, 2010; Davies and Asthon 1999; Wright et al. 2005). Comparative analyses have examined variation in size at onset of maturity in terms of variation among species, ecological groups, and forests. One consistent pattern is that trees that attain a large stature (i.e., have a large asymptotic maximum height) are larger at onset of maturity either

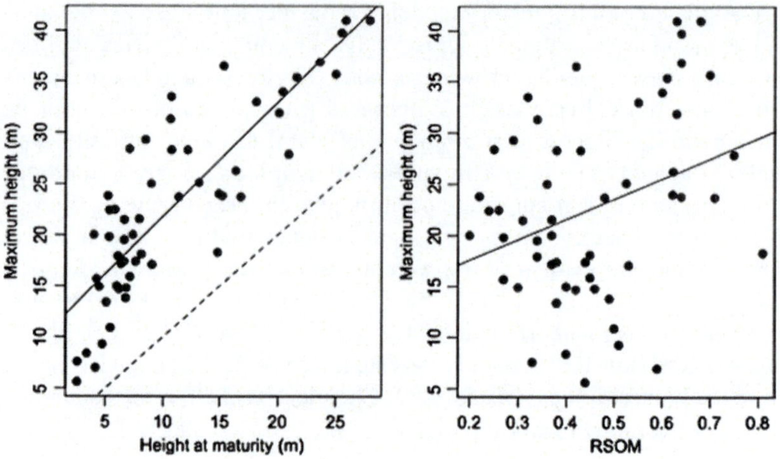

Fig. 2.1 Relationship between tree height at maturity and asymptotic height among 54 tree species, compiling data across studies (mainly of tropical tree species: Thomas 1996b, 2010; Davies and Asthon 1999; Wright et al. 2005). Least-squares linear regression lines are shown as *solid lines*; the *dotted line* indicates a 1:1 relationship for maximum height and height at maturity. Correlations are significant in both cases (absolute height at maturity: $r=0.847$; $P<0.001$; RSOM: $r=0.330$; $P=0.015$)

in absolute or relative terms (Thomas 1996b; Wright et al. 2005). One may also view this pattern in terms of maximum size as the dependent variable: tree species that show a smaller size at onset of maturity (expressed either in absolute terms or as RSOM) reach a smaller maximal height (Fig. 2.1). This pattern is consistent with a systematic cost of reproduction such that early reproduction limits maximum tree size; however, the correlation is relatively weak ($r=0.330$).

4 Size-Dependence of Reproductive Allocation and Effort

Predictions of life-history theory are generally that reproductive allocation should increase with age and size in iteroparous organisms (Williams 1966; Kozlowski and Uchmanski 1987; Iwasa and Cohen 1989). Although it is possible to construct models that predict declines in reproductive allocation with age, scenarios in which this occurs are considered biologically unrealistic (Roff 1992). The size-dependence of reproductive allocation is generally most informatively analyzed using allometric methods (Weiner 1988; Weiner et al. 2009; Cheplick 2005). An allometric relationship between reproductive and somatic biomass with a slope >1 would indicate a pattern of increasing reproductive allocation with size. Data most typically are available for numbers or biomass of reproductive structures as a function of stem diameter. If one assumes tree biomass (M) scales with stem diameter (D) as $M=D^{8/3}$, then life-history theory would correspondingly predict an exponent of $\alpha > 8/3$ for an allometric relationship between reproductive biomass (R) and stem diameter (i.e., $R = D^{\alpha}$) (Thomas 1996a).

Although more information is available for herbaceous plant species (Weiner et al. 2009) – in particular for lifetime reproduction – a number of studies have presented data on reproductive allometry of crop sizes per reproductive event in natural populations of trees. Reported relationships are quite variable, and there is often very high variability among individuals in reproductive output, necessitating large sample sizes to reveal allometric patterns. Several studies have reported values of $\alpha \sim 2$ (Studies reporting exponents in the range of 1.7–2.2 include: Peters et al. 1988; Niklas 1993; Hirayama et al. 2004), and this value has been widely assumed in forest simulation models (Pacala et al. 1993), and in analyses of propagule dispersal kernels (Ribbens et al. 1994; Caspersen and Saprunoff 2005; Muller-Landau et al. 2008). In contrast, several studies of tropical tree species have estimated much higher exponent values: Alvarez-Buylla and Martinez-Ramos (1992) report $\alpha = 3.3$ for the neotropical pioneer *Cecropia obtusifolia*, and Thomas (1996a) found an average $\alpha = 5.3$ among 32 species of sub-canopy tree species in Malaysia. It is possible that the large variability in allometric slopes among studies may be driven by differences in size-dependence access to resources. In particular, vertical gradients in light availability may greatly enhance size-dependence of reproduction in sub-canopy tree species.

In addition to showing greater reproductive output within any given reproductive event, larger trees generally reproduce more frequently than smaller trees. Many studies have documented this pattern in a wide range of species and ecosystems, with examples ranging from Malaysian sub-canopy trees (Thomas and LaFrankie 1993) to tropical alpine dwarf trees (Kudo and Suzuki 2004). Among forest canopy trees positive relationships between size and reproductive frequency have been documented for *Nyssa sylvatica* in New Jersey USA (Cipollini and Stiles 1991), *Bischofia javanica* in the Bonin Islands of Japan (Yamashita and Abe 2002), *Shorea acuminata* in Malaysia (Naito et al. 2008), and *Chrysophyllum lucentifolium* (Sapotaceae) in Brazil (Fonseca et al. 2009). Where analyses have been conducted, size-dependent increases in reproductive frequency have been found to be independent of effects of canopy position or crown illumination (Cipollini and Stiles 1991; Fonseca et al. 2009).

Although most studies have described monotonic increases in reproduction with tree size, apparent decreases in total crop sizes of reproductive structures among very large trees have also been reported. Such a decrease might be expected if declines in net carbon fixation and/or meristem production are not compensated for by increases in reproductive allocation. However, it does not appear that any formal statistical analysis documenting a "hump-shaped" relationship between tree size and reproductive output has been presented (e.g., Greene and Johnson 1994; Gullison et al. 1996; Kainer et al. 2006; Naito et al. 2008), and patterns interpreted as decreases may simply represent increased variability with tree size, as is obvious in many data sets (e.g., Gullison et al. 1996; Kainer et al. 2006). Perhaps the most convincing data showing systematic decreases in reproductive output with tree size has been presented for *Aquilaria* spp. (Thymeleaceae) (Soehartono and Newton 2001), however again there is no formal statistical test. Increased variability in reproductive output among individuals with increasing tree age size in some cases is clearly related to biotic interactions. For example, Kainer et al. (2006) describe large increases in average liana load with tree size in *Bertholletia excelsa* (Lecythidaceae), and negative effects of lianas on reproduction; however, very large (>100 cm dbh) trees with low liana loads showed the highest reproductive output.

There are several reasons why simple allometric analyses of crop sizes do not give an adequate picture of reproductive allocation patterns through tree ontogeny. First, much of the "biomass" of large trees can consist of non-living heartwood tissue. In addition, very large trees of many species show pronounced crown thinning and loss of branches (e.g., Nock et al. 2008), such that stem diameter may not estimate biomass accurately. Plants may be considered as a population of modules (reiterated morphological units, such as a leaf with subtending shoot: Watkinson and White 1985), providing an alternative basis for quantifying size-dependent patterns of reproductive allocation in plants. Studies that have followed this approach have consistently found a pattern of increasing allocation of meristems to reproductive structures with increasing size or age in woody plants (Acosta et al. 1997; Lopez et al. 2001).

Theoretical studies of life history evolution have generally treated reproductive allocation as the total resources invested in reproduction, which corresponds to reproductive effort rather than allocation per se (Bazzaz et al. 2000). The size- and/or age-dependence of reproductive effort in trees, as distinct from reproductive allocation, has received little explicit attention in either theoretical or empirical studies. Reproductive costs other than direct resource investment in reproductive structures should include physiological costs associated with the transport of carbon and nutrients, construction respiration costs, and physiological costs associated with maintenance of turgor during the development of reproductive structures. A similar set of biophysical constraints that pertain to development and physiological function of leaves in tall tree canopies, such as effects of hydraulic path length effects, should pertain to reproductive structures as well. There is thus a strong basis to expect that the non-allocation fraction of reproductive effort (somatic cost of reproduction) should increase with tree size. This reasoning suggests that reproductive effort actually shows stronger increases with tree size than suggested by studies to date based on either allometric or module-based analyses.

4.1 Direct Assessments of Size-Dependent Carbon Partitioning

What is the proportion of carbon fixation allocated (or "partitioned" *sensu* Litton et al. 2007) to reproductive structures, and how does this change with tree size and age? Early studies suggested that this fraction was generally quite low (e.g., <15%: Linder and Troeng 1981; Dick et al. 1990a, b), supporting a conclusion that reproductive allocation was unlikely to be an important determinant of age-related growth declines (Ryan et al. 1997). The allometric data cited above do not give sufficient information to evaluate carbon partitioning, as the amount of carbon fixed and allocated to growth of other tissues (or to storage) has not generally been quantified. While relevant studies are scarce, Hirayama et al. (2008) present data showing increasing biomass allocation to reproduction with tree size in three species of Japanese *Quercus* ($P<0.01$; linear model analysis of data for mast fruiting year presented in their Table 2.1), with annual biomass allocation to reproduction of the largest trees sampled in each species estimated at 32–79% during a mast year. In another recent study Genet et al. (2010) provide estimates based on a chronosequence of *Fagus sylvatica* stands, with assessments of woody tissues growth, and non-structural carbohydrates in addition to reproductive structures. The proportion of annual fixed carbon in reproductive structures increased from ~5% or less in stands <25 years in age, to ~40% in stands >120 years. There is thus recent evidence that carbon allocation to reproduction increases substantially with tree size/age, and is, at least episodically, much higher than previously suggested.

5 Moncarpic Trees

Plants that show a single bout of reproduction shortly followed by senescence and death are termed monocarpic[1]. It has been incorrectly reported in recent papers (Poorter et al. 2005; Read et al. 2006) that there are only four genera and ~30 species of monocarpic trees known, but this is a serious underestimate. There are at least six arborescent palm genera that include strictly monocarpic species (Dransfield et al. 2008a): *Arenga* (20 species, all hapaxanthic, 4 monocarpic), *Caryota* (13 species, all hapaxanthic, 10 monocarpic), *Corypha* (6 species, all monocarpic), *Metroxylon* (6 of 7 species hapaxanthic, 5 monocarpic), *Wallichia* (10 species, all hapaxanthic, 1 monocarpic), *Raphia* (20 species, all hapaxanthic, roughly 1/3 monocarpic), and *Tahina* (monotypic), the last discovered in Madagascar in 2007 (Dransfield et al. 2008b). In addition, two other arborescent palm genera show a hapaxanthic but not strictly monocarpic habit: *Eugeissona* (6 species), and *Nannorrhops* (monotypic). Among dicotyledonous trees, monocarpy has been noted in *Cerberiopsis* (Apocynaceae: 1 of 3 species monocarpic) (Read et al. 2006, 2008), *Spathelia* (Rutaceae: ~15 species, at least 2 monocarpic) (Rodrigues 1962; Groppo et al. 2008), and *Tachigali* (Leguminosae: Caesalopiniodeae; ~20 species, mostly monocarpic) (Foster 1977; Poorter et al. 2005). The large tropical genera

[1]Trees in which individual stems die following reproduction are termed hapaxanthic; in the case of palms that form stem clusters many groups are hapaxanthic, but only species showing a single-stemmed habit are strictly monocarpic (Dransfield et al. 2008a).

Fig. 2.2 Examples of the ~100 known species of monocarpic trees, in which a single episode of reproduction is followed by rapid senescence and death. (**a**) *Tachigali versicolor* (Leguminosae: Caesalopiniodeae), a common tree at Barro Colorado Island, Panama; (**b**) *Cerberiopsis candelabra* (Apocynaceae) endemic to New Caledonia and forming extensive monospecific stands; (**c**) *Tahina spectabilis* (Dransfield et al. 2008b) recently discovered in northwestern Madagascar. Photo credits: (**a**) Robin Foster; (**b**) Jennifer Read; (**c**) John Dransfield

Strobilanthes and *Mimulopsis* (both Acantheaceae) also include monocarpic species; while many species are multi-stemmed shrubs, some grow to be sizeable single-stemmed trees (e.g., *Mimulopsis arborescens*). Additional monocarpic species occur in montane forests in Africa but have received almost no study: these include *Oreacanthus mannii*, *Acanthopale decempedalis*, *Isoglossa glandulosa*, *I. nervosa* (all Acanthaceae), and *Plectranthus insignis* (Lamiaceae) (Duncan Thomas and David Kenfack, *personal communication*). Subtropical examples of monocarpy include several species of rosette trees of the genus *Echium* (Boraginaceae) (Böhle et al. 1996). There are thus likely over 100 species of monocarpic trees in at least 15 genera (Fig. 2.2). This tally does not include the many examples of short-lived tropical trees that reproduce continuously (or nearly so) after the onset of reproduction, and so approach a monocarpic life history.

Monocarpic tree species provide an important set of examples in which the physiological impacts of reproduction completely determine future plant development. Although there are very few formal comparative studies (Poorter et al. 2005; Read et al. 2006, 2008), monocarpic trees appear to be characterized by systematic differences from co-occurring polycarpic trees. Characteristics of monocarpic tree species include: high resource investment in reproductive structures; rapid growth; high growth response to increased light availability; low mortality rates, at least at certain life history stages; and an early age (though not size) at reproduction (Poorter et al. 2005). It has been noted that reproduction in at least some monocarpic tree species is triggered by damage. *Cerberiopsis candelabra* is characteristically a mass-flowering species: entire stands will flower and die synchronously (Read et al. 2008). However, damaged individuals of *C. candelabra* will commonly flower out of synchrony (Read et al. 2006).

Some monocarpic trees (such as *Corypha* spp.) have a single non-branched stem that eventually produces a terminal inflorescence (i.e., the Holttum architectural model of Hallé et al. 1978). However, this is not the case in most monocarpic trees, and meristem limitation alone does not appear to be a sufficient mechanism in most cases. An early proposed mechanism for the coordinated reproduction and senescence observed in monocarpic plants was the "nutrient drain hypothesis": that plant resources are so drained by reproductive structures that insufficient resources remain to sustain other physiological functions (Molisch 1938). However, physiological studies of monocarpic species (mainly agronomic crops) indicate that reproduction does not entirely consume all labile carbon or other resources (Kelly and Davies 1988). A more recent formulation of the nutrient drain hypothesis is that nutrient and carbon sinks related to reproductive development show a feedback with vegetative sinks that disrupts future vegetative development (Kelly and Davies 1988). It has also, more speculatively, been hypothesized that senescence in moncarpic plants may be induced by a signal (e.g., a "death hormone" produced in reproductive structures) that initiates whole-plant senescence (Leopold 1961; Noodén et al. 1997). No such substance has been identified, and it is not clear why plants would evolve such a process. There are some cases in which the nutrient drain hypothesis, even in modified form, does not appear to be a viable explanation (Kelly and Davies 1988). A role for biotic agents has also been suggested. Read et al. (2006) cored *C. candelabra* trees at the time of flowering, and found wood to be consistently characterized by a dark-stained, watery rot compared with clean, pale wood in non-reproductive individuals. This observation suggests that in *C. candelabra* a biotic agent having severe detrimental effects may trigger reproductive onset, or that trees lose rapidly the capacity for pathogen defense following reproduction.

In summary, cases of monocarpic reproduction in trees offer several potentially important insights into effects of reproduction on tree growth and physiology more generally. First, reproduction clearly is capable of causing complete loss of physiological function (though the mechanism is unclear). Monocarpic trees appear to have arisen multiple times from polycarpic ancestors. One would therefore, from an evolutionary perspective, expect to find intermediate cases where reproduction has drastic, though not fatal, effects. Second, the "nutrient drain hypothesis", that death following reproduction in monocarpic plants is driven directly by allocation of all available resources to reproduction, has not been supported. It follows that reproductive costs can substantially outweigh the direct effects of resource allocation. Third, recent evidence that monocarpic trees show unusual combinations of high growth yet low mortality compared to co-occurring polycarpic species (Poorter et al. 2005) suggests that iteroparous reproduction in general results in costs in terms of slower growth and/or higher mortality compared to what is possible given biophysical constraints alone. Fourth, there are intriguing but anecdotal observations that death in at least some monocarpic tree species may be linked to biotic factors associated with reproduction.

6 Empirical Evidence for the "Costs of Reproduction" in Polycarpic Trees

The development of reproductive structures on a plant can affect a wide range of growth and physiological processes (Table 2.1), with implications for current and future survivorship and reproduction. Most, but not all, of these effects are consistent with the hypothesis of a tradeoff based on costs of reproduction to current or future growth. Observed effects on vegetative growth range from reductions in diameter increment and shoot elongation, to increased shoot mortality and meristem initiation. Reductions in leaf size and changes in leaf morphology, chemistry, and physiology have also been widely observed.

Table 2.1 Observed effects of reproduction on tree functional traits based on correlations within or among woody plants

	Effect	References
Growth:	Reduced stem diameter increment	Table 2.2
	Reduced shoot elongation	Table 2.3
	Reduced leaf initiation	Dick et al. (1990a)
	Increased shoot mortality	Gross (1972)
Morphology:	Reduced leaf size	Tappeiner (1969), Tuomi et al. (1982), Caesar and MacDonald (1984), Chapin and Moilanen (1991), Karlsson (1994), Miyazaki et al. (2002), Leal and Thomas (2003)
	Increased crown transparency	Innes (1994)
	Reduced leaf mass per area	Caesar and MacDonald (1984), Miyazaki et al. (2002)
Gas-exchange:	Reduced net photosynthesis	Karlsson (1994), Karlsson et al. (1996), Obeso et al. (1998), Miyazaki et al. (2002), Wheelwright and Logan (2004)
Leaf and shoot chemistry:	Reduced leaf N	Newell (1991), May and Killingbeck (1992), Karlsson (1994), Leal and Thomas (2003)
	Reduced stem N	Newell (1991)
	Increased leaf C:N ratio	Leal and Thomas (2003)
	Reduced leaf chlorophyll content	Leal and Thomas (2003)
	Increased chlorophyll A:B ratio	Leal and Thomas (2003)
	Reduced labile C	Newell (1991), Miyazaki et al. (2002), Ichie et al. (2005)

6.1 Correlation vs. Causation in Analysis of Reproductive Costs

The majority of data related to cost of reproduction in woody plants has been based on correlations observed in nature. In some cases a negative correlation between growth and reproduction may not indicate causation: in particular, if an environmental factor has a positive effect on growth but a negative impact on reproduction (or vice-versa) this can produce a negative correlation that might be spuriously interpreted as a tradeoff (Bell 1980). A recent example of this has been documented in Californian oak species in which rainfall is positively affects diameter increment but negatively affects acorn production (Knops et al. 2007). However, environmental factors commonly have positive effects on both growth and reproduction (e.g., Woodward et al. 1993; Despland and Houle 1997), a pattern that would in general obscure actual tradeoffs. Ideally analyses that control for major environmental drivers or experimental approaches should be used. Unfortunately, it is essentially always the case that the full set of important environmental drivers of tree growth and reproduction are not known a-priori. Experimental approaches are logistically challenging with trees, and may produce artifacts: for example, bud removal experiments may result in wounding responses.

Much of the data pertaining to reproductive effects on tree growth and physiology are based on observations and correlations that do not take into account environmental correlations; the review and meta-analyses presented thus may confound environmental correlations with direct reproductive effects. However, it is likely that environmental factors are at least equally likely to affect growth and reproduction in the same as in opposite directions; thus, environmental influences are most likely to result in increased variance within and (particularly) among studies. In this regard a pooled analysis (and especially a formal meta-analysis) is much less likely to be influenced by environmental covariance patterns than is any given study considered individually. In the following sections I also specifically compare results to available experimental studies and to studies in which attempts have been made to control for environmental factors.

6.2 Meta-analyses: Literature Search and Analytic Methods

For effects on which there is a relatively extensive pertinent literature (>7 species and >7 separate publications), I present meta-analyses of available data to assess the generality of the hypothesized patterns. The sample of studies utilized is based on tracing back references in prior publications, in addition to a systematic search of electronic databases. Meta-analyses of relationships are based on a pooled analysis of correlation coefficients, following Hunter and Schmidt (1990) (see also Field 2001). The Hunter-Schmidt statistic ($\overline{r}$) is simply a weighted mean correlation calculated as:

$$\overline{r} = \frac{\sum_{i=1}^{k} n_i r_i}{\sum_{i=1}^{k} n_i}$$

where r_i and n_i are the observed correlation and sample size of the ith study, and the summations are made among k studies. The significance is tested using a Z statistic ($\overline{r}$ divided by its standard error), with the associated probability calculated using the standard normal distribution (Field 2001). This is a "random effects" analysis, and so does not assume uniformity of relationships across studies. In many cases correlation coefficients were directly presented; in others they were calculated from original data or from data digitized from figures. In some cases reproduction was treated as a categorical variable, and correlations were calculated with non-reproductive stems or individuals scored as zero, and reproductives as 1. Analyses are restricted to publications that present data on trees that have not been subjected to breeding programs. There is, for example, a large literature on relationships between growth, physiology, and fruit production in alternate-bearing (climacteric) apples and other fruit trees. These agronomic taxa have been specifically bred for extreme expression of reproductive traits (and often for reduced stature), and therefore might be expected to show extreme tradeoffs between growth and reproduction.

6.3 *Dendrochronological Studies*

The diameter growth of trees in the temperate zone (and, as is increasingly being documented, of many trees in the tropics) is recorded by growth rings demarcated by latewood formed prior to the winter or dry season quiescent period. Diameter (or basal area) increment can be allometrically scaled to, and generally provides a good approximation of, tree biomass growth. Given that the reproductive output of trees generally shows high inter-annual variability, long-term records of tree reproductive output, in conjunction with dendrochronological analyses of tree ring series, provide a valuable source of information to detect and quantify the effects of reproduction on tree growth.

Meta-analysis results for studies examining tree ring increment in relation to reproductive output reveal a consistent pooled negative correlation (Table 2.2). The Hunter-Schmidt $\overline{r}$ (–0.230) is significantly less than zero at $P<0.001$. Negative patterns are also found for conifers and angiosperms considered separately: for conifers the Hunter-Schmidt $\overline{r}$ = –0.271 ($P<0.001$), and for angiosperms the Hunter-Schmidt $\overline{r}$ = –0.134 ($P<0.001$). The lower value for angiosperms is largely attributable to low correlations found for California oaks (Knops et al. 2007), which represent 5 of 7 samples. The mean correlation is also significantly negative

Table 2.2 Summary of dendrochronological studies included in meta-analysis examining the correlation of reproductive output (variously measured across studies) with tree ring increment across years

Species	Location	Years	r	Reference
Abies grandis	BC Canada	28	–0.317	Eis et al. (1965)
Abies lasiocarpa	Washington USA	28	**–0.044**	Woodward et al. (1993)
Abies lasiocarpa	Oregon USA	28	**–0.222**	Woodward et al. (1993)
Dacrydium cupressinum	New Zealand	33	**–0.304**	Norton and Kelly (1988)
Nothofagus truncata	New Zealand	33	**–0.303*****	Monks and Kelly (2006)
Picea abies	Norway	40	**–0.648*****	Selas et al. (2002)
Pinus banksiana	Quebec Canada	24	0.310	Despland and Houle (1997)
Pinus sylvestris	Spain	11	–0.140	Alonso et al. (2007)
Pseudotsuga menziesii	BC Canada	28	–0.765***	Eis et al. (1965)
Tsuga mertensiana	Washington USA	28	**–0.122**	Woodward et al. (1993)
Tsuga mertensiana	Oregon USA	28	**–0.163**	Woodward et al. (1993)
Quercus lobata	California USA	13	**0.048**	Knops et al. (2007)
Quercus douglasii	California USA	13	**0.007**	Knops et al. (2007)
Quercus agrifolia	California USA	13	**–0.107**	Knops et al. (2007)
Quercus chrysolepis	California USA	13	**0.159***	Knops et al. (2007)
Quercus kelloggii	California USA	13	**0.078**	Knops et al. (2007)

Correlation values listed in ***bold*** are partial correlations that correct for climate effects (or comparable analyses based on indexing). The Hunter-Schmidt $\overline{r} = -0.225 \pm 0.021$ ($P = 0.0001$)
Significant trends for studies considered individually are indicated by asterisks: *$P<0.05$; **$P<0.01$; ***$P<0.001$

if one considers only studies that have included climate variables as covariates: Hunter-Schmidt $\overline{r} = -0.208$ (P<0.001). Although the data analyzed include prominent examples of null results reported in the literature (e.g., Despland and Houle 1997; Alonso et al. 2007; Knops et al. 2007), these studies tended to be of shorter duration and thus have reduced influence on the overall calculations. The finding of a highly significant meta-analysis result, in spite of the fact that relatively few studies are significant considered individually (Table 2.2), is common in cases where sample sizes of individual studies are low and/or the signal to be detected is either weak or variable among studies (Hedges and Olkin 1985). Detection of prevailing trends under these conditions is a main motivation for meta-analysis generally.

Several additional relevant studies were noted but not included in the meta-analysis due to either a small series length, insufficient reporting of data, or lack of access to some older publications. Consistent with the studies in Table 2.2, these studies generally find negative correlations between tree ring increment and reproductive output (generally fruit or cone production): for *Abies balsamea* in New Brunswick, Canada (Morris 1951), *Pinus sylvestris* and *Picea abies* in Sweden (Eklund 1954, 1957), *Pinus ponderosa* in Colorado, USA (Linhart and Mitton 1985), *Pseudotsuga menziesii* in BC, Canada (El-Kassaby and Barclay 1992), and *Fagus crenata* in Japan (Hoshino et al. 2008 – notably based on a 48 year time series). A few additional null results have also been reported: for *Pinus ponderosa* in Washington State USA (Daubenmire 1960), *Pinus sylvestris* in Poland

(Chalupka et al. 1976), and another (much shorter-duration) study of *Fagus crenata* in Japan (Yasumura et al. 2006).

Dendrochronological studies coupled with long series of reproductive measurements also provide a valuable avenue for analyses of time-lagged effects of reproduction, and of the interactions between climate variability, reproduction, and tree growth. A number of studies have employed auto-regressive moving average (ARIMA) models or similar techniques to examine lagged effects of reproduction on growth. For example, Woodward et al. (1993) detected lagged effects of coning on growth over the following 2 years in two montane conifer species. However, other studies have found effects only in the year of fruit or cone production (e.g., Selas et al. 2002).

Several studies have simultaneously considered climatic variability and reproduction as predictors of long-term tree growth patterns. Woodward et al. (1993) examined relationships for two conifers (*Abies lasiocarpa* and *Tsuga mertensiana*) between cone counts and the residuals of a growth index value that was adjusted both for size trends (as per standard dendrochronological indexing methods) and temperature and precipitation data. They report consistently negative correlations based on this approach (Table 2.2). Monks and Kelly (2006) found that the partial negative correlation between growth and reproduction in *Nothofagus truncata* was strengthened when climate variables were included in a predictive model. In contrast, Knops et al. (2007) found in five species of Californian oaks (*Quercus* spp.) that inclusion of precipitation as a predictor resulted in partial correlations between growth and reproduction not different from zero (and one case of a small apparent positive partial correlation between growth in reproduction). In the case of Californian oaks, dry conditions during the early growing season appear to inhibit growth while enhancing pollination success (Knops et al. 2007), a pattern that seems likely to be system-specific.

6.4 Reproductive Allocation and Shoot Extension Within Tree Crowns

Negative correlations between shoot extension growth and production of reproductive structures have been widely reported in the literature. The first observations on this phenomenon appear to date to Morris (1951); most often data have been recorded in terms of annual shoot length increment of internodes produced the year following a given reproductive event (e.g., Gross 1972; Fox and Stevens 1991), but the specific measures of reproduction and growth vary among studies.

The meta-analysis of available data provides remarkably consistent support for a general negative trend between shoot or branch growth and the quantity or mass of reproductive structures (generally fruit or cones) present (Table 2.3). All reported correlations are negative, four of eight being significant considered individually. The Hunter-Schmidt $\bar{r}$ (–0.341) is significantly less than zero at $P<0.001$. Additional studies that did not present simple correlations (or data allowing calculations) have

Table 2.3 Summary of studies included in meta-analysis examining the correlation of stem increment and reproductive output within tree crowns

Species	Location	N	r	Reference
Alnus hirsuta var. *sibirica*	Japan	206	–0.270**	Hasegawa and Takeda (2001)
Alnus hirsuta var. *sibirica*	Japan	49	–0.630***	Hasegawa and Takeda (2001)
Betula alleghaniensis	Ontario Canada	15	–0.518	Gross (1972)
Betula papyrifera	Ontario Canada	15	–0.859	Gross (1972)
Pinus radiata	Australia	6	–0.295	Cremer (1992)
Pinus strobus	Ontario Canada	26	–0.170	Leal and Thomas (2003)
Pseudotsuga menziesii	California USA	10	–0.838***	Tappeiner (1969)
Salix alaxensis	Alaska USA	92	–0.232*	Fox and Stevens (1991)

The Hunter-Schmidt mean $r = -0.341 \pm 0.013$ ($P = 0.0000015$)
Significant trends for studies considered individually are indicated by asterisks: *$P<0.05$; **$P<0.01$; ***$P<0.001$

likewise generally found negative relationships (Henriksson and Ruohomäki 2000), although null results have also been reported (e.g., Bañuelos and Obeso 2004). A number of studies also describe substantial variation among populations and sites, such that negative relationships between reproduction and shoot extension are present in some populations but not others, possibly related to local variation in soil resources or other factors (e.g., Obeso 1997).

As in the case of dendrochronological studies of reproductive effects, the analysis in Table 2.3 is based on correlative data. Thus positive responses of both reproduction and growth to environmental heterogeneity might be expected to mask an underlying tradeoff (Obeso 2002; Karlsson et al. 2006). In addition, physiological integration should also generally act to obscure patterns at the shoot or branch level (Bañuelos and Obeso 2004). Experimental approaches, primarily based on removal of reproductive buds, have also been employed to test for reproductive costs at the branch and whole-tree level. Fox and Stevens (1991) in a bud removal experiment failed to find experimental evidence for negative effects of reproduction on growth in *Salix alaxensis*, in spite of a weak negative correlation between extension growth and reproduction. A follow-up experiment found a compensatory response to removal of either reproductive or vegetative buds, leading to the suggestion that bud removal is not a valid experimental approach to detect or measure reproductive costs due to artifacts related to wounding responses (Fox 1995). Nevertheless, similar studies have detected increased vegetative growth following reproductive bud removal in other woody plants (Houle 2001; Karlsson et al. 2006).

6.5 Reproduction and Shoot Demography Within Tree Crowns

Meristems are generally genetically predetermined to develop as vegetative or reproductive structures, and there is thus a potential for reproduction to limit future growth and development through meristem limitation. Reproduction may also

Fig. 2.3 Crown dieback following heavy production of catkins in *Betula alleghaniensis* (From Gross 1972: used by permission)

sufficiently drain resources to result in subsequent crown dieback. Substantial crown dieback following episodes of very high seed production was observed in *Betula alleghaniensis* by Gross (1972): individuals with high a high proportion of catkins per terminal bud site showed progressively lower shoot growth in the subsequent year, and 3 of 4 individuals with >95% of terminal buds bearing catkins showed branch dieback (Fig. 2.3). This pattern was observed very extensively in the study region, with an estimated 20–50 cm loss of the upper crown common following an intensive regional reproductive episode (Gross 1972). Studies of reproductive effects on shoot demography are particularly logistically difficult in trees, and this remarkable early research has not been replicated to date. A few studies have been conducted on small-statured, accessible trees (e.g., Maillette 1987), but have not detected significant negative effects of reproduction on shoot demography.

6.6 *Leaf-Level Gas-Exchange, Chemistry and Morphology*

Proximity of reproductive structures could either act to enhance net photosynthesis by reducing feedback inhibition, or to reduce carbon uptake due to nutrient allocation effects. Enhanced photosynthesis in leaves near reproductive structures has been commonly found in herbaceous plants and cultivated fruit trees (e.g. DeJong 1986; Urban et al. 2003). However, reported proximity effects of reproductive structures on net photosynthesis in non-cultivated trees appear to be generally negative (Karlsson et al. 1996; Obeso et al. 1998; Miyazaki et al. 2002; Wheelwright and Logan 2004). This may be due to relatively higher nutrient limitation in natural as compared to cultivated systems. Patterns of leaf chemistry appear to be consistent with fruits and seeds constituting a significant local sink for nutrients, in particular nitrogen. Reduced leaf N and reduced chlorophyll content in leaves that develop in proximity to reproductive structures has been reported in *Pinus strobus* by Leal and Thomas (2003), who also report a large increase in the chlorophyll a:b ratio. Such an increase is a predicted consequence of chlorophyll degradation, since chlorophyll b is converted to chlorophyll a prior to further breakdown (Folley and Engel 1999; Hörtensteiner 2006).

Ontogenetic patterns of leaf traits with tree size provide additional indirect evidence for effects of reproduction. Ontogenetic declines in the area of individual leaves have been widely observed in comparisons of saplings vs. mature trees (Thomas and Ickes 1995). However, the first leaves produced in tree ontogeny are typically very small, likely as a result of constraints imposed by seed resources. Average leaf size thus reaches a peak at an intermediate tree size. In a number of tree species this peak occurs approximately at the size of reproductive onset (Alvarez-Buylla and Martinez-Ramos 1992; Thomas and Ickes 1995), though in others the peak clearly falls earlier in development (Reich et al. 2004; Panditharathna et al. 2008; Thomas 2010). If biophysical processes, such as hydraulic effects on leaf turgor, were directly driving changes in leaf size, an increase in size early in ontogeny would not be expected; nor would one predict any association of this pattern with reproductive onset.

While many published studies addressing ontogenetic change in leaf gas-exchange have compared seedlings or saplings vs. mature trees (Bond 2000; Thomas and Winner 2002), there have been surprisingly few efforts to characterize gas-exchange over the entire course of tree ontogeny (Niinemets 2002; Nabeshima and Hiura 2008). In a recent effort to address this gap, I found that light-saturated photosynthesis and leaf N reach a peak at intermediate tree sizes closely approximating the size at onset of maturity in three species of temperate deciduous trees (Fig. 2.4). In contrast, leaf traits that reflect investment in structure and defense, including leaf mass per area, leaf thickness, leaf tissue density, and leaf C content increased monotonically with tree size. Qualitatively similar patterns were found in the three tree species examined (Thomas 2010).

6.7 Belowground Processes

Agronomic studies of fruit trees suggest that there is commonly a very strong tradeoff between root growth and fruit development. For example, in alternate-bearing apple trees, allocation in non-fruiting years is largely directed to root growth; root pruning is widely used to induce reproduction in non-fruiting years (Ferree et al. 1991). Very few studies of non-cultivated trees have been undertaken. A masting event in

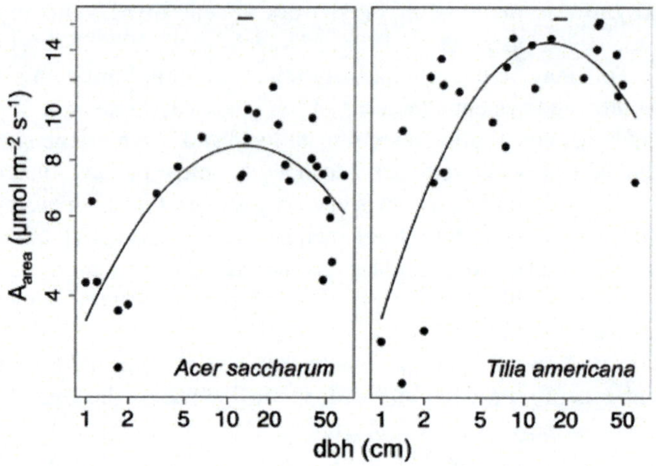

Fig. 2.4 Light saturated photosynthesis (A_{area}) as a function of stem diameter at 1.3 m height (dbh) in *Acer saccharum* and *Tilia americana*. In both cases the peak in A_{area} is closely coincident with the size (dbh) at reproductive onset (14.3 cm in *A. saccharum* and 19.8 cm in *T. americana*: shown by horizontal bars ± 95% CI). *Dotted line* is a 2nd-order polynomial fit to log-log transformed data (Modified from Thomas 2010)

Dryobalanops aromatica (Diptercarpaceae) did not result in detectable declines in root carbohydrate levels (Ichie et al. 2005).

6.8 Reproduction Effects That Don't Fit into a "Costs" Framework

There are a variety of physiological consequences of reproductive activity that do not fit easily into a "costs" framework. Reproductive structures serve as strong sinks for carbon, and in some cases photosynthetic rates of leaves near to reproductive structures can be enhanced, as a consequence of local carbon sink strength reducing feedback inhibition effects (Gifford and Evans 1981), and other factors (Urban et al. 2003). Reproductive structures can also supply much of their own carbon demands through photosynthesis (Bazzaz et al. 1979; Aschen and Pfanz 2003) particularly through refixation of CO_2 generated by construction respiration. Estimates of the proportion of the total C requirements internally generated by developing fruit average ~ 16%, but few measurements have been made on trees (Aschen and Pfanz 2003). Photosynthesis of reproductive structures and carbon refixation are both considered cases of adaptations of plants to offset reproductive costs (Obeso 2002). Development of reproductive structures will also have impacts on biomechanical support, within-crown shading, and plant hormonal balance that may or may not qualify as "costs".

6.9 Sexual Dimorphism of Functional Traits in Dioecious Trees

An additional form of "natural experiment" for evaluating the effects of reproduction on growth patterns and tree functional traits is to compare the sexes of dioecious tree species. The production of pistillate flowers, fruit and seed generally entails a much higher allocation of biomass and mineral nutrients (especially N) than does the production of staminate flowers alone; likewise female cones of conifers require greater investment, particularly of N, than do male cones (McDowell et al. 2000). Thus, the costs of reproduction have generally been predicted (and widely observed) to be higher in pistillate individuals. There have been several recent reviews that focus in whole or in part on sexual dimorphism in plants (Dawson and Geber 1999; Obeso 2002; Case and Ashman 2005); here I review some main results and specifically highlight studies that examine sex differences in ontogenetic patterns.

Dioecious trees (and dioecious plants generally) commonly show pronounced differences between the sexes in leaf-level physiological traits, such as photosynthetic capacity, leaf N, water use efficiency, and tissue water relations (Dawson and Geber 1999; Case and Ashman 2005). A now-classic result is the finding that female trees differ in responses to water stress, with males showing more conservative water use strategies under xeric conditions (Dawson and Bliss 1989). This pattern is a predicted consequence of evolved response to higher reproductive costs of water and nutrient supply to developing seeds and fruits, and has been found in numerous woody dioecious plant species (Dawson and Bliss 1989; Dawson and Ehleringer 1993; Dawson and Geber 1999; Espirito-Santo et al. 2003; Dudley 2006). Female trees, as a consequence, commonly show greater drought sensitivity than males, and may be differentially distributed to more mesic habitats (e.g., Dawson and Ehleringer 1993; Dudley 2006). One may infer from these results that female reproductive function in non-dioecious species must have comparably large impacts on whole-tree function; moreover, increasing reproductive allocation with tree size should result in strong size x sex interactive effects on drought sensitivity. Along these lines, strong interactive effects of sex and size on diameter increment responses to precipitation have recently been found in a dendrochronological study of *Juniperus thurifera* in Spain (Rozas et al. 2009).

The slope of the relationship between growth rate and tree size (following the peak of the differential form of the growth curve) has been considered a signature characteristic of age-related declines tree growth (e.g., Bond et al. 2007). A strong role for reproductive effort in driving age-related decline tree growth would thus be indicated not simply by differences between the sexes in average growth rates, but specifically differences in the slope of this decline. Wheelwright and Logan (2004) provide one of the only pertinent data sets available, on the dioecious tree *Ocotea tenera*, which shows both lower growth rates among female trees, and a much steeper decline in growth with increasing size in female trees (Fig. 2.5). A similar pattern has also been reported for the dioecious shrub *Rhamnus alpina* (Bañuelos and Obeso 2004). Higher growth and/or reduced mortality of male trees late in ontogeny has also been inferred from an elevated

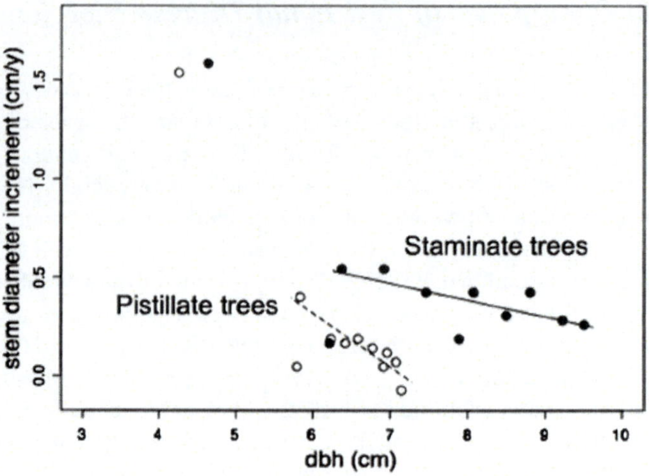

Fig. 2.5 Patterns of decline in diameter increment in males (staminate: *solid symbols*) vs. females (pistillate: *open symbols*) in a natural population of the tropical tree *Ocotea tenera*. Linear regressions for the later phase of growth (excluding the first two annual measurements in each case) are shown: for males (*solid line*) $y = 1.063–0.085x$; for females (*dotted line*) $y = 1.954–0.273x$. Slopes differ significantly between sexes (ANCOVA $P < 0.001$). Data are re-analyzed from Fig. 2.1 of Wheelwright and Logan (2004)

ratio of male to female trees observed in large size classes of tropical dioecious trees (Thomas and LaFrankie 1993).

6.10 Interactions Between Reproduction and Biotic Factors

Although costs of reproduction have figured centrally in analyses of sexual dimorphism in plants, males and females may also show differential evolutionary responses to other factors. In particular, male reproductive success depends on pollen transfer; there may, for example, be strong selection for increased male height in trees with wind-dispersed pollen or height-dependent access to pollinators (Lloyd and Webb 1977). Higher growth rates in males (e.g., Fig. 2.5), may thus potentially be related to sexual selection as well as direct physiological costs. In addition, male and female plants may show differential levels of and tolerance of herbivory, with slow-growing females generally being better defended (Jing and Coley 1990; Ågren et al. 1999; Verdu et al. 2004). There may also be effects of reproduction on growth and physiology mediated by herbivores or pathogens that primarily attack reproductive structures themselves. For example, the ash flower gall mite (*Eriophyes fraxiniflora*) attacks male flowers in many species of *Fraxinus*, but can cause extensive deformation of both flowers and leaves (Wawrzynski and Ascerno 1990). Such sex-specific effects in dioecious taxa suggest similar large secondary effects of reproduction more generally that are mediated by biotic agents.

7 Synthesis and Some Implications for Future Studies

The body of research reviewed is remarkably consistent in indicating that the "costs of reproduction" in trees are not trivial. Eight independent sources of data support this contention: (1) Among species, trees that show a smaller size at onset of maturity tend to show a smaller maximum size; (2) As is the case with most other organisms, various sources of data suggest that reproductive allocation generally increases with tree size; (3) Monocarpic trees provide a set of clear examples in which reproduction catastrophically impacts future growth and development, and comparisons with polycarpic trees suggest that repeated reproduction carries a consistent cost in terms of growth potential; (4) Dendrochonological analyses indicate an overall pattern of negative correlations between tree growth and reproduction, and in some cases of carry-over effects of reproduction that persist for >1 year; (5) Shoot extension growth within tree crowns is generally reduced when in close proximity to reproductive structures, (6) There are likewise frequently local negative effects of reproductive structures on leaf size, N content, and photosynthetic capacity within tree crowns; (7) Leaf size, N content, and photosynthetic capacity commonly increase with tree size early in ontogeny, but show decreases only after reproductive onset; (8) In dioecious tree species, females generally show reduced growth, and growth analyses indicate that females can show much steeper reductions in growth growth rate with tree size than do males.

Can such costs of reproduction explain, at least in part, the general pattern of age-related declines in growth in trees? The meta-analyses presented above (Tables 2.2 and 2.3) indicate pronounced negative effects of reproduction on tree growth, but do not directly assess whether these negative effects increase with tree ontogeny. However, two lines of reasoning suggest that reproductive effects must increase with tree ontogeny, and likely do contribute to age-related declines in growth. Trees universally show a long period of pre-reproductive growth during which negative effects of reproduction cannot be present (Fig. 2.1). Therefore negative effects of reproduction on growth must at a minimum be expressed as a "step function", being present only after reproductive onset. In addition, the widespread pattern of increasing reproductive output and frequency with tree size and age suggests that the negative consequences of reproduction on growth also must generally increase with tree size and age. It has previously been argued that because tree reproduction is intermittent, reproductive cost effects are unlikely to explain any continuous decline in growth (Ryan et al. 1997). However, studies reviewed here have found evidence for carry-over effects of reproduction in prior years on current growth (Woodward et al. 1993) and shoot survivorship (Gross 1972).

How large is the effect of ontogenetic increases in reproductive allocation on tree growth? A recent study of *Fagus sylvatica* by Genet et al. (2010) indicates an increase in annual carbon allocation of from <5% in young trees to ~40% in trees >120 years, with a concomitant decrease in carbon allocation to stem growth. However, the direct mechanistic relevance of these data depend on whether the growth of large trees is carbon limited. Recently, a pattern of increasing non-structural carbohydrate concentrations in old trees (Sala and Hoch 2009) and observations of little or no growth enhancement to elevated CO_2 in mature trees (Körner et al. 2005) have both been interpreted as indicating a lack of carbon limitation. However, the *Fagus* stands studied by Genet et al. (2010) showed no age-related trend in non-structural carbohydrate concentration. Moreover, since concentrations of N and other mineral nutrients are generally higher in reproductive than somatic tissues, reproductive allocation could impact growth even in "carbon-saturated" trees. As noted above, observed physiological effects of leaf proximity to reproductive structures commonly include reduced leaf N, chlorophyll content, and photosynthetic capacity, all suggesting strong impacts of N allocation.

A direct test of the effects of tree reproduction on age-related trends in growth and physiology is in theory possible if one could experimentally inhibit reproduction through all or part of a tree's lifespan, and compare patterns to a control population. Experimental approaches for inhibiting reproduction could range from bud removal experiments to genetic modification methods, which have particular promise in *Populus* (Brunner and Nilsson 2004). This type of experiment has been utilized extensively in studies aimed at quantifying the costs of reproduction in short-lived herbaceous plants (e.g., Reekie and Bazzaz 1987a, b, c), and some woody shrub species (e.g., Houle 2001; Karlsson et al. 2006). While comparable experiments on trees appear never to have been undertaken, they may be feasible in short-lived, small-statured species. Candidate tree species for such experiments might include small-statured ephemeral pioneer trees, such as members of the genera *Trema* (Ulmaceae), *Clerodendrum* (Verbenaceae), and *Carica* (Caricaceae), and *Salix* (Salicaceae) or *Populus* (Salicaceae) in the temperate zone. Such experiments would also offer a valuable platform to better understand physiological mechanisms, including the relative importance of carbon limitation.

Dendrochronological studies linked to long-term records of tree reproduction – and where possible with experimental manipulations – will likely continue to present the best opportunities to understand effects of reproduction on age-related trends in tree growth and functional biology. Four types of dendrochronological studies stand out as offering great potential. First, additional analyses that link climate variability to trends in both tree reproduction and increment growth across a variety of species are essential. Work to date has not specifically sought to quantify the possible role of reproduction in driving age-related changes in growth patterns, and the common practice of removing age-related trends prior to analysis is likely to obscure such patterns. Studies of this type would be particularly novel and potentially revealing in tropical regions, where consistent annual rings are much more common than has commonly been believed (Worbes 2002). Second, studies of growth pattern differences between the sexes of dioecious species offer an important and under-explored avenue for understanding reproduction in relation to age-related trends in growth and physiology. Third, monocarpic trees provide a unique potential for comparative studies on age-related growth and physiology, in that there are no possible effects of reproduction until the end of the life cycle. It would be particularly interesting to examine whether patterns of age-related growth and physiology differ between monocarpic and polycarpic members of such genera as *Tachigali* and *Cerberiopsis*. Finally, there is a near complete absence of studies examining relationships between reproduction and root growth and functional biology in non-cultivated trees.

In all future dendrochronological studies it will be important to statistically partition effects among alternative causal processes, including direct climate effects and reproduction (Knops et al. 2007). In addition, it would be useful to partition "age-related" trends in tree growth to explicit age vs. size effects, in datasets where there is appreciable orthogonal variation in age vs. size. As an example of this approach, in black spruce (*Picea mariana*) declines in diameter increment through ontogeny, and of the capacity to show release responses following harvests, were principally attributable to tree age rather than size (Thorpe et al. 2007).

In terms of research applications, reproductive effects on tree growth have long been considered a potential breeding target for enhancing growth and yield (e.g., Ledig and Linzer 1978; El-Kassaby and Barclay 1992; Strauss et al. 1995). Breeding for trees with low reproductive effort is likely to be a much easier objective than breeding to reduce effects of biophysical limitations to growth, since the objective is essentially a loss of function. There is also considerable interest in breeding for reproductive sterility in genetically engineered trees, to limit possible genetic pollution effects (Strauss et al. 1995; van Frankenhuyzen and Beardmore 2004; Brunner et al. 2007). From an environmental management perspective, a better understanding of the links between growth and reproduction of trees is important in the context of anthropogenic climate change and other major perturbations of forest ecosystems. Reproductive development of trees is commonly more sensitive to climatic variables (and to different climatic variables) than is growth (e.g., Woodward et al. 1993; Despland and Houle 1997; Knops et al. 2007), and reproduction may likewise show stronger responses to rising CO_2 (e.g., LaDeau and Clark 2001). Global change impacts on tree reproductive output may thus have greater ecological impacts on forest community interactions and biodiversity than changes in total biomass accumulation (e.g., Ostfield et al. 1996; Jones et al. 1998; Curran and Leighton 2000).

In conclusion, reproductive effects should be looked at as one among a suite of processes that impact patterns of age-related growth and functional biology of trees. Given the centrality of reproduction and reproductive costs in life-history theory, research to better understand the physiology and comparative ecology of reproductive effects, and their interaction with biophysical limitations to tree growth, should be a priority for future research.

Acknowledgments I thank Michael Ryan, Lourens Poorter, and Jacob Weiner for helpful reviews of the manuscript, and John Dransfield, Robin Foster, David Kenfack, Jenny Read, and Duncan Thomas for input on and/or photos of monocarpic trees.

References

Acosta FJ, Delgado JA, Lopez F, Serrano JM (1997) Functional features and ontogenetic changes in reproductive allocation and partitioning strategies of plant modules. Plant Ecol 132:71–76

Ågren J, Danell K, Elmqvist T, Ericson L, Hjältén J (1999) Sexual dimorphism and biotic interactions. In: Geber MA, Dawson TE, Delph LF (eds.) Gender and sexual dimorphism in flowering plants. Springer, Berlin, pp 217–246

Alonso CM, Valladares F, Camarero JJ, Lopez Arias M, Serrano M, Rodrıguez JA (2007) The uncoupling of secondary growth, cone and litter production by intradecadal climatic variability in a Mediterranean Scots pine forest. For Ecol Manage 253:19–29

Alvarez-Buylla ER, Martinez-Ramos M (1992) Demography and allometry of *Cecropia obtusifolia*, a neotropical pioneer tree – an evaluation of the climax-pioneer paradigm for tropical rain forests. J Ecol 80:275–290

Andrewartha HG, Birch LC (1954) The distribution and abundance of animals. University of Chicago Press, Chicago, 782 pp

Aschen G, Pfanz H (2003) Non-foliar photosynthesis – a strategy of additional carbon acquisition. Flora 198:81–97

Bañuelos M-J, Obeso J-R (2004) Resource allocation in the dioecious shrub *Rhamnus alpinus*: the hidden costs of reproduction. Evol Ecol Res 6:1–17

Bazzaz FA, Carlsson RW, Harper JL (1979) Contribution to reproductive effort by photosynthetic flowers and fruits. Nature 279:554–555

Bazzaz FA, Chiariello NR, Coley PD, Pitelka LF (1987) Allocating resources to reproduction and defense. Bioscience 37:58–67

Bazzaz FA, Ackerly DD, Reekie EG (2000) Reproductive allocation and reproductive effort in plants. In: Fenner M (ed.) Seeds: the ecology of regeneration in plant communities, 2nd edn. CAB International, Oxford, pp 1–37

Bell G (1980) The costs of reproduction and their consequences. Am Nat 116:45–76

Birrell J (1987) Rights in the medieval forest: disputes and conflicts in the thirteenth century. Past Present 117:22–49

Bloom A, Chapin FS, Mooney HA (1985) Resource limitation in plants – an economic analogy. Ann Rev Ecol Syst 16:363–392

Böhle U-R, Hilger HH, Martin WF (1996) Island colonization and evolution of the insular woody habit in *Echium* L. (Boraginaceae). Proc Natl Acad Sci USA 93:11740–11745

Bond BJ (2000) Age-related changes in photosynthesis of woody plants. Trends Plant Sci 5:349–355

Bond BJ, Czarnomski NM, Cooper C, Day ME, Greenwood MS (2007) Developmental decline in height growth in Douglas-fir. Tree Physiol 27:441–453

Brunner AM, Nilsson O (2004) Revisiting tree maturation and floral initiation in the poplar functional genomics era. New Phytol 164:43–51

Brunner AM, Li JY, DiFazio SP, Shevchenko O, Montgomery BE, Mohamed R, Wei H, Ma C, Elias AA, VanWormer K, Strauss SH (2007) Genetic containment of forest plantations. Tree Genet Genomes 3:75–100

Caesar JC, Macdonald AD (1984) Shoot development in *Betula papyrifera*. V. Effect of male inflorescence formation and flowering on long-shoot development. Can J Bot 62:1708–1713

Case AL, Ashman T-L (2005) Sex-specific physiology and its implications for the cost of reproduction. In: Reekie EG, Bazzaz FA (eds.) Reproductive allocation in plants. Academic, New York, pp 129–158

Caspersen JP, Saprunoff M (2005) Seedling recruitment in a northern temperate forest: the relative importance of supply and establishment limitation. Can J For Res 35:978–989

Chalupka W, Giertych M, Krolikowski Z (1976) The effect of cone crops in Scots pine on tree diameter increment. Arbor Kornickie 21:361–366

Chapin FS, Moilanen L (1991) Nutritional controls over nitrogen and phosphorus resorption from Alaskan birch leaves. Ecology 72:709–715

Chapin FS, Bloom AJ, Field CB, Waring RH (1987) Plant responses to multiple environmental factors. Bioscience 37:49–57

Cheplick GP (2005) The allometry of reproductive allocation. In: Reekie EG, Bazzaz FA (eds.) Reproductive allocation in plants. Academic, New York, pp 97–128

Cipollini ML, Stiles EW (1991) Costs of reproduction in *Nyssa sylvatica*: sexual dimorphism in reproductive frequency and nutrient flux. Oecologia 86:585–593

Clark JS (1991) Disturbance and tree lift history on the shifting mosaic landscape. Ecology 72:1102–1118

Cohen D (1968) A general model of optimal reproduction in a randomly varying environment. J Ecol 56:219–228

Cohen D (1976) The optimal timing of reproduction. Am Nat 110:801–807

Cremer KW (1992) Relations between reproductive growth and vegetative growth of *Pinus radiata*. For Ecol Manage 52:179–199

Curran L, Leighton M (2000) Vertebrate responses to spatiotemporal variation in seed production of mast-fruiting Dipterocarpaceae. Ecol Monogr 70:101–128

Daubenmire R (1960) A seven-year study of cone production as related to xylem layers and temperature in *Pinus ponderosa*. Am Midl Nat 64:187–193

Davies SJ, Asthon PS (1999) Phenology and fecundity in 11 sympatric pioneer species of *Macaranga* (Euphorbiaceae) in Borneo. Am J Bot 86:1786–1795

Dawson TE, Bliss LC (1989) Patterns of water use and tissue water relation in the dioecious shrub, *Salix arctica*: the physiological basis for habitat partitioning between the sexes. Oecologia 79:332–343

Dawson TE, Ehleringer JR (1993) Gender-specific physiology, carbon isotope discrimination, and habitat distribution in boxelder, *Acer negundo*. Ecology 74:798–815

Dawson TE, Geber MA (1999) Dimorphism in physiology and morphology. In: Geber MA, Dawson TE, Delph LF (eds.) Gender and sexual dimorphism in flowering plants. Springer, Berlin, pp 175–215

DeJong TM (1986) Fruit effects on photosynthesis in *Prunus persica*. Physiol Plant 66:149–153

Despland E, Houle G (1997) Climate influences on growth and reproduction of *Pinus banksiana* (Pinaceae) at the limit of the species distribution in Eastern North America. Am J Bot 84:928–937

Dick JM, Jarvis PG, Leakey RRB (1990a) Influence of male cones on early season vegetative growth of *Pinus contorta* trees. Tree Physiol 6:105–117

Dick JM, Leakey RRB, Jarvis PG (1990b) Influence of female cones the vegetative growth of *Pinus contorta* trees. Tree Physiol 6:151–163

Dransfield J, Uhl NW, Asmussen CB, Baker WJ, Harley MM, Lewis CE (2008a) Genera palmarum: the evolution and classification of palms. Royal Botanical Gardens, Kew, UK, 732 pp

Dransfield J, Rakotoarinivo M, Baker WJ, Bayton RP, Fisher JB, Horn JW, Leroy B, Metz X (2008b) A new coryphoid palm genus from Madagascar. Bot J Linn Soc 156:79–91

Dudley L (2006) Ecological correlates of secondary sexual dimorphism in *Salix glauca* (Salicaceae). Am J Bot 93:1775–1783

Eis S, Garman EH, Ebell LF (1965) Relation between cone production and diameter increment of Douglas-fir (*Pseudotsuga menziesii* (Mirb.) Franco), Grand fir (*Abies grandis* (Dougl.) Lindl.) and western white pine (*Pinus monticola* Dougl.). Can J Bot 43:1553–1559

Eklund B (1954) Variation in the widths of the annual rings in pine and spruce due to climatic conditions in northern Sweden during the years 1900–1944 (in Swedish). Medd Statens Skogsforsknings Inst 44:5–150

Eklund B (1957) The annual ring variations in spruce in the centre of northern Sweden and their relation to climatic conditions (in Swedish). Medd Statens Skogsforsknings Inst 47:1–53

El-Kassaby YA, Barclay HJ (1992) Cost of reproduction in Douglas-fir. Can J Bot 70:1429–1432

Elser JJ, Bracken MES, Cleland EE, Gruner DS, Harpole WS, Hillebrand H, Ngai JT, Seabloom EW, Shurin JB, Smith JE (2007) Global analysis of nitrogen and phosphorus limitation of primary producers in freshwater, marine and terrestrial ecosystems. Ecol Lett 10:1135–1142

Espirito-Santo MM, Madeira BG, Neves FS, Faria ML, Fagundes M, Fernandes GW (2003) Sexual differences in reproductive phenology and their consequences for the demography of *Baccharis dracunculifolia* (Asteraceae), a dioecious tropical shrub. Ann Bot 91:13–19

Ferree DC, Myers SC, Schupp JR (1991) Root pruning and root restriction of fruit trees-current review. Acta Hort ISHS 322:153–166

Field AP (2001) Meta-analysis of correlation coefficients: a Monte Carlo comparison of fixed- and random-effects methods. Psychol Meth 6:161–180

Folley P, Engel N (1999) Chlorophyll b to chlorophyll a conversion precedes chlorophyll degradation in *Hordeum vulgare* L. J Biol Chem 274:21811–21816

Fonseca MG, Vidal E, dos Santo FAM (2009) Intraspecific variation in the fruiting of an Amazonian timber tree: implications for management. Biotropica 41:179–185
Foster RB (1977) *Tachigalia versicolor* is a suicidal neotropical tree. Nature 268:624–626
Fox JF (1995) Shoot demographic responses to manipulation of reproductive effort by bud removal in a willow. Oikos 72:283–287
Fox JF, Stevens GC (1991) Costs of reproduction in a willow: experimental responses vs. natural variation. Ecology 72:1013–1023
Genet H, Bréda N, Dufrêne E (2010) Age-related variation in carbon allocation at tree and stand scales in beech (*Fagus sylvatica* L.) and sessile oak (*Quercus petraea* (Matt.) Liebl.) using a chronosequence approach. Tree Physiol 30:177–192
Gifford RM, Evans LT (1981) Photosynthesis, carbon partitioning, and yield. Ann Rev Plant Physiol Plant Mol Biol 32:485–509
Gleeson SK, Tilman D (1992) Plant allocation and the multiple limitation hypothesis. Am Nat 139:1322–1343
Gower ST, McMurtrie RE, Murty D (1996) Aboveground net primary production decline with stand age: potential causes. Trends Ecol Evol 11:378–382
Gradowski T, Thomas SC (2006) Phosphorus limitation of sugar maple growth in central Ontario. For Ecol Manage 226:104–109
Gradowski T, Thomas SC (2008) Responses of canopy trees and saplings to P, K, and lime additions in *Acer saccharum* under high N deposition conditions. Tree Physiol 28:173–185
Greene DF, Johnson EA (1994) Estimating the mean annual seed production of trees. Ecology 75:642–647
Groppo M, Pirani JR, Salatino MFL, Blanco SR, Kallunki JA (2008) Phylogeny of Rutaceae based on two noncoding regions from cpDNA. Am J Bot 95:985–1005
Gross HL (1972) Crown deterioration and reduced growth associated with excessive seed production by birch. Can J Bot 50:2431–2437
Gullison RE, Panfil SN, Strouse JJ, Hubbell SP (1996) Ecology and management of mahogany (*Swietenia macrophylla* King) in the Chimanes forest, Beni, Bolivia. Bot J Linn Soc 122:9–34
Hallé F, Oldeman RAA, Tomlinson PB (1978) Tropical trees and forests: an architectural analysis. Springer-Verlag, Berlin, 441 pp
Harper JL, White J (1974) The demography of plants. Ann Rev Ecol Syst 5:419–463
Hasegawa S, Takeda H (2001) Functional specialization of current shoots as a reproductive strategy in Japanese alder (*Alnus hirsuta* var. *sibirica*). Can J Bot 79:38–48
Hedges LV, Olkin I (1985) Statistical methods for meta-analysis. Academic, Orlando, 369 pp
Hemborg AM, Karlsson PS (1998) Altitudinal variation in size effects on plant reproductive effort and somatic costs of reproduction. Ecoscience 5:517–525
Henriksson J, Ruohomäki K (2000) Assessing costs of reproduction in mountain birch: the importance of considering the modular level. Ann Bot 86:503–510
Hirayama D, Itoh A, Yamakura T (2004) Implications from seed traps for reproductive success, allocation and cost in a tall tree species *Lindera erythrocarpa*. Plant Species Biol 19:185–196
Hirayama D, Nanami S, Itoh A, Yamakura T (2008) Individual resource allocation to vegetative growth and reproduction in subgenus *Cyclobalanopsis* (*Quercus*, Fagaceae) trees. Ecol Res 23:451–458
Hörtensteiner S (2006) Chlorophyll degradation during senescence. Ann Rev Plant Biol 57:55–77
Hoshino Y, Yonenobu H, Yasue K, Nobori Y, Mitsutani T (2008) On the radial-growth variations of Japanese beech (*Fagus crenata*) on the northernmost part of Honshu Island, Japan. J Wood Sci 54:183–188
Houle G (2001) Reproductive costs are associated with both the male and female functions in *Alnus viridis* ssp *crispa*. Ecoscience 8:220–229
Hunter JE, Schmidt FL (1990) Methods of meta-analysis: correcting error and bias in research findings. Sage Publications, Thousand Oaks, 592 pp
Ichie T, Tanaka K, Kitahashi Y, Koike T, Nakashizuka T (2005) How does *Dryobalanops aromatica* supply carbohydrate resources for reproduction in a masting year? Trees 19:704–711

Innes JL (1994) The occurrence of flowering and fruiting on individual trees over 3 years and their effects on subsequent crown condition. Trees 8:139–150

Iwasa Y (2000) Dynamic optimization of plant growth. Evol Ecol Res 2:427–455

Iwasa Y, Cohen D (1989) Optimal growth schedule of a perennial plant. Am Nat 133:480–505

Jing SW, Coley PD (1990) Dioecy and herbivory: the effect of growth rate on plant defense in *Acer negundo*. Oikos 58:369–377

Jones CJ, Ostfeld RS, Richard MP, Schauber EM, Wolff JO (1998) Chain reactions linking acorns to gypsy moth outbreaks and Lyme disease risk. Science 279:1023–1026

Kainer KA, Wadt LHO, Gomes-Silva DAP, Capanu M (2006) Liana loads and their association with *Bertholletia excelsa* fruit and nut production, diameter growth and crown attributes. J Trop Ecol 22:147–154

Karlsson PS (1994) Photosynthetic capacity and photosynthetic nutrient-use efficiency of *Rhododendron lapponicum* leaves as related to nutrient status, leaf age and branch reproductive status. Func Ecol 8:694–700

Karlsson PS, Olsson L, Hellstrom K (1996) Trade-offs among investments in different long-shoot functions — variation among mountain birch individuals. J Ecol 84:915–921

Karlsson PS, Andersson M, Svensson BM (2006) Relationships between fruit production and branching in monocarpic shoot modules of *Rhododendron lapponicum*. Ecoscience 13:396–403

Kelly MO, Davies PJ (1988) The control of whole-plant senescence. CRC Crit Rev Plant Sci 7:139–173

King D, Roughgarden J (1982) Graded allocation between vegetative and reproductive growth for annual plants in growing seasons of random length. Theor Popul Biol 21:1–16

Knops JMH, Koenig WD, Carmen WJ (2007) Negative correlation does not imply a tradeoff between growth and reproduction in California oaks. Proc Natl Acad Sci USA 104:16982–16985

Körner C, Asshoff R, Bignucolo O, Hättenschwiler S, Keel SG, Peláez-Riedl S, Pepin S, Siegwolf RTW, Zotz G (2005) Carbon flux and growth in mature deciduous forest trees exposed to elevated CO_2. Science 309:1360–1362

Kozlowski J, Uchmanski J (1987) Optimal individual growth and reproduction in perennial species with indeterminate growth. Evol Ecol 1:214–230

Kudo G, Suzuki S (2004) Flowering phenology of tropical-alpine dwarf trees on Mount Kinabalu, Borneo. J Trop Ecol 20:563–571

Kutsch WL, Wirth C, Kattge K, Nöllert S, Herbst M, Kappen L (2009) Ecophysiological characteristics of mature trees and stands – consequences for old-growth forest productivity. In: Wirth C, Gleixner G, Heimann M (eds.) Old growth forests: function, value and fate. Springer, Berlin, pp 57–69

LaDeau SL, Clark JS (2001) Rising CO_2 levels and the fecundity of forest trees. Science 292:95–98

Leal DA, Thomas SC (2003) Vertical gradients and tree-to-tree variation in shoot morphology and foliar nitrogen in an old-growth *Pinus strobus* L. stand. Can J For Res 33:1304–1314

Ledig FT, Linzer DIH (1978) Fuel crop breeding. Chem Tech 8:18–27

Leopold AC (1961) Senescence in plant development. Science 134:1727–1732

Levins R (1968) Evolution in changing environments. Princeton University Press, Princeton, 132 pp

Linder S, Troeng E (1981) The seasonal course of respiration and photosynthesis in strobili of Scots pine. For Sci 27:267–276

Linhart YB, Mitton JB (1985) Relationships among reproduction, growth rates, and protein heterozygosity in ponderosa pine. Am J Bot 72:181–184

Litton CM, Raich JW, Ryan MG (2007) Carbon allocation in forest ecosystems. Glob Change Biol 13:2089–2109

Lloyd DG, Webb CJ (1977) Secondary sex characters in plants. Bot Rev 43:177–216

Lopez F, Fungairiño S, de las Heras P, Serrano J, Acosta F (2001) Age changes in the vegetative vs. reproductive allocation by module demographic strategies in a perennial plant. Plant Ecol 157:12–21

Maillette L (1987) Effects of bud demography and elongation patterns on *Betula cordifolia* near the tree line. Ecology 68:1251–1261
Mäkelä A (1985) Differential games in evolutionary theory: height growth strategies of trees. Theor Popul Biol 27:239–267
May JD, Killingbeck KT (1992) Effects of preventing nutrient resorption on plant fitness and foliar nutrient dynamics. Ecology 73:1868–1878
McDowell SCL, McDowell NG, Marshall JD, Hultine K (2000) Carbon and nitrogen allocation to male and female reproduction in Rocky Mountain Douglas-fir (*Pseudotsuga menziesii* var. *glauca*, Pinaceae). Am J Bot 87:539–546
Miyazaki Y, Hiura T, Kato E, Funada R (2002) Allocation of resources to reproduction in *Styrax obassia* in a masting year. Ann Bot 89:767–772
Molisch H (1938) The longevity of plants. Science Press, Lancaster, 101 pp
Monks A, Kelly D (2006) Testing the resource-matching hypothesis in the mast seeding tree *Nothofagus truncata* (Fagaceae). Austral Ecol 31:366–375
Morris RF (1951) The effect of flowering on the foliage production and growth of balsam fir. For Chron 27:40–57
Muller-Landau HC, Wright SJ, Calderon O, Condit R, Hubbell SP (2008) Interspecific variation in primary seed dispersal in a tropical forest. J Ecol 96:653–667
Nabeshima E, Hiura T (2008) Size-dependency in hydraulic and photosynthetic properties of three *Acer* species having different maximum sizes. Ecol Res 23:281–288
Naito Y, Kanzaki M, Numata S, Obayashi K, Konuma A, Nishimura S, Ohta S, Tsumura Y, Okuda T, Lee SL, Muhammad N (2008) Size-related flowering and fecundity in the tropical canopy tree species, *Shorea acuminata* (Dipterocarpaceae) during two consecutive general flowerings. J Plant Res 121:33–42
Newell EA (1991) Direct and delayed costs of reproduction in *Aesculus californica*. Ecology 79:365–378
Niinemets U (2002) Stomatal conductance alone does not explain the decline in foliar photosynthetic rates with increasing tree age and size in *Picea abies* and *Pinus sylvestris*. Tree Physiol 22:515–535
Niklas KJ (1993) The allometry of plant reproductive biomass and stem diameter. Am J Bot 80:461–467
Nock CA, Caspersen JP, Thomas SC (2008) Large ontogenetic declines in intra-crown leaf area index in two temperate deciduous tree species. Ecology 89:744–753
Noodén LD, Guiamét JJ, John I (1997) Senescence mechanisms. Physiol Plant 101:746–753
Norton DA, Kelly D (1988) Mast seeding over 33 years by *Dacrydium cupressinum* Lamb. (rimu) (Podocarpaceae) in New Zealand: the importance of economies of scale. Func Ecol 2:399–408
Obeso JR (1997) Costs of reproduction in *Ilex aquifolium*: effects at tree, branch and leaf levels. J Ecol 85:159–166
Obeso JR (2002) The costs of reproduction in plants. New Phytol 155:321–348
Obeso JR, Álvarez-Santullano M, Retuerto R (1998) Sex-ratios, size distributions, and sexual dimorphism in the dioecious tree *Ilex aquifolium* (Aquifoliaceae). Am J Bot 85:1602–1608
Ostfield RS, Jones CG, Wolff JO (1996) Of mice and mast: ecological connections in eastern deciduous forests. Bioscience 46:323–330
Pacala SW, Canham CD, Silander JA (1993) Forest models defined by field measurements I. The design of a northeastern forest simulator. Can J For Res 23:1980–1988
Panditharathna PAKAK, Singhakumara BMP, Griscom HP, Ashton MS (2008) Change in leaf structure in relation to crown position and size class for tree species within a Sri Lankan tropical rain forest. Botany 86:633–640
Paoli GD, Curran LM (2007) Soil nutrients limit fine litter production and tree growth in mature lowland forest of Southwestern Borneo. Ecosystems 10:503–518
Peters R, Cloutier S, Dubr D, Evans A, Hastings P, Kaiser H, Kohn D, Sarwer-Foner B (1988) The allometry of the weight of fruit on trees and shrubs in Barbados. Oecologia 74:612–616

Poorter L, Zuidema PA, Peña-Claros M, Boot RGA (2005) A monocarpic tree species in a polycarpic world: how can *Tachigali vasquezii* maintain itself so successfully in a tropical rain forest community? J Ecol 93:268–278
Rackham O (1980) Ancient woodland: its history, vegetation and uses in England. Edward Arnold, London, 402 pp
Read J, Sanson GD, Jaffré T, Burd M (2006) Does tree size influence timing of flowering in *Cerberiopsis candelabra* (Apocynaceae), a long-lived monocarpic rain-forest tree? J Trop Ecol 22:621–629
Read J, Sanson GD, Burd M, Jaffré T (2008) Mass flowering and parental death in the regeneration of *Cerberiopsis candelabra* (Apocynaceae), a long-lived monocarpic tree in New Caledonia. Am J Bot 95:558–567
Reekie EG, Bazzaz FA (1987a) Reproductive effort in plants. 1. Carbon allocation to reproduction. Am Nat 129:876–896
Reekie EG, Bazzaz FA (1987b) Reproductive effort in plants. 2. Does carbon reflect the allocation of other resources? Am Nat 129:897–906
Reekie EG, Bazzaz FA (1987c) Reproductive effort in plants. 3. Effect of reproduction on vegetative activity. Am Nat 129:907–919
Reekie EG, Bazzaz FA (1992) Cost of reproduction as reduced growth in genotypes of two congeneric species with contrasting life histories. Oecologia 90:21–26
Reich A, Holbrook NM, Ewel JJ (2004) Developmental and physiological correlates of leaf size in *Hyeronima alchorneoides* (Euphorbiaceae). Am J Bot 91:582–589
Ribbens E, Silander JA, Pacala SW (1994) Seedling recruitment in forests – calibrating models to predict patterns of tree seedling dispersion. Ecology 75:1794–1806
Rodrigues WA (1962) Arvore hapaxanta na flora Amazonica (in Portuguese). Inst Nacl Pesqu Amaz, Manaus, Brasil, Pub. 14
Roff DA (1992) The evolution of life histories: theory and analysis. Chapman and Hall, New York
Rozas V, DeSoto L, Olano JM (2009) Sex-specific, age-dependent sensitivity of tree-ring growth to climate in the dioecious tree *Juniperus thurifera*. New Phytol 182:687–697
Ryan MG, Binkley D, Fownes JH (1997) Age-related decline in forest productivity: pattern and process. Adv Ecol Res 27:213–262
Sala A, Hoch G (2009) Height-related growth declines in ponderosa pine are not due to carbon limitation. Plant Cell Environ 32:22–30
Selas V, Piovesan G, Adams JM, Bernabei M (2002) Climatic factors controlling reproduction and growth of Norway spruce in southern Norway. Can J For Res 32:217–225
Shaw EB (1940) Geography of mast feeding. Econ Geogr 16:233–249
Soehartono T, Newton AC (2001) Reproductive ecology of *Aquilaria* spp. in Indonesia. For Ecol Manage 152:59–71
Strauss SH, Rottmann WH, Brunner AM, Sheppard LA (1995) Genetic engineering of reproductive sterility in forest trees. Mol Breed 1:5–26
Tanner EVJ, Vitousek PM, Cuevas E (1998) Experimental investigation of nutrient limitation of forest growth on wet tropical mountains. Ecology 79:10–22
Tappeiner JC (1969) Effect of cone production on branch, needle, and xylem ring growth of Sierra Nevada Douglas-fir. For Sci 15:171–174
Thomas SC (1996a) Reproductive allometry in Malaysian rain forest trees: biomechanics vs. optimal allocation. Evol Ecol 10:517–530
Thomas SC (1996b) Relative size at reproductive onset in rain forest trees: a comparative study of 37 Malaysian species. Oikos 76:145–154
Thomas SC (2010) Photosynthetic capacity peaks at intermediate size in temperate deciduous trees. Tree Physiol 30:555–573
Thomas SC, Ickes K (1995) Ontogenetic changes in leaf size in Malaysian rain forest trees. Biotropica 27:427–434
Thomas SC, LaFrankie JV (1993) Sex, size, and inter-year variation in flowering among dioecious trees of the Malayan rainforest understory. Ecology 74:1529–1537

Thomas SC, Winner WE (2002) Photosynthetic differences between saplings and adult trees: an integration of field results by meta-analysis. Tree Physiol 22:117–127

Thorpe HC, Thomas SC, Caspersen JP (2007) Residual-tree growth responses to partial stand harvest in the black spruce (*Picea mariana*) forest. Can J For Res 37:1563–1571

Tuomi J, Neimela P, Mannila R (1982) Resource allocation on dwarf shoots of birch (*Betula pendula*) – reproduction and leaf growth. New Phytol 91:483–487

Urban L, Le Roux X, Sinoquet H, Jaffuels S, Jannoyer M (2003) A biochemical model of photosynthesis for mango leaves: evidence for the effect of fruit on photosynthetic capacity of nearby leaves. Tree Physiol 23:289–300

Vadeboncoeur MA (2010) Meta-analysis of fertilization experiments indicates multiple limiting nutrients in northeastern deciduous forests. Can J For Res 40:1766–1780

van Frankenhuyzen K, Beardmore T (2004) Current status and environmental impact of transgenic forest trees. Can J For Res 34:1163–1180

Verdu M, Garcia-Fayos P, Gleiser G (2004) Mites attack males of the sexually polymorphic tree *Acer opalus* more harmfully and more often. Func Ecol 18:592–597

Watkinson AR, White J (1985) Some life-history consequences of modular construction in plants. Phil Trans R Soc Lond B 313:31–51

Wawrzynski RP, Ascerno ME (1990) Effects of ash flower induced by *Eriophyes fraxiniflora* on tree vitality. J Arbor 16:182–185

Weiner J (1988) The influence of competition on plant reproduction. In: Lovett- Doust J, Lovett-Doust L (eds.) Reproductive plant ecology: patterns and strategies. Oxford University Press, Oxford, pp 228–245

Weiner J, Thomas SC (2001) The nature of tree growth and the age-related decline in forest productivity. Oikos 94:374–376

Weiner J, Campbell LG, Pino J, Echarte L (2009) The allometry of reproduction within plant populations. J Ecol 97:1220–1233

Wheelwright NT, Logan BA (2004) Previous-year reproduction reduces photosynthetic capacity and slows lifetime growth in females of a neotropical tree. Proc Natl Acad Sci USA 101:8051–8055

Williams GC (1966) Natural selection, the costs of reproduction, and the refinement of Lack's principle. Am Nat 100:687–690

Woodward A, Silsbee DG, Schreiner EG, Means JE (1993) Influence of climate on radial growth and cone production in subalpine fir (*Abies lasiocarpa*) and mountain hemlock (*Tsuga mertensiana*). Can J For Res 24:1133–1143

Worbes M (2002) One hundred years of tree-ring research in the tropics – a brief history and an outlook to future challenges. Dendrochronologia 20:217–231

Worley AC, Harder LD (1996) Size-dependent resource allocation and costs of reproduction in *Pinguicula vulgaris* (Lentibulariaceae). J Ecol 84:195–206

Wright SJ, Jaramillo MA, Pavon J, Condit R, Hubbell SP, Foster RB (2005) Reproductive size thresholds in tropical trees: variation among individuals, species and forests. J Trop Ecol 21:307–315

Yamashita N, Abe T (2002) Size distribution, growth and inter-year variation in sex expression of *Bischofia javanica*, an invasive tree. Ann Bot 90:599–605

Yasumura Y, Hikosaka K, Hirose T (2006) Resource allocation to vegetative and reproductive growth in relation to mast seeding in *Fagus crenata*. For Ecol Manage 229:228–233

Zhang DY (2006) Evolutionarily stable reproductive investment and sex allocation in plants. In: Harder LD, Barrett SCH (eds.) Ecology and evolution of flowers. Oxford University Press, Oxford, pp 41–60

Chapter 3
The Earliest Stages of Tree Growth: Development, Physiology and Impacts of Microclimate

Daniel M. Johnson, Katherine A. McCulloh, and Keith Reinhardt

Abstract Although the germinant and young seedling life stage of trees has the highest mortality rates, our understanding of their fundamental physiology is surprisingly limited in naturally-occurring environments. Field measurements for this life stage are particularly important because of the uniquely harsh environment these small trees experience: their proximity to the soil surface results in an exceptionally dynamic microclimate with large daily shifts in such parameters as temperature, humidity and radiation. The impact of these changing conditions is discussed with respect to seedling survival. Our current knowledge of anatomical and morphological responses of early seedlings to variation in light and water conditions is also reviewed. For example, research has shown that differences in light intensity result in divergence in vessel diameters and number in *Eucalyptus grandis* seedlings. Using these data, we calculated the theoretical hydraulic conductivity for high and low light-grown seedlings. This comparison showed that after 7 weeks the differences in xylem anatomy would result in the high light seedlings having two times the theoretical leaf-specific hydraulic conductivity of the low light seedlings. Future research in this field should include developing new techniques specific to the unique fragility and size of seedlings, determining the adaptive nature of different cotyledon structures, and better quantifying the physiological and developmental characteristics of early germinants.

D.M. Johnson (✉)
Department of Environmental and Plant Biology, Ohio University, Athens, OH 45701, USA
e-mail: johnsod6@ohio.edu

K.A. McCulloh
Department of Wood Science and Engineering, Oregon State University, Corvallis, OR 97331, USA

K. Reinhardt
Department of Biological Sciences, Idaho State University, Pocatello, ID 83209, USA

F.C. Meinzer et al. (eds.), *Size- and Age-Related Changes in Tree Structure and Function*, Tree Physiology 4, DOI 10.1007/978-94-007-1242-3_3,

Keywords Drought • Facilitation • Mortality • Photosynthesis • Radiation • Seedling • Water relations

1 Introduction

1.1 Population Dynamics and Seedlings

The primary determinant of vegetation patterns and species distributions on a global scale is aridity gradients (e.g., Engelbrecht et al. 2007), and moisture regimes (either via precipitation changes or temperature changes and related evaporative demands) have shifted over last 40 years (van Mantgem et al. 2009) and are predicted to continue to change. Species distributions can shift over time through adult mortality and less subsequent regeneration (shrinking) or through seedling establishment beyond current distribution boundaries (expansion). Therefore, a critical component of our ability to predict future species distributions is an understanding of the mechanisms of seedling establishment, which requires a fundamental knowledge of seedling physiology. In fact, the life stage (other than the seed stage) with the greatest mortality, by far, is the germinant seedling stage (Leck and Outred 2008) and young seedling performance may be more important than seed germination for determination of species distributions (e.g. Figueroa and Lusk 2001).

Few studies have addressed the ecophysiology of naturally-occurring current year seedlings (newly emerged) even though this life stage has a high mortality rate (e.g. Alvarez-Buylla and Martinez-Ramos 1992; Whitmore 1996; Masaki and Nakashizuka 2002). Possible reasons for the lack of emphasis on germinant seedling physiology include the difficulty in recognizing and documenting the presence of natural, newly emerged seedlings and the even more difficult task of acquiring physiological measurements (Cui and Smith 1991; Germino and Smith 2001). In general, the coupling of physiological ecology (e.g. carbon gain, water relations) to population biology (e.g. seedling abundance and survival) appears to be an important omission in studies of plant biology and ecology (Ackerly and Monson 2003). Given that the highest mortality of all life stages often occurs during the newly-emerged seedling stage, emphasis on current-year seedling studies may be particularly important as an early and sensitive indicator of community and ecosystem health.

Recent literature suggests that the newly-emerged seedling stage is becoming more in focus as an important life stage for understanding limitations to natural forest regeneration (e.g. Smith et al. 2003; Matthes and Larson 2006; Miina and Saksa 2006). In the southern Appalachian Mountains, there appears to be a consistently high mortality (approximately 95%) of new *Abies fraseri* seedlings on a yearly basis (Johnson and Smith 2005), similar to measurements on *Abies lasiocarpa* in the Rocky Mountains of the western US (Cui and Smith 1991; Germino and Smith 2000, 2001). Studies of first year survival of tropical tree seedlings have also reported high seedling mortality. In 7 out of 9 of the species that were measured at Barro Colorado Island Panama (Augspurger 1984) germinant mortality was between 69% and 100%

and in 50 out of 75 species studies at La Selva, Costa Rica, mortality was greater than 75% (Li et al. 1996). However, studies of the mechanisms of establishment in current-year seedlings are rare in the literature. For example, most studies on *Abies fraseri* seedlings have focused only on the abundance of considerably older, established seedlings, without addressing specific factors influencing the high mortality of newly establishing seedlings (Nicholas et al. 1992; Smith and Nicholas 2000).

The goals of this review are to document the microclimatic conditions of newly emerged seedlings and to discuss young seedling structure and function. This review will focus on woody tree species in particular and will attempt to provide a framework for important omissions and future directions for research on young seedlings. The purpose of this review is not to cover in detail the topics of seedling shade tolerance or seed mass/survival relationships, both of which have been thoroughly covered in the current literature (Moles and Westoby 2004a; Kitajima and Myers 2008).

1.2 Working Definitions of Life Stages

There appears to be a large discrepancy in the heights, diameters or ages of woody plants that are considered seedlings or saplings. For example, trees of up to 2.7 m have been classified seedlings (Whitmore 1996), whereas some researchers reserve the term for plants that still have seed reserves present (Fenner 1987). Throughout this chapter, we will use the term "germinant" or "emergent" for a seedling that is in its first growing season. Our use of the term "seedling" will refer to a young plant in its first few years of growth, which is a more liberal use of the term than many ecologists use (e.g., see Garwood 1996). We also use the phrase "established seedling" to indicate a seedling that has survived beyond its first season of growth.

1.3 Population Range Shifts and Reforestation

Worldwide changes in plant community distributions have been observed over the last decade (Walther et al. 2005) and further shifts are predicted with climate change (e.g. Iverson and Prasad 2001). However, most models predicting future patterns of community distribution rely on correlations between current climate and species current location (a species' "bioclimate envelope," Pearson and Dawson 2003). To develop mechanistic models of species and community distribution under climate change scenarios it will be critical to understand the constraints on seedling establishment under natural conditions. For example, recent studies in tropical forests have shown that emergent seedling drought tolerance is correlated strongly with the occurrence of that species along moisture gradients (Engelbrecht et al. 2007; Kursar et al. 2009). These types of studies and data could be used for predictions of future species distributions under changing moisture regimes. Although any complete picture of forest regeneration and migration would include all life stages (e.g. adult mortality, flower and/or seed production, pollination), forest regeneration and migration into new areas ultimately depends on new seedlings establishing in the understory or beyond the current boundary of the community. Seedling establishment is critical for reforestation after clearcutting, fire or other major disturbance. Successful establishment and subsequent growth beneath canopy gaps is critical for stand regeneration as well, and for preventing encroachment by shrub species that could out-compete understory tree seedlings (White et al. 1985; Pauley and Clebsch 1990).

1.4 Causes of Seedling Mortality

Compared to studies focusing on older age classes, relatively few studies have addressed the causes of mortality during the early establishment phase of new seedlings. In certain communities, herbivory and/or trampling by animals appears to be a significant factor contributing to the mortality of young seedlings (Ehrlen 2003; Moles and Westoby 2004a). In other systems, drought is likely the primary cause of seedling death (e.g. Cui and Smith 1991; Moles and Westoby 2004b; Negussie et al. 2008). In tropical moist forests, pathogens are frequently responsible for a large portion of young seedling mortality (Augspurger 1984), especially in dense shade. Small seedlings are particularly susceptible to uprooting due to frost heaving in areas with frequent frosts (Goulet 1995). Additional causes of mortality can include burial by substrate (e.g. sand, Maun 1998) and litterfall (Clark and Clark 1989; Alvarez-Claire and Kitajima 2009) as well as fire (Plumb 1980; Grace and Platt 1995; Edwards and Krockenberger 2006).

Reductions in carbon gain, which may result in reduced root growth (and possibly desiccation) or even "carbon starvation" (McDowell et al. 2008) can also lead to seedling mortality. For example, current-year seedlings of *A. fraseri* had reduced photosynthetic capacity and sunlight under closed canopies (as compared to open canopies), which also corresponded to increased mortality (Johnson and Smith 2005). Lack of photosynthesis in low light environments (i.e. under closed canopies) appeared to be the primary cause of mortality for *Quercus pyrenaica,* as well (Rodriguez-Calcerrada et al. 2010). In the cases of *A. lasiocarpa* and *Picea engelmannii*, Cui and Smith (1991) found that the high mortality (>90%) in first-year seedlings in the forest understory appeared symptomatically to be due to desiccation, possibly enhanced by a lack of root growth and/or mycorrhizal infection. Additionally, Sacchi and Price (1992) observed high mortality rates in young seedlings of *Salix lasiolepis* in response to limited soil water availability. Carbon gain and survival were also reduced in individual seedlings growing in microsites with either too much or too little sunlight (Germino and Smith 2000; Johnson et al. 2004). This finding of too much or too little sky exposure (degree of canopy openness) was attributed to both positive and negative effects related to sunlight exposure and nighttime sky exposure. Sky exposure was influenced by the overstory tree canopy, plus the vegetation and inanimate structures at ground level. Earlier studies pointed to the possibility that a lack of photosynthetic carbon gain in less favorable microsites (too little or too much sunlight) led to decreased root growth and desiccation-related mortality (Knapp and Smith 1982; Cui and Smith 1991). Seedling photosynthesis declined in response to low minimum leaf temperatures (predawn) as well as high sunlight the following morning, associated directly with the amount of sky exposure of the seedling microsite (Germino and Smith 2000; Maher et al. 2005). As a result, seedling survival in many species appears strongly associated with microsite facilitation provided by surrounding structures such as overhead canopy, tree islands, and ground vegetation, all of which substantially reduce seedling sky exposure (Hättenschwiler and Smith 1999; Germino and Smith 2001).

The amount of sky exposure strongly influences seedling growth and survival, as well as provides a quantitative measure of microsite and canopy openness. In addition to sunlight exposure and photosynthetic effects, the accompanying exchange of longwave (thermal) radiation emitted from the cold night sky is an important forcing variable for nighttime leaf temperatures, often driving leaf temperature well below air temperature (5–7°C, Jordan and Smith 1995). Other studies have also reported large decreases in downwelling radiation from cold night skies due to vegetative overstories (McLaughlin et al. 1990; Örlander 1993), and cloud patterns such as found at southern Appalachian sites can generate large increases in longwave radiation of up to 100 W m^{-2} on nights with fully overcast skies (Feygel'son and Tsvang 1974; Harrison et al. 1993). In addition, cloud immersion in high-altitude southern Appalachian forests has been shown to increase understory sunlight penetration and photosynthesis and decrease transpirational water loss in young seedlings (Johnson and Smith 2008). Substantial decreases in seedling temperatures occur on clear nights and are followed by high sunlight levels in the morning, causing long-term declines (up to several days) in photosynthetic carbon gain (low temperature photoinhibition and/or downregulation of photosynthesis) and, ultimately, reductions in root growth that have been associated with desiccation and mortality in new seedlings (e.g. Knapp and Smith 1982; Ball et al. 1991; Ball 1994; Ball et al. 1997; Cui and Smith 1991; Germino and Smith 2000).

2 Seedling Environments

First-year seedlings experience unique stresses compared to other life stages and are more vulnerable to both biotic and abiotic stresses, compared to older life stages. The unique stresses experienced by seedlings are due to their small size and the fact that they have not escaped the microclimate of the ground surface. The surface environment is dynamic, and can feature large diurnal swings in every environmental parameter (temperature, humidity, radiation, etc.). In contrast to larger, established saplings and adult trees, which have escaped the climate near the ground, emergent seedlings are entirely contained in the near-surface climate at the whole plant level and do not have height or branch complexity to escape the surface environment. Thus, seedling physiological ecology is affected at the whole plant level.

Seedlings emerge from the soil along a continuum of microsites, and it would be difficult to address all situations. Therefore, the approach in this section will be to compare the extremes of possible germinating environments common to seedlings.

The microsites described in the following sections will be a dark, forest understory microsite and an exposed, open canopy microsite, such as might be found for a seedling establishing in an alpine ecotone or during a typical old field succession (this could also be synonymous with forest/shade and pioneer seedling types).

Ultimately, seedling growth and survival depends on adequate carbon capture and processing, avoidance of hydraulic failure (i.e. desiccation), and escaping biota-induced mortality (e.g. trampling, herbivory). Many factors directly influence the carbon and water balance, and thus the survival, of young seedlings.

2.1 Near-Surface Incident Radiation

Incident radiation loads on seedlings depend on both radiation intensity and duration. For first year seedlings, there is a delicate balance between too little versus too much sunlight (Germino and Smith 1999; Johnson and Smith 2005). In situations where there is too little sunlight, seedlings may perish because they cannot escape infestation by pathogens on the soil surface and/or seedlings may experience carbon "starvation", where metabolic needs are not met by adequate carbon assimilation (Givnish 1988). Too much sunlight, especially in combination with low temperatures, can result in decreased carbon gain via photoinhibition (Ball et al. 1991; Germino and Smith 1999; Maher et al. 2005). The optimal incident sunlight regime for a seedling seems to be species-specific, and may co-vary with other factors such as proximity to neighboring plants or rocks, microtopography, and aspect (see further discussion below).

Shaded understory environments can be very dark, with less than 5% of the total radiation at the top of the canopy reaching the forest floor (Baumgartner 1952; Geiger 1965; Lee 1987; Endler 1993; Grant 1997; Johnson and Smith 2005). Net radiation may be less than 21 W m^{-2} day^{-1} (<1 MJ m^{-2} day^{-1}), with the instantaneous quantum flux of photosynthetically active radiation (PAR, ~400–700 nm) averaging between 10 and 20 μmol m^{-2} s^{-1} in deep forest shade. Total daily PAR at the ground is often <20 mol m^{-2} day^{-1}, most of which occurs during occasional sunflecks (Young and Smith 1983; Lee and Downum 1991; Combes et al. 2000; Johnson and Smith 2005; Navrátil et al. 2007). Consequently, in the understory near the forest floor, sunlight intensities are rarely above the light compensation threshold for photosynthesis to occur, even for shade-adapted species. This low light availability can strongly impact carbon gain, and thus, allocation to root, stem, and lateral branch development. Seedlings in such light environments may not gain enough carbon for survival before exhausting carbon reserves (Kohyama 1983; Kobe 1997; Givnish 1988; Walters and Reich 1999; Myers and Kitajima 2007). Furthermore, much of the carbon balance of a newly emerged seedling in a forest understory is dictated by ephemeral periods if sunlight. During these periods, seedlings must quickly adjust to a changing light environment to maximize carbon gain. The ability to respond to sunflecks by quickly upregulating photosynthetic mechanisms (i.e. the photosynthetic "induction period", Chazdon and Pearcy 1986) may be critically important for seedlings inhabiting dark understories where sunflecks are the primary source of light for carbon gain. Little research has addressed the different responses of photosynthetic tissues (cotyledonous vs. primary leaves) to variable sunlight conditions, though there is some evidence that cotyledons may be more efficient at harvesting sunlight for photosynthesis (Johnson and Smith 2005; Reinhardt et al. 2009). It should also be highlighted that some species are remarkably adapted to low light environments, and seedlings of these species are able to persist for many years in such environments with very little growth, until opportunities for substantial positive carbon balance occur (often referred to as "release" in the literature), via dieback of canopy trees (e.g. Kobe et al. 1995).

On the other extreme, seedlings establishing into open areas of an ecotone (for example, beside a tree island at the alpine-treeline ecotone, into an old-field meadow, or into a large canopy gap) can encounter high radiation loads that affect the whole plant (compared to just a few leaves or branches for plants that are taller and/or with a more complex branch architecture). For many hours of the day in an open-canopy alpine scenario, instantaneous whole-seedling radiation exposure may be >1060 W m^{-2}, or >2000 μmol m^{-2} s^{-1} (PAR), with daily radiation loads >30 MJ m^{-2} day^{-1} (>900 mol m^{-2} PAR per day; Lee and Downum 1991; Jordan and Smith 1994; Larcher 2003; Reinhardt and Smith 2008). In some situations, such as in snowmelt concavities or in depressions of light-colored ground where there is substantial reflected sunlight, instantaneous irradiance has been reported to be even greater, exceeding 3500 μmol m^{-2} s^{-1} (Day et al. 1989; Germino and Smith 2000). Such prolonged exposure to high radiation loads can cause increased leaf temperatures, photorespiration, and high-light photoinhibition. These problems with high radiation loads can be exacerbated if there was frost on the leaves the previous night. Frost occurrence on leaves typically happens via a combination of low air temperatures and radiative cooling (e.g. Ball et al. 1991; Jordan and Smith 1994, 1995; Germino and Smith 1999, 2000; Maher et al. 2005).

2.2 *The Effect of the Boundary Layer on Heat and Mass Fluxes*

Boundary layer effects are paramount to understanding radiation (heat) and mass fluxes, including convection and conduction dynamics, of newly emerged seedlings. In some cases, boundary layers protect small seedlings from harsh environmental conditions and/or large diurnal variations in the environment (e.g. temperature, sunlight intensity, humidity, wind). The boundary layer can be described as the zone of air covering a surface where convection (wind) is both laminar (vs. turbulent) and still (due to viscosity) (Nobel 2005). Conceptually, what is considered the "boundary layer" is a scaling issue and can vary by many orders of magnitude, depending on the system. For example, whereas the boundary layer of a forest is usually considered the area below the canopy (10^1–10^2 m), the boundary layer of an individual leaf surface might be on the scale of 10^{-4}–10^{-3} m. The boundary layer near the ground in open areas is on the order of 10^{-3}–10^1 m, depending mostly on surface roughness and wind speed (boundary layer decreases with greater roughness and wind velocity), and is commonly about 5–10 cm over bare ground (Morgan and Rickson 1995). Therefore, small seedlings are often entirely within the near-surface boundary layer.

Within the near-surface boundary layer, heat fluxes of seedlings are greatly reduced relative to taller plants. This difference is because thermal mixing is minimized due to the lack of wind and turbulent eddies, and thermal transfer is dominated by conduction. For example, even with wind velocities at 1 m height in excess of 5 m s^{-1}, wind velocities below 0.5 m height rarely exceed 1 m s^{-1}, with most reported, and modeled, measurements at ground-level less than 0.5 m s^{-1} (Geiger 1965; Elias et al. 1989; Jordan and Smith 1994; Tappeiner and Cernusca 1998). Deacon (1953) and Geiger (1965) showed that the minimum wind velocity for turbulence at 1 m height above the ground is 0.4 m s^{-1} for short grass cover and about 0.8 m s^{-1} for bare ground. As wind velocity decreases logarithmically with decreasing height (Bagnold 1941), the wind velocity in the boundary layer does not induce turbulent mixing the majority of the time. This underscores that newly germinated, and small, seedlings are entirely within the boundary layer of the ground.

The consequences of residing within the boundary layer of still air can be beneficial (protective) or detrimental (exposing). In forest understories, air temperature profiles are more uniform than above the canopy, and have less diurnal variation. The uniformity of these profiles in the understory is because air movement inside the thick forest is low compared to outside the forest, and the understory area is, effectively, one entire boundary layer. Regardless of how large the diurnal range of air temperature is over the course of 24 h, which can be greater than 30°C in temperate forest ecosystems, forest temperatures near the ground in summer typically remain between 15°C and 25°C for the whole day (see Fig. 3.1; similar to the range higher in the canopy), mostly due to trapping and re-radiation of longwave radiation (Geiger 1965; Larcher 2003).

In exposed microsites outside the forest, boundary layer dynamics can be much more variable. In the absence of ground-level convection (as in Fig. 3.1, which was based on a very low wind speed of 0.5 m s^{-1}), leaf temperatures are commonly ~2–6°C above air temperature (Tranquillini 1979; Hadley and Smith 1987; Germino and Smith 2001; Johnson and Smith 2008), which is well within the limits of tolerance of leaf biochemistry. However, there are reports of leaf temperatures of plants near the ground exceeding air temperature by as much as 23°C inside a krummholz mat in open meadows in the alpine (Hadley and Smith 1987). Midday air temperatures at the ground surface in open-canopy microsites are commonly between 40°C and 50°C, and in some instances can exceed 80°C, especially on dark surfaces (Turner 1958; Tranquillini 1979; Alexander 1987). Prolonged exposure to air temperatures greater than 45°C can cause "heat girdling" of stem cortex tissue in succulent seedlings, resulting in seedling death (see Tranquillini 1979; Alexander 1987 and references therein). Because small seedlings are entirely within the surface boundary layer, with little leaf area to provide mutual shading, there are few structural mechanisms to ameliorate this environment. Moderately high leaf temperatures can lead to heat stress of photosystem II, increased respiration, as well as increased

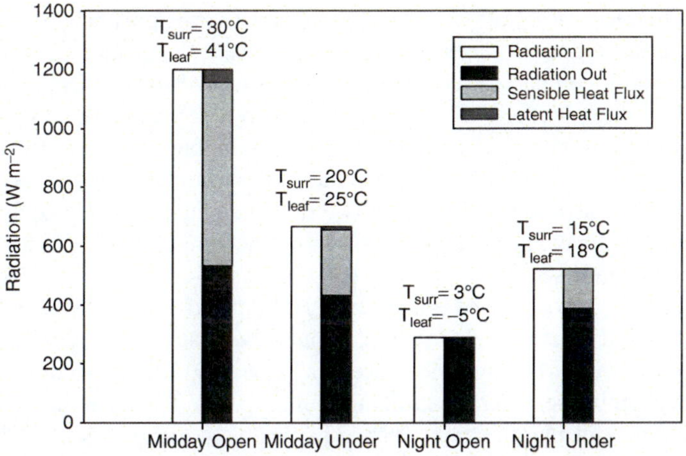

Fig. 3.1 Comparison of seedling energy balance due to microsite. Calculated energy balances of seedlings during midday in open canopy (Midday Open) and understory (Midday Under) microsites, as well as nighttime open canopy (Night Open) and understory (Night Under) microsites. Energy balance calculations are based on the equation: $\text{Radiation}_{in} = \text{Radiation}_{out} + G$ (Sensible heat flux) + L (Latent heat flux). Assumptions: leaf absorptance = 0.6, leaf emmissivity = 0.97, leaf size = 5 cm, wind velocity = 0.5 m s^{-1}, leaf width = 5 cm, leaf shape = flat, horizontal leaf angle (angle of incidence = 0°). Variables: relative humidity = 30% (Midday Open), 80% (Midday Under), 10% (Nighttime Open), 95% (Nighttime Under); midday stomatal resistance = 2500 s m^{-1}; nighttime stomatal resistance = 10000 s m^{-1} (no transpiration)

photorespiration, all of which negatively impact seedling carbon balance. Biochemical heat stress of small seedlings may be exacerbated by decreased plant water status and the inability of juvenile root systems to adequately transport water to leaves for transpirational cooling. Root systems of first year seedlings are usually shallow (on the order of centimeters), and not fully developed with few lateral roots for tap-rooted species (Alexander 1987; Greenwood et al. 2008).

At night in open-canopy microsites, evening near-surface temperatures start off relatively warm compared to higher above the ground (due to soil re-radiation of absorbed longwave radiation during the day), but by early morning (pre-dawn) the temperature near the ground is the coldest section in the air profile. This cold temperature is primarily due to the lack of wind in this part of the profile. Such a situation allows for effective radiative cooling, resulting in seedling leaf temperatures that are commonly 2–4°C below air temperature, and occasionally as much as 8°C below air temperature (Fig. 3.1; Hadley and Smith 1987; Jordan and Smith 1994). Leaf temperature depressions can promote frost formation even when air temperatures are above freezing. Thus, frost development on seedlings is likely to occur more frequently than on taller, adult plants and can occur at the whole plant level, not just on individual leaves.

Mass (e.g. CO_2, H_2O) fluxes of seedlings inside the near-surface boundary layer are also different compared to taller plants, though trends may be different between CO_2 and H_2O. Soil respiration, lack of wind, and trapping of CO_2 by overstory vegetation all act to concentrate CO_2 at the ground relative to upper heights, generating typical CO_2 concentrations of about 450–500 ppm near the ground during evening and morning hours, compared to common ambient values of about 380 ppm (Kimmins 1987; Larcher 2003; Nobel 2005). Greater CO_2 levels at ground level increase the gradient between internal leaf and air CO_2 concentrations, allowing for a greater driving force for CO_2 movement and potentially, photosynthesis. However, the photosynthetic advantage due to increased CO_2 gradients may not be realized in young seedlings due to immature root, leaf, and plant architecture systems. While increased leaf-level photosynthesis under elevated CO_2 concentrations has been observed in seedlings of tropical broadleaved tree species (Holtum and Winter 2003) and temperate conifer trees (Reinhardt and Germino, unpublished data), other researchers have reported no difference, or slight decreases, in photosynthesis in seedlings in both seedbed and greenhouse experiments (Tingey et al. 2007). These decreases in photosynthesis and overall plant carbon balance in seedlings under elevated CO_2 levels may have been due to seedlings altering allocation patterns vs. overall photosynthesis, as well as indirect adverse affects of elevated CO_2 on photosynthesis, caused by drier soils and reduced seed quality. Compared to in the understory, in open-canopy microsites the concentration of ground-level CO_2 is much less and may not differ appreciably from CO_2 concentrations at a height of 1 m (~380–400 ppm), especially in the absence of high soil respiration and/or thick herbaceous mats.

Similar to CO_2, H_2O vapor concentrations close to the ground also are greater compared to above-ground heights typical of saplings and trees (>1 m), for the same reasons associated with boundary layer as mentioned above. Evaporation of surface soil moisture and transpiration from low growing plants such as seedlings, mosses, and small herbs ("surface evapo-transpiration") is minimized within the soil boundary layer, and primarily controlled by conduction. In the absence of ground-level (<5 cm) wind and/or radiation, near-surface air humidity is "trapped" at ground level and is much greater compared to higher above the surface for many hours of the day. Sunlight irradiance acts to increase boundary layer conductance and promote turbulence, even at near surface elevations, especially in spatially heterogeneous understory sunlight environments. Nevertheless, surface humidity in forests is greater, and with less diurnal fluctuation, than higher in the canopy for all hours of a typical day. For both deciduous and conifer temperate forests, ground-level relative humidity averages about 80% for a clear summer day, and typically ranges from about 50% to 95% during the day (Geiger 1965; Elias et al. 1989; Tappeiner and Cernusca 1998; Larcher 2003). At heights near and above the forest canopy, daily relative humidity averages about 60%, commonly fluctuating from 30% to 90% during the day. For a conifer plantation in Europe, near surface air humidity was >80% for 14 h of the day, while at 10 m relative humidity was above 80% only 8 h of the day (Geiger 1965).

When air humidity is high, H_2O gradients between inside the leaf (assumed relative humidity near 100%) and the air are smaller (as is the case in still air). Consequently, water fluxes are minimal because the driving force for transpiration is reduced. In mesic forests, near-surface conditions in dark understories can be wet and cool 24 h a day. Such conditions in the southern Appalachian Mountains, USA led to low seedling water stress (water potentials never less negative than –0.75 MPa) for all times of the day (Johnson and Smith 2005; Reinhardt et al. 2009). However, in the absence of protective snowpack, evergreen seedlings in a mesic forest may be exposed to extended periods of freezing conditions during the winter.

2.3 Importance of Ecological Facilitation to Seedling Environment

Although the presence of neighboring vegetation can result in competitive interactions that can adversely affect seedling survival, many species benefit from facilitative interactions. Many of the stressful environmental conditions mentioned above can be ameliorated to some degree by neighboring structures, such as tree canopies, herbaceous species, rocks, or even topographical depressions that provide some degree of protection against exposure (Ball et al. 1991; Berkowitz et al. 1995; Greenlee and Callaway 1996; Callaway and Walker 1997; Hastwell and Facelli 2003; Hughes et al. 2009). Such "facilitation" has been shown to protect young seedlings from high intensity sunlight exposure, radiative cooling at night, low temperature photoinhibition, and damaging winds (e.g. Hadley and Smith 1987; Germino and Smith 1999; Maher et al. 2005; Maher and Germino 2006). In commercial forestry, seedlings are often planted in close association with "nurse trees" which can result in substantially enhanced seedling survival and/or growth (Schlesinger and Williams 1984; Ashton et al. 1997). In natural systems, facilitation plays a large role in primary succession and in ecotone migration (e.g. Connell and Slayter 1977). For example, Germino and Smith (1999) showed that emergent seedling abundance in the alpine ecotone in Wyoming was greatest in microsites located partially under krummholz mats (40–80% sky exposure), and that nighttime warming treatments (simulating protective herbaceous cover) increased *Abies lasiocarpa* seedling photosynthesis by about 400%, and *Picea engelmannii* seedling photosynthesis by about 150%. Furthermore, Maher et al. (2005) demonstrated that *Pinus albicaulis* seedlings had less negative water potentials (less water stress) and greater maximal quantum efficiency (F_V/F_M) in microsites with overstory vegetation cover. Belowground ecological facilitation also has been shown to increase seedling growth and survival. Between 90% and 95% of all plants benefit from associations of mycorrhizal fungae with their root systems. Such associations usually occur within days or weeks of germination (Janos 1980; See and Alexander 1996), and effectively increase the surface area of root systems (100- to 1000-fold) allowing for greater water uptake by plants (Smith and Read 2008). Furthermore, mycorrhizal infection of seedlings results in increased growth rates and greater nutrient uptake in many species (e.g., Janos 1980; Allsop and Stock 1995).

Apparently, proximity to neighboring vegetation can affect the degree of mycorrhizal infection of roots. Hasselquist et al. (2005) showed that colonization of spruce and fir seedling roots by mycorrhizae were fourfold greater in seedlings adjacent to tree islands compared to seedlings ~7 m away from tree islands, and that colonization increased seedling water potentials. Dickie et al. (2005) found that seedlings had greater degrees of mycorrhizal infection near forest edges, but that there was an optimal distance that minimized shading while still providing benefits of a high degree of mycorrhizal colonization. Beyond facilitating macro-nutrient acquisition, recent research suggests that there is also an exchange of carbon, nutrients, and water between facilitating (host) plants and seedlings, occurring on time scales of weeks (He et al. 2006). In stressful environments where carbon gain is limited, this extra source of carbon may be crucial for seedling survival.

3 Early Seedling Development

Seedlings are entirely dependent on seed energy reserves until cotyledon or leaf machinery is developed enough for photosynthetic carbon capture and until radicle development has progressed to be able to supply the photosynthetic tissues with water and nutrients. The function of cotyledons ranges from purely storage of reserves for growth (e.g., *Phaseolus*, *Quercus*; de Vogel 1980; Kitajima 2003) to photosynthetic tissue that is physiologically indistinguishable from that of true leaves. Photosynthetic cotyledons can possess many of the same traits as mature leaves, including stomata and differentiated mesophyll tissue (e.g. Sasaki and Kozlowski 1969). It is noteworthy that the cotyledons of gymnosperms species are always green, even with no exposure to light, whereas eudicot species require light for cotyledon chlorophyll development (Niklas 2008).

Generally, the cotyledons maintain the same function for all seedlings of a species, and all function in the same capacity. However, it has been documented that certain members of the Piperaceae have one cotyledon that functions in storage and transfer of reserves from within the seed coat and one that emerges (Bell 1991) and may be photosynthetic (Leck and Outred 2008). Both types of cotyledons exist in most forest types (Garwood 1996), although boreal and subalpine forests (primarily conifer-dominated), are populated with species with primarily photosynthetic cotyledons. Additionally, small-seeded species tend to have photosynthetic cotyledons (Fenner and Kitajima 1999). Cotyledon size and number can vary dramatically, even within a single family. For example, the number of cotyledons in Pinaceae seedlings ranges from 3 to 18 (Lubbock 1892) and cotyledon size ranges from several mm to over 7 cm (Griffin 1962). Most eudicots simply have two cotyledons (as their name would indicate) and the majority of gymnosperms have between two and six (Leck and Outred 2008). In general however, eudicot cotyledons have a much greater diversity in morphology and function than do those of gymnosperms (Ji and Yi 2003; Niklas 2008).

The structural and functional differences between cotyledon and primary leaves in a tree species have been studied most comprehensively in *Abies fraseri* (see Table 3.1). Johnson et al. (2005) found that in cotyledons and primary needles of *A. fraseri* from the same light environment, there were distinct differences in cotyledon vs. needle structure as well as in light processing properties.

Table 3.1 Anatomical and physiological characteristics of *Abies fraseri* cotyledons and primary needles (Numbers in parentheses are standard errors; data compiled from Johnson et al. 2005 and Reinhardt et al. 2009)

	Leaf thickness	Leaf area palisade (%)	Depth of maximum [Chl]	Stomatal distribution	A_{SAT}	Apparent quantum efficiency	Light saturation point
Cotyledon	355 (11)	14.7 (1.7)	325	Adaxial	2.47 (0.08)	0.031 (0.004)	96.3 (1.1)
Primary needle	331 (24)	24.2 (0.7)	280	Abaxial	2.50 (0.19)	0.014 (0.003)	244.3 (74.2)

Cotyledons of *A. fraseri* had significantly longer palisade cells than those of primary needles (ca. 56 versus 46 μm, respectively), but had less palisade per unit cross sectional area (24% in primary needles, 15% in cotyledons). Additionally, chlorophyll concentrations in cotyledons were shifted toward the abaxial surface as compared to primary needles, as interpreted from chlorophyll fluorescence. This combination of characteristics resulted in deeper propagation of red and blue light into cotyledons of *A. fraseri* compared to primary needles. Reinhardt et al. (2009) found that there were no significant differences in maximum photosynthesis in primary needles versus cotyledons in *A. fraseri* but that the light required to saturate photosynthesis was two to four times greater in primary needles than in cotyledons. In addition, the apparent quantum yield was two times greater in cotyledons than primary needles. Cotyledons of *A. fraseri* appear to be more effective, as compared to primary needles, at utilizing very low levels of sunlight, which would be typical of a closed-canopy forest understory.

The stresses that seedlings are exposed to in the germinating environment are initially experienced entirely by the embryonic tissues and directly affect the tissues that grow from them. In general, many seedlings show a remarkable degree of phenotypic plasticity across environmental gradients (e.g. Greenwood et al. 2008), although certain traits display little variation even when there is substantial environmental variation (see Day and Greenwood 2011). In the rest of this section, we review changes in the structure of organs and allocation of biomass to shifts in the availability of two main resources to seedlings: light and water.

Little work has examined the effect of light regime on the structure of cotyledons, but it has been shown that chlorophyll concentration increases in cotyledons of *Pinus pinaster* germinated and allowed to develop in white light (as compared to dark-grown seedlings, Cánovas et al. 1991) and in cotyledons of *Tabebuia rosea* grown in high light (as compared to low light, Kitajima and Hogan 2003). It is noteworthy that cotyledon photosynthetic potential may be inversely related to cotyledon thickness (Kitajima 1992), which would be the opposite of the trend in true leaves. For true leaves of young seedlings, a meta-analysis of five separate studies with varying light conditions revealed no trends in differences in specific leaf area (SLA; leaf area per mass) or the leaf mass ratio (ratio of leaf to total mass of the plant) for woody species (Wright and Westoby 2001). When different species were compared at a common size, though, all species had greater SLA in low vs. high light (Reich et al. 1998), which was also observed across leaf size in three liana and two tree congenerics as well as in seedlings of *Tabebuia rosea* (Kitajima and Hogan 2003; Cai et al. 2008).

At the level of the whole seedling, allocation patterns tend to change with differences in light regime. It has been widely observed that species increase shoot mass relative to whole plant or root mass in low vs. high light (Smith 1976; Reich et al. 1998). What has not received as much attention is the anatomical changes in the wood associated with the changes in allocation. In a classic study, Denne (1972) found that under normal growing conditions tracheid diameter and wall thickness increased towards the root tip in *Picea* and *Pinus* seedlings, and that transferring them to shorter day lengths (decreasing integrated daily light) caused the tracheids to narrow throughout the plant due to slower expansion rates. In *Picea* this result was also observed when comparing high and low light intensity treatments rather than day length (Denne 1974). However, Denne (1976) found that after an initial decline in the expansion rate as a response to a decline in light, which lasted approximately 2–3 weeks, the photosynthetic efficiency increased and then tracheid dimensions and number increased again, as well.

Unlike for *Picea*, for *Eucalyptus grandis* seedlings grown at constant low vs. high intensity light regimes, differences in vessel diameter and number continued for weeks (Doley 1979). Using data from Doley (1979) on changes in vessel lumen area, number and total vessel area: leaf area ratios over time, we used the Hagen-Poiseuille equation (Tyree and Zimmermann 2002) to calculate theoretical values of hydraulic conductivity and leaf-specific conductivity (conductivity/leaf area) for seedlings growing in each light regime (Fig. 3.2). Each value of hydraulic function was calculated relative to the maximum day-7 value (7 days after germination, a date at which seedling sizes were not significantly different between treatments, Doley 1978), which for both cases was for the low light intensity seedlings. In fact, the seedlings growing in low light intensity had greater theoretical hydraulic- and leaf-specific conductivity than did those at high light at day 7 and day 14. This trend was reversed by day 21 after which the seedlings growing in high light intensity outpaced the low intensity seedlings in both measures of hydraulic efficiency. By day 49, the high-light seedlings had nearly 600 times the theoretical hydraulic conductivity that the low-light seedlings had had on day 7, and more than four times the value of the low light seedlings of the same age. These increases were caused by the addition of more and wider vessels to the xylem relative to the low light intensity seedlings. Increases in leaf area resulted in smaller differences in hydraulics when compared on a leaf-specific scale (Fig. 3.2b). However, by day 49 the leaf-specific conductivity of the high light grown plants was >11 times the day-7 value of the low light seedlings, and more than double when comparing hydraulic function of the low light plants on day 49.

The functional consequences of these patterns (Fig. 3.2) are apparent when the values from the two panels are compared. For the high light seedlings, the increase in the theoretical hydraulic conductivity between days 21 and 28 is insufficient to compensate for the increase in leaf area, and the leaf-specific value does not increase. The low light grown seedlings exhibit a drop in leaf-specific conductivity because

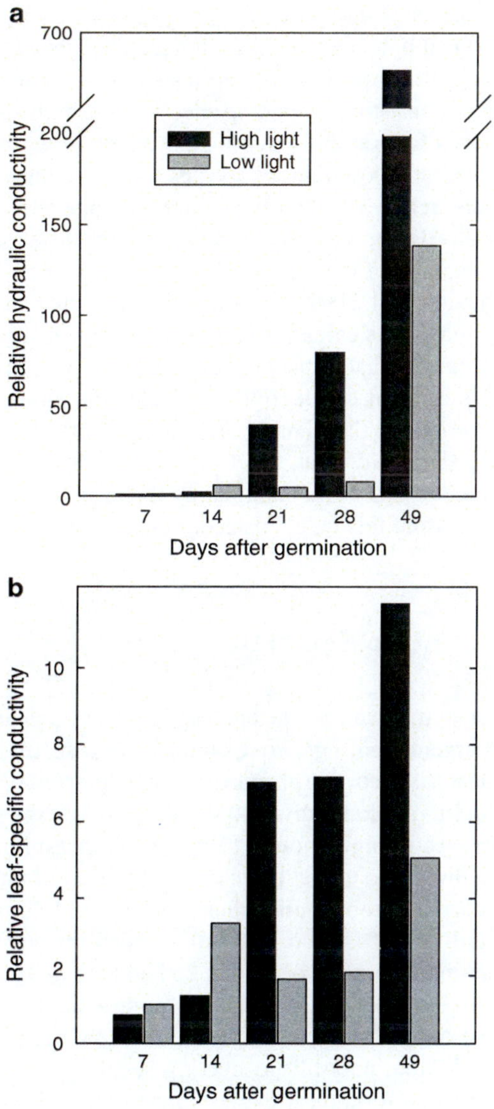

Fig. 3.2 The change in theoretical values of hydraulic conductivity (**a**) and leaf-specific hydraulic conductivity (**b**) over the first 49 days after germination in *Eucalyptus grandis* seedlings grown under high (12.6 mol m^{-2} day^{-1} photosynthetically active radiation) and low (2.8 mol m^{-2} day^{-1}) light intensity. Hydraulic values are relative to the maximum value for either growth treatment on day 7 (which was the low light value for both panels). Hydraulic values were calculated from anatomical data in Doley (1979)

the hydraulic conductivity did not increase in concert with the leaf expansion. Unfortunately, the data are not available to determine if this disconnect between the hydraulic supply and leaf area resulted in limited carbon assimilation due to stomatal closure.

A number of studies have examined the effect of limited water supply on root and shoot structure. In particular, Padilla et al. (2007) showed that when water is less available, individuals of three Mediterranean shrub species tend to elongate their roots relative to individuals of the same species growing under conditions of greater soil moisture. This response to grow deeper roots can enhance seedling survival in habitats with prolonged droughts where deeper soil layers may maintain more moisture than the top layers (Padilla and Pugnaire 2007). When congeneric pairs consisting of one species from a high rainfall environment and one species from a low rainfall environment were grown under common and favorable conditions, most low-rainfall species had lower specific root length (length per mass of roots) relative to their high-rainfall congeneric (Wright and Westoby 1999; Nicotra et al. 2002). The root systems of the low-rainfall species also showed greater elongation of the main axis of the root system, but fewer and shorter side branches (Nicotra et al. 2002). Although the theoretical axial hydraulic conductivity of the root systems were calculated and no differences were found between the congenerics, it is unclear what impact the thicker roots (i.e., lower specific root length) would have on the radial conductivity and thus the entire hydraulic conductance of the root network. As the plant hydraulic conductance is proportional to the photosynthetic rate (Hubbard et al. 2001; Meinzer and Grantz 1990; Sperry et al. 1993), plants would benefit from maximizing their root conductance, yet the tools and techniques to measure this value in seedlings are not currently developed.

4 Future Research

Due to their small size and fragile nature, it is difficult to measure the physiology of emergent seedlings. For example, it is difficult to clamp young seedlings into traditional gas exchange chambers without destroying the seedling. In addition, it typically requires several individual seedlings inside the same chamber to achieve a signal to noise ratio high enough to make the measurements reliable. Consequently, many seedlings end up being sacrificed for a set of physiological measurements, which is especially problematic when working with limited numbers of naturally-occurring seedlings. It is also difficult to perform water potential measurements, due to the small diameter of the cut end and the difficulty in recognizing the balance pressure. Furthermore, measures of hydraulic conductance on emergent seedlings are nonexistent because our current methods are primarily designed for woody tissues. In addition, naturally-occurring seedlings may only be available only every few years due to natural cycles in reproductive output, so most researchers rely on planted seedlings (germinated in the greenhouse) or greenhouse studies, which may not be reflective of naturally-occurring seedling microclimate and physiology.

However, there are promising strategies and new methods being developed that could be easily adapted to emergent seedling physiology. Techniques that could be adapted to the study of young seedlings include chlorophyll fluorescence for studies of photosynthetic potential (or photoinhibition); using smaller chambers for measuring gas exchange of single cotyledons (e.g. LICOR Arabidopsis chamber); and the rehydration kinetics method (Brodribb and Holbrook 2003), previously used for leaf hydraulic conductance, could be used to measure seedling hydraulic conductance. There are also tools that are independent of size: tissue osmometry for measuring solute concentrations, use of Lockhart-based analyses of growth, and studies of gene expression and gene product inactivation (e.g. aquaporins) with inhibitory chemicals.

Overarching questions about which little is known and that are likely to be fruitful in the near future include the following. (1) What are the adaptive nature of different cotyledon structures? Much work has been done on classification of seedling types (e.g. Garwood 1996) and on relationships between cotyledon or seed size/mass and survivorship (Kitajima 1996; Bonfil 1998; Moles and Westoby 2004a). However, we know very little about the adaptive significance of cotyledon shape (and why cotyledons are often very different from primarily leaves), orientation, internal structure, stomatal distribution, cuticular qualities, etc. An energy balance approach incorporating these cotyledon parameters along with the very low wind speeds found near the ground might be particularly elucidative. (2) What determines the timing of the transition between cotyledonous and the first true leaves, and is this transition a time of greater mortality? Does carbon balance or water status determine whether or not young seedlings flush a new set of primary leaves or is it more closely related to environmental conditions (e.g. temperature degree days, red:far-red ratios, etc.) or determined entirely by genetic cues? (3) What are the hydraulic properties of young seedlings and how are they affected by different types of stress? Measurements of hydraulic conductivity and conductance of various parts of emergent seedlings and their vulnerability to hydraulic dysfunction during drought or freeze-thaw conditions could be particularly informative, especially for predictions of future species or community migration under climate change. A greater understanding of how seedling hydraulics impacts carbon gain and seedling survival will allow more accurate predictions of seedling establishment and population range shifts under different predicted scenarios of future climates.

Acknowledgements The authors would like to thank Kaoru Kitajima for a thoughtful and thorough review that resulted in a much stronger chapter. We are also grateful to Drs. Frederick Meinzer and Barbara Lachenbruch, as well as the entire AvH group for critical discussion of ideas.

References

Ackerly DD, Monson RK (2003) Waking the sleeping giant: the evolutionary foundations of plant function. Int J Plant Sci 164(Suppl):S1–S6

Alexander RR (1987) Ecology, silviculture, and management of the Engelmann spruce-subalpine fir type in the central and southern Rocky Mountains. Agric. Handb. No. 659. USDA Forest Service, Washington, DC, 144 pp

Allsop N, Stock WD (1995) Relationship between seed reserves, seedling growth and mycorrhizal responses in 14 related shrubs (Rosidae) from a low-nutrient environment. Funct Ecol 9:248–254

Alvarez-Buylla ER, Martinez-Ramos M (1992) Demography and allometry of *Cecropia obtusifolia,* a neotropical pioneer tree – an evaluation of the climax-pioneer paradigm for tropical rain forests. J Ecol 80:275–290

Alvarez-Claire S, Kitajima K (2009) Susceptibility of tree seedlings to biotic and abiotic hazards in the understory of a moist tropical forest in Panama. Biotropica 41:47–56

Ashton PM, Gamage S, Gunatilleke IAUN, Gunatilleke CVS (1997) Restoration of a Sri Lankan rain forest: using carribean pine *Pinus carribaea* as a nurse for establishing late-successional tree species. J Appl Ecol 34:915–925

Augspurger CK (1984) Seedling survival of tropical tree species: interactions of dispersal distance, light-gaps and pathogens. Ecology 65:1705–1712

Bagnold RA (1941) The physics of blown sand and desert dunes. Methuen, London, 265 pp

Ball MC (1994) The role of photoinhibition during tree seedling establishment at low temperatures. In: Baker NR, Bowyer JR (eds.) Photoinhibition of photosynthesis: from molecular mechanisms to the field. BIOS Scientific Publishers, Oxford, pp 365–377

Ball MC, Hodges VS, Laughlin GP (1991) Cold-induced photoinhibition limits regeneration of snow gum at treeline. Funct Ecol 5:663–668

Ball MC, Egerton JJG, Leuning R, Cunningham RB, Dunne P (1997) Microclimate above grass adversely affects spring growth of seedling snow gum. Plant Cell Environ 20:155–166

Baumgartner A (1952) Die Strahlungsbilanz in einer Fichentendickung. Forstw C 71:337–349

Bell AD (1991) Plant form: an illustrated guide to flowering plant morphology. Oxford University Press, Oxford

Berkowitz AR, Canham CD, Kelly VR (1995) Competition vs. facilitation of tree seedling growth and survival in early successional communities. Ecology 76:1156–1168

Bonfil CZ (1998) The effects of seed size, cotyledon reserves and herbivory on seedling survival and growth in *Quercus rugosa* and *Q. laurina* (Fagaceae). Am J Bot 85:79–87

Brodribb TJ, Holbrook NM (2003) Stomatal closure during leaf dehydration, correlation with other leaf physiological traits. Plant Physiol 132:2166–2173

Cai Z-Q, Poorter L, Han Q, Bongers F (2008) Effects of light and nutrients on seedlings of tropical *Bauhinia* lianas and trees. Tree Physiol 28:1277–1285

Callaway RM, Walker LR (1997) Competition and facilitation: a synthetic approach to interactions in plant communities. Ecology 78:958–1965

Cánovas FM, Canton FR, Gallardo F, Garcia-Gutierrez A, de Vicente A (1991) Accumulation of glutamine synthetase during early development of maritime pine (*Pinus pinaster*) seedlings. Planta 185:372–378

Chazdon RL, Pearcy RW (1986) Photosynthetic response to light variation in rain-forest species. Oecologia 69:529–531

Clark DB, Clark DA (1989) The role of physical damage in the seedling mortality regime of a neotripcal rain forest. Oikos 55:225–230

Combes D, Sinoquet H, Varlet-Grancher C (2000) Preliminary measurement and simulation of the spatial distribution of the morphogenetically active radiation (MAR) within an isolated tree canopy. Ann For Sci 57:497–511

Connell JH, Slayter RO (1977) Mechanisms of succession in natural communities and their role in community stability and organisation. Am Nat 111:119–144

Cui M, Smith WK (1991) Photosynthesis, water relations and mortality in *Abies lasiocarpa* seedlings during natural establishment. Tree Physiol 8:37–46

Day ME, Greenwood MS (2011) Regulation of ontogeny in temperate conifers. In: Meinzer FC, Lachenbruch B, Dawson TE (eds.) Size- and age-related changes in tree structure and function. Springer, New York

Day TA, DeLucia EH, Smith WK (1989) Influence of cold soil and snowcover on photosynthesis and leaf conductance in two rocky mountain conifers. Oecologia 80:546–552

de Vogel EF (1980) Seedlings of dicotyledons: structure development, types, descriptions of 150 woody Malesian taxa. Center for Agricultural Publishing and Documentation, Wageningen

Deacon EL (1953) Vertical profiles of mean wind speed in the surface layers of the atmosphere. Geophys Mem Lond 11:91

Denne MP (1972) A comparison of root- and shoot-wood development in conifer seedlings. Ann Bot 36:579–587

Denne MP (1974) Effects of light intensity on tracheid dimensions in *Picea sitchensis*. Ann Bot 38:337–345

Denne MP (1976) Effects of environmental change on wood production and wood structure in *Picea sitchensis* seedlings. Ann Bot 40:1017–1028
Dickie IA, Schnitzer SA, Reich PB, Hobbie SE (2005) Spatially disjunct effects of co-occurring competition and facilitation. Ecol Lett 8:1191–1200
Doley D (1978) Effects of shade on gas exchange and growth in seedlings of *Eucalyptus grandis* Hill Ex. Maiden. Aust J Plant Physiol 5:723–738
Doley D (1979) Effects of shade on xylem development in seedlings of *Eucalyptus grandis* Hill Ex. Maiden. New Phytol 82:545–555
Edwards W, Krockenberger A (2006) Seedling mortality due to drought and fire associated with the 2002 El Niño event in a tropical rain forest in north-east Queensland, Australia. Biotropica 38:16–26
Ehrlen J (2003) Fitness components versus total demographic effects: evaluating herbivore impacts on a perennial herb. Am Nat 162:796–810
Elias P, Kratochvilova I, Janous D, Marek M, Masarovicova E (1989) Stand microclimate and physiological activity of tree leaves in an oak-hornbeam forest. Trees 4:227–233
Endler JA (1993) The color of light in forests and its implications. Ecol Monogr 63:1–27
Engelbrecht BMJ, Comita LS, Condit R, Kursar TA, Tyree MT, Turner BL, Hubbell SP (2007) Drought sensitivity shapes species distribution patterns in tropical forests. Nature 447:80–82
Fenner M (1987) Seedlings. New Phytol 106:35–47
Fenner M, Kitajima K (1999) Seed and seedling ecology. In: Pugnaive F, Valladares F (eds.) Handbook of functional plant ecology. Marcel Dekker, New York, pp 589–621
Feygel'son YM, Tsvang LR (1974) In: Feygel'sonYe M, Tsvang LR, Feygel'son Ye M, Tsvang LR (eds.) Heat transfer in the atmosphere, translation of teploobmen v atmosfere. Nauka Press, Moscow
Figueroa JA, Lusk CH (2001) Germination requirements and seedling shade tolerance are not correlated in a Chilean temperate rain forest. New Phytol 152:483–489
Garwood NC (1996) Functional morphology of tropical tree seedlings. In: Swaine MD (ed.) The ecology of tropical tree seedlings. UNESCO and The Parthenon Publishing Group, Paris, pp 59–129
Geiger R (1965) The climate near the ground, 3rd edn. Harvard University Press, Cambridge, MA, 611 pp
Germino MJ, Smith WK (1999) Sky exposure, crown architecture, and low-temperature photoinhibition in conifer seedlings at alpine treeline. Plant Cell Environ 22:407–415
Germino MJ, Smith WK (2000) Differences in microsite, plant form, and low-temperature photoinhibition in alpine plants. Arct Antarct Alp Res 32:388–396
Germino MJ, Smith WK (2001) Relative importance of microhabitat, plant form and photosynthetic physiology to carbon gain in two alpine herbs. Funct Ecol 15:243–251
Givnish TJ (1988) Adaptation to sun vs. shade: a whole-plant perspective. Aust J Plant Physiol 15:63–92
Goulet F (1995) Frost heaving of forest tree seedlings: a review. New For 9:67–94
Grace SL, Platt WJ (1995) Effects of adult tree density and fire on the demography of pregrass stage juvenile longleaf pine (*Pinus palustris* Mill.). J Ecol 83:75–86
Grant RH (1997) Partitioning of biologically active radiation in plant canopies. Int J Biometeorol 40:26–40
Greenlee JT, Callaway RM (1996) Abiotic stress and the relative importance of interference and facilitation in montane bunchgrass communities in western Montana. Am Nat 148:386–396
Greenwood MS, O'Brien CL, Schatz JD, Diggins CA, Day ME, Jacobson GL, White AS, Wagner RG (2008) Is early life-cycle success a determinant of the abundance of red spruce and balsam fir? Can J For Res 38:2295–2305
Griffin JR (1962) Intraspecific variation in *Pinus sabiniana* Dougl. PhD Thesis, University of California, Berkeley, 274 p
Hadley JL, Smith WK (1987) Influence of krummholz mat microclimate on needle physiology and survival. Oecologia 73:82–90

Harrison EF, Minnis P, Barkstrom B, Gibson GG (1993) Radiation budget at the top of the atmosphere. In: Gurney RJ, Foster JL, Parkinson CL (eds.) Atlas of satellite observations related to global change. Cambridge University Press, Cambridge, pp 19–38

Hasselquist NJ, Germino MJ, McGonigle TC, Smith WK (2005) Variability of *Cenoccocum* infection and its ecophysiological significance for young conifers at alpine-treeline. New Phytol 165:867–873

Hastwell GT, Facelli JM (2003) Differing effects of shade-induced facilitation on growth and survival during the establishment of a chenopod shrub. J Ecol 91:941–950

Hättenschwiler S, Smith WK (1999) Seedling occurrence in alpine treeline conifers: a case study from the central Rocky Mountains, USA. Acta Oecol 20:219–224

He XH, Bledsoe CS, Zasoski RJ, Southworth D (2006) Rapid nitrogen transfer from ectomycorrhizal pines to adjacent ectomycorrhizal and arbuscular mycorrhizal plants in a California oak woodland. New Phytol 170:143–151

Holtum JAM, Winter K (2003) Photosynthetic CO_2 uptake in seedlings of two tropical tree species exposed to oscillating elevated concentrations of CO_2. Planta 218:152–158

Hubbard RM, Stiller V, Ryan MG, Sperry JS (2001) Stomatal conductance and photosynthesis vary linearly with plant hydraulic conductance in ponderosa pine. Plant Cell Environ 24:113–121

Hughes NM, Johnson DM, Akhalkatsi M, Abdaladze O (2009) Characterizing Betula litwonii seedling microsites at the alpine-treeline ecotone, central Greater Caucasus Mountains, Georgia. Arct Antarct Alp Res 41:112–118

Iverson LR, Prasad AM (2001) Potential changes in tree species richness and forest community types following climate change. Ecosystems 4:186–199

Janos DP (1980) Vesicular-arbuscular mycorrizae affect lowland tropical rain forest plant growth. Ecology 61:151–162

Ji XB, Yi NG (2003) The seedling types of dicots and their evolutionary relationships. Acta Phytotakinomica Sin 41:447–464

Johnson DM, Smith WK (2005) Refugial forests of the southern Appalachians: photosynthesis and survival in current-year *Abies fraseri* seedlings. Tree Physiol 25:1379–1387

Johnson DM, Smith WK (2008) Cloud immersion alters microclimate, photosynthesis and water relations in *Rhododendron catawbiense* and *Abies fraseri* seedlings in the southern Appalachian Mountains, USA. Tree Physiol 28:385–392

Johnson DM, Germino MJ, Smith WK (2004) Abiotic factors limiting photosynthesis in seedlings of *Abies lasiocarpa* and *Picea engelmannii* at alpine timberline. Tree Physiol 24:377–386

Johnson DM, Smith WK, Vogelmann TC, Brodersen CR (2005) Leaf architecture, incident light direction, and mesophyll fluorescence profiles inside a broadleaf, conifer needle and cotyledon. Am J Bot 92:1425–1431

Jordan DN, Smith WK (1994) Energy-balance analysis of nighttime leaf temperatures and frost formation in a sub-alpine environment. Agric For Meteorol 71:359–372

Jordan DN, Smith WK (1995) Microclimate factors influencing the frequency and duration of growth season frost for subalpine plants. Agric For Meteorol 77:17–30

Kimmins JP (1987) Forest ecology. Macmillian Publishing Company, New York, 531 pp

Kitajima K (1992) Relationship between photosynthesis and thickness of cotyledons for tropical tree species. Funct Ecol 6:582–589

Kitajima K (1996) Cotyledon functional morphology, patterns of seed reserve utilization and regeneration niches of tropical tree seedlings. In: Swaine MD (ed.) The ecology of tropical tree seedlings. UNESCO and The Parthenon Publishing Group, Paris, pp 193–210

Kitajima K (2003) Impact of cotyledon and leaf removal on seedling survival in three tree species with contrasting cotyledon functions. Biotropica 35:429–434

Kitajima K, Hogan KP (2003) Increases of chlorophyll a/b ratios during acclimation of tropical woody seedlings to nitrogen limitation and high light. Plant Cell Environ 26:857–865

Kitajima K, Myers JA (2008) Seedling ecophysiology: strategies towards achievement of positive carbon balance. In: Leck MA, Parker VT, Simpson RL (eds.) Seedling ecology and evolution. Cambridge University Press, UK, pp 172–188

Knapp AK, Smith WK (1982) Factors influencing understory seedling establishment of Engelmann spruce (*Picea engelmannii*) and subalpine fir (*Abies lasiocarpa*) in southeast Wyoming. Can J Bot 60:2753–2761

Kobe RK (1997) Carbohydrate allocation to storage as a basis of interspecific variation in sapling survivorship and growth. Oikos 80:226–233

Kobe RK, Pacala SW, Silander JA Jr, Canham CD (1995) Juvenile tree survivorship as a component of shade tolerance. Ecol Appl 5:517–532

Kohyama T (1983) Seedling stage of two subalpine *Abies* species in distinction from sapling stage: a matter-economic analysis. J Plant Res 96:49–65

Kursar TA, Engelbrecht BMJ, Burke A, Tyree MT, El Omari B, Giraldo JP (2009) Tolerance to low leaf water status of tropical tree seedlings is related to drought performance and distribution. Funct Ecol 23:93–102

Larcher W (2003) Physiological plant ecology, 4th edn. Springer, New York, 513 pp

Leck MA, Outred HA (2008) Seedling natural history. In: Leck MA, Parker VT, Simpson RL (eds.) Seedling ecology and evolution. Cambridge University Press, UK, pp 17–55

Lee DW (1987) The spectral distribution of radiation in two neotropical rainforests. Biotropica 19:161–166

Lee DW, Downum KR (1991) The spectral distribution of biologically active solar radiation at Miami, Florida, USA. Int J Biometeorol 35:48–54

Li M, Lieberman M, Lieberman D (1996) Seedling demography in undisturbed tropical wet forest in Costa Rica. In: Swaine MD (ed.) The ecology of tropical tree seedlings. UNESCO and The Parthenon Publishing Group, Paris, pp 285–304

Lubbock J (1892) A contribution to our knowledge of seedlings. Appleton and Company, New York, 646 pp

Maher EL, Germino MJ (2006) Microsite differentiation among conifer species during seedling establishment at alpine treeline. Ecoscience 13:334–341

Maher EL, Germino MJ, Hasselquist NJ (2005) Interactive effects of tree and herb cover on survivorship, physiology, and microclimate of conifer seedlings at the alpine tree-line ecotone. Can J For Res 35:567–574

Masaki T, Nakashizuka T (2002) Seedling demography of *Swida controversa*: effect of light and distance to conspecifics. Ecology 83:3497–3507

Matthes U, Larson DW (2006) Microsite and climatic controls of tree population dynamics: an 18 year study on cliffs. J Ecol 94:402–414

Maun MA (1998) Adaptations of plants to burial in coastal sand dunes. Can J Bot 76:713–738

McDowell N, Pockman WT, Allen CD, Breshears DD, Cobb N, Kolb T, Plaut J, Sperry J, West A, Williams DG, Yepez EA (2008) Mechanisms of plant survival and mortality during drought: why do some plants survive while others succumb to drought? New Phytol 178:719–739

McLaughlin SB, Anderson CP, Edwards NT, Roy WK, Layton PA (1990) Seasonal patterns of photosynthesis and respiration of red spruce saplings from two elevations in declining Southern Appalachian stands. Can J For Res 20:485–495

Meinzer FC, Grantz DA (1990) Stomatal and hydraulic conductance in growing sugarcane: stomatal adjustment to water transport capacity. Plant Cell Environ 13:383–388

Miina J, Saksa T (2006) Predicting regeneration establishment in Norway spruce plantations using a multivariate multilevel model. New For 32:265–283

Moles AT, Westoby M (2004a) Seedling survival and seed size: a synthesis of the literature. J Ecol 92:372–383

Moles AT, Westoby M (2004b) What do seedlings die from and what are the implications for evolution of seed size? Oikos 106:193–199

Morgan RPC, Rickson RK (1995) Slope stabilization and erosion control: a bioengineering approach. E & F Spon, London, 274 pp

Myers JA, Kitajima K (2007) Carbohydrate storage enhances seedling shade and stress tolerance in a neotropical forest. J Ecol 95:383–395

Navrátil M, Spunda V, Markova I, Janous D (2007) Spectral composition of photosynthetically active radiation penetrating into a Norway spruce canopy: the opposite dynamics of the blue/red spectral ratio during clear and overcast days. Trees 21:311–320

Negussie A, Aerts R, Gebrehiwot K, Muys B (2008) Seedling mortality causes recruitment limitation of *Boswellia papyrifera* in northern Ethiopia. J Arid Environ 72:378–383

Nicholas NS, Zedaker SM, Eagar C (1992) A comparison of overstory community structure in three southern Appalachian spruce–fir forests. Bull Torrey Bot Club 119:316–332

Nicotra AB, Babicka N, Westoby M (2002) Seedling root anatomy and morphology: an examination of ecological differentiation with rainfall using phylogenetically independent contrasts. Oecologia 130:136–145

Niklas KJ (2008) Embryo morphology and seedling evolution. In: Leck MA, Parker VT, Simpson RL (eds.) Seedling ecology and evolution. Cambridge University Press, UK, pp 103–129

Nobel PS (2005) Physicochemical and environmental plant physiology, 3rd edn. Academic Press/Elsevier, Burlington, 567 pp

Örlander G (1993) Shading reduces both visible and invisible frost damage to Norway spruce seedlings in the field. Forestry 66:27–36

Padilla FM, Pugnaire FI (2007) Rooting depth and soil moisture control Mediterranean woody seedling survival during drought. Funct Ecol 21:489–495

Padilla FM, de Dios MJ, Pugnaire FI (2007) Early root growth plasticity in seedlings of three Mediterranean woody species. Plant Soil 296:103–113

Pauley EF, Clebsch EEC (1990) Patterns of *Abies fraseri* regeneration in a Great Smoky Mountains spruce–fir forest. Bull Torrey Bot Club 117:375–381

Pearson RG, Dawson TP (2003) Predicting the impacts of climate on the distribution of species: are bioclimate envelope models useful? Glob Ecol Biol 12:361–371

Plumb TR (1980) Response of oaks to fire. In: Plumb TR (technical coordinator) Proceedings of the symposium on the ecology, management, and utilization of California oaks; 1979 June 26–28; Claremont, CA. Gen. Tech. Rep. PSW-44. Berkeley, CA: U.S. Department of Agriculture, Forest Service, Pacific Southwest Forest and Range Experiment Station, pp 202–215

Reich PB, Tjoelker MG, Vanderklein DW, Buschena C (1998) Close association of RGR, leaf and root morphology, seed mass and shade tolerance in seedlings of nine boreal tree species grown in high and low light. Funct Ecol 12:327–338

Reinhardt K, Smith WK (2008) Impacts of cloud immersion on microclimate, photosynthesis and water relations of Abies fraseri (Pursh.) Poiret in a temperate mountain cloud forest. Oecologia 158:229–238

Reinhardt K, Johnson DM, Smith WK (2009) Age-class differences in Fraser fir (*Abies fraseri*) photosynthesis and water relations. Can J For Res 39:1–5

Rodriguez-Calcerrada J, Cano FJ, Valbuena-Carabaña M, Gill L, Aranda I (2010) Functional performance of oak seedlings naturally regenerated across microhabitats of distinct overstorey canopy closure. New For 39:245–259

Sacchi CF, Price PW (1992) The relative roles of abiotic and biotic factors in seedling demography of Arroyo willow (*Salix lasiolepis*: Salicaceae). Am J Bot 79:395–405

Sasaki S, Kozlowski TT (1969) Effects of cotyledon and hypocotyl photosynthesis on growth of young pine seedlings. New Phytol 69:493–500

Schlesinger R, Williams RD (1984) Growth responses of black walnut *Juglans nigra* to interplanted trees. For Ecol Manage 9:235–243

See LS, Alexander IJ (1996) The dynamics of ectomycorrhizal infection of *Shorea leprosula* seedlings in Malaysian rain forests. New Phytol 132:297–305

Smith CJ (1976) Adaptations to environmental change in *Picea sitchensis* seedlings – changes in photosynthesis, growth and substrate availability. Ann Bot 40:1003–1015

Smith GF, Nicholas NS (2000) Size- and age-class distributions of Fraser fir following balsam woolly adelgid infestation. Can J For Res 30:948–957

Smith SE, Read DJ (2008) Mycorrhizal symbiosis, 3rd edn. Academic, London, 605 pp

Smith WK, Germino MJ, Hancock TE, Johnson DM (2003) Another perspective on the altitudinal occurrence of alpine tree lines. Tree Physiol 23:1101–1113

Sperry JS, Alder NN, Eastlack SE (1993) The effect of reduced hydraulic conductance on stomatal conductance and xylem cavitation. J Exp Bot 44:1075–1082

Tappeiner U, Cernusca A (1998) Model simulation of spatial distribution of photosynthesis in structurally differing plant communities in the central Caucasus. Ecol Model 113:201–233

Tingey DT, Lee EH, Phillips DL, Rygiewicz PT, Waschmann RS, Johnson MG, Olszyk DM (2007) Elevated CO2 and temperature alter net ecosystem C exchange in a young Douglas fir mesocosm experiment. Plant Cell Environ 30:1400–1410

Tranquillini W (1979) Physiological ecology of the Alpine timberline: tree existence at high altitudes with special reference to the European Alps. Springer, New York, 137 pp

Turner H (1958) Maximaltemperaturen oberflachennaher Bodenschichten an der alpinen Waldgrenze. Wetter Leben 10:1–12

Tyree MT, Zimmermann MH (2002) Xylem structure and the ascent of sap. Springer Verlag, Berlin

van Mantgem PJ, Stephenson NL, Byrne JC, Daniels LD, Franklin JF, Fulé PZ, Harmon ME, Larson AJ, Smith JM, Taylor AH, Veblen TT (2009) Widespread increase of tree mortality rates in the western United States. Science 323:521–524

Walters MB, Reich PB (1999) Research review: low-light carbon balance and shade tolerance in the seedlings of woody plants-do winter deciduous and broad-leaved evergreen species differ? New Phytol 143:143–154

Walther G-R, Berger S, Sykes MT (2005) An ecological footprint of climate change. Proc R Soc B 272:1427–1432

White PS, Mackenzie MD, Busing RT (1985) Natural disturbance and gap phase dynamics in southern Appalachian spruce–fir forests. Can J For Res 15:233–240

Whitmore TC (1996) A review of some aspects of tropical rainforest seedling ecology with suggestions for further enquiry. In: Swaine MD (ed.) The ecology of tropical tree seedlings. UNESCO and The Parthenon Publishing Group, Paris, pp 3–39

Wright IJ, Westoby M (1999) Differences in seedlings growth behaviour among species: trait correlations across species and trait shifts along nutrient compared to rainfall gradients. J Ecol 87:85–97

Wright IJ, Westoby M (2001) Understanding seedlings growth relationships through specific leaf area and leaf nitrogen concentration: generalizations across growth forms and growth irradiance. Oecologia 127:21–29

Young DR, Smith WK (1983) Effect of cloudcover on photosynthesis and transpiration in the subalpine understory species *Arnica latifolia*. Ecology 64:681–687

Part III
Integration of Structure and Function Across Ontogeny

Chapter 4
Regulation of Ontogeny in Temperate Conifers

Michael E. Day and Michael S. Greenwood

Abstract Through their life stages, long-lived forest trees must adapt to challenges resulting from vast changes in size and external environment. Trees accomplish this by producing new tissues and growth habits adapted to specific life stages by pluripotent meristems. The morphology and physiology of these new tissues are determined by complex interactions of the external and internal environments of the tree, and gene expression both within meristems and during differentiation of cells produced by meristems. The trajectories of various age-related changes are both inter-related and independent at various scales, and, for the majority of species, are not purely physio-mechanical responses to increased tree height. Understanding the relationships between tree developmental state, tree size and its environment requires both a whole-tree approach integrated through time to encompass life stage strategies, and a molecular approach to understand the cues, transduction pathways and epigenetic mechanisms that regulate whole-life ontogeny.

1 Introduction

Forest trees are not only long lived but undergo immense changes in size over their life spans. Therefore, they must adapt not only to substantial variation in microclimate and edaphic conditions, but also must cope with the demands of vast increases in size and complexity of their vegetative bodies. Continual growth throughout the tree lifetime derives from its pluripotent meristems, the shoot and root apical meristems (SAMs and RAMs, respectively) and the cambia that provide for growth in girth.

M.E. Day (✉) • M.S. Greenwood
School of Forest Resources, University of Maine, Nutting Hall,
Orono, ME 04469–5755, USA
e-mail: daym@maine.edu

F.C. Meinzer et al. (eds.), *Size- and Age-Related Changes in Tree Structure and Function*, Tree Physiology 4, DOI 10.1007/978-94-007-1242-3_4,

Pluripotent meristems are capable of producing new tissue within a genetically determined range of cell types and continual replacement of meristem initials. The daughter cells produced by the meristem initials retain developmental plasticity until differentiating into specific cell types such as parenchyma, epidermal tissues, xylem or phloem (Steeves and Sussex 1989). As ontogeny (whole-tree development) progresses, the behavior of meristems and tissue maturational processes that regulate the properties of the new plant organs undergo changes that are life-stage specific. For example, the frequency of reproductive organs increases and foliage often becomes more massive. Of particular interest to foresters, forest managers and others is the ontogenetic course of stem wood productivity, which is a principal factor in determining optimum age for harvest as well as a basis for estimating rates of carbon sequestration in forests.

The intricate relationships between size, age and ontogenetic life-stage results in significant challenges to understanding the trajectories of morphological and physiological changes over a tree's lifetime. A germinant competes for light and nutrients in a very restricted area both horizontally and vertically, but, as it increases in size, a tree exploits a habitat that expands exponentially in both dimensions. In parallel, exponential increases in size require modification in strategies for transport of water, nutrients, and carbohydrates, communication between plant tissues, and addressing static and dynamic forces. Here we review our current understanding of how temperate conifers cope with both changes in the external environment faced by their roots and shoots, as well the consequences of their ever more complex vegetative structure.

In a competitive and often physically hostile world, physiological and morphological adaptations that serve the individual well in any particular life stage may prove mal-adapted to survival in another life-stage. For instance, relatively low shoot-to-root ratios (S:R) in seedlings, which must rely on the upper soil strata for water, may not be adaptive in later life-stages when canopy competition for light is the key to survival. Size-related trajectories of changes in heritable traits such as height growth provide evidence for the life-stage independence of conifer ecophysiological strategies. For example, the correlation between family rankings in height growth between first year germinants (where differences among families are highly significant) with those of mature trees in the same families is nearly zero (Lambeth 1980). Many tree improvement programs have attempted to recognize individuals with superior growth potential in germinant/seedling stages in order to decrease generational turnover times. Even though heritable physiological traits have been identified in many of these studies, a lack of correlation between life-stages has made predictive efforts largely unsuccessful (Pharis et al. 1991; Greenwood and Volkaert 1992). Ecophysiological attributes tested in this regard include photosynthetic rates and quantum efficiency, water use efficiency, chlorophyll content, specific leaf mass, shoot-root ratios, metabolic efficiency and hormone levels. These evaluations have been made on a wide variety of temperate conifers including *Picea abies* (Larsen and Wellendorf 1990), *Pinus sylvestris* (Nilsson and Anderson 1987), *Pinus resinosa* (Carter et al. 1990), and *Picea mariana* (Sulzer et al. 1993).

We suggest that there are two broad categories of mechanisms that regulate age- (or size-) related changes in tree growth habits. The best studied, though certainly not yet fully understood, group of mechanisms are the physical constraints that occur through the interaction of the external environment with the morphological and physiological attributes of a tree. These biophysical constraints can be viewed as limitations placed on tree growth, development and/or productivity by a mechanistic interaction of the controlling factor(s) with a species' phenotypic possibilities. For example, the role of hydraulic biophysical constraints in determining the height of the tallest trees has been examined by Yoder et al. (2004), Koch et al. (2004), McDowell et al. (2005), Domec et al. (2008), Woodruff et al. (2008), and is discussed in several chapters in this volume (McDowell et al. 2011; Mencuccini et al. 2011; Scholz et al. 2011; Woodruff and Meinzer 2011).

The second group of mechanisms acts through the behavior of meristems and/or the differentiating cells they produce to develop tissue and organs that exhibit age-, life-stage- or size-specific evolutionarily adaptive properties. These mechanisms may result from size-independent changes in meristem behavior or be entrained by "cues" in the external environment or within the tree itself followed by transduction of signals to the meristem and/or developing cells. The results are phenotypic changes that either react to or anticipate potential challenges associated with a particular life-stage. In many cases these pathways act to circumvent or minimize biophysical constraints, or they may have evolved to counteract life-stage specific challenges to survival.

Meristematic behavior is controlled by complex interactions of patterns of gene expression in response to environmental and organismal cues (Poethig 1990). Day et al. (2002) proposed a classification for trends in life-stage attributes based on the pathways controlling meristematic behavior: extrinsic, intrinsic and complex, potentially synergistic, interactions (Table 4.1). Extrinsic control is characterized by reversible responses to cues external to the meristem. This would include not only the environment external to the tree, but also cues originating from the previously developed tissues and organs of the tree. Extrinsic pathways affect development of new cells and tissues only after their production by meristematic initials. They are readily reversible as external cues or stimuli change, but may or may not show reversibility in already differentiated tissues. Morphological and physiological attributes of sun-adapted and shade-adapted foliage are well known examples of extrinsic control.

In contrast, intrinsic controls involve epigenetic changes in meristematic gene expression and tend to be relatively irreversible. The cues that initiate intrinsic pathways may originate within the meristem itself or may be regulated by external cues operating through transduction pathways. Cues internal to meristems would involve gene induction and epigenetic changes in genes analogous to aging in animals, where genetic changes result from mechanisms such as telomere shortening, histone modification or DNA-methylation. While internal epigenetic "time-keeping" appears to occur in the meristems of some annual species, the mechanisms have not been elucidated, and their relevance to long-lived perennials is uncertain. However, intrinsic meristem behaviors entrained by cues external to the meristem do appear to be important drivers of age-related trends in conifers, as identified by reciprocal grafting among trees of various age-classes (Table 4.1). Intrinsic behaviors, such as the phase-change to reproductive competency, are generally irreversible by normal

Table 4.1 Summary of factors affecting meristem behavior in red spruce

	Traits regulated by:	
Life stage:	Intrinsic factors	Extrinsic factors
Germinant	Acicular needles Chlorophyll development High root shoot ratios Neoformed growth No reproductive competence	None apparent
Seedling	Shade-tolerance response Low foliar LMA High branch density No reproductive competence High stomatal sensitivity to VPD	Neoformed growth (expressed under high light, moisture) Photosynthetic capacity
Sapling	Shade-tolerance (early sapling) or shade-avoidance responses Increased LMA, decreased needle L:W Declining R:S ratio Branch density declines Some reproductive potential	Increasing LMA, decreasing needle L:W Increased apical control Precocious reproduction under stress Photosynthetic capacity
Young-mature	High LMA, decreasing needle L:W Full reproductive competence Apical control maximized?	High LMA, decreasing needle L:W Maximum apical control Photosynthetic capacity
Old growth	Maximum LMA, minimum L:W Full reproductive competence Water use efficiency increases? Carbon allocation strategy?	Maximum LMA, minimum L:W Apical control declines Water use efficiency increases? Carbon allocation strategy?

Intrinsic factors are relatively permanent ontogenetic changes in meristem behavior. Extrinsic mechanisms are the responses of differentiating cells to the environment external to the meristem, including increased tree size. Attributes with both intrinsic and extrinsic regulation will appear in both columns. Question marks indicate potential but unconfirmed attributes. See text for details

external conditions, or may, as with branching habit, be only slowly or incompletely reversible (Greenwood et al. 2010).

The results of tree- and organ-level studies suggest that many age-related trends are complex interactions of extrinsic and intrinsic control pathways. For example, increasing needle width is a robust, nearly universal, age-related trend in temperate conifers. However, reciprocal grafting studies, where scions from donors of different age classes are grafted onto rootstocks of the same age groups, suggest complex control pathways. In red spruce, grafted scions retain much of the needle width attributes of their donor age class, but also assume the character of their rootstock to a substantial degree (Fig. 4.1).

In addition to ontogenetic trajectories, evolutionarily successful, long-lived species also respond plastically to external (and internal) stresses that may occur seasonally, gradually over extended periods of time, or stochastically at any life-stage. The separation of more ephemeral plastic changes and true ontogenetic trajectories presents a

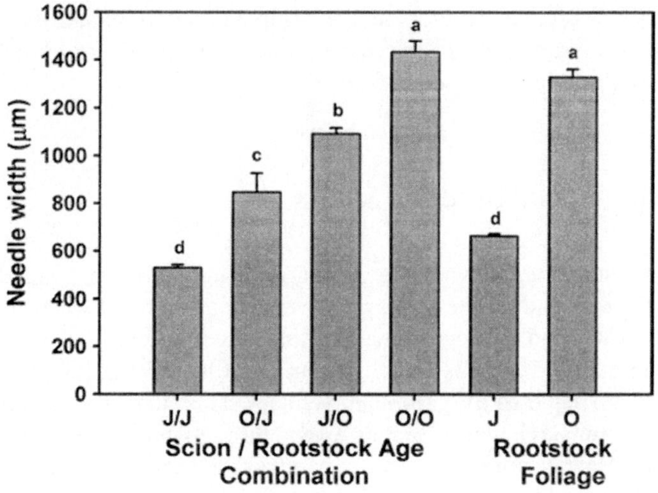

Fig. 4.1 Needle widths of old (125 y+) scions on young (5 y) rootstock and young scions on old rootstock 3 years post-grafting. The right hand columns show needle widths on non-grafted rootstock branches for comparison. The intermediate needle widths of scions suggest that this characteristically age-related attribute is regulated by both intrinsic and intrinsic pathways. Letters indicate significant differences with Tukey's HSD test at $\alpha = 0.05$, $n = 20$

major challenge for physiologists studying tree aging processes, as they may appear phenotypically mixed. Extrinsically controlled developmental pathways may show substantial phenotypic plasticity to environmental variables such as light intensity, water availability and mineral resources, while intrinsic aspects and would tend to exhibit a lower degree of plasticity, but follow ontogenetic trends that have proved successful for the species in the past. Another factor critical to our understanding of ontogenetic trajectories is the almost inevitable relationship between life-stage and tree size, which may cloud our analysis of intrinsic and extrinsic mechanisms. At the whole-tree level, reciprocal grafting and clonal approaches have provided valuable insights into potential regulatory mechanisms and illuminated new directions for research. We are now entering a new and exciting phase, where genomics can be coupled with whole-tree physiology to explore age-related changes in gene expression that can be ultimately associated with meristem gene regulation and environmental and within-tree cues, and the transduction pathways that link them.

2 Regulation of Conifer Vegetative Ontogeny

The conifer ontogenetic trajectory or progression of life-stages includes phase-change or maturation, terms more appropriately used to describe the often dramatic and more temporally restricted transition between vegetative and reproductive development. Other life-stage transitions will exhibit more gradual and less defined boundaries. This section describes the morphological and physiological changes associated with vegetative development in temperate conifers and their potential controlling pathways. Control of the transition from juvenile (pre-reproductive) to reproductive maturity is discussed in Sect. 3.

2.1 Germinant and Seedling Stages

The germinant stage begins with the extrusion of the embryonic radicle beyond the seed coat and lasts until the root system secures a source of water that permits further plant growth and the development of true leaves (as opposed to cotyledons), (Leck and Outred 2008). If this establishment process is not accomplished before seed resources are depleted the germinant perishes. This stage may occupy only the first few post-germination weeks, or may extend through much or all of the first year of growth. In temperate conifers, intrinsic pathways appear to dominate most ontogenetic characteristics and trajectories of germinants (Table 4.1). Germinants and young seedlings show little plastic response to extrinsic environmental conditions. For most angiosperm germinants, light cues through phytochrome red:far-red sensing and blue-light reception are critical to initiate development of photosynthetic systems and other seedling de-etiolation processes. However, in conifers these processes are usually uncoupled from light perception, with germinants developing chlorophyll and photosynthetic capabilities in the dark (Bogorad 1950; Shinya et al. 2001; Matthews 2005). Pools of phytochrome analogs have been identified in *Pinus elliottii* and *Pseudotsuga menziesii* seedlings, although at far lower concentrations than typical of angiosperm seedlings, and the physiological ramifications of these low concentrations are not understood (Burgin et al. 1999).

In conifers early patterns in R:S allocation appear to be substantially under intrinsic control. That is, species differences in early growth allocation are strongly maintained despite variation in resource availability (Day et al. 2005; Greenwood et al. 2008a), although response to above and below-ground resource ratios may modify intrinsic shoot-to-root ratios in second and third year seedlings (Day et al. 2005). Greenwood et al. (2008a) demonstrated that shoot-to-root ratios (S:R) in red spruce (*Picea rubens* Sarg.) and balsam fir (*Abies balsamea* Mill.) did not vary with soil volumetric water content of 15–75% during the first 105 days of growth. The timing of strict intrinsic control in germinants is highly species-specific. S:R and root length in balsam fir germinants and first-year seedlings was unaffected by light environments of 10% and 70% ambient sunlight during their first growing season (210 days). In contrast, red spruce responded to the higher light environment, and presumably greater evaporative demand, by increasing both S:R and root length after 90 days of growth. Consistent with the concept that early developmental patterns in conifers are "pre-programmed" by intrinsic mechanisms, most species exhibit limited or no plasticity in physiological response to varying light environments. Landhäusser and Lieffers (2000) found very limited photosynthetic plasticity in seedlings of cool temperate and boreal conifers growing in different field light environments, and Lavigne et al. (2001) reported no gas exchange (photosynthetic or stomatal conductance) responses to experimentally manipulated light-environments in second-year balsam fir seedlings.

Most temperate conifers appear to retain a large degree of intrinsic control of foliar morphology throughout their seedling stage. Conifer seedling foliage from all needle-leaved species has been consistently characterized as having the highest specific leaf area (SLA, $cm^2\ g^{-1}$), lowest needle width, greatest length-to-width ratio and a pronounced acicular form (Steele et al. 1989; Day et al. 2001; Greenwood et al. 2009). In pines, seedlings have acicular primary needles borne individually on stems with typical fascicular bundles developing subsequently. Many of these foliar morphological characteristics are partially retained in the foliage produced by juvenile meristems grafted into the tops of mature understock for many growing seasons (Fig. 4.1), suggesting a genetic "memory" for these traits that is not readily and completely altered by external factors, a characteristic of intrinsic mechanisms (Day et al. 2001; Rebbeck et al. 1992; Greenwood et al. 2010).

Strong intrinsic regulation is consistent with temperate conifer reproductive strategies. Temperate conifers cannot be characterized as either "r-type" seed producers (such as aspen) that produce massive quantities of widely dispersed seeds, or as typical "k-type" producers (such as oaks) with large seeds containing robust supplies of resources. In addition, conifer seeds do not exhibit the extended seed dormancy associated with survival in seed banks and mechanisms for detecting favorable germination conditions.

Conifer germinants and young-seedlings follow adaptive strategies based on intrinsic programming which have been proven successful in previous generations. Key components of the juvenile strategy are high area-based photosynthetic rates compared to later life-stages (Bond 2000) and relatively low per area foliar structural costs (Day et al. 2001) and neoformed growth (Greenwood et al. 2008a; Powell 2009), all attributes being adaptive for a life-stage having needles with a high probability of loss to herbivory or physical damage, or obsolescence from competitive or mutual shading. Little is known about comparative photosynthetic rates between first-year germinants and older seedlings. In red spruce (*Picea rubens* Sarg.) and balsam fir (*Abies balsamea* (L.) Mill.) germinants, Zazzaro (2009) found 40–50% higher rates of light-saturated photosynthesis in germinants than second-year seedlings grown in simulated understory light environments. Such high differences in the germinant-seedling transition are consistent with both a strategy of maximizing the capture of resources needed for establishment in the germinant stage and the rapid changes in foliar morphology documented to occur during the earliest seedling years (Day et al. 2001; Steele et al. 1989).

2.2 Seedling to Small-Sapling Stage

The transition from early seedling to sapling stages is generally marked by an enhanced interaction between meristems and extrinsic factors (Table 4.1). In conifers with preformed growth habits this transition is often associated with a departure from neoformed growth and formation of a resting bud with some or all of the second-year foliar and branch primordia preformed inside (Greenwood et al. 2009). In shade-intolerant species with high relative growth rates the sapling stage may last only a few growing seasons, while in shade-tolerant species growing under an existing canopy this stage may extend for several decades (Seymour 1992). In this life-stage resource allocation and crown development appear to be highly contingent on the light environment. When developing under an existing canopy, shade-tolerant seedlings such as red spruce and balsam fir and mid-tolerant species such as white pine show little height growth, in contrast to individuals growing in the open. This shade-tolerance response enhances the ability of seedlings and small saplings to survive extended periods in the understory in anticipation of future canopy disturbance by maximizing light capture and minimizing respiratory costs. Balsam fir develops a crown with greatly shortened internodes and its longest branches near the apex, assuming a "candelabra" appearance (Davis 1991). White pine (*Pinus strobus*) similarly minimizes orthotropic (upright) stem growth with most allocation going to extensive plagiotropic (horizontal) branching, which can exceed 90% of sapling height (O'Connell and Kelty 1994). The mechanisms underlying control of crown form (extension growth as opposed to branchiness) in these shade-tolerant species is not understood but grafting studies suggest complex extrinsic-intrinsic pathways in red spruce (Greenwood et al. 2010). Light intensity, red:far-red wavelengths and blue light wavelengths are all potential environmental cues for the shade-tolerance response.

In seedlings and saplings of shade-intolerant angiosperms, the ratio of red to far-red light wavelengths (R:FR) is a determinant of stem elongation. R:FR is decreased by transmission and reflection from the foliage of neighbors or an existing overhead canopy. The lower R:FR is sensed by phytochrome, which releases stem growth from genetic inhibition (Franklin and Whitelam 2005). The result is the classic shade-avoidance response of rapid stem elongation and reduced allocation to branching and stem diameter. Gilbert et al. (2001) demonstrated that this shade avoidance response is far greater in shade-intolerant than shade-tolerant hardwoods. De la Rosa et al. (1998) enhanced stem growth in Scots pine (*Pinus sylvestris*) seedlings by 20% by providing far-red enriched sidelight. In their study overall biomass decreased by 15% in treated seedlings, presumably from lower relative allocation to light-harvesting foliage. Similar enhancements in stem allocation and height growth have been reported for black spruce (*Picea mariana*), white spruce (*P. glauca*) and jack pine (*Pinus banksiana*) seedlings by Hoddinott and Scott (1996). Jack pine, their most intolerant species, more than doubled height growth with a 21% relative increase in stem biomass. Ritchie (1997) reported evidence for a phytochrome response for Douglas-fir (*Pseudotsuga menziesii*) seedlings. In *Picea omorika*, an endemic pioneer from the Balkan region, Tucić et al. (2006) found that the response of epicotyl elongation to FR was steady and significant through the first 20 weeks post-germination but increased by 60% between weeks 20 and 30, suggesting a increasing sensitivity to environment during ontogeny. Potential reception/regulation of stem elongation by blue light wavelengths in conifers has received little attention, but is suggested by indirect evidence in Scots pine (Sarala et al. 2007).

The high RGR potential of seedling and sapling apical meristems has been demonstrated by comparing the growth of grafted scions from juvenile and mature donors onto juvenile rootstock (Rebbeck et al. 1992; Day et al. 2001; Greenwood et al. 2010). Studies that have evaluated multiple life stages have generally found that photosynthetic rates in conifer seedlings or saplings tend to be higher than those reported for later stages (Bond 2000; Day et al. 2001). An exception is the deciduous conifer *Larix laricina* in which Greenwood et al. (1989) found greater photosynthetic rates and chlorophyll content in older individuals. This may be related to carbon economics of the deciduous habit, a hypothesis that needs evaluation in other conifers with annual leaf turnover.

Red spruce juveniles exhibit greater whole-tree hydraulic conductance than adults (Greenwood et al. 2010) and are able to photosynthesize at higher rates when soil moisture is adequate. However, with low stem hydraulic capacitance, they also show greater stomatal sensitivity to leaf-to-air vapor pressure deficits (VPD) than adult trees (Day 2000). As seedling and saplings in the forest understory are usually subjected to less preconditioning by high VPD than adults, this observation suggests that stomatal sensitivity may be an intrinsic adaptive characteristic.

2.3 Late Sapling to Young-Mature (Pole) Stage

These stages are characterized by a decrease in branch density (branches per unit stem length) and greater relative allocation to height growth (Powell 2009). In shade-intolerant conifers, and shade-tolerant species growing under open conditions, trees transition to this developmental state within the first few years of life. In shade-tolerant conifer seedlings growing under an existing canopy, this stage usually is initiated by an increase in resource availability due to canopy disturbance. Height growth greatly increases during this stage. Foliage becomes more massive, with needle width and LMA significantly increasing as a function of both extrinsic and intrinsic factors (Fig. 4.1), with parallel trends in both sun-adapted and shade-adapted foliage (Day et al. 2001). The latter portion of this life-stage encompasses the "phase change" between purely vegetative growth and reproductive competence (Table 4.1). Both male and female strobili begin to form in most species, and reproductive allocation can be readily enhanced by flower stimulation treatments (Owens and Colangeli 1989; Caron and Powell 1989). As discussed later, bodies of evidence suggest that reproductive maturation is both a function of behavioral changes within the meristem itself (Hackett 1985; Poethig 1990; Greenwood and Hutchison 1993), and changes within the tree external to the meristem (Owens and Colangeli 1989) or external to the tree (Caron and Powell 1989).

Because of greater allocation of resources to stemwood production, this stage has been commonly described as the period of greatest mean annual increment in commercial tree species (Smith et al. 1997). With height growth at maximal rates, proportional cambial increment is required for structural integrity (King 1990; 2011).

From an evolutionary standpoint this allocation to height growth defines the tree's period of maximum vegetative competitiveness.

For most conifers the transition into this life stage marks a point of no return. Relative allocation to non-photosynthetic tissue is greatest, and results in a resource sink that must be continually supported by allocation to maintenance respiration. Trees overtopped by competitors in the "race to the upper canopy" are likely to perish from the effects of a negative carbon balance (Niinemets 2006). Contrary to the more general probability of mortality decreasing with ontogeny in trees until maximum ages are reached (Harcombe 1987), Kneeshaw et al. (2006) found that once shade-tolerant conifers leave the seedling or early sapling stage the probability of mortality increases during the period of transition to the upper canopy. They ascribed this to the inability of trees beyond that transition to reassume a crown form adaptive to shaded habitats with horizontally spreading branches and minimal height growth. Hemlocks (*Tsuga* spp.), the most shade-tolerant of temperate conifers, are an exception to this rule and often survive multiple release and suppression periods in their journey into the canopy.

The cue or cues that initiate vegetative transitions to this life-stage are poorly understood, and may act asynchronously with reproductive competence (see below). Numerous studies have linked the growth rates of sapling and pole-sized trees to light environment and release from competition, and there is indirect evidence that R:FR sensing may be involved (Harper et al. 2009). However the phytochrome pathway has not been experimentally isolated from the effects of light fluence and, perhaps, blue-wavelength growth stimulation in conifers. Once initiated, crown formation during this life-stage appears to be largely directed by the leader SAM through apical control mechanisms (Wilson and Gartner 2002; Cline and Harrington 2007). These studies suggest that the leader SAM limits carbohydrate availability to subtending branches by creating strong cambial sinks through auxin production, maximizing growth in the primary stem. In this model the leader SAM is intrinsically entrained by an extrinsic condition or combination of cues. The vigorous elongation growth may result from a combination of hormones and resource balance unique to this life stage. The results of a 7 year reciprocal grafting study with red spruce supports this model, and further suggests that similar cues influence lateral branch terminal buds. Greenwood et al. (2010) found enhanced elongation growth and less branch density in juvenile scions grafted into the crowns of pole-sized trees compared to the crown forms of natural juveniles and juvenile-juvenile self-grafts. Further evidence supporting hormone-mediated control by the leader comes from decreasing apical control and greater light-environment control of growth responses with distance down the stem from the SAM (Baldwin et al. 2000; Mäkelä and Vanninen 2001).

Trees in this middle-age life stage may benefit from their substantial stem hydraulic capacitance as most temperate conifer species maintain several years, even decades, of hydraulically functional xylem tissue. In more xeric climates, such as the Intermountain West in North America, conifers have been shown to exhibit greater stomatal sensitivity to leaf-to-air vapor pressure deficit (VPD) with increasing age (Hubbard et al. 1999) and stronger midday depression in stomatal conductance (Yoder et al. 1994). In contrast, red spruce in the more mesic northeastern North

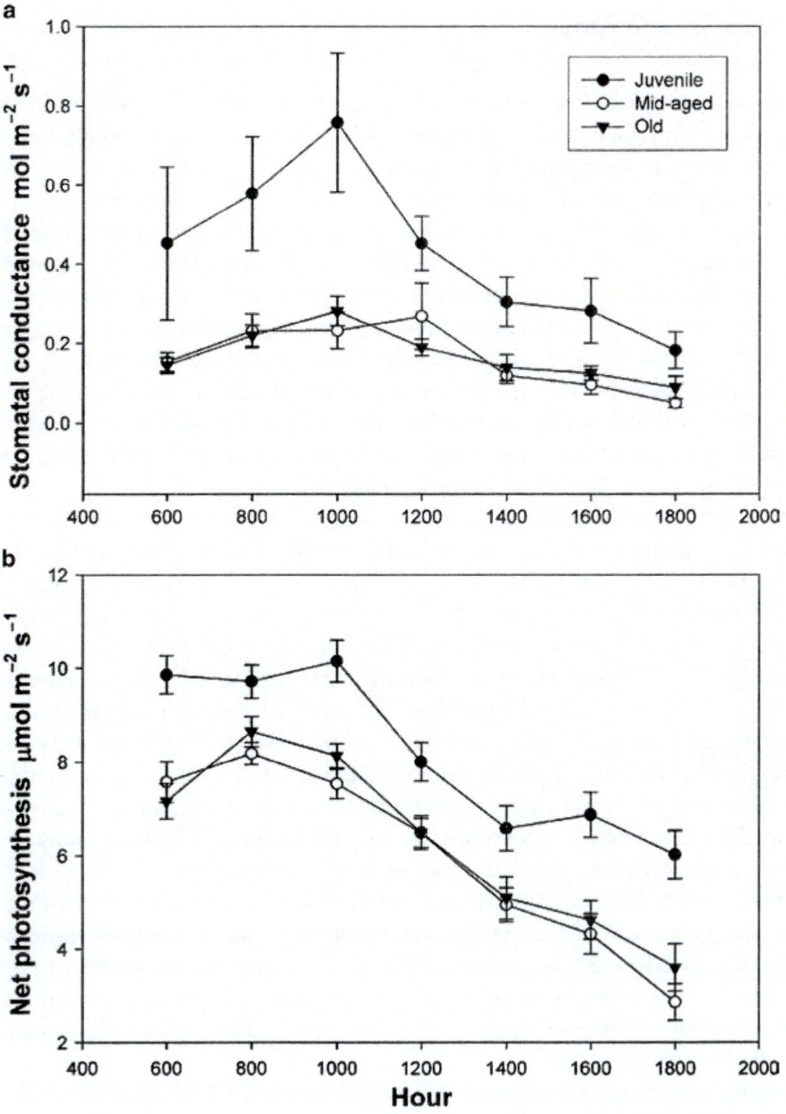

Fig. 4.2 Diurnal trends in gas exchange attributes between foliage from juvenile 10 y / 1 m (–•–), mid-aged 54 y / 10 m (–■–), and old 130 y / 21 m (–▼–) red spruce: A assimilation of CO_2, and B stomatal conductance to water vapor. Mid-aged and old trees were not different at any time and age-x-hour interactions were not different ($P>F = 0.55$ for A, and 0.99 for g_s) between age classes. ($n=30$). Age-classes refer to mean age of study trees. Trends calculated on a projected area basis showed the same patterns and statistical relationships as those on a mass basis (Data from Adams 2006)

American Acadian Forest shows the greatest sensitivity to VPD in seedlings and small saplings with limited stem capacitance (Day 2000) and similar patterns of midday depression across age classes, with no difference between middle aged and old trees (Fig. 4.2).

2.4 Old-Growth Stage

The transition to the old or "old-growth" stage (Table 4.1), most noticeable as a marked change in growth habit of the leader SAM, appears more gradually than that between seedling to young sapling and middle aged trees. In this transition, leader elongation declines, apical control lessens, and needles become gradually more massive with increases in width and leaf mass per area (LMA), and continued decrease in needle length-to-width ratios (Steele et al. 1989; Bond 2000; Day et al. 2001; Apple et al. 2002; Greenwood et al. 2010; Ishii 2011). In red spruce, grafting of old scions onto juvenile rootstock suggests that a persistent generic "memory" for old tree needle anatomical characteristics may be maintained indefinitely in apical meristems (Day et al. 2001). With decreased apical control, the top of the crown becomes flattened and height growth becomes minimal (Fig. 4.3). Reproductive output generally remains high for many decades or even centuries after the old life-stage is reached (Powell 2009; Greenwood et al. 2010).

The gradual transition into this stage is also associated with a decline in the mean annual increment of stemwood. Until recently it was widely believed that total tree productivity and/or the productivity per unit of foliage decreased dramatically during this stage. In red spruce stemwood production per unit leaf area decreases by over 50% between 60 and 120 years old (Day et al. 2001; Seymour and Kenefic 2002). However, recent findings have cast doubt on this paradigm. Sillett et al. (2010) provide evidence that peak rates of productivity may continue in redwoods and eucalypts that have reached their old life-stage and near maximum height growth. For lodgepole pine, Sala and Hoch (2009) report that old trees have greater non-structural carbohydrate contents, suggesting that their growth is not limited by net primary production. In a 7 year reciprocal grafting study in red spruce (Greenwood et al. 2010), the mid-aged to old change in growth habit is not associated with a decline in meristem vigor. Meristems from old-growth trees grafted onto juvenile rootstock are just as vigorous as those from mid-aged trees that exhibit peak MAI, although they exhibit the old-tree characteristics of greater branch density and less extension growth of lower order (main) branches. Conversely, scions from mid-aged and old donors on mid-aged and old rootstocks displayed identical vigor.

Although studies have suggested decreasing rates of photosynthesis with age in conifers (Bond 2000; Day et al. 2001; Steppe et al. 2011), it is potentially misleading to draw broad conclusions on productivity from these results for several reasons. Most studies of age-related trends in conifer photosynthesis have relied on spot measurements of maximum photosynthetic rates (A_{max}). Few studies compare repeated diurnal measurements by age class. For example, Yoder et al. (1994) found A_{max} to be greater in old compared to younger ponderosa pine and lodgepole pine (*Pinus contorta* Dougl. Ex Loud.) in the early morning, but became significantly lower during the mid-day, a result of age-related variation in stomatal limitations. Similarly, Day et al. (2001) reported a trend to lower rates in 120 y old compared to 60 y old red spruce, based on spot measurements. However, subsequent diurnal tracking of the same population shows these two age classes

Fig. 4.3 Mid-aged, 60 y, (*left*) and old, 150 y, red spruce crowns showing the difference in growth habit between age-classes. *Insets* show extension growth of upper-crown lateral shoots (From Greenwood et al. 2010)

to have statistically identical diurnal trends in A_{max} (Fig. 4.2). A better approach compares photosynthetic capacity (carboxylation and/or electron transport rates), which is less subject to variation from environmental conditioning prior to measurement (Niinemets 2002). Photosynthetic capacity is indicative of resource allocation to assimilatory systems and generally represents a balance with longer-term carbon demands (Paul and Foyer 2001). In red spruce, measurements of photosynthetic capacity corroborate the trends reported from diurnal tracking (Adams 2006). A still better approach includes a broader view of age- or size-related tends in carbon economy (Körner 2003), which couples net primary production with carbon requirements. However, this has been evaluated in only a few conifer species. In red spruce (Fig. 4.4) as in western North American pines (Sala and Hoch 2009; Sala et al. 2011) non-structural carbohydrate concentrations suggest old trees are not carbon resource limited. Carbon economy may relate to photosynthetic rates and capacities via both short- and long-term feedback mechanisms (Paul and Foyer 2001), which have been experimentally demonstrated in diverse conifers (Myers et al. 1999; Equiza et al. 2006).

Reciprocal grafting studies suggest that photosynthetic rates and capacities are under extrinsic control. Grafting studies on a wide range of conifers including sugi (*Cryptomeria japonica* D. Don) (Matsuzaki et al. 2005), Scots pine (*Pinus sylvestris* L.) (Mencuccini et al. 2005), Douglas-fir (Bond et al. 2007), and red spruce (Ward 2005; Adams 2006) have universally indicated that the extrinsic (to the meristem) environment is the principle factor governing age-related differences in photosynthetic rates. However, grafting of old scions onto juvenile rootstock has indicated a persistent meristematic "physiological memory" may be retained in some species, at least over several post-grafting growing seasons (Rebbeck et al. 1992; Day et al. 2001).

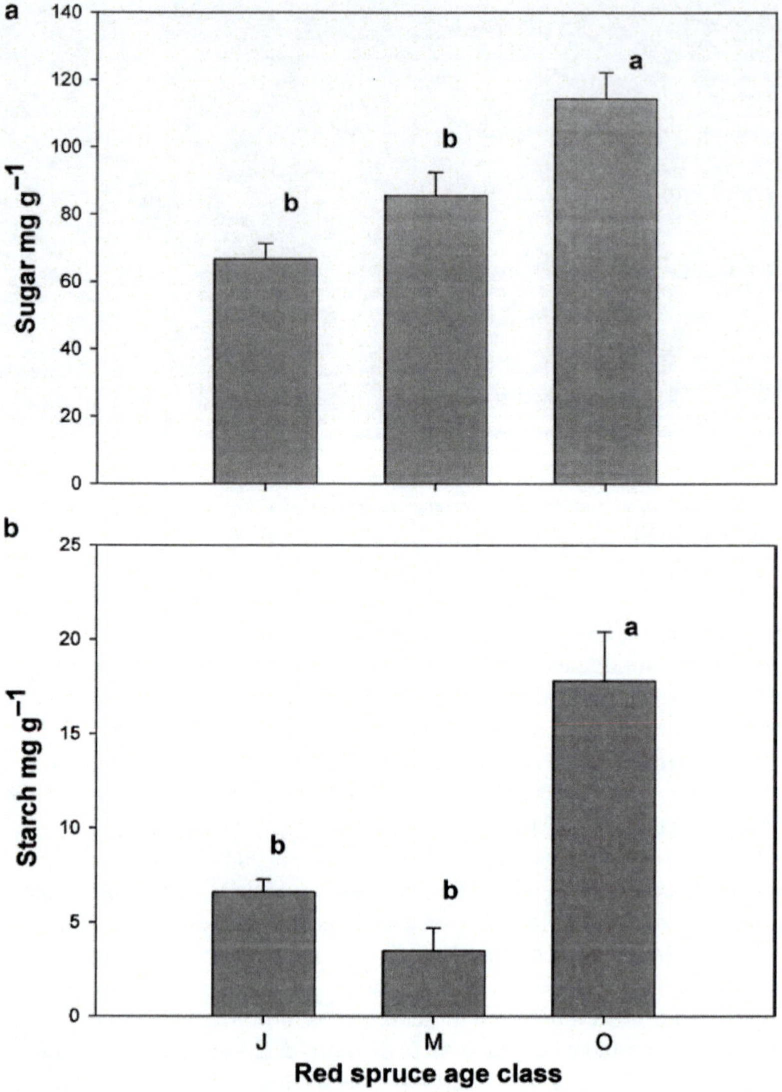

Fig. 4.4 Midday nonstructural carbohydrate concentrations (*sugar panel a and starch panel b*) in the foliage of red spruce from juvenile, mid-aged, and old trees. Higher NSC concentrations are indicative of "excess" photosynthetic capacity over growth sink demands. Bars give standard errors and letters designate different classes for percent starch (Tukey's HSD $\alpha = 0.05$, $n = 30$) (Data from Adams 2006)

A decrease in $\delta^{13}C$ with tree age has been reported for such diverse species as redwood (Koch et al. 2004), Douglas-fir (Woodruff et al. 2008), lodgepole pine (Yoder et al. 1994), sugi (Matsuzaki et al. 2005), loblolly pine (Drake et al. 2010), and red spruce (Greenwood et al. 2008b), and is reviewed in this volume (McDowell et al. 2011). This phenomenon has often been suggested as evidence that old trees are more likely to experience stomatal limitations to assimilation. This may be true in very tall trees where xylem resistance and gravitational effects may reduce water potential well over −2 MPa in a 100 m tree (McDowell et al. 2002, 2005, 2011; Domec et al. 2008), but its importance for 10 and 20 m trees in mesic habitats is questionable. Alternatively, the lower carbon isotope discrimination values can be interpreted as reflecting greater water use efficiency and a more conservative gas exchange strategy. This interpretation is supported by data from red spruce, where whole-tree conductance was equal between 10 and 12 m high mid-aged and 20–25 m old trees, while carbon isotope discrimination was significantly lower in the older individuals (Greenwood et al. 2008b). This strategy may be particularly important in more xeric or drought-prone environments (e.g., Drake et al. 2010). A more conservative water use strategy is consistent with the continuing trend of decreasing needle surface to volume ratio from mid-aged to old trees (Day et al. 2001; Ishii 2011).

The emerging paradigms for carbon dynamics offers new insight on the issue of age-related trends in productivity and provides a much more complex picture of age-related "decline". If our past perception based on stemwood production proves to be an inadequate metric (Sillett et al. 2010), the transition to the old life stage may represent changed allocation priorities rather than decreasing productivity. An alternative hypothesis might include increased carbon costs of tissues in old trees. Wood scientists have long observed that old trees have denser wood with thicker cell walls that are not only more costly to produce in terms of cellulose (Panshin and de Zeeuw 1980), but also have a greater proportional concentration of lignin with a construction respiration cost of nearly two times that of cellulose (Penning de Vries et al. 1974). Lachenbruch et al. (2011) have suggested that age-related trends in radial patterns in the xylem are regulated by the interaction of developmental, structural-mechanical and hydraulic factors. They view the relative importance of intrinsic and the two extrinsic factors in regulating wood properties as a function of ontogenetic state of the cambium. Likewise, the costs of foliar tissue appear greater in old conifers which are more robust with greater lignin content and waxy cuticles (Steppe et al. 2011). In Douglas-fir Apple et al. (2002) found an increase in density and sclerenchyma tissue as trees age, and in red spruce Greenwood et al. (2008a) report a 28% decline in internal air space in the denser needles of old trees. Other untested possibilities include greater allocation to branch construction, secondary defensive compounds, reproduction, reproductive structures and carbon reserves.

3 Regulation of the Vegetative-Reproductive Transition and Reproductive Effort

3.1 Intrinsic Regulation of Precocity and Fecundity

Flowering in conifers is influenced by a variety of factors, both extrinsic and intrinsic to the meristem, including genetics and maturational state associated with increased size. The ability of a woody plant meristem to become determined as a reproductive structure (reproductive competence) may not occur for decades in some conifer species, but occurs relatively early in others. There are obvious advantages to ontogenetic limitations to flowering, where reproductive competence is delayed while seedlings and saplings compete for space and resources (Thomas 2011). During the transition to reproductive competence intrinsic changes within the meristem interact with the extrinsic effects of increased size. This includes increased distance from the root system, which will affect water stress and availability of root-produced growth regulators. Even though flowering is uncommon in the first few years of development, precocious flowering has been observed in many species.

There are distinct differences in precocity within the genus *Pinus*. About 25% of 5 year-old jack pine (*Pinus banksiana* Lamb.) saplings in a plantation produced flowers (Wright 1964), and nursery-grown seedlings have flowered in their first year (Rudolph 1979). In contrast loblolly pine (*Pinus taeda* L.) saplings in wide-spaced plantations rarely flower before 10–years of age (Wright 1964; Dorman and Zobel 1973), and strobili are totally absent on nursery seedlings. In addition, since jack pine, a boreal species, grows more slowly than loblolly pine, its flowering begins on much smaller individuals. There are also some extreme examples of precocious flowering within the genus *Pinus*, where both ovulate and staminate strobili have been observed on germinant stage seedlings less than a year old, but at very low frequencies (Smith and Konar 1969). This suggests that intrinsic epigenetic mechanisms that suppress gene expression needed for reproductive development until a critical size and maturation state is reached can, on rare occasions, be overcome.

In later life-stages, fecundity is not only under genetic control, but is also strongly influenced by the external environment. Fecundity is highly variable among individual genotypes within a species. Variation in flowering among 17 clones of red spruce is shown in Table 4.1. The ramets of each clone resulted from grafted scions taken from mature trees, selected from desirable growth characteristics, and were 1.5–2.5 m tall about 8 years after grafting. Considering the male and female flowering response of the untreated control trees in the rightmost two columns, note that there are orders of magnitude of variation among the clones. In addition, both water stress and exogenous applications of the hormone gibberellin affect clonal response. Similar variation has been described among 18 clones in a loblolly pine seed orchard (Schmidtling 1983). Genetic effects accounted for 50% of the variation, and there were significant clone by year interactions. Thus fecundity is highly heritable among conifers.

3.2 *Effect of Size and Meristem Maturation: Grafting Studies*

Further evidence for the regulatory roles of both size and maturation state on both precocious flowering and fecundity comes from grafting studies, where the flowering response of scions of different maturation states are grafted onto rootstocks ranging from juvenile to mature. Seed orchard establishment usually involves grafting scions from mature select trees onto juvenile rootstock. Few strobili are produced by either juvenile or mature scions for several years following grafting (e.g. Greenwood et al. 1989), probably until a minimum size is attained. Forcing growth under greenhouse conditions greatly reduces the time to flowering in fruit trees and conifers as well, indicating that size, not chronological age, is a determinant of reproductive competence (Hackett 1985).

Once flowering commences, mature scions of most conifers generally flower more prolifically than juvenile scions. Scions from 12 year old donors of loblolly pine produced 6 and 129-fold more female and male strobili respectively than scions from 1 year-old donors of the same half- sib families (Greenwood 1984). Grafted scions of eastern larch (*Larix laricina* (Du Roi) K. Koch, a precocious species) from juvenile and mature donor trees of the same half-sib families produced few strobili 2 years after grafting, even when subjected to optimal conditions in a heated greenhouse. However, in the second and third growing seasons, juvenile scions actually produced more female strobili than mature scions, but the latter produced more male strobili. The total number of strobili was slightly higher on juvenile scions. The different responses of juvenile scions between loblolly pine and larch may be due to larch being more reproductively precocious, increasing in fecundity with increased size much more rapidly than loblolly pine. Again this may be an adaptation to slower growth rates in the boreal range of larch, or perhaps the much shorter return interval between destructive crown fires in boreal ecosystems. As conifer saplings become reproductively mature, female strobili appear first, and the male to female ratio gradually increases over time (Greenwood and Hutchison 1993). Thus the time when the male to female sex ratio stabilizes may be a better indicator of full reproductive maturity than female fecundity.

Grafting juvenile scions into the tops of mature trees (topworking), used in attempts to achieve earlier fecundity (Greenwood and Schmidtling 1981), can also further inform us about the relative roles of size and meristem maturation in reproductive competence. Juvenile scions of loblolly pine topworked into the crowns of mature seed orchard trees produced very few strobili 2–3 years after grafting, although most of the neighboring branches of the understock were producing male strobili.

Like loblolly pine, virtually no strobili are found on open-grown red spruce regeneration 10–12 years old, which also appear juvenile in terms of foliar characteristics. We followed reproductive development during a reciprocal grafting study, where both juvenile and mature scions were grafted onto juvenile rootstock as well as topworked into the upper crown of reproductively mature trees (Greenwood et al. 2010). In topworked grafts female and male strobilus production by juvenile scions was eightfold less than that observed on mature scions, even seven growing seasons after grafting. In addition, juvenile foliar characteristics were retained on juvenile grafts. Scions from all ages of donors produced few strobili on juvenile root stock after 2–3 years, but those that were observed occurred on mature scions. Thus meristem maturation appears to be a major determinant in reproductive competence in conifers, with increased size playing a lesser but significant role.

3.3 *Effects of Shoot Growth Habit and Budset on Reproductive Development*

Discussion of all the intrinsic and extrinsic factors that affect shoot development in relation to flowering is beyond the scope of this review, but some environmental and hormonal effects are informative about how size and maturation affect bud development in relation to reproductive development.

The grafting studies described above clearly indicate that increased size does not completely overcome suppression of flowering in juvenile scions on mature rootstock nor conversely that decreased size suppresses the greater tendency of mature scions to flower on juvenile rootstock. This raises the question of the nature of the mechanisms, both intrinsic and extrinsic, which regulate flowering.

The fact that GA_3 promotes flowering in seedlings of the Cupressaceae and Taxodiaceae suggests that failure of juvenile trees to flower may be due to lack of sufficient levels of gibberellins (Kato et al. 1959). However, only $GA_{4/7}$ has been found to promote flowering in the Pinaceae, and its promotive effects have only been demonstrated on sexually mature trees (Zimmermann et al. 1985). Endogenous $GA_{4/7}$ or other GAs may play a critical roles in either initiation or differentiation of strobili through the *SOC1* transcription factor pathway, but their regulatory pathway in conifers is still unclear at present (e.g. Smith 1998). Furthermore, hypotheses suggesting that the primary determinant for flowering requires minimal levels of specific GAs or other growth regulators that are lacking in juvenile plants seem unlikely to be true, since one would expect that prolifically flowering rootstock branches would export GAs to neighboring topworked juvenile scions in sufficient quantities to largely offset the effects of maturation. An alternative hypothesis is that flowering is extrinsically suppressed by root-produced inhibitory substances (perhaps cytokinins), which will be diluted with increased distance from the root system. This may play a role in the delay in flowering observed when mature scions are grafted onto juvenile rootstock. Smith and Greenwood (1995) showed that exogenously applied cytokinins inhibited flowering in black spruce, while root pruning had a promotive effect, possibly by decreasing the supply of root-produced cytokinins. Thus, the limited effect of rootstock on overcoming the effect of maturation state on flowering is probably due to the effects of plant growth substances, which, in turn, may be regulated by environmental factors such as drought. However, empirical studies suggest that the intrinsic maturation state of the meristem dominates these effects.

Meristem response to treatments that stimulate flowering is highly variable and is strongly influenced by crown position; female flowering is most abundant in the upper quadrant of the live crown, while male flowering is most abundant in the lower quadrants. Irrigation significantly decreased female flowering in a red spruce seed orchard, while having little effect on male flowering (Table 4.2), and lack of irrigation enhanced the effect of exogenously applied $GA_{4/7}$ in promoting female flowering. GA application significantly promoted female production, but the smaller increase in male flowering was not significant.

In addition, the intrinsic sensitivity of meristems to growth regulators or other flower stimulation treatments may change with meristem maturation. $GA_{4/7}$ significantly

Table 4.2 Variation in flowering response among 17 clones in a red spruce clonal seed orchard, with and without GA foliar sprays and irrigation

	Irrigated				Not irrigated			
	GA spray		No GA		GA spray		No GA	
Clone	♀	♂	♀	♂	♀	♂	♀	♂
1	133	134	2	49	243	229	1	4
2	19	3	1	2	38	3	1	5
3	106	117	9	385	152	143	7	321
4	**11**	**1**	**1**	**1**	**27**	**3**	**11**	**147**
5	137	347	3	129	168	443	34	127
6	**2**	**20**	**1**	**5**	**147**	**10**	**5**	**13**
7	59	2	2	1	101	1	8	32
8	42	68	2	216	76	147	3	58
9	94	195	1	10	119	1	2	53
10	**168**	**251**	**21**	**293**	**160**	**168**	**82**	**385**
11	85	142	17	174	337	329	48	310
12	**266**	**437**	**0**	**36**	**420**	**1,451**	**5**	**188**
13	45	23	6	241	166	235	4	35
14	64	15	6	106	107	12	2	46
15	26	319	2	175	104	394	7	24
16	**5**	**11**	**2**	**21**	**12**	**12**	**1**	**18**
17	2	20	1	5	173	10	5	123
MEAN	75±25	126±34	5±2	110±29	173±35	210±86	13±5	113±29

Two blocks received irrigation (2–3 cm/week June through September) and two did not. Ramets of each clone within each block either received a $GA_{4/7}$ spray applied weekly for 8 weeks beginning in mid-May or was an untreated control

enhanced the ratio of female to male strobili on scions from 1 year-old eastern larch donor trees grafted onto juvenile rootstock, but this ratio gradually declined the older the donor tree, and $GA_{4/7}$ had no effect at all on the sex ratio in scions from donors older than 33 years (Eysteinsson and Greenwood 1993). When loblolly pine scions on juvenile rootstock from donors aged 1,4, 8 and 12 years were grown under a 24 h photoperiod were suddenly subjected to 12 h ambient photoperiod (when moved outside in the spring), they temporarily set bud and many of the trees later developed strobili later in the summer, and pollen was shed with no intervening dormancy period. This treatment was significantly more effective with increasing scion donor age, with the oldest scions producing fourfold more strobili than the youngest. In addition, the younger the scions began to flush earlier and produced more vegetative growth cycles which resulted in longer shoots. These observations provided evidence that meristems of different maturation states respond differently to flower enhancing treatments and also suggest an inverse relation between the tendency to produce vegetative growth and ability to flower.

An inverse relationship between conditions which favor the development of vegetative organs and ability to flower has been proposed for jack pine and loblolly pine (Larson 1961; Greenwood 1981). Also, irrigation significantly promoted height increment in red spruce and at the same time inhibited the female strobilus production, suggesting a negative tradeoff between formation of reproductive and vegetative shoot primordia. Additionally, in loblolly pine, female strobili occur in the upper quadrant of the crown on the most vigorous shoots, which can flush several times during the growing season, whereas male strobili occur lower down on less vigorous shoots that flush only once. Thus the timing of bud set, which is affected by both maturation and the external environment, may be critical in terms of the initiation and differentiation of reproductive primordia.

3.4 Genetic Regulation of Reproductive Competence

Recent advances in understanding the regulation of flowering and budset also contribute to our understanding of how reproductive competence may be regulated. For example, *Populus* orthologs of the genes that make up the CO/FT regulatory module, that is necessary for regulation of flowering by day length in *Arabidopsis*, also control the timing of flowering as well as seasonal growth cessation in *Populus* (Bohlenius et al. 2006). This module consists of two genes (An et al. 2004; Huang et al. 2005), *CONSTANS (CO)* and *FLOWERING LOCUS T(FT)*. In *Arabidopsis, CO*, whose expression is modulated by the circadian clock and day length, encodes a putative transcription factor under long days which in turn upregulates *FT*. *FT* then produces a transcription factor that is translocated to the shoot apex, where it binds to the promoter for *APTALA1 (AP1)*, a floral meristem identity gene. In turn, *AP1* orchestrates up- and downregulation of an array of organ identity and gibberellin metabolism genes, which specify the formation of floral organs (Kaufmann et al. 2010). Thus the *FT* gene product appears to meet the criteria for a mobile florigen (a flower-inducing substance).

In *Populus tremuloides,* when *PtCO2* expression peaks before the end of the day then *PtFT1* expression increases; if *PtCO2* peaks in early darkness then there is no change in *PtFT1* expression, and budset occurs. Budset occurs under longer photoperiods in provenances that occur at higher latitudes, and is associated with low levels of both FT and CO expression towards the end of the day.

Suppression of *PtFT1* expression may prevent precocious flowering in *Populus.* Male *P. tremula x tremuloides* hybrid stem segments transformed via *Agrobacterium tumefaciens* with *35 S::PtFT1*(a *Populus* homolog of FT linked to a promoter gene so that the gene is constitutively expressed) initiated flowerlike structures within 4 weeks, compared with a normal flowering time of 8–20 years. This would suggest that suppression of this gene occurs in younger plants, accounting for the failure to flower. Indeed the expression of wild type FT in *Populus* is relatively low in younger plants (Fig. 4.5).

Could *PtFT1* expression play a key role in conifer ontogeny? In *Populus PtFT1* expression steadily increases from age 2 to 6 years, suggesting that reduced reproductive competence in younger trees may be due to reduced expression of this gene. Gyllenstrand et al. (2007) have shown that expression of a homolog of *FT* in Norway spruce (*Picea abies* L. Karst.), *PaFT4,* plays a role in photoperiodically-induced bud set in Norway spruce seedlings, so some form of the CO/FT regulatory module

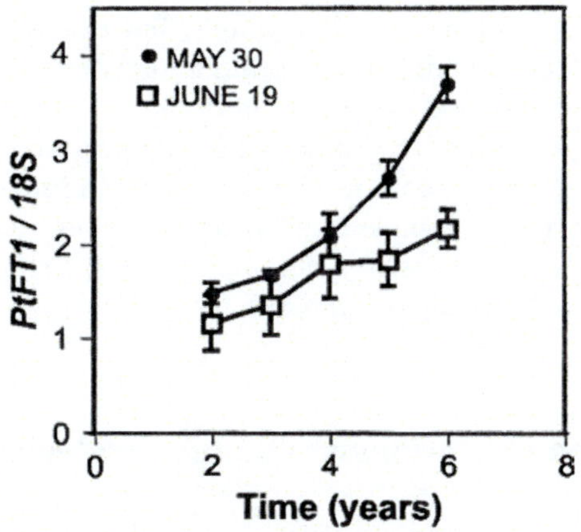

Fig. 4.5 The expression of PtFT1 in ramets of a single field-grown clone of *Populus trichocarpa x tremuloides* relative to simultaneous expression of the ribosomal 18 S gene. This clone flowers for the first time between age 5–6 years (Graph from supporting online material from Bohlenius et al. 2006. www.sciencemag.org/cgi/content/full/1126038/DC1)

appears to be active in conifers as well as angiosperms. These findings give rise to the hypothesis that changes in expression of an FT homolog, as well as CO homologs, could play a role in maturational changes. More specifically, the intrinsic suppression of flowering in juvenile plants may due to the epigenetic suppression of FT expression, which continues even when juvenile scions are grafted onto reproductively competent rootstock.

Another possibility is that juvenile shoots are incapable of responding to the floral stimulus. Juvenile scions do not flower even in the close proximity of flowering shoots of the understock. This is in contrast to grafting studies with photoperiodically-induced herbaceous plants where the floral stimulus can be transmitted from the induced scion throughout the non-induced rootstock and vice versa (e.g. Lang 1965). This would imply that the FT gene product may be translocated only short distances in conifers. In the crowns of reproductively competent conifers flowering shoots are often found in close proximity to non-flowering shoots, suggesting that reproductive competence occurs on a shoot-by-shoot basis, perhaps due to localized variation in production or distribution of flowering stimulus. The hypothesis that localized expression of the FT gene product is repressed in juvenile shoots on both juvenile and mature rootstock could be tested by comparison of expression of the CO/FT module homologs in juvenile and mature scions grafted onto juvenile and mature rootstock. This would confirm whether or not local FT expression over time is intrinsically regulated within juvenile and mature shoots independent of tree size. These studies should be initially carried out on *Populus* as well as coniferous species, which will provide a broader picture of how the CO/FT module works in woody species.

Focus on the CO/FT regulatory module is desirable because of the obvious importance of its gene products in the regulation of flowering and timing of budset in *Populus*. A more complete understanding of how budset and flowering is regulated in woody plants has many potential applications. For example, if the period of maximum growth potential exhibited by seedling meristems could be prolonged by genetic selection or genetic engineering in desirable species, yields could be increased. Also, manipulation of reproductive development to induce early flowering for breeding purposes or increase fecundity in seed and fruit orchards would be of great value. Finally, a more complete understanding of how the timing of budset and bud break is regulated will assist in developing strategies to assure adaptation of critical woody species to changing climate.

4 Life Stages and Evolutionarily Stable Strategies

Forest scientists have traditionally addressed age-related changes in form and function by considering stresses and restrictions placed on trees by the interaction of their morphology and physiology with physiomechanical and physiochemical constraints that occur with increased size. This approach has worked well for the interesting and quite seductive question of what limits the heights of the world's tallest trees and has provided many insights into the processes of age-related change. However, these advances have fallen short of providing a holistic paradigm for ontological trends in the majority of tree species, which exhibit characteristics of old life-stage behavior at a fraction of the height obtained by the tallest species. Understanding age-related trends in the vast majority of species and biomes will require a different perspective, a whole-life perspective with a focus on maintaining a species through successful reproduction, i.e. an evolutionarily stable strategy (ESS) perspective (e.g. Vincent et al. 1996). Based on game theory, an ESS maintains a species presence by preventing a competitor from displacing it from its niche. In this regard, we would suggest Ecologically Stable Strategy as an ecophysiologically more appropriate term that incorporates both biotic and abiotic challenges to the survival of a species.

An overview of the life-stage approach summarizes how the ecologically stable perspective could be applied to the typical temperate conifers considered in this review. With minimal seed reserves and limited ability at sensing their environment, germinants generally develop following intrinsic genetic programming, reflecting selection of strategies that have proved successful for past generations in their particular habitat. If establishment is successful, the intrinsic programming is modified by extrinsic conditions with the meristem creating a crown structure that maximizes light capture under shade or extension growth under high light intensity. Seedling and sapling foliage exhibits minimal structural investment (high SLA, low density), reflecting a greater probability of loss to herbivory and mutual shading by new foliar cohorts. With a minimal capacity to buffer carbohydrate sources and sinks, investment in photosynthetic capacity is maximized. Shade-adapted crown structure will be abandoned when higher R:FR or more intense light is detected, and saplings to middle aged individuals maximize allocation to extension growth of the main stem, maintained by strong apical control. Suppression by being overtopped during this stage is usually fatal, so allocation to carbohydrate reserves is minimized in favor of height growth (Iwasa et al. 1985). When individuals are successful in obtaining upper canopy status, allocation to extension growth can be safely minimized and resources redirected to enhanced branchiness and a more conservative allocation strategy. Foliage, now with a much lower probability of functional obsolescence by competition or mutual shading, is thicker, denser and better protected by investment in cuticular waxes, sclereids, and secondary defensive metabolites. Seed cones are often protected by heavy excretions of resins. Concurrently, foliar trends result in increased water use efficiency (assimilation/transpiration) by decreasing surface to volume ratios and increasing WUE through stomatal regulation. A spreading crown, with lower branch extension and more densely packed foliage provides for maximizing light capture, development of new photosynthetically efficient foliage and, perhaps, better occupation of physical growing space to the disadvantage of competitors. As an effective ESS, species-level reproductive efforts are maximized by longevity of individuals in the upper canopy.

Evolutionarily-ecologically stable strategy theory also provides a potential explanation as to why most trees assume the structural attributes of an old life-stage (Fig. 4.3) on reaching upper canopy status. Continuing growth above co-occurring trees increases the risk of damage from physical disturbance agents, which is likely to increase exponentially for emergent individuals that no longer are protected from extreme winds by canopy boundary layers and cannot rely on mutual support of surrounding trees to reduce bending stress. In addition, additional height growth requires not only a greater resource investment in stem extension, but a concurrent, and much greater overall, biomass investment in girth to support the increased static and dynamic loads on the stem (King 1990; 2011) and the increasing costs of maintaining hydraulic efficiency (Midgley 2003). Both increased risks and resource costs suggest a conflict between continual height growth (runaway competition) and maximizing long-term reproductive effort. Continually strong photosynthetic capacity and increased NSC reserves and water use efficiency are also suggestive of a conservative strategy during the old life-stage. Models of interspecific interactions suggest that an EES of maximizing height growth would be limited to environments with abundant water and nutrient resources and low rates and intensities of physical disturbance, i.e. environments where evolutionary success is primarily driven by competition (Falster and Westoby 2003).

5 Overall Conclusions

Meristem behavioral changes associated with the progression through life stages include the transition from neoformed to preformed shoot, induction of shade-avoidance and shade-tolerance growth strategies, increases and declines in shoot growth potential, changes in foliar and branching habits, and the onset of reproductive competence. Accumulating evidence at many levels suggests that they are complex functions of intrinsic changes (meristem maturation) and extrinsic effects of increased size (such as water stress or hormone balance) and external environment (such as red:far-red irradiance ratios). It is also clear that the time frames of various ontogenetic processes may both act independently and interactively depending on the trajectory involved. To date, most physiological research on these questions has occurred at the tree- or organ-level. Research approaches at these levels have often been confounded by the correlation between ontogeny, tree size and external environment. However, studies that control for size and environment using reciprocal grafting and clonal approaches have provided many insights in to the mechanisms regulating ontogeny. Our current understanding also indicates that many of these life-stage- (or size-) related trends result from epigenetically induced changes in gene expression patterns within meristems. In order to fully understand how these changes are regulated, molecular approaches are now essential, and will serve to overcome the many logistical difficulties of experimental manipulation of organisms as large as mature trees. Remarkable progress has been recently reported at the molecular level on the regulation of bud development and flowering in *Populus*, which provides a template for understanding these processes in conifers.

A challenge to tree physiologists is identifying the cues and control pathways for inter-stage transitions. For example, there is strong circumstantial evidence that R:FR light and, likely, the phytochrome extrinsic-intrinsic pathway functionally controls the seedling to sapling transition. The evidence becomes much weaker for environmental control over allocation during the sapling and middle aged pole stages, and is non-existent for the middle aged to old life-stage transition. Genomic techniques and advances in understanding conifer carbon dynamics provide promise for understanding developmental regulation by revealing patterns of gene expression and resource allocation that become active during certain life-stages and are dominant in others. Of particular potential are recently described genes associated with light quality, neighbor sensing and branchiness in Arabidopsis (Finlayson et al. 2010). However, physiologists must be mindful of integrating molecular approaches with an understanding of whole-tree function and the roles of ontological trajectories in EES life strategies.

Acknowledgements This material is based on work supported by the National Science Foundation under Grant No.IBN-0110142. (Any opinions, findings and conclusions or recommendations expressed here are those of the authors and do not necessarily reflect the views of the National Science Foundation). Additionally, the authors' research on age-related trends has been supported by the USDA Forest Service, Northeastern States Research Cooperative and the Maine Agricultural and Forest Experiment Station. We are indebted to the Northern Research Station USDA Forest Service for permitting research in their long-term silvicultural studies, to the many graduate and undergraduate students who have made these studies possible, and to Barb Lachenbruch and the anonymous reviewers for their input on earlier drafts of the manuscript. This is Maine Agricultural and Forest Experiment Station publication number 3141.

References

Adams SL (2006) A physiological examination of the age-related decline in photosynthesis in *Picea rubens*. School of Forest Resources/Maine University, Orono, p 75

An H, Roussot C, Suarez-Lopez P, Corbesier L, Vincent C, Pineiro M, Hepworth S, Mouradov A, Justin S, Turnbull C, Copland G (2004) CONSTANS acts in the phloem to regulate a systemic signal that induces photoperiodic flowering of *Arabidopsis*. Development 131:3615–3626

Apple MK, Tiecotter K, Snow M, Young J, Soeldner A, Phillips D, Tingey D, Bond BJ (2002) Needle anatomy changes with increasing age in Douglas-fir. Tree Physiol 22:129–136

Baldwin VC Jr, Peterson KD, Clark A III, Ferguson RB, Strub M, Bower DR (2000) The effects of spacing and thinning on stand and tree characteristics of 38 year-old loblolly pine. For Ecol Manage 137:91–102

Bogorad L (1950) Factors associated with the synthesis of chlorophyll in the dark in seedlings of *Pinus jeffreyi*. Bot Gaz 111:221–241

Bohlenius H, Huanh T, Charbonnel-Campa L, Brunner A, Jansson S, Strauss SH, Nilsson O (2006) *CO/FT* regulatory module controls timing of flowering and seasonal growth cessation in trees. Science 312:1040–1043

Bond BJ (2000) Age-related changes in photosynthesis of woody plants. Trends Plant Sci 5:349–353

Bond BJ, Czarnomski N, Cooper C, Day ME, Greenwood MS (2007) Developmental decline in height growth in Douglas-fir. Tree Physiol 27:441–453

Burgin MJ, Casal JJ, Whitelam CG, Sánchez RA (1999) A light-regulated pool of phytochrome and rudimentary high-irradiance responses under far-red light in *Pinus elliottii* and *Pseudotsuga menziesii*. J Exp Bot 50:831–836

Caron GE, Powell GR (1989) Patterns of seed-cone and pollen-cone production in young *Picea mariana* trees. Can J For Res 19:359–364

Carter KK, Adams GW, Greenwood MS, Nitschke P (1990) Early selection in jack pine. Can J For Res 20:285–291

Cline MG, Harrington CA (2007) Apical dominance an apical control in multiple flushing of temperate woody species. Can J For Res 37:74–83

Davis WC (1991) The role of advance regeneration of red spruce and balsam fir in east central Maine. In: Simpson CM (ed.) Proceedings, conference on natural regeneration management. Forestry Canada Maritimes Region, Fredericton, pp 157–168

Day ME (2000) Influence of temperature and leaf-to-air vapor pressure deficit on net photosynthesis and stomatal conductance in red spruce (*Picea rubens*). Tree Physiol 20:47–53

Day ME, Greenwood MS, White AS (2001) Age-related changes in foliar morphology and physiology in red spruce and their influence on declining photosynthetic rates and productivity with tree age. Tree Physiol 21:1195–1204

Day ME, Greenwood MS, Diaz-Sala C (2002) Age- and size-related trends in woody plant shoot development: regulatory pathways and evidence for genetic control. Tree Physiol 22:507–513

Day ME, Scheldbauer JL, Livingston WH, Greenwood MS, White AS, Brissette JC (2005) Influence of seedbed, light-environment and elevated night temperature on growth and carbon allocation of pitch pine (*Pinus rigida*) and jack pine (*Pinus banksiana*) seedlings. For Ecol Manage 205:59–71

de la Rosa TM, Aphalo PJ, Lehto T (1998) Effects of far-red light on the growth, mycorrhizas and mineral nutrition of Scots pine seedlings. Plant Soil 201:17–25

Domec JC, Lachenbruch B, Meinzer FC, Woodruff DR, Warren JM, McCulloh KA (2008) Maximum height in a conifer is associated with conflicting requirements for xylem design. Proc Natl Acad Sci USA 105:12069–12074

Dorman KW, Zobel BJ (1973) Genetics of loblolly pine. USDA For Serv Res Paper 19:1–21

Drake JE, Raetz LM, Davis SC, DeLucia EH (2010) Hydraulic limitation not declining nitrogen availability causes the age-related photosynthetic decline in loblolly pine (*Pinus taeda* L.). Plant Cell Environ 33:1756–1766

Equiza MA, Day ME, Jagels R, Li X (2006) Photosynthetic down-regulation in the conifer *Metasequoia glyptostroboides* growing under continuous light: the significance of carbohydrate sinks and paleo-ecophysiological implications. Can J Bot 84:1453–1461

Eysteinsson T, Greenwood MS (1993) Effects of maturation and gibberellin A4/7 on flowering and branching characteristics of *Larix laricina*. Can J For Res 23:14–20

Falster DS, Westoby M (2003) Plant height and evolutionary games. Trends Ecol Evol 18:337–343

Finlayson SA, Krishnareddy SR, Kebrom TH, Casal JJ (2010) Phytochrome regulation of branching in Arabidopsis. Plant Physiol 152:1914–1927

Franklin KA, Whitelam GC (2005) Phytochromes and shade-avoidance responses in plants. Ann Bot 96:169–175

Gilbert IR, Jarvis PG, Smith H (2001) Proximity signal and shade avoidance differences between early and late successional trees. Nature 411:792–795

Greenwood MS (1981) Reproductive development in loblolly pine II. The effect of age, gibberellin plus water stress and out-of-phase dormancy on long shoot growth behavior. Am J Bot 68:1184–1190

Greenwood MS (1984) Phase change in loblolly pine: shoot development as a function of age. Physiol Plant 61:518–522

Greenwood MS, Hutchison KW (1993) Maturation as a developmental process. In: Ahuja MR, Libby WJ (eds.) Clonal forestry I: genetics and biotechnology. Springer, New York, pp 14–33

Greenwood MS, Schmidtling RC (1981) Regulation of catkin production. In: Franklin EC (ed.) Pollen management handbook. USDA Forest Service Agriculture Handbook No. 587, pp 20–25

Greenwood MS, Volkaert HA (1992) Morphophysiological traits as markers for the early selection of conifer genetic families. Can J For Res 22:1001–1008

Greenwood M, Hooper CA, Hutchison KW (1989) Maturation in larch. I. Effect of age on shoot growth, foliar characteristics and DNA methylation. Plant Physiol 90:406–412

Greenwood M, O'Brien CL, Schatz JD, Diggins CA, Day ME, Jacobson GL, White AS, Wagner RG (2008a) Is early life cycle success a determinant of the abundance of red spruce and balsam fir. Can J For Res 38:2295–2305

Greenwood MS, Ward MH, Day ME, Adams SA, Bond BJ (2008b) Age-related trends in red spruce foliar plasticity in relation to declining productivity. Tree Physiol 28:255–232

Greenwood MS, Day ME, Berlyn GP (2009) Regulation of foliar plasticity in conifers: developmental and environmental factors. J Sustain Forest 28:48–62

Greenwood MS, Day ME, Schatz J (2010) Separating the effects of tree size and meristem maturation on shoot development of grafted scions of red spruce (*Picea rubens* Sarg.). Tree Physiol 30(4):459–468

Gyllenstrand N, Clapham D, Kallman T, Lagercrantz U (2007) A Norway spruce *FLOWERING LOCUS T* homolog is implicated in control of growth rhythm in conifers. Plant Physiol 144:248–257

Hackett WP (1985) Juvenility, maturation, and rejuvenation in woody plants. Hortic Rev 7:109–155

Harcombe PA (1987) Tree death: cause and consequence. Bioscience 37(8):557–568

Harper G, O'Neill M, Fielder P, Newsome T, Delong C (2009) Lodgepole pine growth as a function of competition and canopy light environment within aspen dominated mixedwoods of central interior British Columbia. For Ecol Manage 257:1829–1838

Hoddinott J, Scott R (1996) The influence of light quality and carbon dioxide enrichment on the growth and physiology of seedlings of three conifer species I. Growth responses. Can J Bot 74:383–390

Huang T, Bohlenius H, Eriksson S, Parcy F, Nilsson O (2005) The mRNA of the *Arabidopsis* FT gene moves form leaf to shoot apex and induces flowering. Science 309:1694–1696

Hubbard RM, Bond BJ, Ryan MG (1999) Evidence that hydraulic conductance limits photosynthesis in old *Pinus ponderosa* trees. Tree Physiol 19:165–172

Ishii H (2011) How do changes in leaf/shoot morphology and crown architecture affect growth and physiological function of large, old trees? In: Meinzer FC, Lachenbruch B, Dawson TE (eds.) Size- and age-related changes in tree structure and function. Springer

Iwasa Y, Cohen D, Leon JA (1985) Tree height and crown shape as results of competitive games. J Theor Biol 112:279–298

Kato Y, Fukuhara N, Kobayashi R (1959) Stimulation of flower bud differentiation of conifers by gibberellin. J Jpn For Soc 41:309–311

Kaufmann K, Wellmer F, Muino J, Ferrier T, Wuest S, Kumar V, Serrano-Mislata A, Madueno F, Krajewski P, Meyerowitz E, Angenent G, Reichmann J (2010) Orchestration of floral initiation by APETALA1. Science 328:85–89

King DA (1990) The adaptive significance of tree height. Am Nat 135:809–828

King DA (2011) Size-related changes in tree proportions and their potential influence on the course of height growth. In: Meinzer FC, Lachenbruch B, Dawson TE (eds.) Size- and age-related changes in tree structure and function. Springer

Kneeshaw DD, Kobe RK, Coates KD, Messier C (2006) Sapling size influences tolerance ranking among southern boreal tree species. J Ecol 94:471–480

Koch GE, Sillett SC, Jennings GM, Davis SD (2004) The limits of tree height. Nature 428:851–854

Körner C (2003) Carbon limitation in trees. J Ecol 91:4–17

Lachenbruch B, Moore JR, Evans R (2011) Radial variation in wood structure and function in woody plants, and hypotheses for its occurrence. In: Meinzer FC, Lachenbruch B, Dawson TE (eds.) Size- and age-related changes in tree structure and function. Springer

Lambeth CC (1980) Juvenile-mature correlations in Pinaceae and implications for early selection. For Sci 26:571–580

Landhäusser SM, Lieffers VJ (2000) Photosynthesis and carbon allocation of six boreal tree species grown in understory and open conditions. Tree Physiol 21:243–250

Lang A (1965) Physiology of flower initiation. In: Ruhland W (ed.) Encyclopedia of plant physiology, vol Old Series 15. Springer, Berlin, pp 1380–1535

Larsen JB, Wellendorf H (1990) Early test in *Picea abies* full sibs by applying gas exchange, frost resistance and growth measurements. Scand J For Res 5:369–380

Larson PR (1961) Influence of date of flushing on flowering in *Pinus banksiana*. Nature 192:82–83

Lavigne MB, Little CHA, Major JE (2001) Increasing the sink: source balance enhances photosynthetic rate of 1 year-old balsam fir foliage by increasing allocation of mineral nutrients. Tree Physiol 21:417–426

Leck MA, Outred HA (2008) Seedling natural history. In: Leck MA, Parker VT, Simpson RI (eds.) Seedling ecology and evolution. Cambridge University Press, Cambridge, pp 17–55

Mäkelä A, Vanninen P (2001) Vertical structure of Scots pine crowns in different age and size classes. Trees 15:385–392

Matsuzaki J, Norisada M, Kodaira J, Suzuki M, Tange T (2005) Shoots grafted into the upper crowns of tall Japanese cedar (*Cryptomeria japonica* D. Don) show foliar gas exchange characteristics similar to those of intact shoots. Trees 19:198–203

Matthews S (2005) Phytochrome evolution in green and nongreen plants. J Hered 96:197–204

McDowell NG, Phillps N, Lunch CK, Bond BJ, Ryan MG (2002) An investigation of hydraulic limitation and compensation in large, old Douglas-fir trees. Tree Physiol 22:763–774

McDowell NG, Licata J, Bond BJ (2005) Environmental sensitivity of gas exchange in different-sixed trees. Oecologia 145:9–20

McDowell NG, Bond BJ, Hill LT, Ryan MG, Whitehead D (2011) Relationships between tree height and carbon isotope discrimination. In: Meinzer FC, Lachenbruch B, Dawson TE (eds.) Size- and age-related changes in tree structure and function. Springer

Mencuccini M, Martínez-Vilalta J, Vanderklein D, Hamid HA, Korakaki E, Lee S, Michiels B (2005) Size-mediated ageing reduces vigor in tall trees. Ecol Lett 8:1183–1190

Mencuccini M, Hölttä T, Martinez-Vilalta J (2011) Comparative criteria for models of the vascular transport system of tall trees. In: Meinzer FC, Lachenbruch B, Dawson TE (eds.) Size- and age-related changes in tree structure and function. Springer

Midgley JJ (2003) Is bigger better in plants? The hydraulic costs of increasing size in trees. Trends Ecol Evol 18:5–6

Mirov MT (1956) Photoperiod and flowering of pines. For Sci 2:328–332

Myers DA, Thomas RB, DeLucia EH (1999) Photosynthetic responses of loblolly pine (*Pinus taeda*) needles to experimental reduction in sink demand. Tree Physiol 19:235–242

Niinemets Ü (2002) Stomatal conductance alone does not explain the decline in foliar photosynthetic rates with increasing tree age and size in *Picea abies* and *Pinus sylvestris*. Tree Physiol 22:515–535

Niinemets Ü (2006) The controversy over traits confering shade-tolerance in trees: onogenetic changes revisited. J Ecol 94:464–470

Nilsson JE, Anderson B (1987) Performance in freezing tests and field experiments of fill-sib families of *Pinus sylvestris* (L). Can J For Res 17:140–1347

O'Connell BM, Kelty MJ (1994) Crown architecture of understory and open-grown white pine (*Pinus strobus* L.) saplings. Tree Physiol 14:89–102

Owens JN, Blake MD (1985) Forest tree seed production. Canadian Forestry, Service Chalk River, 161p

Owens JN, Colangeli AN (1989) Promotion of flowering in western hemlock by gibberellin $A_{4/7}$ applied at different stages of bud and shoot development. Can J For Res 19:1051–1058

Panshin AJ, de Zeeuw C (1980) Textbook of wood technology, 4th edn. McGraw-Hill, New York

Paul MJ, Foyer CH (2001) Sink regulation of photosynthesis. J Exp Bot 52:1383–1400

Penning de Vries FWT, Brunsting AHM, Van Laar HH (1974) Products requirements and efficiency of biosynthetic processes: a quantitative approach. J Theor Biol 45:339–377

Pharis RP, Yeh FC, Dancik BP (1991) Superior growth potential in trees: what is its basis, and can it be tested for at an early age? Can J For Res 21:368–374

Poethig RS (1990) Phase change and the regulation of shoot morphogenesis in plants. Science 250:923–930

Powell GR (2009) Lives of conifers. The Johns Hopkins University Press, Baltimore, 276 p

Rebbeck J, Jesen KF, Greenwood MS (1992) Ozone effects on grafted mature and juvenile red spruce: photosynthesis, stomatal conductance, and chlorophyll concentration. Can J For Res 23:450–456

Ritchie GA (1997) Evidence for red: far red signaling and photomorphogenic growth response in Douglas-fir (*Pseudotsuga menziesii*) seedlings. Tree Physiol 17:161–168

Rudolph TD (1979) Female strobili on 12-month old jack pine. Tree Planters Notes 30:24–26

Sala A, Hoch G (2009) Height-related growth declines in ponderosa pine are not due to carbon limitation. Plant Cell Environ 32:22–30

Sala A, Fouts W, Hoch G (2011) Carbon storage in trees: does relative carbon supply decrease with tree size? In: Meinzer FC, Lachenbruch B, Dawson TE (eds.) Size- and age-related changes in tree structure and function. Springer

Sarala M, Taulavuori K, Taulavuori E, Karhu J, Laine K (2007) Elongation of Scots pine seedlings under blue light depletion is independent of etiolation. Environ Exp Bot 60:340–343

Schmidtling RC (1983) Genetic variation in fruitfulness in a loblolly pine (*Pinus taeda* L.) seed orchard. Silvae Genet 32:76–81

Scholz FG, Phillips NG, Bucci SJ, Meinzer FC, Goldstein G (2011) Hydraulic capacitance: biophysics and functional significance of internal water sources in relation to tree size. In: Meinzer FC, Lachenbruch B, Dawson TE (eds.) Size- and age-related changes in tree structure and function. Springer

Seymour RS (1992) The red spruce-balsam fir forest of Maine: evolution of silvicultural practice in response to stand development patterns and disturbances. In: Kelty MJ, Larson BC, Oliver CD (eds.) The ecology and silviculture of mixed-species forests. Kluwer Academic Publishers, Boston, pp 217–244, 287 p

Seymour RS, Kenefic LS (2002) Influence of age on growth efficiency of *Tsuga canadensis* and *Picea rubens* trees in mixed-species, multi-aged northern conifer stands. Can J For Res 32:2032–2042

Shinya M, Kojima K, Igasaki T, Azumi Y, Shinohara K (2001) Inhibition of the light-independent synthesis of chlorophyll in pine cotyledons at low temperature. Plant Cell Physiol 42:868–872

Sillett SC, Van Pelt R, Koch GW, Ambrose AR, Carroll AL, Antoine ME, Mifsud BM (2010) Increasing wood production through old age in tall trees. For Ecol Manage 259:976–994

Smith RF (1998) Effects of stem injections of gibberellin $A_{4/7}$ and paclobutrazol on sex expression and the within-crown distribution of seed and pollen cones in black spruce (*Picea mariana*). Can J For Res 28:641–651

Smith RF, Greenwood MS (1995) Effects of gibberellin $A_{4/7}$, root pruning, and cytokinins on seed- and pollen cone production in black spruce (*Picea mariana* (Mill.) B.S.P.). Tree Physiol 15:457–466

Smith WH, Konar RN (1969) Initiation of ovulate strobili in cotyledon-stage seedlings of *Pinus elliottii*. Can J Bot 47:624–626

Smith DM, Larson BC, Kelty MJ, Ashton PMS (1997) The practice of silviculture, applied forest ecology. Wiley, New York

Steele MJ, Coutts MP, Yeoman MM (1989) Developmental changes in Sitka spruce as indices of physiological age. I. Changes in needle morphology. New Phytol 113:367–375

Steeves TA, Sussex IM (1989) Patterns in plant development. Cambridge University Press, Cambridge

Steppe K, Niinemets Ü, Teskey RO (2011) Tree size- and age-related changes in leaf physiology and their influence on carbon gain. In: Meinzer FC, Lachenbruch B, Dawson TE (eds.) Size- and age-related changes in tree structure and function. Springer

Sulzer AM, Greenwood MS, Livingston WH, Adams G (1993) Early selection of black spruce using physiological and morphological markers. Can J For Res 23:657–664

Thomas SC (2011) Age-related changes in tree growth and functional biology: the role of reproduction. In: Meinzer FC, Lachenbruch B, Dawson TE (eds.) Size- and age-related changes in tree structure and function. Springer

Tucić B, Dučić J, Pemac D (2006) Phenotypic responses to early signals of neighbor proximity in *Picea omorika*, a pioneer conifer tree. Basic Appl Ecol 7:443–454

Vincent TL, Van MV, Geh BS (1996) Ecological stability, evolutionary stability and the EES maximum principle. Evol Ecol 10:567–591

Ward MH (2005) Age-related trends in red spruce needle anatomy and their relationship to declining productivity. School of Forest Resources/University of Maine, Orono, p 97

Wilson BE, Gartner BL (2002) Effects of phloem girdling in conifers on apical control of branches, growth allocation and air in wood. Tree Physiol 22:347–353

Woodruff DR, Meinzer FC (2011) Size-dependent changes in biophysical control of tree growth: the role of turgor. In: Meinzer FC, Lachenbruch B, Dawson TE (eds.) Size- and age-related changes in tree structure and function. Springer

Woodruff DR, Meinzer FC, Lachenbruch B, Johnson DM (2008) Coordination of leaf structure and gas exchange along a height gradient in a tall conifer. Tree Physiol 29:261–272

Wright JW (1964) Flowering age of clonal and seedling trees as a factor in choice of breeding system. Silvae Genet 13:21–27

Yoder BJ, Ryan MG, Waring RH, Schoettle AW, Kaufmann MR (1994) Evidence of reduced photosynthetic rates in old trees. For Sci 40:513–527

Zazzaro S (2009) Plasticity in response to changing light environment for red spruce and balsam fir. Ecology and environmental sciences program, University of Maine, Orono, 74 p

Zimmermann RH, Hackett WP, Pharis RP (1985) Hormonal aspects of phase change and precocious flowering. In: Pharis RP, Reid DM (eds.) Encyclopedia of plant physiology. Springer, Berlin, pp 79–115

Chapter 5
Radial Variation in Wood Structure and Function in Woody Plants, and Hypotheses for Its Occurrence

Barbara Lachenbruch, John R. Moore, and Robert Evans

Abstract Woody stems, branches, and roots display large progressive changes in structure and function from the pith outwards, a pattern known as the core/outerwood or juvenile/mature wood pattern. For example, the typical radial pattern (TRP), characterized mostly for commercial softwoods, includes increases in wood density by a factor of up to 1.5, tracheid width by a factor of 2, and tracheid length by a factor of up to 4. These large changes will have substantial effects on the success or failure of the plant. Our knowledge of the drivers and the universality of the typical radial pattern are minimal in spite of 150 years of research that has been concentrated mostly on commercial softwoods. This chapter (1) provides a brief description of the nature and magnitude of the structural and functional changes along the radius of woody stems, (2) clarifies whether ring number from the pith or size best explain the typical radial pattern, and (3) introduces and briefly evaluates hypotheses for the functional significance of the radial patterns, with developmental, hydraulic, and mechanical hypotheses. The chapter indicates major knowledge gaps in different biomes and plant types, and provides examples of systems in which research advances could be made. An improved knowledge of why wood develops into the structures we observe will provide a basis for better predictions of woody plant behavior across environments. It is also essential for tree breeders in developing adapted stock for different environments, and foresters and wood processors in predicting end-product quality.

B. Lachenbruch (✉)
Department of Wood Science and Engineering, Oregon State University, Corvallis, OR 97331, USA
e-mail: Barb.Lachenbruch@oregonstate.edu

J.R. Moore
Forest Products Research Institute, Edinburgh Napier University, Edinburgh EH10 5DT, Scotland, UK

R. Evans
CSIRO Materials Science and Engineering, Clayton, VIC, Australia

F.C. Meinzer et al. (eds.), *Size- and Age-Related Changes in Tree Structure and Function*, Tree Physiology 4, DOI 10.1007/978-94-007-1242-3_5,

1 Introduction

Throughout their lives, animals reinvent themselves by replacing their cellular structure at various rates. Juvenile characteristics are lost even in bone. Trees and other woody plants, on the other hand, do not reabsorb their structural elements, so that their entire morphological history is recorded in the wood as the cells die. Apart from the deposition of materials to form heartwood, block conduits, and increase the resistance to decay by microorganisms, little else happens to the original wood structure throughout the life of the plant – which can sometimes be thousands of years. Therefore, the wood properties at a specific location within a plant reflect any plastic response to the environment superimposed on the plant's ontogenetic stage at the time when the wood was formed. The anatomical, chemical, physical and mechanical properties of wood vary considerably within plants; these variations exist between different plant parts (such as main stems, branches, and roots) and within any given plant part. If these wood properties reflect adaptations to the plant's needs, then the study of their variation may help us to identify the most important factors that through natural selection have shaped plants, and especially their xylem, at different stages of their lives.

As a plant grows, a given point within it will experience changes in both its external environment and its internal physiology (Day et al. 2002; Day and Greenwood 2011). It is likely that plants experience shifts in the relative importance of various hydraulic and mechanical demands that vary systematically with the position in the plant and with time. Firstly, a small plant will experience the environment at a finer scale than will a larger plant with respect to many factors (such as water and light availability). Secondly, a small plant will also have a lesser ability to buffer its interactions with the environment (such as transient high evaporative demand, short-term low temperatures, and localized disturbance from herbivory or stem breakage) because of its lesser biomass for storage (Scholz et al. 2011), and its smaller spatial extent for averaging patchy resources and disturbances (Johnson et al. 2011). Lastly, any location in a plant will experience changing hydraulic and mechanical demands as the biomass distal and/or proximal to it changes in extent (through growth, reproduction or breakage) and/or physiology (such as stomatal behavior or root uptake capacity). Therefore, it is likely that species have evolved mechanisms to produce xylem that functions appropriately for the stresses encountered at any given stage of growth. Because xylem serves a number of roles, there will be tradeoffs among the many potential roles that the xylem can play. For example, Meinzer et al. (2009) showed that species with low capacitance (water storage) and dense wood tend to have high daily maximum xylem tensions and structural adaptations that help them resist embolism at these tensions. In contrast, species with higher capacitance and less dense wood are able to use the stored water to avoid the transient high maximum xylem tensions, and they have wood that is less resistant to embolism at the high tensions (which will rarely occur). Other examples for specific situations include the within-plant tradeoff of specific hydraulic conductivity and resistance to embolism (Domec et al. 2008, 2009) and the between-plant tradeoff of basic density and vulnerability to embolism (Hacke et al. 2001; Rosner et al. 2008; Meinzer et al. 2010), or decreases in specific conductivity that can occur if compression wood is produced (Spicer and Gartner 1998; Mayr and Cochard 2003).

In this chapter we concentrate on the radial variation in wood structure and properties within main stems (i.e. differences in the xylem from the pith toward the bark) of angiosperm and gymnosperm trees, which are generally more pronounced than variation within the same growth ring along the stem (Zobel and van Buijtenen 1989). The radial patterns of wood structure may reflect the changing hydraulic and mechanical demands placed upon woody plants as they grow, thus providing a relatively simple case study in structure and function. Because the structure is recorded in the wood, one can infer the function with assumptions, as elaborated here. We acknowledge that patterns may vary by species and environment, and even along the compass directions of the same annual ring (e.g. Lachenbruch et al. 2010), but this variation may illustrate, rather than take away from, the thesis that the variation has a functional significance. The focus on radial variation is a good starting place because of the relatively large amount of information available, and because it can be of economic importance if we can manipulate it. Moreover, to a first approximation, the vertical pattern of wood properties (from the tip downward to the base) is similar to the radial pattern (from the pith outward) allowing use cambial age information from a basal disk to estimate vertical distributions of properties.

Within a tree, the wood located in the region close to the pith, which generally exhibits relatively large radial gradients in properties, is commonly referred to as 'corewood' owing to its location near the pith or 'juvenile wood' on account of the young age of the vascular cambium at the time that it was formed, whereas wood located outside this zone is often referred to as 'outerwood' or 'mature wood' (Fig. 5.1, and Cown 1992; Zobel and Sprague 1998). Note that the terms 'juvenility' and 'maturity' in wood are unrelated to the same terms used to describe the ability of a plant to reproduce. Some of the earliest observations of radial variation in wood properties were reported by Sanio (1872) who found that the length of tracheids in Scots pine (*Pinus sylvestris*) stems and branches increased from pith towards the bark reaching a final value in a quasi-asymptotic manner. The typical radial pattern of variation (called TRP in this chapter) is discussed in more detail in Sect. 2.2, but briefly, it involves rapid inter-annual changes in characteristics such as cell length, lumen diameter, cell wall thickness, microfibril angle in the S2 layer of the cell wall (MfA, measured with respect to the cell's axial direction) and density in the first few years of tree growth. Eventually, a cambial age or distance from the pith is reached at which these radial changes are extremely gradual or are swamped by variation caused by the environment. The age of transition between corewood and outerwood, which is very dependent on the analysis methods used and the characteristic under consideration, is often reported as somewhere between 5 and 20 years (Bendtsen 1978; although Abdel-Gadir showed the age ranged from 15 to 38 in a sample of *Pseudotsuga menziesii*), after which if there are still continued changes in traits, they are very slight and are not of notable economic value. The time scale over which this occurs can be very long indeed.

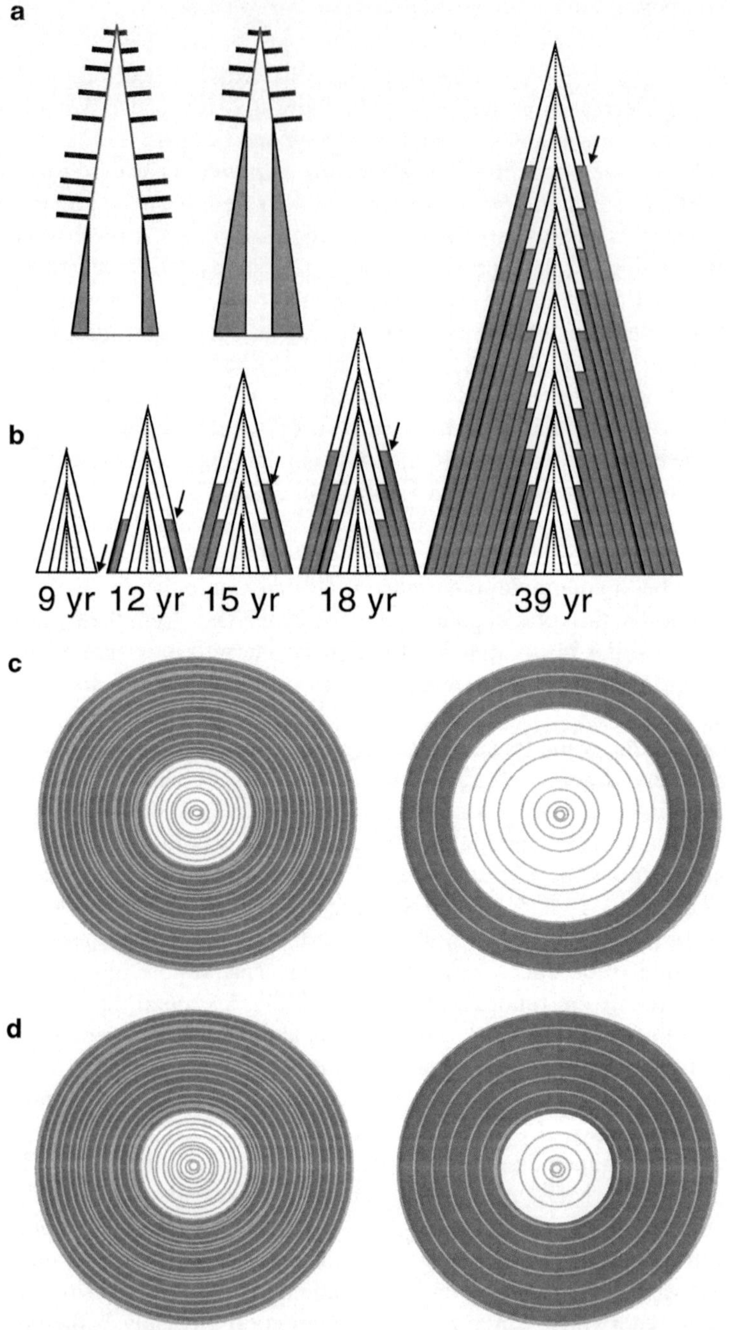

Fig. 5.1 Mechanistic hypotheses for the location of the corewood/outerwood transition. Corewood is *white*, and outerwood is *gray*. (**a**) Location determined by the position of the live crown. (**b**) Location determined by cambial age. Shown schematically for transition age of 9 years, with each sheath of the tree representing 3 years' growth. At any stage of tree growth, corewood is produced at cambial age of ≤9 years and outerwood is produced at cambial age of >9 years (*arrow*). (**c**) Location determined by cambial age. Shown schematically for transition age of 9 years for a slow-grown and a fast-grown tree. (**d**) Location determined by diameter

For example, Gerry (1916) reported that tracheid length in Douglas-fir (*Pseudotsuga menziesii*) was still increasing slowly in the 400th growth ring from the pith, while Baas et al. (1986) found that tracheids in *Pinus longaeva* were still increasing in length after more than 2,000 years, with no sign of leveling off. These typical radial patterns are not seen in all species, a fact that may provide us with plant material for use in comparative studies to better understand why (and under which circumstances) the TRP exists.

Although published studies on wood anatomy date back almost 150 years, the research concerning radial variation is not well-distributed taxonomically: it has been concentrated in the conifers that have an abrupt transition from earlywood to latewood (such as *Pinus taeda*, *P. radiata*, and *Pseudotsuga menziesii*), which include most of the valuable timber species that are used in construction. In these species, knowledge of the radial variation has been used in tree breeding programs and has helped silviculturalists anticipate the wood quality they are producing. It has also been very useful for processing and utilizing the lumber. In contrast, we have much less information on all of the following: conifers with a gradual transition (such as members of the Cupressaceae, and several genera in Pinaceae such as *Abies*, *Picea*, and *Tsuga*), hardwoods with diffuse-porous wood that is often reputed to be relatively uniform across the cross-section (such as *Alnus rubra* or hybrid poplar) and which are not used to the same extent for strength properties, hardwoods with ring-porous wood structure which tend to produce an earlywood band of constant width, and thus whose many properties are largely tied to variation in the width of the latewood band, and tropical hardwoods, including not only trees, but woody plants that have phases with different degrees of mechanical support (such as lianas and hemi-epiphytes). Data are also rare on shrubs, whose wood density is often high and whose growth form is often distinct from that of a tree. In general, the TRP is much less marked in hardwood trees than conifer trees.

Moreover, there have been very few adaptive explanations of the TRP. The early literature on wood variation mainly focused on anatomical characteristics that could be examined through transmitted light microscopes and on mechanical properties of boards cut from different radial positions within a tree. Quite sophisticated methods, some automated, are now available for characterizing radial variation, and unlike most of the work undertaken during the twentieth century, there is now careful physiological work in addition to anatomical and mechanical work characterizing radial changes in structure and function. Much of this chapter is concerned with what we can infer from knowing whether the radial variation in function is more closely related to ring number from the pith (cambial age) or size (more strictly referred to as radius or overall diameter). We use the term ‘size’ rather than stem radius because the tree’s functions (related to allometry, physiology, and/or mechanics) depend on its three-dimensional form rather than a simple linear or quadratic function of radius. For example, tree size is related to traits such as how deep is the canopy and at what range of water potentials does the tree transpire (for water needs), the extent to which the canopy’s mass is offset (for bending stresses), or the stem’s taper (for axial load).

This chapter has three main objectives:

- To provide a brief description of the nature and magnitude of the typical structural and/or functional changes along the radial direction of a stem at any height (the TRP). We will use breast height as the standard height in order to reduce the number of factors discussed, but most of this discussion would pertain to locations higher or lower in the bole as well.
- To clarify whether ring number from the pith or size best explains the TRP in self-supporting trees that are not subjected to any unusual mechanical environments.
- To briefly introduce and evaluate hypotheses for the functional significance of the structural variation exhibited in the TRP. We also suggest plant material/environment systems in which these hypotheses can be further tested.

If the evaluations proposed here improve our knowledge of why wood develops into the structures we observe, we will have a good basis for better understanding the woody plant habit and predicting its behavior in various environments. Moreover, knowledge of the extent and location of variation in wood properties within a stem is of importance to foresters and wood processors as it has major implications for end-product quality.

2 Radial Variation's Terminology, Patterns, and Effects on Wood Utilization

In this section, we review the terminology, location, and structure of corewood and outerwood and its effects on wood utilization. We then briefly discuss three mechanistic theories to explain the typical radial pattern (TRP) that is observed in selected wood properties.

2.1 Terminology

The terminology used to describe the wood found at different positions within a tree has been the source of much debate (e.g. Larson 1969; Amarasekara and Denne 2002; Burdon et al. 2004). In the wood technology and wood science literature, wood located near the pith is commonly referred to as juvenile wood, and wood found further from the pith as mature wood. However, Larson (1969) argued that the term juvenile wood is a misnomer as it can only adequately describe the type of wood formed in young trees and not the wood formed higher up in the stem of older trees. He stated that it would be appropriate to refer to this type of wood from a positional perspective as 'core wood' or from a developmental perspective as 'crown-formed wood'. In fact, the term 'crown-formed wood' has been in existence since at least the 1930s, when it was used by Trendelenburg (1935, cited in Paul 1957) to describe the core of low density wood in a tree, and reflects the effect that proximity to the active crown has on the properties of the wood that is formed. 'Crown-formed wood' has the disadvantage of not distinguishing between wood that is truly in close proximity to foliage and/or terminal meristems, and wood that is instead near long branches, which may not be influenced by crown-derived hormones. While the terms 'juvenile wood', 'core wood' and 'crown-formed wood' have often been used synonymously, Amarasekara and Denne (2002) proposed that the term 'crown-formed wood' should be used to describe fluctuations in wood structure associated with the size of the crown, which are superimposed upon the trends due to cambial age. There are also overlaying patterns resulting from the effects of gravity and wind (e.g. Chiba 2000). These perspectives were advanced further by Burdon et al. (2004) who developed two separate concepts: juvenility versus maturity to describe axial variation in wood properties; and corewood versus outerwood to describe variations from pith to bark. They argued that these terms should be used in conjunction with each other, rather than as alternatives. For example, wood located near the pith at the base of the tree would be referred to as juvenile corewood, while further up the tree it would be referred to as mature corewood. We will use the term corewood to refer to the wood located near the pith at all heights in a tree where there is considerable radial variation in wood properties, and outerwood to refer to the wood exterior to the core. We will also use the term typical radial pattern (TRP) to describe the radial variation at any given height.

2.2 *The Typical Radial Pattern (TRP)*

Radial variation exists not only in the length of cells within a stem, but in most other cell characteristics. In softwoods, growth rings near the pith usually consist of a large proportion of earlywood tracheids (which have larger diameter and thinner cell walls than latewood tracheids), that gradually transitions to a larger proportion of latewood tracheids with increasing ring number from the pith. Earlywood tracheid diameter increases with cambial age and together these two properties control most of the variation in density, which is the most widely studied physical wood property. (Unless stated otherwise, we will use the term density to mean basic density, which is dry mass per unit green volume.) Tracheid shape and length have relatively minor effects on density, and the density of the cell wall itself is almost constant at approximately 1,530 kg m^{-3} (reviewed in Siau 1984). Automated x-ray densitometers have greatly aided in data acquisition on radial density patterns. For conifers in the Pacific Northwest of North America, mean ring-level density generally follows a pattern where there is an initial decrease for the first 5–20 growth rings outward from the pith, followed by a rapid increase until a maximum value is reached asymptotically (Kennedy 1995). The typical pattern in many hard pines, including *Pinus radiata* and *P. taeda*, as well as a number of other species is to have no initial decrease in density, just a rapid increase in density with ring number, followed by a leveling off (Turnbull 1948, reviewed in Zobel and van Buijtenen 1989). With the advent of rapid x-ray diffraction technology such as SilviScan (Evans et al. 1999), within-tree variation in

MfA has been studied in great detail for a number of species. In softwoods, MfA generally declines rapidly in the first few growth rings out from the pith before it approaches a minimum value asymptotically (Cave and Walker 1994; Mansfield et al. 2009); however, other radial patterns do exist.

There is no universal position within a stem at which wood properties reach a quasi steady-state value; the progression from corewood to outerwood is usually gradual, and the steady-state values as well as the pattern themselves can vary between wood properties, as well as between individual trees and sites (Cown et al. 1991; Fabris 2000). Therefore, for convenience, the extent of the corewood zone (i.e. the number of rings from the pith or the area that it occupies at a given height in a tree) can be defined by an arbitrarily assigned threshold in a systematically changing property. In many cases, the number of rings from the pith has been used to identify the extent of the corewood zone, but in other cases, distance from the pith has been used (see Sect. 3). In southern pines (primarily *Pinus taeda* and *Pinus elliottii*) its extent is commonly based on visual assessments of latewood percentage (Larson et al. 2001). Its extent has also been delimited using densitometry (which gives an estimate of wood density that is in turn related to mechanical properties) (e.g. Abdel-Gadir et al. 1993; Fabris 2000), or fiber or tracheid length (e.g. Shiokura 1982; Mansfield et al. 2009). More recently, microfibril angle has been used to identify corewood because of its influence on wood stiffness, shrinkage potential and dimensional stability (e.g. Cown et al. 2004).

The radial extent of corewood differs by species, between individuals within a species, by height within the stem, and with the property used to define it (Zobel and Sprague 1998). For example, in *Pinus taeda*, the juvenile core is generally taken to consist of the inner ten growth rings (Larson et al. 2001), whereas in lodgepole pine (*Pinus contorta*) Mansfield et al. (2009) found that the transition from corewood to outerwood occurred at 31, 18 and 15 years based on the analysis of radial profiles of wood density, tracheid length and MfA, respectively. For a particular property the age at which the transition from corewood to outerwood occurs within a species appears to be under a degree of genetic control (e.g. Gapare et al. 2006). Irrespective of the property or properties used to define the extent of the corewood zone, the change from corewood to outerwood is usually not abrupt. Almost all woods of self-supported woody plants show a transition from corewood to outerwood, which can be called the transition zone (Zobel and Sprague 1998). The discussion of this zone, or the properties of the wood contained within it, adds little to our insights while adding to the complexity of an already uncertain concept and, therefore, will not be pursued further in this chapter.

Differences between corewood and outerwood have been studied for many species, particularly conifers. Zobel and Sprague (1998) summarized the general characteristics of corewood and outerwood in both conifers and hardwoods. Here we present what we are calling the typical radial pattern (TRP), which is most likely biased by the larger number of studies on softwoods than hardwoods, and on traits of traditional importance to wood processors and tree breeders (such as wood density and microfibril angle) rather than on other physical and biological traits (such as critical strain, or peak transpiration/sapwood area). This pattern is typical for self-supported plants;

those plants that are supported externally for some or all of their life will not necessarily conform to the TRP, nor will those that have other unusual mechanical loads such as from a prevailing wind, or from self-weight if the whole stem is leaning. As more growth forms, habitats, and environments are studied, many of the generalizations about the TRP may well shift significantly. The main TRP discussed in this paper includes the following differences in key characteristics of corewood compared to outerwood:

- Shorter cells with smaller diameter and thinner cell walls.
- Smaller contribution (in conifers) of the S2 wall layer to the double cell wall thickness.
- Lower specific gravity. In some pine species (mostly hard pines) the specific gravity of outerwood can be as much as double that of corewood, but 10–20% differences are more common.
- High microfibril angle (40–50° is common for softwoods such as pines; 25–35° is more common for hardwoods such as eucalypts, poplar and birch).
- Low modulus of elasticity and strength due to lower specific gravity and higher microfibril angle. Specific stiffness (the ratio of modulus of elasticity to specific gravity) is also generally lower in corewood.
- High spiral grain angle (the angle of the longitudinal cells with respect to the stem's axis). In many species spiral grain reaches a maximum in the first few rings from the pith, and then declines to a low value in the outerwood.
- Higher lignin and lower cellulose content.
- Higher incidence of reaction wood, i.e. compression wood in softwoods and tension wood in hardwoods.
- Greater propensity for distortion when wood is dried because of higher longitudinal shrinkage (associated with high microfibril angle), spiral grain angle and incidence of reaction wood.
- Greater susceptibility to internal checking (Ball et al. 2005).

These properties listed above have been studied because of their effect on wood utilization. The following characteristics of corewood have been found by tree physiologists:

- Lower specific conductivity (Mencuccini et al. 1997; Spicer and Gartner 2001; Domec et al. 2005).
- Lower vulnerability to embolism in conifers (e.g. Domec and Gartner 2002, 2003; Domec et al. 2005; Rosner et al. 2006).

In addition, gene expression studies are showing that there are differential activities and products produced by the cambium at different ages; essentially, rather than explaining why the physical and physiological traits exist, these studies are providing a new set of traits that follow the TRP. For example, Cato et al. (2006) showed that the expression of many genes of developing tissues have monotonic gradients from the cambial zone of corewood to that of outerwood in *Pinus radiata*. Similarly, Paiva et al. (2008) sampled *Pinus pinaster* trees along a height (and thus cambial age) gradient. They showed that the relative expression of genes for processes of importance in the corewood region (a cluster of 71 genes) was higher in cambial material from the top than the bottom of the tree, and that the relative expression of genes for processes of importance in the outerwood region (a cluster of 53 genes) was higher at the bottom than the top of the tree.

There is usually markedly less radial variation in wood anatomy within hardwoods than within softwoods (Panshin and deZeeuw 1980; Zobel and Sprague 1998), although Wiemann and Williamson (1988) found that tropical pioneer angiosperm species can have very high radial variation in wood density. Diffuse-porous species have patterns that are roughly similar to those in softwoods, and generalized curves were presented by Maeglin (1987), showing that fiber length, cell diameter and wood density increased with ring number from the pith while MfA decreased. A study in *Alnus rubra* showed, however, that some diffuse-porous species have no significant variation in density from pith to bark, in spite of an increase in vessel proportion and a decrease in fiber proportion (Gartner et al. 1997). In ring-porous species (those species with markedly wider vessels at the beginning than the end of the growth ring), there is a very large difference between earlywood and latewood densities. This fact coupled with the tendencies to have narrower growth rings at higher ring number from the pith but to maintain approximately constant earlywood width, results in ring-porous hardwood species tending to decrease in density from the pith outward (Lei et al. 1996; Woodcock and Shier 2002).

Several characteristics can be classed as exhibiting a radial pattern, but which are actually integrative values that depend not only on the currently produced structure, but on the structure elsewhere in the tree. One example is leaf-specific conductivity, which is a measure of how conductive the xylem is relative to the leaf area that it supplies. In most trees, the corewood has lower leaf-specific conductivity than the outerwood (Zimmermann 1978; Ewers and Zimmermann 1984; Domec et al. 2005). However, leaf-specific conductivity at a particular location depends on both the conductivity at that location and all of the leaf area distal to that point. Another example is the growth stress at a location (the force per unit cross-sectional area which the xylem exerts on its surrounding wood). Longitudinal growth stresses display strong radial gradients, with the outer layers (xylem in the outer part of the radius) contracting and inner layers (xylem within about the inner 2/3 of the radius) expanding upon removal from surrounding tissues (Jacobs 1945; Boyd 1950; Archer 1986). These values, however, depend on the characteristics of the wood that had already been laid down: the growth stress exerted by the newest growth layer will result in a different stress profile within the stem depending on what the previous profile had been (Fournier et al. 1990).

2.3 *Effects of Radial Variation on Wood Utilization*

Much of the interest in the radial variation in wood properties and the size of the corewood zone arises from their importance to the wood processing industry. As a result there is a substantial body of literature on both the effects of different silvicultural practices on the size of the corewood zone (particularly the percentage of log volume occupied by corewood) as well as on the effects of corewood on the performance of different products for a large number of tree species, mostly commercially-important softwoods. We highlight some of the major corewood/outerwood differences and the implications for manufacturers; interested readers are directed to the reviews by Zobel and van Buijtenen (1989), Zobel and Sprague (1998), and Larson et al. (2001). Without genetic and silvicultural intervention, the proportion of corewood in the commercial forest resource worldwide will inevitably continue to increase for several reasons. The proportion of timber originating from plantations has increased substantially in the last few decades but outerwood in many cases does not yield much higher prices than corewood. Therefore, to maximize the rate of return on investment in land and silviculture, there has been a move toward harvesting trees much younger than in the past, and also in increasingly using the trees that are removed during thinning operations.

Corewood is generally regarded as being of lower quality than outerwood for most structural uses, although it is actually preferred for certain products such as medium density fiberboard and for some types of paper. Zobel and Sprague (1998) caution that the large number of publications focusing on the adverse properties of corewood have led to it being labeled as "bad wood", whereas they argue that from a utilization perspective it should be considered as "different wood" with advantages of its own. The relatively low density of juvenile softwood affects pulp and paper properties mainly through the greater collapsibility of the tracheids. Kraft pulp handsheets produced from corewood have higher tensile strength but lower tear strength than those produced from outerwood. Printing paper produced by the thermo-mechanical process from young radiata pine (*Pinus radiata*) has high structural uniformity, optical performance and surface smoothness (Burdon et al. 2004). The strength properties of medium density fiberboard panels produced from the corewood of black spruce (*Picea mariana*) were found to be superior to those of panels produced from outerwood taken from the same trees (Shi et al. 2005).

Corewood does, however, present real problems for sawmills producing structural timber. For these uses, the low density and high microfibril angle of corewood results in timber with low modulus of elasticity (low stiffness) and bending strength (Bendtsen and Senft 1986), an often high incidence of compression wood (Barger and Ffolliott 1976; Timell 1986), and a longitudinal shrinkage as much as ten times greater than that of outerwood (Voorhies and Groman 1982). This results in poor dimensional stability, meaning high shrinkage or swelling in response to changes in ambient humidity. These strong property gradients lead to warping and twisting of timber that contains both corewood and outerwood. Such dimensional instability can cause significant serviceability problems in timber structures, even if it only affects a small number of the pieces of timber within a structure. Numerous studies have shown that engineered wood products such as oriented strand board have poorer performance if they are made from corewood (e.g. Cloutier et al. 2007);

however, corewood can be placed in certain locations in some engineered wood products where it will not adversely affect overall performance (e.g. in the middle layers in laminated veneer lumber or glulam beams; e.g. Kretschmann et al. 1993). Corewood also tends to have a higher proportion of knots by virtue of its location in the tree and geometry; these knots negatively impact grade recovery and strength, but are not the subject of this chapter.

3 Developmental Controls of the Typical Radial Pattern

Gradual radial changes in many xylem properties are the rule, rather than the exception, in most woody plant species that have been investigated, but what mechanism is responsible for these changes? Are the patterns ontogenetically fixed, or are they plastic, triggered by factors that vary as the plant increases in size? The radial pattern may be of adaptive value regardless of whether the control is ontogenetic or plastic because natural selection can favor a similar radial pattern through either developmental route. Knowledge of the nature of the driver, however, can tell us whether the variation we observe is genetic or is related to the environment that the plant lives in, and can also give us tools to use if we want to alter the pattern of wood development. In this section, we review terminology that make more explicit the meaning of 'fixed' and 'plastic' control, then present three hypotheses for the mechanism responsible for the TRP. While we acknowledge that it is likely that a combination of fixed and plastic mechanisms is involved, for simplicity we will examine the potential mechanisms separately. We then focus on rings from the pith vs. diameter (a special case of 'age vs. size') and ask why it may be useful to understand which factor is more important, and then give examples of studies that supported the importance of either ring number or diameter for determining the TRP. The examples show that for both conifers and hardwoods, in some cases age is more strongly correlated with a trait and in other cases it is size. It is not surprising to find such variability in the control (fixed or plastic) or the type of factors (age or size) that describe the development of the TRP in the woody plants: given the physiological, developmental, and mechanical interactions that occur within woody plants, one would not expect an extremely simple set of rules to orchestrate the development of adaptive patterns in all trees and under all circumstances.

Day and colleagues (Day et al. 2002; Day and Greenwood 2011) define three models of potential pathways controlling age-related changes in the development of plant shoots. *Intrinsic control* results from internal factors (those within the meristem itself) that result from programmed changes in gene expression (the typical meaning of 'ontogenetically fixed' or 'developmentally controlled'). Environmental (plastic) control can be broken down into two types. *Extrinsic control* results from external factors (those beyond the meristem—but that can include environmental as well as within-plant signals) that act on the products of the meristem such that they develop differently (for example, the effect of light on developing leaf tissue). In

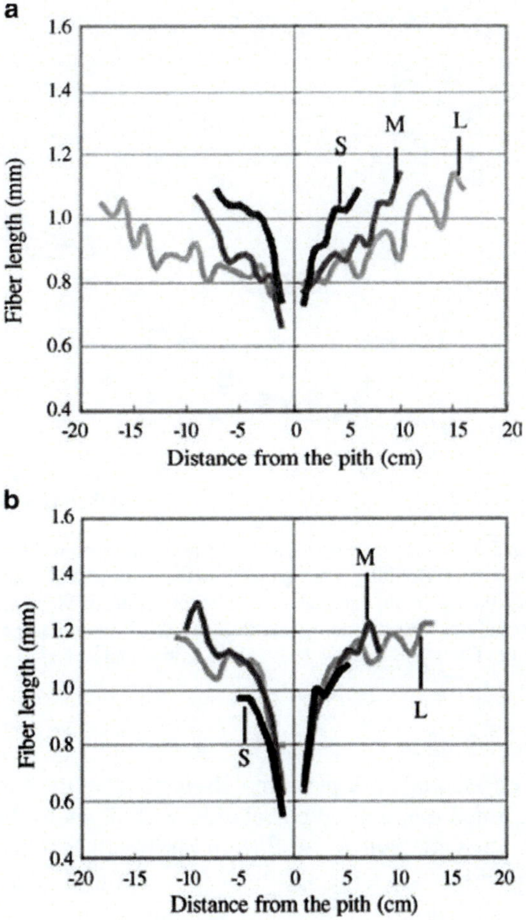

Fig. 5.2 Radial distribution of fiber length at breast height from small (S), medium (M), and large (L) diameter plantation grown trees. (**a**) *Eucalyptus grandis* (14 years old, mean DBH of 21.7, 28.0, and 37.9 cm, respectively, n=6 trees/size class) showing dependence of fiber length on cambial age, not stem diameter. (**b**) *Acacia auricuriformus* (11 years old, mean DBH of 14.1, 19.1, and 25.8 cm, respectively, n=4 trees/size class) showing dependence of fiber length on stem diameter, not cambial age (From Kojima et al. 2009. Used with permissions)

this case, the meristem remains inherently the same over time, but the signals change. *Extrinsic-intrinsic control* results from changes in gene expression that are induced by environmental factors. Extrinsic-intrinsic control appears to be more common than extrinsic controls.

Although the theory makes clear distinctions among these controls, in practice we are not always able to discern which type of control is acting, and so our ability to distinguish among these types of control is somewhat blurred. An ontogenetically fixed (intrinsically controlled) pattern could be manifested by a tight correlation of a trait with cambial age but a poor correlation with diameter (e.g. Fig. 5.2a), as discussed by Day et al. (2002). A fixed pattern could also be manifested by the reverse, a tight correlation with diameter but a poor correlation with cambial age (e.g. Figs. 5.2b and 5.3). In the first case, the cambium's algorithm would be triggered by a time-related metric such as the number of dormancy periods experienced, and in the second case, by a diameter-related algorithm

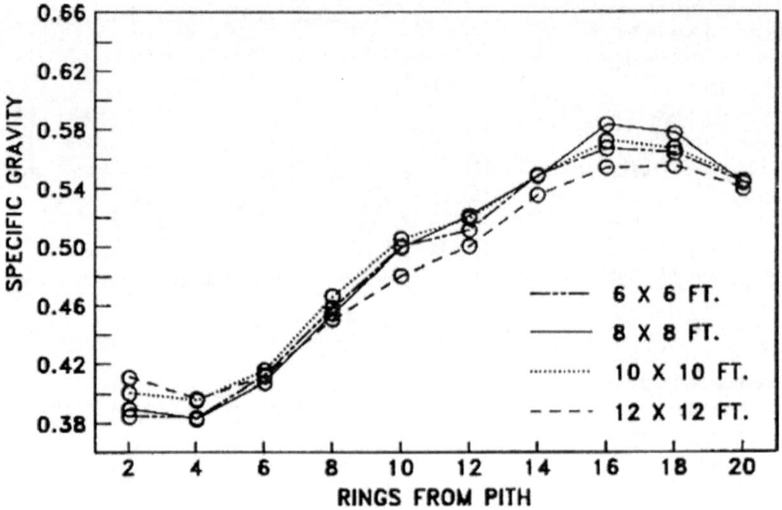

Fig. 5.3 Influence of tree spacing on the radial pattern of wood specific gravity at breast height in *Pinus taeda* (loblolly pine) growing in the Piedmont of South Carolina showing the dependence of specific gravity on cambial age, not stem diameter. Mean diameter at breast height (and height) of trees at the closest spacing was 20.3 cm (20.9 m) and at the widest spacing was 28.7 cm (22.5 m) (From Clark and Saucier 1989. Used with permissions)

such as number of periclinal divisions that the cambium has undergone. It should be noted that a good correlation with diameter but a poor correlation with cambial age could also result from plastically-controlled development, if the growth algorithm was triggered by a force acting beyond the cambium itself, such as a bending stress. In this case, the magnitude of the force depends on factors such as the amount of biomass distal to the xylem sample, as well as its geometric distribution. Unlike a size-related pattern that is ontogenetically controlled, a plastically-controlled one could be altered with manipulations of the driving force, such as staking plants in the example in which the TRP is driven by changes in bending stresses.

To investigate the extent of ontogenetic (intrinsic) vs. plastic (intrinsic/extrinsic) control of wood development, Lenz et al. (2010) estimated the heritability of eight traits that follow the TRP for rings 1–16 in *Picea glauca*. If a trait was highly heritable, then its control was most likely ontogenetic, and likewise if it had low heritability, then its control was most likely plastic. Near the pith, there was little evidence to support ontogenetic control: none of the traits were strongly heritable, with the highest heritability values there being 0.3–0.4 (for earlywood radial cell diameter and earlywood and latewood MfA). Heritability then increased markedly for earlywood traits from the first to the 16th growth ring for five of the eight traits studied (wood density, specific fiber surface, cell wall thickness, tangential cell diameter, radial cell diameter), with heritability at ring

16 exceeding 0.7 for three of these traits. The authors inferred that ontogenetic control becomes more important for these traits at increasing ring number. Similar research on other factors and in other taxa would be very valuable for determining the degree to which the patterns can be changed through genetic selection vs. through environment.

3.1 Mechanistic Hypotheses Explaining the Location of the Corewood/Outerwood Transition

Proposed mechanisms controlling the TRP are the number of rings from the pith (pattern ontogenetically fixed), diameter (usually treated as a plastic response, but as shown above, could also be ontogenetically fixed), and the location with respect to the position of the live crown (plastic). We refer to these hypotheses as mechanistic, as distinct from the adaptive hypotheses (Sect. 4) that explain the functional advantages of such radial variation.

3.1.1 Transition Occurs in Response to Ring Number from the Pith (Cambial Age)

This hypothesis proposes that the transition from corewood to outerwood is related to the age of the cambial initials (synonymous with the terms 'cambial age', and 'number of growth rings from the pith') at a particular height within the stem (Fig. 5.1b, c); the TRP is intrinsically controlled. At young cambial ages (in a young tree or higher up the stem in an older tree), the products of the cambial initials have xylem properties that are characteristic of corewood (e.g. short cells with narrow diameter and relatively high MfA). As a tree grows outward, the older cambial initials at the base of the tree are the first to reach the transition from corewood to outerwood as shown by these properties, while younger cambial initials higher in the tree are still producing corewood (Fig. 5.1b) but will be reaching the transition in the future. As discussed in more detail in Sect. 3.3, there are many examples in which ring number is a better predictor of the trajectory of the TRP than is stem diameter, in both softwoods (e.g. Herman et al. 1998; Clark and Saucier 1989) and hardwoods (e.g. DeBell et al. 1998).

3.1.2 Transition Occurs in Response to Stem Diameter

This hypothesis proposes that the transition from corewood to outerwood is related to stem diameter (Fig. 5.1d). One possible explanatory mechanism for this is that the cambial cells in a small-diameter stem would need to have a much higher proportion of anticlinal divisions (those that produce tangentially-aligned cambial cells) than would those in a larger stem to produce a new radial increment. However, true anticlinal divisions are rare; cells generally divide pseudo-transversely (i.e. rather than having a cell divide into two thin cells of the same length, the cell divides into two shorter cells, one above the other, but with some overlap of the bottom and top of the resulting cells). This mode of cell division would depress the apparent tendency for cambial initials to increase in length with increasing cambial age, and would explain the gradual increase in cell dimensions as the cambium develops a lesser radius of curvature at larger sizes (discussed in DeBell et al. 1998). Indeed, Bannan (1967) found that tracheids arising from pseudotransverse divisions are considerably shorter and therefore the radial pattern of tracheid length can be interpreted in terms of the relative frequency of pseudotransverse divisions.

However, diameter is actually a one-dimensional trait that serves as an index of the size of the tree (see King 2011), in which the size also encompasses its mass, height, center of mass, foliage quantity, and other factors. While the signal to transition from to corewood to outerwood could be fixed such as the total number of cell divisions in the cambium (which would give an index of diameter) or the current number of divisions/number of cells in the cambium (which would give an index of the cambium's radius of curvature, and thus its distance from the pith), it could also be plastic, triggered by a factor such as self-weight, compression, tension, or torsion sensed in the cambial zone or hormones produced in a specific location (such as crown or root).

3.1.3 Transition Occurs in Response to Location with Respect to the Live Crown

The *crown position* hypothesis proposes that crown-produced auxins favor production of corewood, and their absence favors production of outerwood (Fig. 5.1a and Larson 1969). Larson (1969) further states that xylem in the crown region has a greater proportion of earlywood and thinner cell walls overall. There are no definitive studies that are strongly supportive of this hypothesis, although some are mildly supportive (e.g. Amarasekara and Denne 2002; Gartner et al. 2005).

Crown position may well be an important factor for determining the radial pattern of xylem traits, and it may follow the mechanism describe here, but it could, alternatively follow other mechanisms. For example, the production of plant growth regulators in the crown or the sensitivity of the cambium to these regulators could be related to factors such as the rate of change in water potential, the amount of available water storage, or the quantity of sunlight that reaches the bark (Gartner 1996). The crown-position hypothesis is the basis for suggestions that silvicultural treatments which affect the depth of the live crown, such as pruning or manipulation of stand density (i.e. through initial spacing and/or thinning), will lead to different patterns of corewood and outerwood in same-aged trees (see the critical review in Briggs 1995). We do not discuss this hypothesis further in this paper, but it may be a fruitful area of future research.

3.2 *Why Ask Whether Age or Size Is a Better Predictor of Traits?*

As stated before, ring number from the pith is related to the age of the cambium at a given height, and diameter at that location is related to the size of the tree, so these characteristics can represent age and size. Clarification of whether the TRP is more closely related to ring number from the pith or diameter for a particular wood property, species, and environment may give insight to physiologists seeking to understand the factors that have influenced wood structure and function, on both the short-term and the evolutionary time scales. Limits to the inferences, however, include the possibility (a) that age and size are correlated with one another so strongly that we will learn little from knowing with which variable a trait has the stronger correlation, (b) that age (or size) exhibits the stronger correlation with the trait of interest but that size (or age) was a simpler path for natural selection to use, and (c) that an observed correlation with age or size is spurious and the relationship is actually evolutionarily neutral.

Assuming that these caveats are noted, we can address the implications of size vs. age correlations. As an example, if age is more strongly correlated with a morphological trait than is size, we can infer that evolution favored an age-related factor (such as the hydraulic redundancy provided by having numerous growth rings) over some size-related factor (such as the physical stress caused by the canopy in the wind) as the better predictor of appropriate function. Physiologists can then develop and test hypotheses in accordance with age rather than size to further understand plant biology. Knowledge of whether traits are more closely correlated with number of rings from the pith or with diameter can be very useful to forest managers in situations in which the incidence of corewood will affect value or use (Kojima et al. 2009). This knowledge can be used to manipulate tree growth to maximize log value. It can also be used to predict the value of logs knowing the management history, to better schedule when to harvest a stand, and to better target log sales and processing for their appropriate use.

3.3 *Is There a Small Set of Potential Drivers of Age- or Size-Dependent Changes in Xylem Traits?*

It is instructive to list the most apparent similarities and differences one would expect between a small and a large tree of the same age, and a young and an old tree of the same size. If the lists are short, it may be simple to infer the factors of importance for the plant during passage from one size or age to the next. These factors could then be explored with respect to the radial changes in wood properties. While not exhaustive, Table 5.1 illustrates that there are numerous candidate functions on which natural selection may have acted, and that there is no one set of similarities or differences from which we can infer why a certain species and trait is more closely correlated with one of the independent variables (age or size) than with the other.

Table 5.1 Similarities and differences between trees that differ in age but not size, or in size but not age, and their relevance for development of hypotheses to explain radial variation in xylem structure. These attributes do not take into account any mitigating differences in wood structure or function, such as the centrifugal increase in modulus of elasticity and cell length

Small tree compared to large tree of same age[a]	
Similarities	Potential relevance
Both have been subjected to environmental factors for the same length of time (e.g. freeze-thaw cycles, oxidation)	Both may exhibit similar deterioration or changes due to physical and chemical environment
Both have same likelihood of experiencing infrequent extreme events (e.g. fire, extreme wind, ice-storms, pest outbreaks, oxidation)	Both are at similar risk for damage (although magnitude of damage is probably size-related)
Both have had same number of growth and dormancy periods	Both have had same number of cues, if species has mechanism to count periods to assess age
Both may have same number of plies (growth increments), and same number of branch clusters if determinate growth	Both may have similar numbers of growth rings through which water must traverse from inner to outer wood
Both have had same number of years to reach current architecture	Both should have similar morphology and perhaps physiology at same-aged locations, if meristems have age-dependent behavior
Both may have expended reproductive effort	Small tree may have further reduced growth
Differences	*Potential relevance*
Small tree has shorter internodes if determinate growth	Small tree may have more closely spaced branches, foliage, reproduction
Small tree has shorter stem overall	Small tree may have less hydraulic resistance due to force of gravity and due to path length
Small tree has fewer total divisions at meristem over lifetime	Small tree may be more homogeneous genetically and may track the environment more poorly, if meristematic divisions are associated with somatic mutations
Small tree has lower lever arm, smaller crown relative to stem, less self-weight/stem cross-sectional area	Small tree may need less investment in structural support
Small tree has more growth layers per cm of wood	Small tree may have tougher stem if latewood zones retard crack propagation
Small tree may have lower leaf area	Small tree may have lower absolute carbon gain, less leaf redundancy so more vulnerable to leaf loss (from herbivores, pathogens, physical factors)
Small tree may have narrower sapwood, shallower or less dispersed rooting and/or less stored water per leaf area	Small tree may be more subject to drought stress if drought conditions are limiting

(continued)

Table 5.1 (continued)

Young tree compared to old tree of same size[b]	
Similarities	Potential relevance
Both have same physical environment – wind, light, temperature	Both have same spatial patterns of wood properties if species responds plastically to these factors
Both have undergone roughly same number of division in meristems	Both have same spatial pattern of wood properties, if species has intrinsic meristematic controls that change with size (as number of divisions)
Both have same spatial distribution of mechanical stresses in all the axes: axially, radially, and tangentially along bole, and along individual branches and roots (if distribution of elastic properties were the same in both trees)	Both have wood designed to same mechanical safety factor if species can anticipate (probably through intrinsic controls) or react to (through extrinsic controls) magnitude of stresses
Both may have same water potentials (assuming similar leaf area, root area, and specific conductivities)	Both have wood designed to same hydraulic safety factor if species can anticipate (probably through intrinsic controls) or react to (through extrinsic controls) magnitude of drought stress
Both may have roughly same abilities to capture assimilate (assuming similar leaf area, fine root area, and whole-plant architecture)	Both may have similar growth efficiencies, carbon to allocate to defenses, re-growth after damage, etc.
Differences	*Potential relevance*
Young tree has experienced environment for shorter period	Young tree has had less time to deteriorate from ongoing, predictable processes
Young tree is less likely to have experienced any particular environmental perturbation	Young tree is less likely to have been damaged by extreme events
Young tree has had fewer growth and dormancy periods	Young tree is developmentally behind, if development is influenced by years
Young tree has fewer but wider layers ('plies'), and either the earlywood, latewood or both will be present in wider expanses	Young tree may have different stress distribution, access to stored water, risk (due to diff. amount of redundancy in structure), depending on importance of plies to these factors
Young tree has had fewer years to reach current whole-plant architecture	Young tree is more heterogeneous, has less redundancy in structure if determinate growth (ex: loss of one branch whorl causes a greater biomass loss)
Young tree may not have expended reproductive effort	Young tree may have more vigor
Young tree architecture may be more hierarchic, with older tree consisting of repeated hierarchic units	Young tree's productivity possibly less tied to demography of the units, and thus less buffered

[a]Comparisons are of idealized trees that arrive at different sizes for the same age but that grew without constraints related to tree social status (such as being dominant vs. suppressed). For illustration, consider small (5 m) vs. large trees (30 m) of the same age (50 years)

[b]Comparison of idealized trees that require different ages to attain the same size but that grew without constraints related to environment (such as differences in rainfall). For illustration, consider young (10 years) vs. old (50 years) trees of the same size (10 m)

3.4 Examples of Age- or Size-Related Xylem Traits

One type of experiment of use for testing age vs. size dependence is to find or develop plant material with a common age but that differs in size. (It would be redundant to also compare trees of the same size but different age because this situation is also found by dissecting the larger tree back to the point at which it had the same size, but different age, as the smaller tree.) The criterion of same-aged but different sized trees is often met in silvicultural trials that have variable thinning, fertilization, or irrigation. It may also be met in a cohort of trees that have established following a disturbance.

In the conifers, most such studies have shown that ring number from the pith is more closely correlated than is diameter to the values of traits, although there are examples of each outcome (reviewed in Zobel and Sprague 1998, pp. 9–11). For example, in thinning or spacing trials, the transition from corewood to outerwood was more closely related to ring number for wood density in two species of southern pine (*Pinus taeda* and *P. elliottii*, Clark and Saucier 1989) (Fig. 5.3) and for tracheid length in two species of spruce (*Picea mariana* and *P. glauca*, Yang 1994). A similar study with Norway spruce (*Picea abies*) also showed that tracheid length and MfA were more closely related to ring number from the pith, but that tracheid diameter was more closely related to diameter (Saranpaa et al. 2000). Many studies also show no effect of radial growth rate on wood density. Zobel and van Buijtenen (1989), summarizing 59 studies on hard pines, reported that 59% of the studies showed no significant relationship between radial growth rate and wood density, 19% showed a large negative relationship, 15% showed a small negative relationship, and 7% showed a positive relationship. One must be cautious in labeling this a TRP, however, because density can be driven by changes in latewood proportion that could have more to do with tree spacing and foliage area, rather than changes in latewood or earlywood density (e.g. Cown 1973, 1974).

In hardwoods, there are too few studies to conclude which pattern is more common, but as with conifers, there are examples of both age and diameter having the strongest influence. Two projects studied the radial pattern of wood density in tropical hardwoods of different size but the same age: plantation trials in a secondary successional species in Brazil (de Castro et al. 1993), and cohorts of four self-seeded pioneer species in Costa Rica (Williamson and Wiemann 2010). In most cases in both studies, the small-diameter trees had significantly greater slope of the density on radial distance relationship than did the larger-diameter trees of the same cohort. Moreover, there were no significant differences between the densities of the outer wood in the smallest vs. the largest individuals in most of the species-stand combinations, indicating that density was more closely related to cambial age than to diameter. Studies have also compared the radial patterns of fiber length in trees whose radial growth rates were manipulated. In some studies (e.g. *Populus*, DeBell et al. 1998, and two *Eucalyptus* species, Kojima et al. 2009), fiber length was more closely related to cambial age than to diameter (Fig. 5.2b). However, Kojima et al. (2009) studied four taxa (two *Acacia* species and two accessions of one species of *Paraserianthes*, formerly *Albizia*) in which fiber length was more closely related to size than to cambial age (e.g. Fig. 5.2a).

4 Hypotheses for Why Trees Develop Radial Changes in Wood Properties

The TRP is sufficiently common for us to infer that it has evolved through natural selection to help the plants survive and reproduce. This inference is not necessarily correct, of course: patterns can develop randomly, or as incidental by-products of another function upon which selection has acted. An example of an incidental by-product is the color of wood, which is not thought to be of adaptive value itself, but to result incidentally from the materials of which wood is formed and by which it is chemically defended.

In this section we present three hypotheses for the evolutionary drivers of the TRP related to developmental constraints (*D*), maintaining adequate water distribution (*H*), and maintaining structural integrity of the tree subjected to its own body forces as well as external loads such as wind (*M*) (summarized in Table 5.2). We then give examples of comparisons to help us link specific structures and functions with the *H* and *M* constraints. Although we treat these hypotheses separately, it is unlikely that any plant was shaped through evolution solely by one of these types of constraint. The magnitude of different constraints surely vary among species, plant parts (e.g. stems vs. roots), and cambial ages, and in the xylem of individuals in different habitats. As pointed out by Zimmermann et al. (2004) in their review of hydraulic mechanisms, trees are most likely to have evolved ways to use all the available physical and biological avenues to maximize their chances of survival. This includes the ability to respond to both long-term and short-term environmental changes, where 'environment' refers to all external influences including climate, gravity, day length, latitude, terrain, soil properties, competition from other plants and predation—and, of course, silvicultural conditions related to regimes such as spacing, thinning, and pruning. For this reason, making inferences about which constraint was the most important may be difficult: disparate constraints and tradeoffs may dampen any signature of one constraint over another. Thus, we may learn the most about the adaptive significance of wood structure and function through comparisons of wood that developed in extremes of the *H* and *D* constraints—through manipulation of plants, or choosing plants living in extreme hydraulic or mechanical environments.

4.1 Development (D): Production of Sequentially More Optimal Xylem as Cambium Matures

Hypothesis D (developmental constraints) states that the TRP strategy is a by-product of the need for the young cambium to mature before it can produce the most adaptive wood (Table 5.2). This view suggests that a developmental constraint prevents the meristem from producing the 'good' wood from the beginning and that corewood has no particular adaptive value other than that the tree must produce it before its cambium matures to the point of producing the outerwood.

Table 5.2 Adaptive hypotheses for the purposes of the typical radial pattern (TRP) in secondary xylem of woody plants

Purpose of TRP	When young or small	When old or large	Structure and/or function involved
D: Byproduct of the need for the young cambium to mature before it can produce the most adaptive wood[a]	The cambial initials have derived from the apical meristem, where cells are small, and so they are constrained to produce small, short cells with high microfibril angle	By virtue of the number of cambial divisions already made, the initials have had opportunities to 'mature' gradually, making products that are now optimal for the tree	All characteristics, including morphological and physiological
H: To permit higher resistance to embolism in corewood, and to permit higher specific conductivity in outerwood[b]	Need to allow more negative pressure before hydraulic or mechanical failure because young axes have less readily available stored water (and thus, less buffering capacity) than outerwood, because of their smaller mass, and in young plants, smaller root system	Need more water transport capacity to compensate for increased resistance caused by friction over longer path and gravity on taller path	Cell lumen diameter, cell wall thickness, cell length, pit characteristics, MfA
M: To permit a higher critical strain in corewood, and to permit greater strength and stiffness in outerwood[c]	Small diameter trees need to be flexible to bend through large angles in response to applied loads (e.g. from wind), and to have a high critical strain so it won't fracture with this bending. Low stiffness wood also permits more effective stem re-orientation by the growth stresses associated with reaction wood	As trees increase in size, stem needs increased wood strength and stiffness in order to avoid failure from both self-weight from the stress produced when the stem is bent by applied wind or snow loads	MfA and wood density, which in turn affect critical strain, stiffness and strength

[a]Related to developmental (*D*) constraints on type of wood the young cambium can produce
[b]Related to maintaining hydraulic (*H*) integrity of the water distribution network
[c]Related to maintaining structural (*M*) integrity of the tree

This hypothesis can explain all characteristics that change along the radial transect, as long as the outerwood is presumed to have universally superior function over the corewood. The TRP reflects the decreasing intrinsic control and the increasing production of wood that is optimal for the tree (Fig. 5.4).

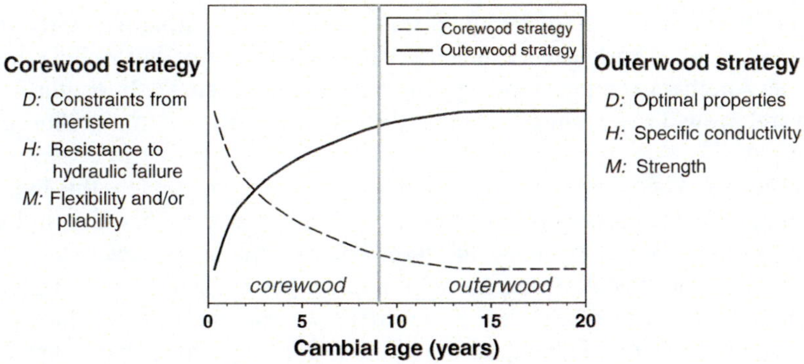

Fig. 5.4 Schematic diagram of hypotheses for why trees develop radial changes in wood properties, showing the changing strategy expressed in the corewood (*dashed curve*, *left axis*) and the strategy expressed in the outerwood (*solid curve*, *right axis*). See Sect. 4 for explanations of the developmental (*D*), hydraulic (*H*), and mechanical (*M*) hypotheses

A general understanding of the vascular cambium supports this hypothesis if the size and shape of the cambial initials constrains the size, shape, and ultrastructure of the cambium's products. The apical meristem lays down cells that differentiate to become the vascular cambium. These vascular cambium cells (the fusiform cambial initials) produce secondary xylem immediately exterior to the growth that had been laid down by the apical meristem. It is logical to expect that these first fusiform initials will be relatively short. Bailey (1920) showed that for branch and stem segments of 13 gymnosperm taxa the tracheid length was about the same length as (or up to 10% longer than) the cambial initials, although he did not survey wood immediately adjacent to the pith. Therefore, these first fusiform initial cells in conifers, if short, may be incapable of producing the extremely long axial cells that are found in outerwood, unless their products are capable of a large amount of intrusive growth. In angiosperms, however, Bailey (1920) reported that fiber tracheids were often more than twice as long as the fusiform initials, showing their ability to exhibit this high intrusive growth.

Hypothesis D is supported by data for some characteristics, particularly for the first several growth rings from the pith, which often appear to be the same size regardless of environment. For example, Christensen-Dalsgaard et al. (2008, discussed below) show a cross-section of a distal root 1 m from the bole, in which they observed that vessel diameters increased markedly for the first 2 mm, even though their calculations show no change in the mechanical strain across the time period during which this wood is developed. From this discrepancy the authors state that these early changes may well result from ontogenetic requirements rather than adaptive responses (either plastic or evolved). They did not estimate hydraulic needs for the small plants, however, so the patterns could have conformed to the hydraulic constraints hypothesis (below). Another example is that it is often noted that growth ring structure often 'takes a few rings to settle down.' For example, ring-porous woods may look diffuse-porous in the first several growth rings, and in conifers it is often much more difficult to distinguish earlywood from latewood in the early rings using either lumen diameter or cell wall thickness as criteria. Another example is the typical pattern of increasing cell lengths in the first several growth rings, suggestive that the cambium can only make short cells near the pith.

Hypothesis D is not supported by several kinds of evidence. First, this hypothesis assumes that the corewood characteristics are not of value. This assumption is violated by physiological evidence that corewood's hydraulic characteristics (such as its high resistance to embolism) are helpful for survival (e.g. Domec and Gartner 2002, see Sect. 4.2), and mechanical evidence that compression wood, which helps a stem re-orient when it is bent, is more common in corewood than outerwood (Pillow and Luxford 1937).

Secondly, structural evidence often shows that the cambium is capable of producing very abrupt radial changes in anatomy, suggesting that the cambial products are not entirely constrained. Compression wood can be detected within days of a stem's being bent (reviewed in Timell 1986, pp. 680–686). When individuals that transition from self-supported to semi- or totally-supported experience a change in their mechanical environment, they often exhibit coincident radial anatomical changes (Ménard et al. 2009). An example is shown with the tropical liana *Condylocarpon guianense*. Rowe and Speck (1996) present a cross-section in which the inner 0.9 mm of xylem is extremely dense and has vessels that average about 50 μm in diameter. Immediately adjacent is xylem that is not dense in which vessels average about 200 μm in diameter. This abrupt change in anatomy implies plasticity with very rapid cambial response.

Lastly, *Hypothesis D* is not supported by data for characteristics that largely are invariant across the radius. Root tracheids provide such an example: in several studies in conifers, the tracheids in the first rings adjacent to the pith are similar in length to those found in the outerwood (Sanio 1872; Matsumura and Butterfield 2001; Peterson et al. 2007). Peterson et al. (2007) reported that the tracheids in the first two growth rings from the pith averaged 3.8–4.7 mm in length for the *Pseudotsuga menziesii* roots examined; the same roots had a mean of 3.8–4.1 mm at 40–41 rings from the pith. Other studies show that conifer root wood produced in the first cm (Matsumura and Butterfield 2001, *Pinus nigra*, *P. radiata*) or the first 8–13 years (Sanio 1872, *Pinus sylvestris*) has very long tracheids. In hardwoods, Sun and Suzuki (2001) reported a lack of radial variation in vessel element and fiber lengths for two of the five mangrove species investigated; the main environmental difference was that the two species that lacked the radial gradient lived in sites with less wind, less force from shifting tides, and less inundated sites than did the other three species. Certain rosette plants also lack a radial gradient in axial cell length (e.g. Carlquist 1962). This research shows that the young cambium is not necessarily canalized into producing short axially-oriented cells. More research is needed to help us tease apart the relative contributions of intrinsic controls vs. plastic responses to environment for the many traits that do exhibit radial changes throughout the secondary xylem.

4.2 Hydraulics (H): Maintaining Adequate Water Distribution as Cause for the TRP

Hypothesis H (hydraulic constraints) states that the TRP strategy permits corewood to withstand high negative pressures and allows the outerwood to have high specific conductivity. There is ample evidence of a tradeoff between resistance to embolism and specific conductivity (Domec and Gartner 2002; Rosner et al. 2008; Domec et al. 2008), depending on the scales compared (Meinzer et al. 2010). The rationale for *Hypothesis H*, detailed below, is that young/small axes have little access to stored water, and they may not have a steady supply of water from their root systems, suggesting plants could benefit from allowing high negative pressures in the corewood before conduits fail mechanically or hydraulically. Large/old axes, in contrast, can buffer the water potentials with stored water and soil water to a greater extent (Scholz et al. 2011). However, as trees grow, resistance to water transport will increase from both the longer and taller path, suggesting that old/large trees could benefit from high specific conductivity in the outerwood (Table 5.2). The TRP reflects the changing strategy from hydraulic safety to efficiency as the tree increases in age/size (Fig. 5.4).

4.2.1 Corewood

First, young/small trees may need to operate with more negative water potentials than old/large trees because they are likely to have less developed root systems and therefore less certainty of access to a steady water supply. Secondly, young/small plants have been shown to have less water storage, and thus to be less capable than larger plants at modulating their water tensions (Scholz et al. 2011). Phillips et al. (2003) showed that young trees can have lower volume of sapwood per leaf area, and that there is a decreasing reliance on stored water with increasing plant size in two coniferous and one hardwood species. Cermak et al. (2007) showed that in old-growth *Pseudotsuga menziesii* trees, most of the stored water comes from stem sapwood, with the largest amount coming from the lower stem. The spikes in tension gradient were buffered to a larger extent lower in the stem (in the outerwood) than higher up (in the corewood). Thirdly, if the corewood is located higher on the stem than is the outerwood, the upper position will experience greater xylem tension because of the resistance from the greater path length, and because of the effect of gravity acting on the taller water column.

Hydraulic failure in the corewood can occur through air-seeding of embolisms or through conduit collapse. In either case, the driving force is a water tension that exceeds some critical value, and plants need be protected from both failure modes simultaneously. Physiological and acoustic measurements in conifers show that corewood (Domec and Gartner 2002; Domec et al. 2009; Rosner et al. 2006, 2008, 2009) and branches (e.g. Domec and Gartner 2002; Mayr et al. 2003; Dunham et al. 2007) are often more resistant to embolism than is outerwood. Anatomical evidence also suggests this higher resistance to embolism and/or collapse in corewood than outerwood. Hacke et al. (2001) calculated a wall reinforcement factor to explain collapse of a simplified conduit, using engineering equations for a plate that has the width of the cell lumen diameter (*t*) and the thickness of the double cell wall (*b*). This wall reinforcement factor, $(t/b)^2$, is positively correlated with conduit collapse, but it is also positively correlated with the resistance to embolism (Hacke et al. 2001). Domec et al. (2009) showed that two coniferous species maintain higher safety factors to air-seeding of their xylem than to collapse, but that the values relative to one another are similar (at about 2:1) throughout the plant. Moreover, these two species had higher $(t/b)^2$ in corewood than in outerwood. The higher MfA of corewood (i.e. running at more of an angle to the cell axis than in outerwood) may also contribute to the higher collapse-resistance of corewood; a model such as that by Innes (1995) could be parameterized to test the sensitivity of collapse to MfA. Moreover, resistance to embolism is positively correlated with structural features that are observed more frequently in corewood than outerwood, such as small pit apertures, low pit frequencies, and narrow tracheids (Pittermann and Sperry 2003; Domec et al. 2006) and low capacitance (water storage, Meinzer et al. 2009; Barnard et al. 2011).

On first glance, some of the research on internal checking appears to run counter to *Hypothesis H*, but in fact, it may not. There has been much research on breakage in xylem in which the wood separates within a growth ring along the radial plane, termed internal checking, intra-annual cracks, frost cracks, and collapse. It has been studied from points of view of wood utilization, material science, dendrochronology, tree nutrition, and tree physiology. Some studies report that these breakages occur more frequently in the inner part of the cross-section than in the outer part, and that they occurred (based on resin or growth intrusions) when the wood was sapwood rather than heartwood (Cherubini et al. 1997; Pang et al. 1999, discussed in Ball et al. 2005). When it occurs in wood that is at moisture contents above the fiber saturation point, the stress is generated by transpiration or wood-drying and is transferred to the inside of the conduit walls where the water tension pulls inward (e.g. Innes 1995; Booker 1999; Grabner et al. 2006). If it pulls to failure of the cell wall, it will cause a check or crack. Rosner et al. (2010) called this stress the tension stress to distinguish it from shrinkage stress (which occurs only in very rare pockets of the stem of living plants where the wood is drier than the fiber saturation point). It is likely that the internal checking, if more prevalent in corewood than outerwood, is caused by additive factors beyond the water tension in the wood, which may not differ significantly from the corewood to the outerwood (e.g. Domec et al. 2009). Through elegant experimentation, Rosner et al. (2010) deduced that the cell wall deformation from tension stress was much lower in corewood than outerwood for all water tensions more negative than about 1 MPa in *Picea abies*. The lesser deformation in the corewood was related to its higher wall reinforcement, $(t/b)^2$. Thus, at a given tension stress, the corewood will deform less than the outerwood. All things being equal, a modeling approach from material science showed that wider growth rings (as are usually found in the corewood) will be more prone to this internal checking (Nairn 2010). Other studies have shown that internal checking is more prevalent in wood with lower earlywood density (Grabner et al. 2006) or with altered cell wall infrastructure (Putoczki et al. 2007) or chemistry (Downes and Turvey 1986).

4.2.2 Outerwood

Old/large trees need to have high specific conductivity (K_s) to meet their water transport demands. As shown above, the corewood will have lower K_s because of the heightened importance in corewood of embolism resistance, and the negative relationship between resistance to embolism and K_s. Specific conductivity, K_s (kg s^{-1} m^{-1} MPa^{-1}), is defined in Eq. 5.1 as

$$K_s = F\,L\,P^{-1} A_s^{-1} \tag{5.1}$$

in which F is water flow rate (kg s^{-1}), L is segment length (m), P is the pressure difference across the segment's length (MPa), and A_s is the segment's sapwood cross-sectional area (m^2). Higher (K_s) in outerwood is almost always observed.

Leaf-specific conductivity (K_l) is similar to K_s but it is normalized by the distal leaf area (A_l, that it supplies with water) rather than by the sapwood area (across which the water flows) (Eq. 5.2):

$$K_l = F\,L\,P^{-1} A_l^{-1} \tag{5.2}$$

Therefore,

$$K_l = K_s A_l^{-1} A_s \tag{5.3}$$

Leaf-specific conductivity (K_l) typically increases greatly with stem diameter (Tyree and Zimmermann 2002) by a factor of 3–10. In an extreme example K_l increased by a factor of 300 from outerwood at the tip to the corewood at the base (Tyree et al. 1991; Tyree and Ewers 1991). Leaf area/sapwood area ($A_l A_s^{-1}$) often (but not always) decreases as trees grow, by a factor of 20% over a height range of 50 m (McDowell et al. 2002). Assume for this exercise that there is no change. If a tree were constructed entirely from wood with the K_s of corewood, then to achieve a 300-fold increase in K_l from its smallest to its largest stem diameter, it would need 300 times more sapwood than observed— i.e. have a basal stem diameter of 6.9 m rather that 0.4 m—or else maintain only 1/300th of its leaf area. Clearly, trees benefit by having an increase in K_s from the corewood to the outerwood.

Data are largely supportive of *Hypothesis H*, that K_s increases from corewood to outerwood (Spicer and Gartner 2001; Domec and Gartner 2002), and that these are related to the tree's changing hydraulic needs as it grows from young/small to old/large. While this hypothesis touches on functional behavior caused by the morphology of cells (length, diameter), inter-conduit pits, and their membranes, it is not centered on MfA or density, both of which are known to change in the TRP. The following mechanical hypothesis relates to MfA and density, which in turn affect wood stiffness and strength.

4.3 Mechanics (M): Maintaining Adequate Structural Integrity as Cause for the TRP

Hypothesis M (mechanical constraints) states that the TRP strategy permits small stems (and branches) to bend without breaking in response to applied loads such as those from wind and snow; provides old/large trees with the strength needed to withstand the forces due to their self-weight as well as from applied loads such as from wind, ice, and snow; and facilitates more effective use of growth stresses to re-orientate the stem (Table 5.2). The strength and rigidity of the stem are functions of its size (diameter) and wood properties (Niklas 1992), and the tree is genetically programmed to adjust both of these in order to maintain adequate mechanical stability for a given situation. A number of studies have investigated the scaling that occurs between diameter and height (or length in the case of branches) to ensure that a tree maintains its structural integrity (e.g. Greenhill 1881; McMahon and Kronauer 1976; King and Loucks 1978; Wilson and Archer 1979; Bertram 1989; Niklas 1994; King 2011). In the case of branches, Bertram (1989) suggested that the more slender form of peripheral branch segments enables them to bend in response to wind and to reduce the associated drag, while the more robust form of non-peripheral segments is due to the need to resist greater bending loads. Material properties also play an important role, particularly if structural integrity is to be maintained using a minimum of material. While some studies have considered material properties in scaling analyses (e.g. Anten and Schieving 2010), in most studies the radial variation in these properties within a tree has been ignored. We hypothesize that the TRP is necessary to enable a tree to change its strategy from compliance to resistance in terms of its response to applied loading as it increases in size (Fig. 5.4).

4.3.1 Corewood

Stems and branches of small diameter can avoid large wind loads by becoming streamlined and therefore presenting a smaller frontal area to the wind (Bertram 1989; Vogel 1989; Steinberg 2002). Not only is this aided by the increased flexibility of peripheral branch segments (Bertram 1989), but also by the increased flexibility of the upper part of the stem in more mature trees (Hedden et al. 1995). Therefore, it is an advantage for the tree to be flexible below a given diameter over the whole height of the tree. The alternative, a stem of small diameter that remains relatively rigid in the wind, would require wood with higher stiffness and strength to avoid breakage due to the larger wind (or other) force acting on it. These concepts can be illustrated by using a simple model where a tree stem or branch is represented as a cantilever beam of uniform diameter (d) and modulus of elasticity (E) along its length subjected to a lateral point load (P) at its free end. The deflection x at any point z along the height (H) of the idealized stem is given by:

$$x(z) = \frac{32P(z^3 - 3Hz^2)}{3\pi Ed^4} \tag{5.4}$$

It is readily apparent from Eq. 5.4 that for a given applied force (*P*), an increase in stem diameter (*d*) results in a reduced tree deflection ($x(z)$). Furthermore, for a given diameter the amount that a tree deflects under an applied load is inversely proportional to the modulus of elasticity (*E*) of the wood. Therefore, wood with a lower value of *E* would allow a tree stem to deflect more. *E* is strongly dependent on microfibril angle and density (e.g. Cave and Walker 1994; Barnett and Bonham 2004; Lachenbruch et al. 2010); wood with a high MfA and low density has low stiffness and in softwoods Cave and Walker (1994) reported that *E* can increase by a factor of 3–5 during the first 30 years of growth.

The large deflections permitted by this strategy will result in an additional bending moment due to the offset mass of the stem and crown; when the tree is small, these second-order effects due to the displaced mass are comparatively small and it is at lower risk of mechanical instability caused by such large deflections. However, for this strategy to be successful the corewood requires a high fracture strain (ε_{max}) to allow the tree to bend through a large angle without rupture of the wood fibers. The strain in the outermost part of the stem depends on the magnitude of the deflection (*x*) as well as the length over which the deflection is acting and the stem diameter (*d*). For the simple case of the cantilever beam with uniform circular cross-section, the surface strain (ε) at the base resulting from a deflection (*x*) at the tip is given by:

$$\varepsilon = \frac{3xd}{2H^2} \tag{5.5}$$

By re-arranging Eq. 5.5, it is apparent that the maximum amount a tree can deflect before failure is governed by the fracture strain of its wood. Lichtenegger et al. (1999) advanced this further and showed theoretically that in order for the tip of a stem or branch to bend by a certain amount without buckling or fracture of the wood requires the optimization of $\varepsilon_{max}^2 E$. While data on the radial variation of fracture strains are rare, particularly for wood in the saturated condition found in a living tree, Reiterer et al. (1999) present data for Norway spruce which show an almost fivefold increase in $\varepsilon_{max}^2 E$ with a change in MfA from 5° up to 30°. Therefore, the high MfA found in the first few growth rings from the pith, would allow corewood to undergo large amounts of deformation with a lower risk of mechanical failure. This strategy is assumed to be further aided by the higher spiral grain angle often found in corewood, which enables considerable energy absorption to occur in young stems as they bend and twist in the wind (Kubler 1991; Skatter and Kucera 1997; Schulgasser and Witztum 2007).

4.3.2 Outerwood

As a tree gets larger, body forces (i.e. those associated with self-weight) will increase, but the wind load will also increase as the crown increases in size and the tree increases in height (larger force × longer moment arm = larger bending moment at the base of the stem) (Niklas 1992). Lichtenegger et al. (1999) showed that for the case of an applied lateral load (F_L), the height that a tree of a given diameter can grow is limited by the strength of its wood (σ_{max}), i.e.

$$H < \frac{2I}{dF_L}\sigma_{max} \tag{5.6}$$

where I is the second area moment of inertia ($= \pi d^4 / 64$ for a beam with circular cross section). This is supported by empirical observations (e.g. Putz et al. 1983) showing that trees which fail by stem fracture in wind storms have less strength (as well as lower wood density and stiffness) than trees that are uprooted. For the case of a vertical compressive force (F_G), Lichtenegger et al. (1999) showed that the maximum height that a vertical column of a given diameter can reach before buckling occurs is a function of *E*, i.e.

$$H < \frac{\pi}{2}\sqrt{\frac{EI}{F_G}} \tag{5.7}$$

Clearly, these two situations do not occur independently. When a tree is displaced by a lateral wind load, there will be an additional bending moment due to the offset mass of the stem and crown, termed the p-delta effect (Megson 2005). This can lead to instability as the p-delta effect increases the deflection of the tree, which in turn further increases the p-delta effect. This is particularly the case for tall, slender trees. In this situation, the tree is on the verge of falling over when the increase in deflection means that the toppling moment from the p-delta effect just exceeds the restoring moment provided by the rigidity of the stem. Increased strength is required to resist this additional moment due to the p-delta effect, and additional rigidity is required to reduce the risk of instability. The first requirement can be satisfied if wood strength increases as tree size increases. This is consistent with measured radial profiles of wood strength that show a substantial increase from pith to bark (e.g. Bendtsen and Senft 1986; Evans II et al. 2000).

When a tree adds new wood to the outside of the stem the rigidity of the stem (*EI*) increases not only because diameter increases, but also because this new material is as stiff or stiffer than material previously laid down (i.e. there is an asymptotic radial increase in wood stiffness up to a maximum value). Because this new material is added furthest from the neutral axis of the stem (assumed here to coincide with the pith), the overall stiffness of the stem is highly dependent on the stiffness of the outermost material (Lundstrom et al. 2008). If the tree only added wood with the same value of *E* as found in the first few growth rings, it would need to add considerably more material to achieve the same overall rigidity. However, there is an upper limit to the value of *E* that can be achieved, which is dependent on the product of the relative proportion of cell wall material in a tracheid and the *E* value of this cell wall material. The relative proportion of cell wall material in a tracheid will be limited by hydraulic conductivity requirements, while the *E* value of this cell wall material will be limited by MfA; once MfA reaches a value of around 10°, further reductions in MfA will only have a relatively small effect on cell wall stiffness. From a biomechanical perspective, it is also unlikely to be advantageous for a tree to keep producing increasingly stiff wood as it grows. This is due to the requirement for strain compatibility in the longitudinal direction within the stem (Archer 1986), which means that the bending stresses due to applied loads such as wind will be resisted by the stiffest material. If the tree continues to add successive increments of increasingly stiff material, then the bending stresses will be transmitted through the outermost thin layer of high stiffness material. This situation can be avoided if E remains approximately constant with increasing radial position after it has reached its maximum value; such a radial profile of wood stiffness will ensure that bending stresses are distributed over a larger area.

The radial profile of wood stiffness typically found in trees also enables trees to more effectively use their growth stresses to re-orient their stem or branches (Alméras and Fournier 2009). Such re-orientations are either a negative gravitropic response (e.g. bending of the stem to correct a lean or to induce upward inclination and/or minimize downward inclination of a branch), or a phototropic response to asymmetric light conditions (Matsuzaki et al. 2006). In the case of tree stems and branches that have finished elongating, the gravitropic response occurs via the internal bending moment generated by the asymmetric pattern of growth stresses in their wood (Archer 1986; Alméras and Fournier 2009). This asymmetric growth stress pattern is due to the presence of reaction wood. For example, in conifers compression wood is formed in response to a lean (Timell 1986). Compression wood has an unusually thick S2 cell wall layer with a very high MfA so that on maturing, the cell will tend to lengthen (Archer 1986; Schulgasser and Witztum 2007), resulting in a longitudinal compressive growth stress which a tree uses to "push" its leaning stem back to the vertical position. The bending moment provided by the addition of a new growth layer is a function of the growth stress differential generated by the asymmetric distribution of reaction wood in this layer and the effective distance that this acts from the neutral axis of the stem (Timell 1986). In turn, the change in curvature that this moment is able to generate is inversely related to the stem diameter and is positively related to the ratio of E in the new growth layer to the effective stiffness of the inner section of the stem (Alméras and Fournier 2009). This ratio increases with an increasing radial gradient in wood stiffness, so that trees with a large radial gradient in wood stiffness are able to use their growth stresses more effectively than trees having a smaller gradient.

While more detailed studies are required to fully test *Hypothesis M*, the principle that the radial changes in wood properties from corewood to outerwood are related to the tree's changing requirements for mechanical support as it grows from young/small to old/large is supported by other published studies (e.g. Bertram 1989; Woodcock and Shier 2002, 2003). Woodcock and Shier (2002, 2003) suggest that radial increases in wood density are characteristic of early-successional species and reflect a pattern of growth in which there is a shift in allocation of resources from growth in height, associated with production of low-specific gravity wood, to greater structural reinforcement of the stem, associated with the production of denser wood.

If trees germinate in the understory, this additional reinforcement may be required as they reach the canopy and are subjected to increasing stresses associated with wind loading. The strongest radial gradients occur in tropical pioneer species that have very low density wood near the pith (Wiemann and Williamson 1988). Woodcock and Shier (2002) suggest for those species in which there is a radial decrease in wood density, that the structural reinforcement can be achieved by diameter growth. This latter argument is also supported by Larjavaara and Muller-Landau (2010) who argue that a large stem of low-density wood can have greater strength at lower construction cost than a thinner stem of high-density wood. However, this conclusion is based on the assumed relationships between wood stiffness, strength and density. Because wood stiffness and strength are functions of both density and MfA (Evans and Ilic 2001; Via et al. 2009; Lachenbruch et al. 2010), radial increases in these two properties could still occur in species showing a radial decrease in density if the reduction in MfA was sufficient.

4.4 Comparative and Experimental Approaches to Inferring Hydraulic and Mechanical Causes of the TRP

The diversity of wood patterns and environments in nature suggest that with careful comparisons, we may be able to infer the relative importance of developmental, hydraulic, and/or mechanical constraints on the TRP. The clearest comparisons for inferring the relationship of xylem structure to a type of constraint may be manipulations that develop, or natural comparisons that display, extreme levels of an environmental factor that one could categorize as hydraulic or mechanical. This approach can be used for different species, genotypes, individuals, and plant parts of individuals. Alternatively, one can start by identifying species, genotypes, individuals, and plant parts that differ in anatomy or properties, and work the other direction to infer drivers for these differences. Both approaches will help us avoid study of patterns in which the structure is a rather neutral reflection of multiple factors resulting in a compromise wood structure (either plastic or fixed but evolved) that bears little clear signature of any one driving force.

Note that these experiments need to be designed to study the radial variation, and not simply the ecological wood anatomy of the trait itself. The former is rarely done (with the exception of work in tree breeding programs for specific cases in commercially-important woods that exhibit ‘problems’ with corewood), whereas the latter is a mainstay in physiological ecology. For example, most studies on within-plant variation in structure and function only characterize the outer wood of a given position (i.e. Dunham et al. 2008; Lintunen and Kalliokoski 2010) rather than the radial variability. Many of the studies only characterize selected factors of importance to either hydraulics or mechanics, but not both, and they may not be the characteristics of most interest for distinguishing among the hypotheses in Table 5.2. For example, Peterson et al. (2007) looked at radial patterns in tracheid length and wood density for trunks, root, and branch of four 65-year old *Pseudotsuga menziesii* trees to learn if there is a TRP in all three plant parts. They concluded there are distinct radial patterns, but they did not ask whether hydraulic or mechanical issues appeared more important in one location than another. Several studies have characterized the TRP of hydraulic and mechanical characteristics, but have not used another factor such as water availability or diameter to tease out the adaptive significance of the observed patterns (e.g. Mencuccini et al. 1997). Therefore, targeted research on within-plant variability is necessary to help us understand why plants produce the observed radial patterns in wood structure and plant function.

4.4.1 Comparisons of Species or Genotypes That Vary in Radial Pattern

One experimental approach is to compare taxa that differ in the radial pattern of a set of wood traits, and ask (correlatively) what characteristics of the species or environment may be responsible for these differences. On a large scale, consider softwoods vs. hardwoods, softwoods having much greater radial variation in wood density, cell length, and microfibril angle than hardwoods. Both taxa can attain similar sizes. At first glance, this suggests that the TRP is unlikely to be related to mechanical considerations. However, the radial pattern of growth stresses also differs greatly between softwoods and hardwoods (Fournier et al. 1990), and in the outer wood, softwoods have much lower growth stresses than hardwoods, both of which could compensate for the mechanical considerations above, again, putting mechanics as a possible adaptive driver of the TRP. On a smaller scale, many tests of the hydraulic and mechanical hypotheses could be undertaken.

There is also much potential in comparing between genotypes that show variation in their radial pattern of a target characteristic. Clonally-reproduced plant material can provide replicate individuals within a treatment. One could study the correlations of the TRP of wood density, MfA, propensity to produce reaction wood, or height/diameter ratio, to learn whether they track environment, appear to have intrinsic control, or show signs of doing both simultaneously. There are many excellent and comprehensive studies with wood properties of commercial value, particularly with components of wood density in pines (e.g. Cown and Ball 2001; Cown et al. 2002), that have been interpreted in part to partition the variation into intrinsic ('genetic') and environmental ('phenotypic') variation. Breeding programs that seek to decrease the type and/or quantity of lignin, thereby producing wood with a lower stiffness, could also be used in this type of research. Increasingly, we will be able to use genomics approaches for studying these vegetative changes (Greenwood et al. 1989; Brunner and Nilsson 2004; Willmann and Poethig 2005; Cato et al. 2006; Paiva et al. 2008).

4.4.2 Comparisons of Individuals or Species from Different Environments

Another approach is to compare radial patterns for individuals or species that have contrasting mechanical or hydraulic environments to infer the role of those radial differences on function. These environments can be found or produced experimentally. Mechanical environments may derive from wind, ice loads, self-weight, mechanical stimulation, effects of neighbors, bending, compression, tension, addition of weights, reduction of force with pulleys, inclination at various angles, etc. Likewise, hydraulic environments may derive from drought, freezing, leaf area/sapwood area, root area/leaf area, short vs. long paths for the water transport, wood with low vs. high hydraulic resistance, nutrient status/soil fertility, etc. The plant material can also involve clones (common in forestry, agriculture, and horticulture) to provide closely-related plants that differ in an hydraulic aspect (such as leaf area/sapwood area) or a mechanical aspect (such as the tendency to lean). If groups growing in different hydraulic and mechanical environments have similar radial wood patterns, one can infer that the constraint is intrinsic (such as for earlywood and latewood densities in ten families of 22-year old *Pinus radiata*, grown at seven sites, Cown and Ball 2001). If, for example different hydraulic environments are associated with different radial patterns, one can interpret that the pattern is related to a putative hydraulic function. Another twist to this type of study would be to study radial patterns in monocarpic vs. polycarpic species to ask if there is an effect of reproductive effort on the structure and function; the monocarpic species would have no reproductive effort until the end of their lives (Thomas 2011).

Many observations have been made on individuals or species that differ hydraulically or mechanically from the typical plant. Mature lianas, for example, typically provide little of their own mechanical support, but some individuals may remain self-supporting (e.g. Gartner 1991a, b). A comparison of free-standing and supported individuals of *Croton pullei* indicated that the mechanical environment constrained growth form: if an individual had to support itself, its wood was stiff throughout the radial transect, but if an individual was supported by external objects, there was a radial progression (or even a switch) to having less stiff (Gallenmüller et al. 2001) and less hydraulically conductive wood (Carlquist 1991). The less stiff outerwood contrasts with the pattern typical of softwoods, but the more conductive outerwood is consistent with what is seen in softwoods (Spicer and Gartner 2001).

4.4.3 Within-Plant Comparisons, with Known Environmental or Radial Differences

Yet another approach for inferring whether the TRP is caused more by mechanical or hydraulic constraints is to study radial variation among plant parts within an individual where those plant parts are known to have contrasting mechanical or hydraulic environments. An example in tropical hardwoods compared radial patterns of hydraulic conductivity and longitudinal strain at several locations in three species of buttressed trees (Christensen-Dalsgaard et al. 2008). Theoretical conductivity was estimated from measurements of vessel frequency and vessel diameter on wood samples from the upper trunk (above the buttress), from the buttress near its top where it attaches to the trunk (proximal buttress), and from the buttress near its base and far from the trunk (distal buttress). They estimated the strain that would have been present when the wood was developing for the inner, middle, and outer radial positions of the proximal and distal buttresses. The proximal buttress had very high levels of strain that were similar at all radial positions. The distal buttress, in contrast, had very low strain in the inner position, higher strain at the middle position, and much higher strain at the outer position. These strains, however, were still much lower than the strain in the proximal buttresses. (They were estimated at 15% and 35% of the strain values of the middle and outer positions, respectively, of the proximal buttress). In the trunk above the buttresses in all three species, theoretical conductivity increased markedly from the inner to the outer wood of the trunk, as expected for the TRP. Theoretical conductivity showed no meaningful radial change in the proximal buttresses, but increased markedly from the inner through middle and then outer radial position for the distal buttresses. The research suggests that the radial variation in hydraulic characteristics was induced by the mechanical environment, with very different hydraulic patterns at these three locations.

Just as the buttresses can have quite different mechanical demands at different locations and cambial ages, many lianoid and hemiepiphytic species also will have within-plant variation in mechanical demands. Lianas, for example, are often self-supporting when young. This pattern is reflected in large radial changes in wood anatomy (reviewed in Carlquist 1991) including dense wood with small vessels near the pith, and new wood of lower density (Caballé 1998) and dramatically lower stiffness when the plant becomes supported (Rowe and Speck 1996; Ménard et al. 2009). Rowe and Speck (1996) noted that the properties did not progress smoothly from bottom to tip of an axis, giving evidence that the radial changes are functional responses to the mechanical environment, rather developmentally-induced trends.

Domec and Gartner (2002) made a within-plant comparison of safety factors to embolism and buckling in the temperate softwood *Pseudotsuga menziesii*. This study examined trunkwood from 4- and 10-year old saplings and 98-year old trees. Samples from the outer sapwood were tested for vulnerability to embolism, and the authors used the air entry point for calculations of safety factor, which is the inflection point early on the sigmoidal vulnerability curve at which wood has a sudden decrease in conductivity. Density of this outer sapwood was also determined, empirical relationships with density were used to estimate modulus of elasticity (E), and this estimated E was used with tree geometry to estimate the critical buckling height of the tree. The safety factor for the air entry point was low throughout the tree, and hovered near unity, meaning that the trees were on the verge of becoming embolized. In contrast, the safety factor for buckling was much higher (ranging from 1.5 to 10), suggesting that the trees were at greater risk from hydraulic rather than mechanical failure. Moreover, simplistic modeling of corewood in the place of outerwood and outerwood in the place of corewood showed much larger effects of wood type for hydraulic safety factors than for mechanical safety factors. This result is suggestive that the selection pressure is currently higher for hydraulics than mechanics. It may have previously evolved in such a way that the mechanical challenges were largely met.

A further type of within-plant comparison could involve study of the radial patterns within the sequentially-produced ramets of plants such as *Rubus alceifolius*, which make a series of increasingly robust self-supporting shoots that increase markedly in pith area, stem cross-sectional area, and internode length (Baret et al. 2003). As one shoot begins to wither, the next shoot is developing. The interpretation is that the first shoot is water-limited, but that sequentially-produced shoots are less water-limited. The growth culminates with the production of a shoot that initially is self-supporting. It grows very long, and develops vine-like mechanical (and probably hydraulic) characteristics at its developing and arching tip. One could predict that the first shoot will have radial wood properties that are adapted to producing a small upright stem, and each subsequently-produced ramet could have a radial pattern reflective of less hydraulic constraint. Another system of interest could be original axes vs. re-sprouts (such as after coppicing or a burn, e.g. Utsumi et al. 2010). One could expect the re-sprouts to have much more water available/leaf area, and so a much lessened hydraulic constraint. What would this signature be on the radial wood pattern?

5 Conclusions

Woody plants experience very different environments as they get older and larger. In most cases they display pre-set (ontogenetic) or plastic changes in their xylem's structure and function. It is presumed that the observed characteristics of corewood and outerwood mostly represent evolutionary 'solutions' to the particular design tradeoffs facing the individual as it gets older and often larger. This chapter has shown that these solutions may change with size and age, and they are more varied than has been widely understood. Until recently, much of the research on these radial patterns has focused on characteristics of importance for wood utilization, and perhaps for that reason, there was an underlying assumption that corewood, which is more flexible and weaker than outerwood in the axial direction, is the best that a young cambium can make, and that as the cambium matures, it can produce the stiffer, stronger xylem that is of higher quality for the tree as well as for humans. The view in this paper is that the radial progression of structure and function has evolved to increase the chances of survival of the whole plant at all stages of vegetative growth. In some cases the TRP will have evolved more for hydraulic than mechanical purposes and in other cases it will have evolved for mechanical rather than hydraulic purposes. The extraordinary plasticity of the cambium ensures that the tree can respond rapidly to environmental variation by favoring one or more mechanisms over others, thereby greatly altering hydraulic and mechanical functionality. This view is an extension of Larson's (1962) prescient work in which he stated that we can only advance our understanding of wood quality through looking at how the environment affects the plant biology, rather than the wood directly.

This paper shows that the particular radial patterns are variable in hardwoods vs. softwoods, and among growth forms, species, environments, and plant parts. There are sets of traits that are most common for a given category of plants (such as plastic response to the mechanical environment in semi-self-supporting species), but the literature is scant on some of these categories and so our knowledge is quite incomplete. The paper also shows that there are numerous examples of changes that are determined ontogenetically and other examples in which the changes are plastic; both types of example can be found in softwoods and hardwoods. With what is now known, it is not easy to predict whether changes are intrinsically or plastically controlled, with a few exceptions (such as the tendency for MfA and tracheid length in aboveground stems to be under intrinsic control). It is important to note that while this paper focuses on radial changes, there are also changes that occur with height that were considered beyond our scope.

The paper presents three frameworks. The first framework (Sect. 3.1) discusses the possible triggering mechanisms responsible for the production of the radial changes in wood. Elucidation of these proximal mechanisms could be very useful for management of forests to increase the value of logs and wood. The second framework (Sect. 3.3) delineates a range of abiotic, developmental, mechanical, and architectural attributes that are likely to be similar, and another set of attributes that are likely to differ, between same-aged small and large plants, and same-sized young and old plants. This framework shows that no selective factors stand out above others to help us interpret the TRP. The third framework (Sect. 4) introduces three hypotheses for why woody plants have evolved a TRP. Our intention was to illustrate that there are numerous hypotheses for patterns, and, indeed, we would not expect one hypothesis to describe the functioning of a plant at all times. With careful comparisons, we should be able to better elucidate the conditions during which different types of radial variation are adaptive.

These changes in xylem structure and function are not trivial. They make substantial contributions to the success or failure of the plant. We will benefit greatly from an understanding of the physiology and structure of xylem as humans continue to impact the environment; as natural plant ranges are disrupted and more plants are left with maladapted traits in their altered environments; as economic pressures push foresters to both accelerate growth and harvest trees younger; as processing technologies for wood change the material requirements for products; as agricultural practices change in response to economic factors, availability of natural resources, prevalence of competing vegetation, pathogens, and herbivores; as industrial agriculture and horticulture increasingly develop new clonal materials that need to be evaluated; as new uses for plants are adopted such as for biofuels, carbon sequestration, and as chemical factories; and as scientists increasingly relate gene function to development, structure, and function. The appreciation of the diverse functions of xylem and its heterogeneity will help physiologists move forward on many fronts. The vascular cambium that produces xylem is also producing phloem (Mencuccini et al. 2011), and so future work at the whole plant, whole lifespan level will benefit from the study of xylem and phloem together, at developmental, anatomical, and ecophysiological levels.

Acknowledgements We thank Dave Briggs, Rowland Burdon, Dave Cown and an anonymous reviewer for comments on the manuscript, and Dan Ridley-Ellis and other colleagues with whom we had many useful discussions.

References

Abdel-Gadir AY, Krahmer RL, McKimmy MD (1993) Estimating the age of demarcation of juvenile and mature wood in Douglas-fir. Wood Fiber Sci 25:242–249

Alméras T, Fournier M (2009) Biomechanical design and long-term stability of trees: morphological and wood traits involved in the balance between weight increase and gravitropic reaction. J Theor Biol 256:370–381

Amarasekara H, Denne MP (2002) Effects of crown size on wood characteristics of Corsican pine in relation to definitions of juvenile wood, crown formed wood and core wood. Forestry 75:51–61

Anten NPR, Schieving F (2010) The role of wood mass density and mechanical constraints in the economy of tree architecture. Am Nat 175:250–260

Archer RR (1986) Growth stresses and strains in trees. Springer, New York, 240 pp

Baas PR, Schmid R, van Heuven BJ (1986) Wood anatomy of *Pinus longaeva* (bristlecone pine) and the sustained length-on-age increase of its tracheids. IAWA Bull 7:221–228

Bailey IW (1920) The cambium and its derivative tissues, II. Size variations of cambial initials in gymnosperms and angiosperms. Am J Bot 7:355–367

Ball RD, McConchie MS, Cown DJ (2005) Evidence for associations between SilviScan-measured wood properties and intraring checking in a study of twenty-nine 6-year-old *Pinus radiata*. Can J For Res 35:1156–1172

Bannan MW (1967) Anticlinal divisions and cell length in the conifer cambium. For Prod J 17:63–69

Baret SE, Nicolini E, Le Bourgeois T, Strasberg D (2003) Developmental patterns of the invasive bramble (*Rubus alceifolius* Poiret, Rosaceae) in Reunion Island: an architectural and morphometric analysis. Ann Bot 91:39–48

Barger RL, Ffolliott PF (1976) Factors affecting occurrence of compression wood in individual ponderosa pine trees. Wood Sci 8:201–208

Barnard DM, Meinzer FC, Lachenbruch B, McCulloh KA, Johnson DM, Woodruff DR (2011) Climate-related trends in sapwood biophysical properties in two conifers: avoidance of hydraulic dysfunction through coordinated adjustments in xylem efficiency, safety and capacitance. Plant Cell Environ 34:643–654

Barnett JR, Bonham VA (2004) Cellulose microfibril angle in the cell wall of wood fibres. Biol Rev 79:461–472

Bendtsen BA (1978) Properties of wood from improved and intensively managed trees. For Prod J 28:61–72

Bendtsen BA, Senft J (1986) Mechanical and anatomical properties in individual growth rings of plantation-grown eastern cottonwood and loblolly pine. Wood Fiber Sci 18:23–38

Bertram JEA (1989) Size-dependent differential scaling in branches: the mechanical design of trees revisited. Trees 4:241–253

Booker RE (1999) Internal checking and collapse – which comes first? In: 4th IUFRO wood drying conferences, Rotorua, New Zealand, 9–13 August 1994. pp 133–40

Boyd JD (1950) Tree growth stresses, II. The development of shakes and other visual failures in timber. Aust J Appl Sci 1:296–312

Briggs DG (1995) Pruning in relation to forest inventory, wood quality, and products. In: Hanley DP, Oliver CD, Maguire DA, Briggs DG, Fight RD (eds.) Forest pruning and wood quality. College of Forest Resources, University of Washington, Seattle, pp 21–35

Brunner A, Nilsson O (2004) Revisiting tree maturation and floral initiation in the poplar functional genomics era. New Phytol 164:43–51

Burdon RD, Walker JCF, Megraw RA, Evans R, Cown DJ (2004) Juvenile versus mature wood: a new concept, orthogonal to corewood versus outerwood, with special reference to *Pinus radiata* and *P. taeda*. For Sci 50:399–415

Caballé G (1998) Le port autoportant des lianes tropicales: une synthèse des stratégies de croissance. Can J Bot 76:1703–1716

Carlquist S (1962) A theory of paedomorphosis in dicotyledonous woods. Phytomorphology 12:30–45

Carlquist S (1991) Anatomy of vine and liana stems: a review and synthesis. In: Putz FE, Mooney HA (eds) The biology of vines. Cambridge University Press, Cambridge, pp 53–71

Cato CL, McMillan L, Donaldson L, Richardson T, Echt C, Gardner R (2006) Wood formation from the base to the crown in *Pinus radiata*: gradients of tracheid wall thickness, wood density, radial growth rate and gene expression. Plant Mol Biol 60:565–581

Cave ID, Walker JCF (1994) Stiffness of wood in fast-grown softwoods: the influence of microfibril angle. For Prod J 44:43–48

Cermak J, Kucera J, Bauerle WL, Phillips N, Hinckley TM (2007) Tree water storage and its diurnal dynamics related to sap flow and changes in stem volume in old-growth Douglas-fir trees. Tree Physiol 27:181–198

Cherubini P, Schweingruber FH, Forster T (1997) Morphology and ecological significance of intra-annual radial cracks in living conifers. Trees 11:216–222

Chiba Y (2000) Modelling stem breakage caused by typhoons in plantation *Cryptomeria japonica* forests. For Ecol Manage 135:123–131

Christensen-Dalsgaard KK, Ennos AR, Fournier M (2008) Are radial changes in vascular anatomy mechanically induced or an ageing process? Evidence from observations on buttressed tree root systems. Trees 22:543–550

Clark A III, Saucier JR (1989) Influence of initial planting density, geographic location, and species on juvenile wood formation in southern pine. For Prod J 39:42–48

Cloutier A, Ananias R, Ballerini A, Pecho R (2007) Effect of radiata pine juvenile wood on the physical and mechanical properties of oriented strandboard. Holz Als Roh Werkst 65:157–162

Cown DJ (1973) Effects of severe thinning and pruning treatments on the intrinsic wood properties of young radiata pine. NZ J Forest Sci 3:370–389

Cown DJ (1974) Comparison of the effect of two thinning regimes on some wood properties of radiata pine. NZ J Forest Sci 4:540–551

Cown DJ (1992) Corewood (juvenile wood) in *Pinus radiata* – should we be concerned? NZ J Forest Sci 22:87–95

Cown DJ, Ball R (2001) Wood densitometry of ten *Pinus radiata* families at seven contrasting sites: influence of tree age, site and genotype. NZ J Forest Sci 31:88–100

Cown DJ, McConchie DL, Young, GD (1991) Radiata pine wood properties survey. New Zealand Ministry of Forestry, FRI Bulletin No. 50 (revised edition)

Cown DJ, McKinley RB, Ball RD (2002) Wood density variation in 10 mature radiata pine clones. NZ J Forest Sci 32:48–69

Cown DJ, Ball RD, Riddell MJC (2004) Wood density and microfibril angle in 10 *Pinus radiata* clones: distribution and influence on product performance. NZ J Forest Sci 34:293–315

Day ME, Greenwood MS (2011) Regulation of ontogeny in temperate conifers. In: Meinzer FC, Lachenbruch B, Dawson TE (eds) Size- and age-related changes in tree structure and function. Springer, Dordrecht

Day ME, Greenwood MS, Diaz-Sala C (2002) Age- and size-related trends in woody plant shoot development: regulatory pathways and evidence for genetic control. Tree Physiol 22:507–513

de Castro F, Williamson GB, Morales de Jesus R (1993) Radial variation in the wood specific gravity of *Joannesia princeps*: the role of age and diameter. Biotropica 25:175–182

DeBell JD, Gartner BL, DeBell DS (1998) Fiber length in young hybrid *Populus* stems grown at extremely different rates. Can J For Res 28:603–608

Domec J-C, Gartner BL (2002) Age- and position-related changes in hydraulic versus mechanical dysfunction of xylem: inferring the design criteria for Douglas-fir wood structure. Tree Physiol 22:91–104

Domec J-C, Gartner BL (2003) Relationship between growth rates and xylem hydraulic characteristics in young, mature, and old-growth ponderosa pine trees. Plant Cell Environ 26:471–483

Domec J-C, Pruyn M, Gartner BL (2005) Axial and radial profiles in conductivies, water storage, and native embolism in trunks of young and old-growth ponderosa pine trees. Plant Cell Environ 28:1103–1113

Domec J-C, Lachenbruch B, Meinzer FC (2006) Bordered pit structure and function determine spatial patterns of air-seeding thresholds in xylem of Douglas-fir (*Pseudotsuga menziesii*; Pinaceae) trees. Am J Bot 93:1588–1600

Domec J-C, Lachenbruch B, Meinzer FC, Woodruff DR, Warren JM, McCulloh KA (2008) Maximum height in a conifer is associated with conflicting requirements for xylem design. Proc Natl Acad Sci 105(33):12069–12074

Domec J-C, Warren JM, Meinzer FC, Lachenbruch B (2009) Safety factors for xylem failure by implosion and air-seeding within roots, trunks and branches of young and old conifer trees. IAWA J 30:100–120

Downes G, Turvey ND (1986) Reduced lignifications in *Pinus radiata* D. Don. Aust For Res 16:371–377

Dunham SM, Lachenbruch B, Ganio LM (2007) Bayesian analysis of Douglas-fir hydraulic architecture at multiple scales. Trees 21:65–78

Dunham SM, Ganio LM, Gitelman AI, Lachenbruch B (2008) Partitioning variation in Douglas-fir xylem properties among multiple scales via a Bayesian hierarchical model. Tree Physiol 28:1017–1024

Evans R, Ilic J (2001) Rapid prediction of wood stiffness from microfibril angle and density. For Prod J 51:53–57

Evans R, Hughes M, Menz D (1999) Microfibril angle variation by scanning x-ray diffractometry. Appita J 52(5):363–367

Evans JW II, Senft JF, Green DW (2000) Juvenile wood effect in red alder: analysis of physical and mechanical data to delineate juvenile and mature wood zones. For Prod J 50(7/8):75–87

Ewers FW, Zimmermann MH (1984) The hydraulic architecture of balsam fir (*Abies balsamea*). Physiol Plant 60:453–458

Fabris S (2000) Influence of cambial ageing, initial spacing, stem taper and growth rate on wood quality of three coastal conifers. Faculty of Graduate Studies, Department of Forestry, The University of British Columbia, Vancouver, 250 pp.

Fournier M, Bordonne PA, Guitard G, Okuyama T (1990) Growth stress patterns in tree stems: a model assuming evolution with tree age of maturation strains. Wood Sci Technol 24:131–142

Gallenmüller F, Müller U, Rowe N, Speck T (2001) The growth form of *Croton pullei* (Euphorbiaceae) - functional morphology and biomechanics of a neotropical liana. Plant Biol 3:50–61

Gapare WJ, Wu HX, Abarquez A (2006) Genetic control of the time of transition from juvenile to mature wood in *Pinus radiata* D. Don. Ann For Sci 63:871–878

Gartner BL (1991a) Stem hydraulic properties of vines vs. shrubs of western poison oak, *Toxicodendron diversilobum*. Oecologia 87:180–189

Gartner BL (1991b) Structural stability and architecture of vines vs. shrubs of poison oak, *Toxicodendron diversilobum*. Ecology 72:2005–2015

Gartner BL (1996) Does photosynthetic bark have a role in the production of core vs. outer wood? Wood Fiber Sci 28:53–61

Gartner BL, Lei H, Milota MR (1997) Variation in the anatomy and specific gravity of wood within and between trees of red alder (*Alnus rubra* Bong.). Wood Fiber Sci 29:10–20

Gartner BL, Robbins JM, Newton M (2005) Effects of pruning on wood density and tracheid length in young Douglas-fir. Wood Fiber Sci 37:304–312

Gerry E (1916) Fibre measurement studies: a comparison of tracheid dimensions in longleaf pine and Douglas fir, with data on the strength and length, mean diameter and thickness of wall of the tracheids. Science 43:360

Grabner M, Cherubini P, Rozenberg P, Hannrup B (2006) Summer drought and low earlywood density induce intra-annual radial cracks in conifers. Scand J For Res 21:151–157

Greenhill AG (1881) Determination of the greatest height consistent with stability that a vertical pole or mast can be made, and of the greatest height to which a tree of given proportions can grow. Proc Camb Philos Soc 4:65–73

Greenwood MS, Hopper CA, Hutchison KW (1989) Maturation in larch. I. Effect of age on shoot growth, foliar characteristics and DNA methylation. Plant Physiol 90:406–412

Hacke UG, Sperry JS, Pockman WT, Davis SD, McCulloh KA (2001) Trends in wood density and structure are linked to prevention of xylem implosion by negative pressure. Oecologia 126:457–461

Hedden RL, Fredricksen TS, Williams SA (1995) Modeling the effect of crown shedding and streamlining on the survival of loblolly pine exposed to acute wind. Can J For Res 25:704–712

Herman M, Dutilleul P, Avella-Shaw T (1998) Growth rate effects on temporal trajectories of ring width, wood density, and mean tracheid length in Norway spruce (*Picea abies* (L.) Karst.). Wood Fiber Sci 30:6–17

Innes TC (1995) Stress model of a wood fibre in relation to collapse. Wood Sci Technol 29:363–376

Jacobs MR (1945) The growth stresses of woody stems, vol Bulletin No. 28. Commonwealth Forestry Bureau, Australia, 67 p

Johnson DM, McCulloh KA, Reinhardt K (2011) The earliest stages of tree growth: development, physiology and impacts of climate change. In: Meinzer FC, Lachenbruch B, Dawson TE (eds.) Size- and age-related changes in tree structure and function. Springer, Dordrecht

Kennedy RW (1995) Coniferous wood quality in the future: concerns and strategies. Wood Sci Technol 29:321–338

King DA (2011) Size-related changes in tree proportions and their potential influence on the course of height growth. In: Meinzer FC, Lachenbruch B, Dawson TE (eds.) Size- and age-related changes in tree structure and function. Springer, Dordrecht

King D, Loucks OL (1978) The theory of tree bole and branch form. Radiat Environ Biophys 15:141–165

Kojima MH, Yamamoto H, Yoshida M, Okjio Y, Okumura K (2009) Maturation property of fast-growing hardwood plantation species: a view of fiber length. For Ecol Manage 257:15–22

Kretschmann DE, Moody CR, Pellerin RF, Bendtsen BA, Cahill JM McAlister RH, Sharp DW (1993) Effect of various proportions of juvenile wood on laminated veneer lumber. Res Pap FPL-RP-521. Madison, WI: USDA Forest Service, For Prod Lab, 31 p

Kubler H (1991) Function of spiral grain in trees. Trees 5:125–135

Lachenbruch B, Johnson GR, Downes GM, Evans R (2010) Relationships of density, microfibril angle, and sound velocity with stiffness and strength in mature wood of Douglas-fir. Can J For Res 40:55–64

Larjavaara M, Muller-Landau HC (2010) Rethinking the value of high wood density. Funct Ecol 24:701–705

Larson PR (1962) A biological approach to wood quality. Tappi 45:443–448

Larson PR (1969) Wood formation and the concept of wood quality, vol Bulletin 74, Yale University School of Forestry. Yale University, New Haven, 54 p

Larson PR, Kretschmann DE, Clark III A, Isebrands JG (2001) Formation and properties of juvenile wood in southern pines: a synopsis. General Technical Report FPL-GTR-129. Madison, WI: U.S. Department of Agriculture, Forest Service, Forest Products Laboratory. 42 p

Lei H, Milota MR, Gartner BL (1996) Between- and within-tree variation in the anatomy and specific gravity of wood in Oregon white oak (*Quercus garryana* Dougl). IAWA J 17:445–461

Lenz P, Cloutier A, MacKay J, Beaulieu J (2010) Genetic control of wood properties in Picea glauca – an analysis of trends with cambial age. Can J For Res 40:703–715

Lichtenegger H, Reiterer A, Stanzl-Tschegg SE, Fratzl P (1999) Variation of cellulose microfibril angles in softwoods and hardwoods – a possible strategy of mechanical optimization. J Struct Biol 128:257–269

Lintunen A, Kalliokoski T (2010) The effect of tree architecture on conduit diameter and frequency from small distal roots to branch tips in *Betula pendula, Picea abies* and *Pinus sylvestris*. Tree Physiol 30:1433–1447

Lundstrom T, Stoffel M, Stockli V (2010) Fresh-stem bending of silver fir and Norway spruce. Tree Physiol 28:355–366

Maeglin RR (1987) Juvenile wood, tension wood and growth stress effects on processing hardwoods. In: Applying the latest research to hardwood problems: Proceedings of the 15th annual hardwood symposium of the hardwood Research Council, Memphis, 10–12 May 1987, pp. 100–108

Mansfield SD, Parish R, Di Lucca CM, Goudie J, Kang K-Y, Ott P (2009) Revisiting the transition between juvenile and mature wood: a comparison of fibre length, microfibril angle and relative wood density in lodgepole pine. Holzforsch 63:449–456

Matsumura J, Butterfield BG (2001) Microfibril angles in the root wood of *Pinus radiata* and *Pinus nigra*. IAWA J 22:57–62

Matsuzaki J, Masumori M, Tange T (2006) Stem phototropism of trees: a possible significant factor in determining stem inclination on forest slopes. Ann Bot 98:573–581

Mayr S, Cochard H (2003) A new method for vulnerability analysis of small xylem areas reveals that compression wood of Norway spruce has lower hydraulic safety than opposite wood. Plant Cell Environ 26:1365–1371

Mayr S, Rothart B, Damon B (2003) Hydraulic efficiency and safety of leader shoots and twigs in Norway spruce growing at the alpine timberline. J Exp Bot 54:2563–2568

McDowell N, Barnard H, Bond BJ, Hinckley T, Hubbard RM, Ishii H, Kostner B, Magnani F, Marshall JD, Meinzer FC (2002) The relationship between tree height and leaf area: sapwood area ratio. Oecologia 132:2–20

McMahon TA, Kronauer RE (1976) Tree structures: deducing the principle of mechanical design. J Theor Biol 59:443–466

Megson THG (2005) Structural and stress analysis, secondth edn. Elsevier Butterworth-Heinemann, Oxford, 724 p

Meinzer FC, Johnson DM, Lachenbruch B, McCulloh KA, Woodruff DR (2009) Xylem hydraulic safety margins in woody plants: coordination of stomatal control of xylem tension with hydraulic capacitance. Funct Ecol 23:922–930

Meinzer FC, McCulloh KA, Lachenbruch B, Woodruff DR, Johnson DM (2010) The blind men and the elephant: the impact of context and scale in evaluating conflicts between plant hydraulic safety and efficiency. Oecologia 164:287–296

Ménard L, McKey D, Rowe N (2009) Developmental plasticity and biomechanics of treelets and lianas in *Manihot* aff. *quinquepartita* (Euphorbiaceae): a branch-angle climber of French Guiana. Ann Bot 103:1249–1259

Mencuccini M, Grace J, Fioravanti M (1997) Biomechanical and hydraulic determinants of tree structure in Scots pine: anatomical characteristics. Tree Physiol 17:105–113

Mencuccini M, Hölttä T, Martinez-Vilalta J (2011) Comparative criteria for models of the vascular transport system of tall trees. In: Meinzer FC, Lachenbruch B, Dawson TE (eds.) Size- and age-related changes in tree structure and function. Springer, Dordrecht

Nairn JA (2010) Polar shear-lag analysis of a composite of concentric cylinders with longitudinal cracks: application to internal checking in wood. Compos A 41:850–858

Niklas KJ (1992) Plant biomechanics: an engineering approach to plant form and function. University of Chicago Press, Chicago, 607 p

Niklas KJ (1994) Plant allometry: the scaling of form and process. University of Chicago Press, Chicago, 395 p

Paiva JAP, Garces M, Alves A, Garnier-Gere PH, Rodrigues JC, Lalanne C, Porcon S, Le Provost G, Da Silva-Perez D, Brach J, Frigerio J-M, Claverol S, Barres A, Fevereiro P, Plomion C (2008) Molecular and phenotypic profiling from the base to the crown in maritime pine wood-forming tissue. New Phytol 178:283–301

Pang S, Orchard R, McConchie D (1999) Tangential shrinkage of Pinus radiata earlywood and latewood, and its implication for within-ring internal checking. NZ J Forest Sci 29:484–491

Panshin AJ, deZeeuw C (1980) Textbook of wood technology, 4th edn. McGraw-Hill Book Co., New York, 736 p

Paul BH (1957) Juvenile wood in conifers. FPL Report 2094, U.S. Department of Agriculture Forest Service, Forest Products Laboratory, Madison, 6 p
Peterson MG, Dietterich HR, Lachenbruch B (2007) Do Douglas-fir branches and roots have juvenile wood? Wood Fiber Sci 39:651–660
Phillips NG, Ryan MG, Bond BJ, McDowell NG, Hinckley TM, Cermak J (2003) Reliance on stored water increases with tree size in three species in the Pacific Northwest. Tree Physiol 23:237–245
Pillow MY, Luxford RF (1937) Structure, occurrence, and properties of compression wood. USDA Agriculture Technical Bulletin 546, Forest Products Laboratory, USDA FS, 32 p
Pittermann J, Sperry J (2003) Tracheid diameter is the key trait determining the extent of freezing-induced embolism in conifers. Tree Physiol 23:907–914
Putoczki TL, Nair H, Butterfield B, Jackson S (2007) Intra-ring checking in *Pinus radiata* D. Don: the occurrence of cell wall fracture, cell collapse, and lignin distribution. Trees 21:221–229
Putz FE, Coley PD, Lu K, Montalvo A, Aiello A (1983) Uprooting and snapping of trees, structural determinants and ecological consequences. Can J For Res 13:1011–1020
Reiterer A, Lichtenegger H, Tschegg S, Fratzl P (1999) Experimental evidence for a mechanical function of the cellulose microfibril angle in wood cell walls. Philos Mag A 79:2173–2186
Rosner S, Karlsson B, Konnerth J, Hansmann C (2009) Shrinkage processes in standard-size Norway spruce wood specimens with different vulnerability to cavitation. Tree Physiol 29:1419–1431
Rosner S, Klein A, Wimmer R, Karlsson B (2006) Extraction of features from ultrasound acoustic emissions: a tool to assess the hydraulic vulnerability of Norway spruce trunkwood? New Phytol 171:105–116
Rosner S, Klein A, Muller U, Karlsson B (2008) Tradeoffs between hydraulic and mechanical stress responses of mature Norway spruce trunk wood. Tree Physiol 28:1179–1188
Rosner S, Konnerth J, Plank B, Salaberger D, Hansmann C (2010) Radial shrinkage and ultrasound acoustic emissions of fresh versus pre-dried Norway spruce sapwood. Trees 24:931–940
Rowe NP, Speck T (1996) Biomechanical characteristics of the ontogeny and growth habit of the tropical liana Condylocarpon guianense (Apocynaceae). Int J Plant Sci 157:406–417
Sanio K (1872) Uber die Grosse der Holz Zellen bei der gemeinen Kiefer (*Pinus sylvestris*). Jahrb Wisch Bot 8:401–420
Saranpaa P, Pesonen E, Saren M, Andersson S, Siiria S, Serimaa R, Paakkari T (2000) Variation of the properties of tracheids in Norway spruce (Picea abies [L.] Karst). In: Savidge RA, Barnett JR, Napier R (eds.) Cell and molecular biology of wood formation. BIOS Scientific Publishers, Oxford, pp 337–345
Scholz FG, Phillips NG, Bucci SJ, Meinzer FC, Goldstein G (2011) Hydraulic capacitance: biophysics and functional significance of internal water sources in relation to tree size. In: Meinzer FC, Lachenbruch B, Dawson TE (eds.) Size- and age-related changes in tree structure and function. Springer, Dordrecht
Schulgasser K, Witztum A (2007) The mechanism of spiral grain formation in trees. Wood Sci Technol 41:133–156
Shi JL, Zhang SY, Reidl B (2005) Effect of juvenile wood on strength properties and dimensional stability of black spruce medium-density fiberboard panels. Holzforsch 59:1–9
Shiokura T (1982) Extent and differentiation of the juvenile wood zone in coniferous tree trunks. J Jpn Wood Res Soc 28:85–90
Siau JF (1984) Transport processes in wood. Springer, Berlin
Skatter S, Kucera B (1997) Spiral grain: an adaptation of trees to withstand stem breakage caused by wind induced torsion. Holz-als-Roh Werkst 55:207–214
Spicer R, Gartner BL (1998) Hydraulic properties of Douglas-fir (*Pseudotsuga menziesii*) branches and branch halves with reference to compression wood. Tree Physiol 18:777–784
Spicer R, Gartner BL (2001) The effects of cambial age and position within the stem on specific conductivity in Douglas-fir *Pseudotsuga menziesii* sapwood. Trees 15:222–229
Steinberg V (2002) Hydrodynamics: bend and survive. Nature 420:473

Sun Q, Suzuki M (2001) Quantitative character variations of cambial derivatives in mangrove and their functional significance. Trees 15:249–269

Thomas SC (2011) Age-related changes in tree growth and functional biology: the role of reproduction. In: Meinzer FC, Lachenbruch B, Dawson TE (eds.) Size- and age-related changes in tree structure and function. Springer, Dordrecht

Timell TE (1986) Compression wood in gymnosperms, vol I–III. Springer, Berlin, 2150 p

Trendelenburgh R (1935) Schwankungen des Raumgewichts wichtiger Nadelholzer nach Wuchsgebiet, Standort und Einzelstamm. Z Ver Dtsch Ing Bd 79, Nr. 4, s 85–89, 26

Turnbull JW (1948) Some factors affecting wood density in pine stems. S Afr For J 16:22–43

Tyree MT, Ewers FW (1991) The hydraulic architecture of trees and other woody plants. New Phytol 119:345–360

Tyree MT, Zimmermann MH (2002) Xylem structure and the ascent of sap. Springer, Berlin

Tyree MT, Snyderman DA, Wilmot TR, Machado JL (1991) Water relations and hydraulic architecture of a tropical tree (*Schefflera morototoni*). Data, models, and a comparison with two temperate species (*Acer saccharum* and *Thuja occidentalis*). Plant Physiol 96:1105–1113

Utsumi Y, Bobich EG, Ewers F (2010) Photosynthetic, hydraulic and biomechanical responses of *Juglans californica* shoots to wildfire. Oecologia. doi:10.1007/s00442-010-1653-x

Via BK, So CL, Shupe TF, Groom LH, Wikaira J (2009) Mechanical response of longleaf pine to variation in microfibril angle, chemistry associated wavelengths, density, and radial position. Compos A 40:60–66

Vogel S (1989) Drag and reconfiguration of broadleaves in high winds. J Exp Bot 40:941–948

Voorhies G, Groman WA (1982) Longitudinal shrinkage and occurrence of various fibril angles in juvenile wood of young-growth ponderosa pine. Arizona Forestry Notes No. 16, School of Forestry, Northern Arizona University, 18 pp

Wiemann MC, Williamson GB (1988) Extreme radial changes in wood specific gravity in some tropical pioneers. Wood Fiber Sci 20:344–349

Williamson GB, Wiemann MC (2010) Age-dependent radial increases in wood specific gravity of tropical pioneers in Costa Rica. Biotropica 42:590–597

Willmann MR, Poethig RS (2005) Time to grow up: the temporal role of small RNAs in plants. Curr Opin Plant Biol 8:548–552

Wilson BF, Archer RR (1979) Tree design: some biological solutions to mechanical problems. BioScience 29:293–298

Woodcock DW, Shier AD (2002) Wood specific gravity and its radial variations: the many ways to make a tree. Trees 16:437–443

Woodcock DW, Shier AD (2003) Does canopy position affect wood specific gravity in temperate forest trees? Ann Bot 91:529–537

Yang KC (1994) Impact of spacing on width and basal area of juvenile and mature wood in *Picea mariana* and *Picea glauca*. Wood Fiber Sci 26:479–488

Zimmermann MH (1978) Hydraulic architecture of some diffuse-porous trees. Can J Bot 56:286–2295

Zimmermann U, Heike Schneider H, Wegner LH, Haase A (2004) Water ascent in tall trees: does evolution of land plants rely on a highly metastable state? New Phytol 162:575–615

Zobel BJ, Sprague JR (1998) Juvenile wood in forest trees. Springer, Berlin

Zobel BJ, van Buijtenen JP (1989) Wood variation: its causes and control. Springer, Berlin

Chapter 6
Size-Related Changes in Tree Proportions and Their Potential Influence on the Course of Height Growth

David A. King

Abstract As trees grow larger and older they show characteristic changes in their proportions and growth rates. This paper explores the hypothesis that mechanical support costs of growing trees contributes to tree height limitation and general growth decline in aging trees. The chapter reviews scaling hypotheses, including geometric similarity, elastic similarity, stress similarity, and the pipe model, as well as the patterns in tree allometry that have been reported. It then uses the remarkable published record of diameter, volume and height growth, deriving from stem analysis, of one 437-year-old, 83-m-tall noble fir (*Abies procera*) to evaluate the degree to which different scaling models apply at different stages in that tree's life. The analysis suggests that the overall pattern of height growth could be related to tree allometry and wood production rates, but that the tree showed substantial decadal variation around the general patterns. The chapter suggests that many more studies are needed to characterize individual trees, either with known environments or after manipulations that alter allometry, and in different ecosystems beyond the Pacific Northwest and California. Such data will enable scientists to analyze the degree to which these allometric models pertain in different plant types and at different life stages, and will give insight into the role of mechanical and hydraulic support in the limitation of tree growth and productivity.

1 Introduction

As trees grow larger and older they show characteristic changes in their proportions and growth rates. Forest or plantation-grown trees that are exposed to overhead sun typically attain maximum height growth rates early in life followed by prolonged

D.A. King (✉)
Biological and Ecological Engineering, Oregon State University,
Corvallis, OR 97331, USA
e-mail: dkingaz@yahoo.com

F.C. Meinzer et al. (eds.), *Size- and Age-Related Changes in Tree Structure and Function*, Tree Physiology 4, DOI 10.1007/978-94-007-1242-3_6,

declines to very low extension rates that may be terminated by top breakage or dieback (Ishii et al. 2000; Bond et al. 2007). Trunk diameter growth also declines in aging trees, but usually to a lesser extent than height growth. Concurrently, the ratio of trunk diameter to tree height increases, as is necessary to insure mechanical stability and reduce the likelihood of breakage by extreme winds. Trees also appear to have an upper limit to height that depends on species, genotype and environment (Thomas 1996; Ryan and Yoder 1997), which is consistent with this decline in height growth.

Proposed mechanisms for limits to height and general declines in growth in aging trees have emphasized size-related limitations to water transport and other size-related changes in morphology and physiology (Ryan and Yoder 1997; Koch et al. 2004; Woodruff et al. 2004; Domec et al. 2008; Ishii 2011; McDowell et al. 2011; Mencuccini et al. 2011; Sala et al. 2011; Steppe et al. 2011; Woodruff and Meinzer 2011). Reciprocal grafting experiments between seedlings and older trees by Mencuccini et al. (2005) and Bond et al. (2007) have shown that declines in height growth rate are related to increases in tree size rather than meristem age, consistent with the above hypotheses. Foresters have recognized that trees subjected to early suppression and release show subsequent growth rates that are similar to non-suppressed trees of similar size but not age (Spurr and Barnes 1980), an observation that also supports size-related influences on growth.

Other proposed mechanisms that may limit tree height include size-related increases in allocation to other tree parts and processes (Ryan and Waring 1992; Thomas 2011). King (1990) predicted evolutionary limits to height based on the assumption that the fraction of gross production allocated to stem maintenance respiration increases substantially with height. However, direct measurements of respiration do not support this assumption (Ryan et al. 1997; Pruyn et al. 2002).

Tree size may also influence the course of height growth via the scaling of mechanical support costs. The proportions of trunks, branches and crowns are important in enabling trees to withstand mechanical forces, intercept light and transport water to their foliage. The relation between tree proportions and size (allometry) can be viewed as both the result of cumulative past growth and an influence on future growth. As trees increase in size, their interception of light and photosynthetic output generally increase, but so do the construction costs of growing yet taller and broader. The minimal design criterion that a trunk be thick enough to prevent itself from buckling under its own weight requires that trunk biomass increase with the fourth power of tree height (McMahon 1973). This disproportionate increase in support costs requires a slowing in height growth with increasing height (King 2005).

However, there is uncertainty regarding the extent to which allometric constraints may explain observed patterns of height growth. Recent studies have advanced our understanding of mechanical constraints on tree architecture and their influence on the tradeoffs between support costs and photosynthetic benefits (e.g. Givnish 1995; van Gelder et al. 2006; King et al. 2006; Sterck et al. 2006; Anton and Schieving 2010). But these studies have not explicitly considered the influence of allometry on height growth rate. Whereas models of forest growth commonly incorporate allometric relations, these models are designed to predict wood production or forest dynamics

(Shugart 1984; Monserud 2003) and have not emphasized the links between allometry and height growth.

This chapter addresses the potential influence of allometry on the course of height growth by first reviewing past scaling hypotheses and observed patterns in tree allometry. Several simple models of the scaling of wood production and allometry and their predictions of height growth are presented – which may apply to different stages of growth in trees of even-aged forests. A life-long record of the course of growth in stem diameter, volume and height derived from stem analysis of a 437 year-old, 83 m tall noble fir (*Abies procera*) by Herman et al. (1975) is used to evaluate this approach and suggest further directions for research.

2 Models of Tree Allometry

The scaling or relation of trunk dimensions to tree size has been derived from several structural models, which may be viewed as tree scaling hypotheses. These include the pipe model (Shinozaki et al. 1964) and the principles of geometric similarity, elastic similarity and stress similarity (McMahon 1975; McMahon and Kronauer 1976; Wilson and Archer 1979; Bertram 1989; Niklas and Spatz 2000), described as follows:

2.1 Geometric Similarity

Geometrically similar objects have the same proportions, regardless of size, i.e., a small tree is an exact miniature of a large one – though this conceptualization is generally applied only to the stem and branches and not the leaves. Thus, crown width and length and stem diameter at a given fraction of total height are all directly proportional to tree height – and stem volume is proportional to tree height cubed. This model is not necessarily related to any specific mechanical principle, but serves as the simplest null model of tree proportions. Under certain conditions geometric similarity is predicted by the stress similarity model, as described below.

2.2 Elastic Similarity

Elastically similar columns or rods show the same angle of deflection from base to tip due to bending under their own weight when held out at a given angle to the vertical. For rods of a given material that are untapered or with the same proportional taper from base to tip, elastic similarity implies that basal diameter is proportional to total rod length raised to the 3/2 power, i.e. $D \propto L^{3/2}$ (McMahon and Kronauer 1976).

Of more importance for tree stems, the minimum diameter required to prevent a vertical pole from bending over under its own weight due to elastic instability is also proportional to pole height to the 3/2 power. More specifically, for a cylindrical, untapered pole, the critical buckling diameter

$$D_{crit} = 1.42(G\rho / E)^{1/2} H^{3/2} \quad (6.1)$$

(Greenhill 1881), where E is Young's modulus of elasticity, G is the gravitational force per unit mass at Earth's surface (9.8 N kg^{-1}), ρ is the pole material density, H is the pole height, and all quantities are expressed in the same system of units – here Standard International units involving newtons, kilograms, seconds and meters – for both height and diameter. Young's modulus denotes the resistance of wood to stretching or compression parallel to the trunk and is closely associated with resistance to bending. This relation may also be expressed as the buckling height for a pole of given diameter, $H_{crit} = 0.792\ (E/G\rho)^{1/3}\ D^{2/3}$. Note that for dimensional comparisons with living trees, the density and elastic modulus of green wood should be used.

The same proportionalities apply for poles with a given proportional taper (for example having conical or parabola-shaped longisections) or for weighted poles having a given top weight to pole weight ratio (King 1986). However, the diameter scaling constant (1.42 for cylindrical poles) is smaller for tapered poles and larger for weighted poles. The calculation of this constant is mathematically rather challenging and may require iterative techniques, but the fundamental scaling relation of $D_{crit} \propto H^{3/2}$ for poles of a given material and taper can be demonstrated as follows.

Consider a slender vertical pole that is slightly bent to one side and is just at the point of elastic instability where the torque or bending moment due to the offset weight of the pole (T_w) is just matched by the restoring moment (T_r) generated by the strains within the wood – stretching along the grain on the side opposite the direction of the bend and compression on the other side. At the pole base,

$$T_w = \text{pole weight} \times \text{mean horizontal offset.} \quad (6.2)$$

For a pole of given taper, pole weight $\propto D_b^2H$, where D_b is basal diameter and H is again total height.

Regarding the offset, if we assume that the bending pole takes on a parabolic arc as measured from the vertical axis, then the horizontal offset of any point is proportional to the square of its height above the base – as defines a parabola. Hence, both the tip offset and the mean offset of the pole weight (which is a certain fraction of the tip offset) will also be proportional to H^2. In general, a pole at the instability limit will not form a parabolic arc, as this arc is determined by the balance between bending and restoring moments along the whole pole and will vary with stem taper. But if the curvature at the base is assumed constant and the relative shape of the arc is independent of total height H, then the above relationship (mean offset $\propto H^2$) will apply. Thus, relation (6.2) implies that

$$T_w \propto D_b^2 H^3 \quad (6.3)$$

The restoring bending moment or torque generated by the wood near the base is proportional to the cross sectional area ($\propto D_b^2$) × mean strain across either the side in tension or compression ($\propto D_b$ for a given stem curvature) × mean distance of the stretched or compressed fibers from the neutral plane bisecting the stem (also $\propto D_b$). Thus, the resisting torque

$$T_r \propto D_b^4 \tag{6.4}$$

Setting $T_w = T_r$, i.e., $D_b^2 H^3 \propto D_b^4$, thus yields $D_b \propto H^{3/2}$ for poles at the buckling point.

Basically, the great benefit of a thick stem for rigidity is partly offset by its greater weight – while increasing total height also increases weight and substantially increases the offset of a pole with given degree of curvature at the base. So a disproportionate increase in diameter is required to maintain stability as height is increased.

Stem diameter D (as measured at a given fraction of total height) should be proportional to $H^{3/2}$ for trees as well as poles, assuming that the trees being compared have identical wood properties, stem taper, ratio of crown to trunk weight and ratio of actual to critical buckling diameter – i.e., the same buckling safety factor. For such trees, stem volume and biomass ($\propto HD^2$) are proportional to H^4 – rather than the H^3 proportionality expected for geometrically similar trees. This difference in the scaling of biomass has a substantive influence on the scaling of height growth rate, as explored below.

2.3 Stress Similarity

According to concept of stress similarity or uniform stress, trees are designed such that the peripheral compressive and tensile stresses on either side of a bent limb are equal along the length of the limb or among limbs, so that no one cross section is more vulnerable to breakage than any other. Here stress is defined as force per unit cross sectional area. However, stress similarity leads to different predictions depending on whether the stress is due to wind forces or gravitational loads.

This scaling concept is closely related to that of strain similarity and plants may in fact sense and respond to local strains rather than stresses (Coutand and Moulia 2000). However, for wood of given physical properties, strain (relative deformation) is proportional to stress, so similar allometries should result.

A rod of uniform material with a circular cross section will experience maximal stretching and compressive stresses along the lines running along its uppermost and lowermost surfaces, respectively, as it bends under its own weight. These stresses will be constant along the length of the rod if it is tapered such that rod diameter is proportional to the square of the distance from its tip – with the provisos that the bending angle from base to tip is small and the rod is relatively slender. The same taper relation applies to a repeatedly ramifying leafless branch that is self similar – such that the branching pattern remains similar from branch base to the infinitely slender tips (McMahon and Kronauer 1976). This relation also applies to the basal diameter of separate branches of differing length, but similar ramification – **and angle to the vertical**. That is, $d_b \propto l^2$, where d_b is branch basal diameter and l is total branch length.

This rapid increase in diameter with length is required for gravitational stress similarity because stress resistance ∝ diameter^3, but weight ∝ diameter^2 × length and bending moment ∝ diameter^2 × length^2, so making a branch thicker to make it more stress resistant substantially increases its bending moment. However, gravitational stress similarity also predicts that steeply ascending branches will be considerably thinner than horizontal branches because the bending moment is proportional to the horizontal projection of the branch tip from the base – which decreases to zero for a vertical branch. But a glance out the window in winter at the bare crowns of deciduous trees shows little evidence of this pattern and small and intermediate-sized branches are generally much over-designed with respect to breaking under their own weight. Furthermore, tree branches are not as steeply tapered as predicted by this theory (McMahon and Kronauer 1976). Thus, limbs are generally not designed to equalize gravitational stresses.

Gravitational forces do become important as branch size increases or trees lean. In these cases trees develop elliptical limb cross sections which more efficiently resist gravitationally induced bending stresses than do circular cross sections. Large, old spreading limbs often produce reaction wood that is placed under tension in angiosperms or compression in conifers, which keeps branches from drooping excessively as their bending moments increase with growth (Wilson and Archer 1979). Branches also tend to sag increasingly as they grow older and larger (Wilson and Archer 1979), a pattern which results in lower branch angles near the crown base thereby spreading the foliage over a larger area. Severe ice storms greatly increase branch weights and may cause permanent bending in small trees (typically <10 cm dbh), stem breakage in somewhat larger trees and severe branch loss in yet larger trees (Greene et al. 2007).

Stress similarity implies different scaling relations if wind rather than gravity is considered as the stress-generating force. For slow, steady winds causing small deflections to branches that are geometrically similar in foliar outline, the foliage cross sectional area perpendicular to the wind and presumably the wind force are proportional to branch length squared. Thus, the bending moment (= force × distance to the center of force) is proportional to branch length cubed. The resistive bending moment of a limb with a given peripheral stress is proportional to the branch wood cross sectional area × branch diameter, i.e. to diameter cubed. Hence, branch diameter should be directly proportional to branch length, as would occur for branches that are geometrically similar in thickness as well as in foliar distribution.

Among trees of similar crown proportions exposed to winds of the same (low) speeds, stress similarity also predicts geometrically similar stem proportions, i.e. $D_b \propto H$. However, within forests wind speeds increase rapidly with height in the canopy (Quine and Gardiner 2007) and taller trees may bear proportionately larger crowns than their shorter, more crowded neighbors. So, among trees of differing height in a given forest a substantially greater increase in diameter with total height is required to maintain stress similarity. In addition, trees subject to damaging high-velocity winds show much stem bending and crown streamlining that reduces wind forces and bending moments, especially for small, relatively slender trees (Vogel 1984, 1989).

A different relation is predicted for the stem taper of forest trees, as measured below the crown base. Here the bending moment = wind force on the crown x distance below the center of wind force (L), i.e. bending moment $\propto L$. As the resistive bending moment for constant stress is again $\propto D^3$, where D is the stem diameter at distance L below the center of wind force, setting the resistive and wind associated bending moments equal yields $D \propto L^{1/3}$.

Among forest-grown conifers, trunk taper below the crown base often approximates this wind stress-equalizing relation (Büsgen and Münch 1929; Morgan and Cannell 1994). Small slender trees and saplings, which may bend nearly to the ground before breaking, typically show steeper tapers, i.e. $D \propto L^{\geq 0.5}$, particularly among hardwoods (King 1986). Under large deformations a steeper taper (slenderer upper stem) is predicted to equalize breaking stresses in wind bent trees, as the bending moment on the upper stem is much reduced if it arcs over to become nearly parallel to the ground (Leiser and Kemper 1973). Trees sway in response to wind gusts and the taper of the branching system seems well suited to dampen this swaying (Quine and Gardiner 2007). Variation in wind profiles, crown streamlining and dynamic responses to wind gusts makes stress equalization from tree base to twigs across all weather conditions unlikely (Niklas and Spatz 2000).

However, the principle that trees allocate wood so as to move towards stress equalization appears to govern the repair of wounds and other defects (Mattheck and Kubler 1995). Cracks, holes and constrictions to diameter growth all produce high local stress concentrations that can initiate breakage and trees preferentially allocate new wood so as to reduce these dangerous stresses (Mattheck and Kubler 1995; Mattheck 1998).

2.4 Pipe Model

The pipe model is based on an idealized hydraulic view of stem proportions that considers stems and branches to be composed of parallel pipes of constant diameter from trunk base to leaves (Shinozaki et al. 1964). Thus, the cross sectional area of any stem should be equal to the summed cross sectional areas of all of the higher limbs that branch off the stem and to the leaf area borne by those limbs – in young plants or branches which have not yet shed any of their leaves. The pipe model also includes the concept of disused pipes leading to abscised leaves and shed branches, as needed to explain the taper of tree branches and trunks below their foliated parts (Mäkelä 2002; Mäkelä and Valentine 2006).

Although the hydraulic conductivity per unit cross sectional area varies among species (Tyree and Zimmermann 2002), the pipe model may nonetheless yield satisfactory predictions of empirical patterns. Shinozaki et al. (1964) found that this model adequately predicted the within-canopy scaling of stem cross sectional areas. Sapwood cross sectional area at the crown base is highly correlated with foliage biomass, as predicted by the pipe model (Maguire and Hann 1989), though the ratio of sapwood area to foliage biomass varies substantially among species (Pruyn et al. 2005). Within Douglas-fir (*Pseudotsuga menziesii*) trees, Long et al. (1981) found that sapwood cross sectional area was roughly proportional to foliage biomass, whereas heartwood cross sectional increased with distance below the crown base such that overall taper approximated that of the wind stress similarity model. That is, the progressive formation of heartwood allowed trees to meet both hydraulic and support requirements in an efficient way.

3 Observed Patterns in Tree Allometry

As trees cannot read these erudite theories and face varying biotic and abiotic stresses throughout life, they generally do not follow any one model from seedling to old age (Farnsworth and Niklas 1995). However, specific models do apply over restricted size ranges and environmental conditions. In evaluating tree allometries it is important to bear in mind the particular forest type and management practices involved, e.g. all-aged multi-species old growth forests vs. even-aged single-species forests or plantations, with or without thinning treatments.

3.1 Conifers and Dicotyledonous Trees

McMahon (1973) found that for record-sized trees of species of the USA, diameter at breast height (dbh) was roughly proportional to $H^{3/2}$, as predicted by the elastic similarity model – though with substantial scatter. These trees averaged about eight times as thick as the critical buckling diameter for wooden columns of similar heights – but they were chosen by a point system based on trunk circumference in inches plus height in feet plus one-fourth of crown width in feet. This procedure often selects trees of extreme girth, but less than maximum height, and diameter at breast height (dbh) is also inflated by butt swell in such stout trees. King et al. (2009) found that for mature adults of species of widely varying stature in tall, dense tropical forests of Malaysia, dbh (or diameter above the buttress in larger species) was also proportion to $H^{3/2}$. However, these trees were much closer to the critical buckling diameter than the above record-sized trees. Other studies of forest-grown trees of all life stages have found that small-crowned trees of intermediate size may approach buckling limits, but that both the smallest and the largest of trees show greater safety factors (e.g. King 1981; Sterck and Bongers 1998), as described below.

Among old forests with trees of all ages, the relation between trunk diameter and tree height varies with tree height and is in part linked to variation in crown width – which varies among species (King 1996; Sterck and Bongers 1998). Small juvenile trees <10 m. tall often approach geometric similarity in stem allometry, while taller trees up to the height of the overstory canopy lie closer to the elastic similarity model (diameter $\propto H^{3/2}$) (Osunkoya et al. 2007). Diameter often increases yet more steeply with height for upper canopy and emergent trees, particularly among species with broad crowns at full size (King 1996; Poorter et al. 2006). Very old trees often lose height due to top dieback and/or breakage, though such damaged individuals are typically omitted from allometric analyses. Asymptotic models that predict linear relations between diameter and height for the smallest individuals and an asymptotic approach to a maximum height with increasing diameter often provide adequate fits to allometric data (Kira 1978; Thomas 1996).

Among even-aged, monospecific forests and plantations the relation between mean trunk diameter and mean height per stand may approximate the elastic similarity model over an intermediate range in stand age and height. But within any given stand, diameter increases with a higher power of height, reflecting the fact that shorter than average trees have small crowded crowns that are more sheltered from wind than are the larger crowns of stand dominants (e.g. Beck and Della-Bianca 1970; King 1990).

Small saplings seem mechanically over designed compared to larger trees (King 1987; Niklas 1994), but are subject to damage from falling debris (Aide 1987) and mammalian herbivores (Ickes et al. 2003). In young saplings lacking heartwood, the pipe model relationship of sapwood cross sectional area $\propto$ leaf biomass should lead to increasing buckling safety factors with decreasing sapling height – if corresponding changes in Young's modulus of elasticity are small.

Among temperate saplings subject to snow loads, species with spreading evergreen foliage are substantially thicker stemmed than are deciduous species with large leaves and thus sparsely branched crowns – comparing saplings of the same height and crown width (King 1991). But larger trees of these contrasting leaf habits show little difference in allometry. This observation is consistent with the fact that crown area and potential snow and ice loads increase with a substantially lower power of height than does stem weight, making this loading less important with increasing tree size.

Wood properties may vary with stage of development – which is not usually accounted for in allometric theories (Niklas 1993). The juvenile wood produced in the first decade or so is often lower in density and substantially less stiff than adult wood (Lachenbruch et al. 2011) – though these differences tend to be more pronounced in shade intolerant species than in shade tolerant ones (Wiemann and Williamson 1989; Zobel and van Buijtenen 1989). Perhaps more importantly, the presence of rot generally increases as trees age (Van Pelt and Sillett 2008). This pattern makes old trees more susceptible to toppling and breakage (Frelich and Reich 1996) despite their relatively thick stems. Decay of the central heartwood increases the likelihood of breakage by shearing along radial cracks in the remaining peripheral wood (Ossenbruggen et al. 1986), while surface defects cause stress concentrations that initiate breakage under lower overall bending stresses (Mattheck and Kubler 1995).

3.2 *The Contrasting Allometry of Palms*

The allometry of palms differs from that of the dicotyledonous trees and conifers considered heretofore – because palms lack branches and a vascular cambium. As a result, palms show limited increases in crown size as they grow in height (Rich et al. 1995) and cannot add new wood to their stem peripheries. Palms do increase the density and stiffness of their stems by increasing the cell wall thickness and diameter of existing vascular fibers – particularly those fibers near the stem periphery (Rich 1987a). This increase in fiber diameter along with the expansion (and sometimes division) of the surrounding parenchyma cells also causes varying amounts of stem expansion (Waterhouse et al. 1978; Tomlinson 1990). However, the increase in stem diameter as palms grow in height is considerably less than that observed in co-occurring dicotyledonous trees and scarcely occurs at all in columnar palms (Rich et al. 1986). The latter are substantially overbuilt in youth, but not in adulthood. The adults of some palm species exceed the stability limits for cylindrical columns of typical wood properties. Their very dense and stiff peripheral wood, particularly near the stem base, enables these palms to keep their relatively small crowns erect (Rich 1987b).

3.3 *Local Influences on Tree Proportions*

Tree spacing has a pronounced effect on crown width and length and on trunk thickness. Isolated open-grown trees typically bear much broader and longer crowns than do forest-grown trees of similar height. Open-grown oaks may be as broad as they are tall, whereas open-grown conifers tend to be somewhat narrower crowned than angiosperms (Ek 1974). Trembling aspen (*Populus tremuloides*) of even-aged, self-thinning stands have crowns that are only 1/3 of the width of open-grown aspen (Ek 1974; King 1981). Open-grown trees typically bear crowns that extend most of the way to the ground, but the crowns of mature even-aged aspens occupy only the upper third of the stem (King 1981). Such forest-grown aspen have stems that are 2/5 as thick as open-grown aspen of similar height. Thus, the large well-illuminated crowns of open-grown trees are more productive than those of forest-grown trees, but they allocate correspondingly larger masses of wood to their trunks and branches. The fact that the greater productivity of widely spaced trees is offset by greater support costs may explain the observation that height growth is relatively insensitive to tree spacing (Sjolte-Jorgensen 1967), justifying the use of height growth curves to specify site productivity (Spurr and Barnes 1980). However, lower height growth rates have been reported for both widely spaced and very crowded trees (Alexander et al. 1967; Flewelling et al. 2001)

Trees respond to changes in crowding by varying the thickness of their current annual growth rings as viewed in longisection and by varying their ratios of height to diameter growth. The rings of young plantation-grown conifers are initially thickest at the base, but rings produced later in life, when the trees are more crowded, tend to be thicker towards the treetop (Kozlowski 1971, p. 78). After stand thinning, wood allocation initially shifts towards the base, resulting in a temporary increase in stem taper (Kozlowski 1971, p. 110). Bormann (1965) observed that diameter growth at breast height declined to near zero, while height growth continued in *Pinus strobus* trees undergoing increasing suppression by their taller neighbors – a response that prolongs access to light in such trees.

Stem diameter is also related to topographic position and exposure to wind. Trees of wind-exposed ridges are shorter and squatter than those of sheltered slopes (Lawton 1982), whereas trees that are guy-wired to prevent swaying grow less in diameter than freely swaying trees (Jacobs 1954). Holbrook and Putz (1989) found that *Liquidambar styraciflua* saplings that were both guy-wired and shaded from the sides grew nearly twice as fast in height as the open-grown controls, but often buckled under their own weight when their supports were removed. Trees allocate wood preferentially to stem sections experiencing greater bending and Coutand et al. (2009) found that in *Populus*, a single stem flexion activated genes that increased cambial growth over the stressed area in subsequent days. Thus, trees seem able to sense environmental stressors and modify their growth in an adaptive fashion.

In summary, tree allometry varies with tree size, species, crowding by neighbors and exposure to light and wind. However, there are general allometric trends, particularly for trees of specified forest conditions, such as the even-aged stands or plantations for which much information on growth and wood yield has been published. In general, small forest-grown trees (<10 m tall) lie closer to the geometric similarity model, whereas larger trees often approach the elastic similarity model. Yet older trees show even greater increases in diameter vs. height, particularly if they have suffered top breakage or dieback. Site specific patterns include shifts in allometry over time in response to increased crowding by neighbors or increased exposure to light and wind following thinning or natural disturbances.

4 Linking Allometry and the Course of Height Growth

The height growth rates of trees can be related to the factors influencing stem wood production (photosynthesis, respiration and allocation) – and the allocation of that wood between height and diameter growth. These processes all vary as trees grow and age and experience varying competitive relations, which complicates the analysis of the factors influencing height growth. However, allometric growth models employing the simplest of assumptions regarding the relation between stem wood production and tree size are useful in isolating the effects of tree proportions on growth. This approach considers the tree to be a simple, self-growing machine that approximates the allometries and production efficiencies of real trees (King 2005). The following three cases are instructive.

1. Geometric similarity in crown and stem proportions, with stem wood production proportional to crown area. Thus, stem biomass $M \propto H^3$, crown area $\propto H^2$ and hence stem wood production rate, $\mathrm{d}M/\mathrm{d}t \propto H^2$. Such relations might pertain to a dominant tree in a very young stand that can increase its crown size as it grows taller.
2. Geometric similarity in crown proportions, elastic similarity in stem proportions ($M \propto H^4$) and again $\mathrm{d}M/\mathrm{d}t \propto H^2$. Such relations might apply to a dominant tree in an intermediate-aged stand.
3. Elastic similarity in stem proportions ($M \propto H^4$), but a constant wood production rate. This constant wood production might be associated with a tree that maintained the same crown size as it grew taller or a tree that increased its crown size but showed a commensurate decline in stem wood production per unit crown area – due to some combination of declining photosynthetic efficiency and declining allocation of photosynthate to stem wood production (Ryan et al. 1997). Such relations might pertain to an aging tree.

For a specified stem allometry, the height growth rate is given by the product of the rate of stem wood production ($\mathrm{d}M/\mathrm{d}t$) and the height growth increment per unit of wood added to the stem ($\mathrm{d}H/\mathrm{d}M$). That is, $\mathrm{d}H/\mathrm{d}t = \mathrm{d}H/\mathrm{d}M \times \mathrm{d}M/\mathrm{d}t$ by the chain rule of calculus, which is equivalent to

$$\mathrm{d}H / \mathrm{d}t = \left(\mathrm{d}M / \mathrm{d}t\right) / \left(\mathrm{d}M / \mathrm{d}H\right) \tag{6.5}$$

For case 1, $M \propto H^3$, so $\mathrm{d}M/\mathrm{d}H \propto H^2$ and the height growth rate is constant, i.e.

$$\mathrm{d}H / \mathrm{d}t = C \tag{6.6}$$

where the constant C depends on the rate of wood production per unit crown area and the ratio of diameter to height.

One can make sense of this relation without resorting to formal calculus by considering the special case of a solid cube that adds a thin slab of material of given thickness to each side. The rate of mass accumulation is proportional to the area of its sides, i.e. $\mathrm{d}M/\mathrm{d}t \propto H^2$, where H is now the height of the cube. Since the cube maintains geometric similarity as it grows, $M \propto H^3$. Thus the cube shows the same proportional relations for mass growth ($\mathrm{d}M/\mathrm{d}t$) and the material cost of height growth ($\mathrm{d}M/\mathrm{d}H$) as the geometrically similar tree. Since it is adding the same annual growth increment to all sides, it obviously has a constant height growth rate. Of course, a solid cube would grow much more slowly than a tree because its material cost would be much greater than that of the tree.

For case 2, $M \propto H^4$, so $\mathrm{d}M/\mathrm{d}H \propto H^3$ and the height growth rate is inversely proportional to height,

$$\mathrm{d}H / \mathrm{d}t \propto 1 / H \tag{6.7}$$

For case 3, $M \propto H^4$, but $\mathrm{d}M/\mathrm{d}t$ is constant, so

$$\mathrm{d}H / \mathrm{d}t \propto 1 / H^3 \tag{6.8}$$

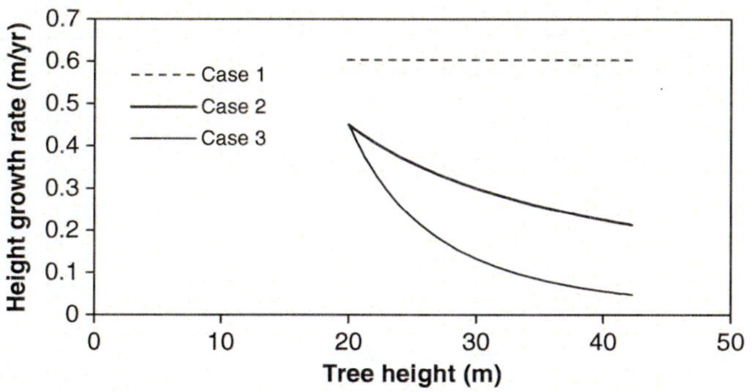

Fig. 6.1 Relation between height growth rate and tree height (H) predicted for Case 1: stem biomass $\propto H^3$ and stem wood production rate $\propto H^2$, Case 2: stem biomass $\propto H^4$ and stem wood production rate $\propto H^2$, and Case 3: stem biomass $\propto H^4$ and constant stem wood production rate, not varying with H. Trees of the three cases are assumed to have identical stem biomasses and stem wood production rates when 20 m tall. Height growth rates are lower at a height of 20 m for cases 2 and 3 because a greater biomass increment is required per height increment when biomass scales with H^4 vs. H^3. The additional divergences in height growth rate as H exceeds 20 m follow the derived scaling relations

Thus, a doubling of height causes the height growth rate to decline to one-eighth of its initial value for this case, as shown in Fig. 6.1.

As cases 1, 2 and 3 may apply approximately to young, intermediate-aged and old stands, respectively, this analysis suggests that the general decline in height growth rate in aging stands could be linked to shifts in allometry and efficiency of stem wood production.

To further examine this premise it is useful to consider the height growth curves derived by foresters to assess site productivity for the even aged stands commonly managed for timber production. An example curve for Douglas-fir was chosen because it extends to a relatively old age for this long-lived species and was based on stem analysis of sample trees, a method considered more reliable than the temporary plots used in earlier American studies (Monserud 1984, 1985). In stem analysis each tree is sectioned at many heights and counts of annual rings are used to determine its height growth history.

As shown in Fig. 6.2a, the height vs. age curve gradually levels off, implying a declining height growth rate. This pattern is shown more clearly in the graph of the height growth rate, which is the derivative of the height curve taken with respect to time (Fig. 6.2b). It is even more useful to plot the height growth rate not against age, but against tree height (Bond et al. 2007). Figure 6.2c shows that the height growth rate peaks as the trees attain a height of 10–12 m and then declines. This decline is nearly linear as tree height surpasses 20 m, while the declines predicted above for either case 2 or case 3 are more curvilinear.

However, the height growth patterns of individual trees are more variable than the smooth curves fit to many trees of varying age and site productivity. The particular functions fit to the measured patterns may also introduce biases, particularly for the more sparsely sampled older trees. The growth pattern derived from a detailed stem

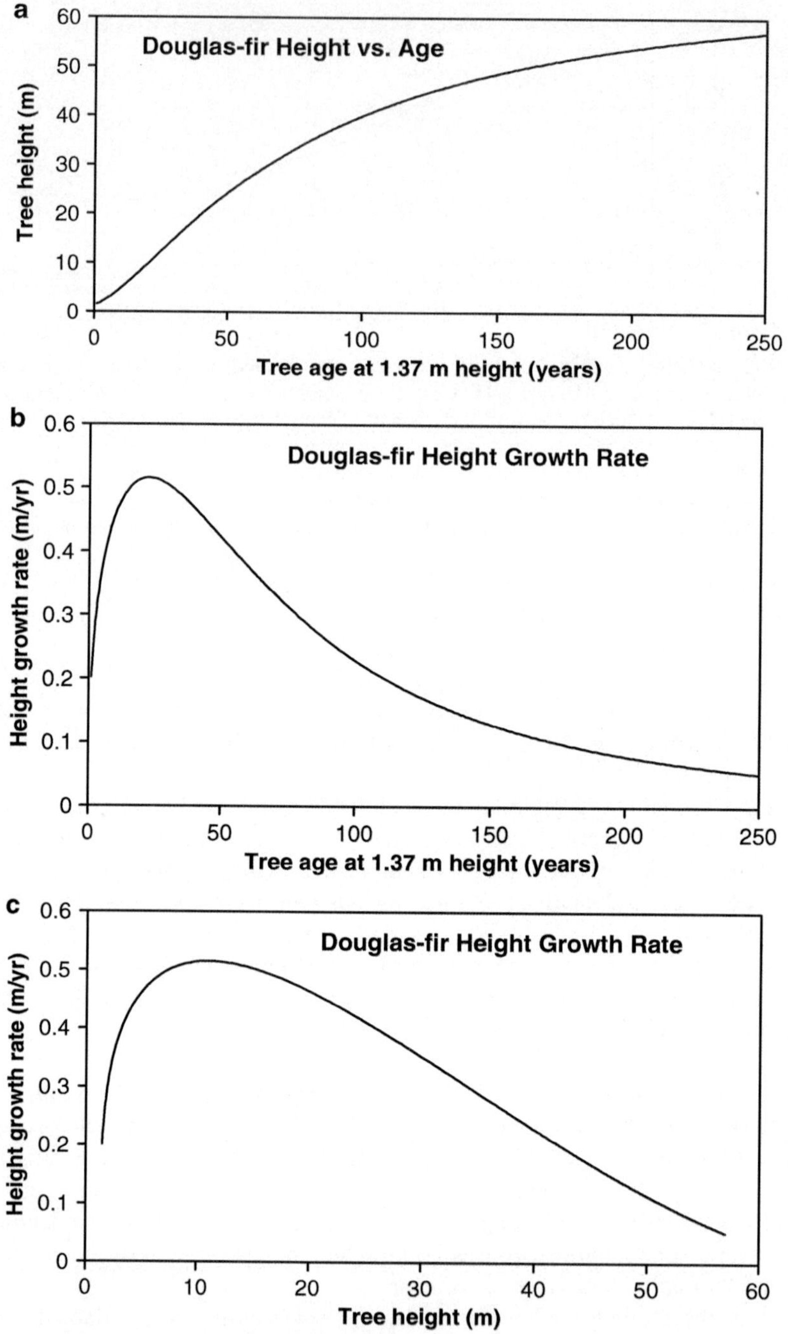

Fig. 6.2 Tree height vs. age (**a**), height growth rate vs. age (**b**) and height growth rate vs. tree height (**c**) for dominant Douglas-fir trees growing in even-aged stands on good sites at higher elevations in the Cascade Mountains of Oregon and Washington, as projected by Curtis et al. (1974), based on data from stands of a wide range in age and site quality

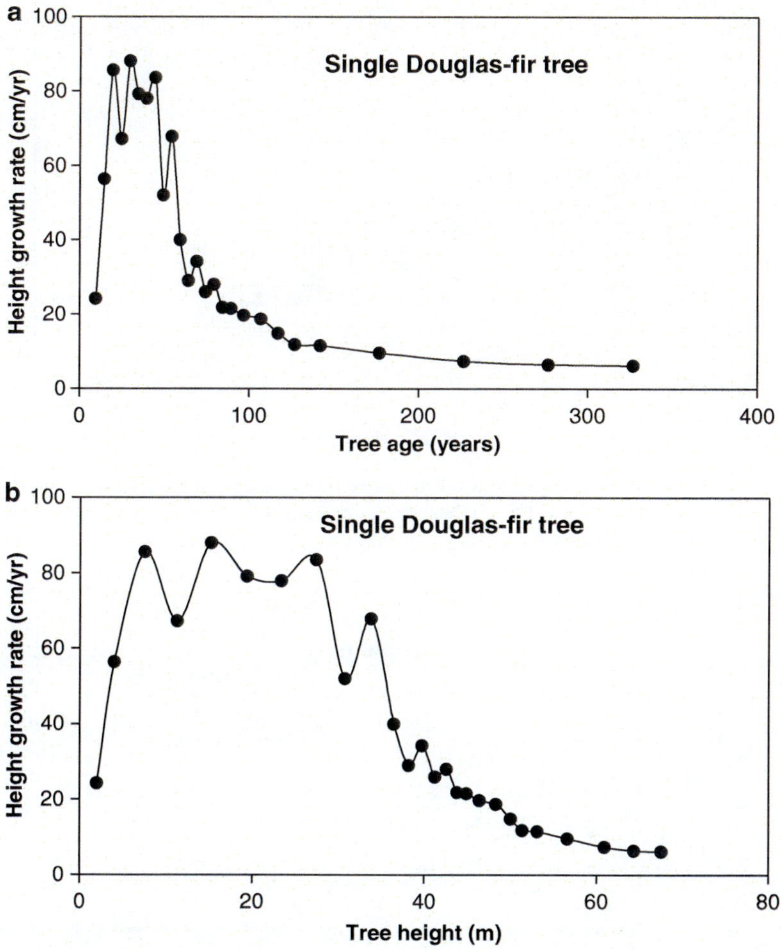

Fig. 6.3 Height growth rate vs. tree age (**a**) and tree height (**b**) for the intensively studied Douglas-fir tree of Winter et al. (2002)

analysis of an old Douglas-fir by Winter et al. (2002) (Fig. 6.3a, b) shows a roughly constant height growth rate over the 8–28 m height range followed by a curvilinear decline.

The curve in Fig. 6.3b is consistent with some combination of changing allometry and wood production efficiency, but more information on allometry and wood production is needed to distinguish these possible mechanisms. Such information was published for an extremely tall noble fir by Herman et al. (1975) to illustrate the method of stem analysis. They sliced the tree into 45 sections and measured the positions of every tenth growth ring over the 437-year life of the tree, as shown in Fig. 6.4. These data can be used to calculate the volume of wood added to the trunk and the corresponding height growth over each 10-year period, as well as the relationship between trunk diameter and tree height over the entire life of the tree.

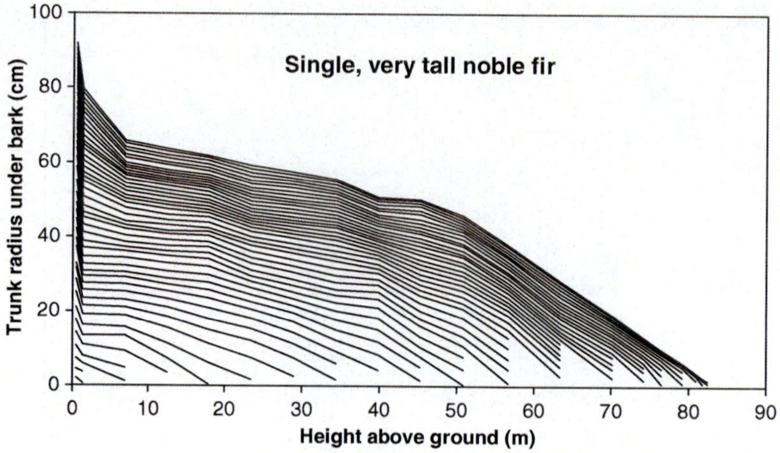

Fig. 6.4 Shape of the stem of a 437 year-old noble fir tree (shown lying on its side) at 10 year intervals, as given by Herman et al. (1975). Note the exaggeration of the scale for trunk radius; the tree was actually quite slender for its height

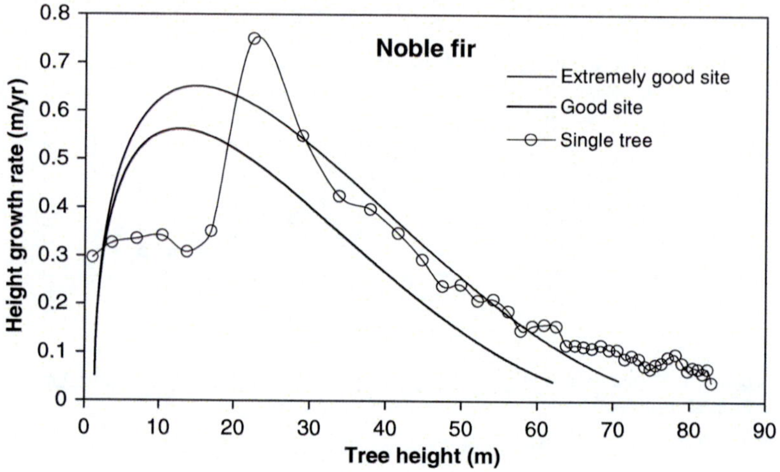

Fig. 6.5 Height growth rate of the noble fir tree sectioned by Herman et al. (1975), as compared to the projected height growth rate of dominant noble fir trees in even-aged stands determined by Herman et al. (1978). Each point shows the sectioned tree's growth rate over a 10-year interval

The height growth rate of this tree is plotted as a function of height in Fig. 6.5, as compared with the projected height growth rates of even-aged noble fir stands on productive and very productive sites. The tree's early height growth rate is lower than projected, followed by a spike in growth at a height of 19–26 m. The growth rate then declines, following the curve for trees on very good sites until reaching a height of 60 m. From here on the observed height growth rate declines more gradually than does the projected growth rate. However, Herman et al. (1978) noted

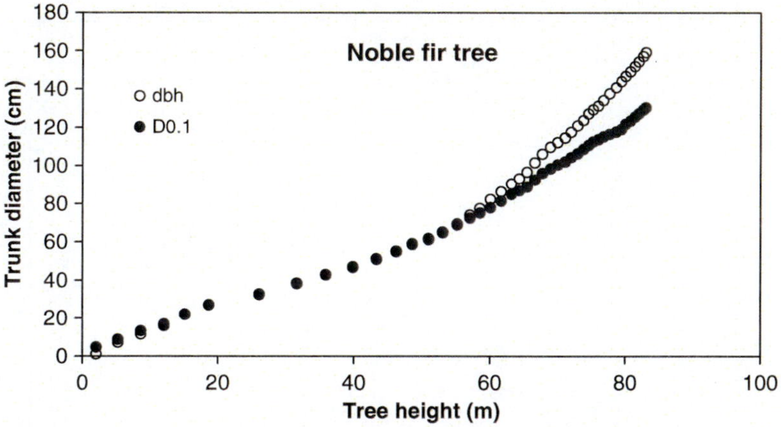

Fig. 6.6 Trunk diameter under the bark at breast height (dbh) and at 1/10th of tree height (*D0.1*) vs. tree height. The points show the height and diameter every 10 years for the tree sectioned by Herman et al. (1975)

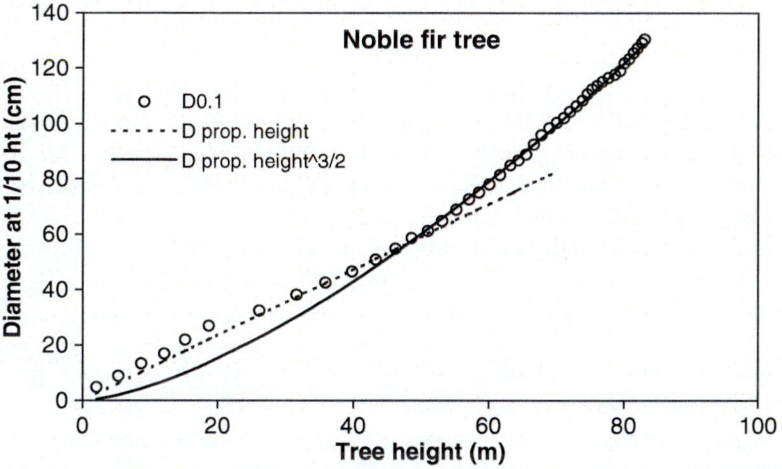

Fig. 6.7 Trunk diameter inside the bark at 1/10th of tree height vs. tree height, as compared to diameter proportional to height (*dotted line*) and diameter proportional to height$^{3/2}$ (*solid line*)

prolonged height growth in very old trees and therefore omitted trees >260 years old from the data used to construct the height growth curves.

This pattern of height growth can be related to the allometry and wood production of the tree over its life. However, diameter measured at breast height is a biased measure of stem thickness due to increasing butt swell (Fig. 6.6) and dbh is a poor predictor of the volumes of large old trees (Van Pelt 2001). Diameter at 1/10th of tree height avoids the region of butt swell and is hence used instead.

Based on diameter at one-tenth of tree height, Fig. 6.7 indicates that the tree was relatively thick-trunked until it reached a height of 19 m, whereas diameter was

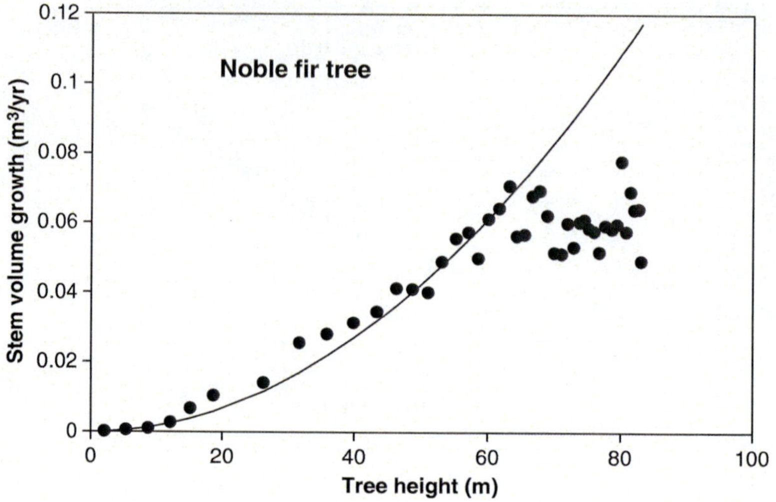

Fig. 6.8 Stem volume growth rate of the noble fir determined from the volumes of the concentric shells of wood shown in cross section in Fig. 6.4. The line indicates a production rate proportional to height squared

directly proportional to height as it grew from 25 to 50 m. The stem showed anomalously low diameter growth during the decade in which the tree grew from 19 to 26 m (Fig. 6.7) and this transient change in allometry is associated with the spike in height growth rate over this interval (Fig. 6.5).

A remarkable feature of this tree is that diameter at 1/10th height was almost exactly proportional to $H^{3/2}$ over the last 300 years of its life, as it grew from 50 to 83 m tall. During this long period, the tree was only about 30% thicker than the minimum diameter that an untapered pole of similar height would require to keep from bending over under its own weight – calculated for the physical properties of green noble fir wood (USDA Forest Products Laboratory 1999).

The stem volume growth rate increased curvilinearly with height for the first 200 years of life and then remained relatively constant as height exceeded 60 m (Fig. 6.8). Because stem diameter $\propto H^{3/2}$ over this latter interval, the prediction of case 3 that $dH/dt \propto 1/H^3$ should apply. The observed height growth rate showed a substantial, but variable decline over this interval (Fig. 6.5). This variation may be due to the variation in volume growth (Fig. 6.8) and transient shifts in stem taper, which would change dM/dH, the biomass increment per unit height growth. However, the observation that the height growth rate for the tree at 80 m was about 2/5 of that at 60 m agrees closely with the prediction of case 3 that the 80 m height growth rate should decline to $(3/4)^3 = 42\%$ of the rate at 60 m.

This analysis suggests that overall patterns of height growth can be related to tree allometry and wood production rates, though trees may show substantial decadal variation about these general patterns (Figs. 6.3 and 6.5). An important caveat is that the example tree considered here was likely selected for illustration by Herman et al. (1975) for its extreme height, which was only 7 m less than the 90 m height of the current tallest known noble fir (Van Pelt 2001). The noble fir trees used for analysis by Herman et al. (1978) were each chosen as the single tallest undamaged dominant tree per 0.1 ha plot within stands dominated by even-aged trees. Although the tree considered here may be atypical, the general tendency of diameter to scale with an increasingly higher power of height as trees grow taller (Observed Patterns section) will cause height growth rates to decline, thereby contributing to the universal decline in height growth rate among aging trees.

5 Discussion

Tree proportions vary with microenvironment and species as reviewed in this chapter. However, tree stems commonly show increases in the ratio of diameter to total height as trees age. This shift in allometry may be particularly apparent in large, long-lived species, such as coast redwood (Fujimori 1977). Among trees of even-aged stands, height growth rates decline with increasing stature after peaking at a young age. This pattern was linked in part to changes in allometry for the example 83-m-tall noble fir considered here. However, other factors may limit the ultimate maximum heights of trees (Koch et al. 2004; Domec et al. 2008).

Whole stand rates of stem wood production and wood production per unit leaf area or light interception commonly decline as stands age (Ryan et al. 1997) and this effect will also reduce height growth rates by increasing the time required to synthesize the wood required for a given height increment. The cause of this decline is still uncertain, though it is likely related in part to water transport (Ryan et al. 2006; Martinez-Vilalta et al. 2007). The extent of the decline is also uncertain, as it is based primarily on forest growth and yield tables that are mostly derived from young to intermediate aged stands, as are commonly managed for wood production (e.g. Curtis et al. 1982). Thorough measurements of wood production in old-growth trees by Sillett et al. (2010) indicate that the world's tallest species, coast redwood, shows little further decline in the efficiency of stem wood production after an initial decline during the first 200 years of growth – on the most productive sites.

For the example tree considered here, gravitational constraints on stem thickness coupled with a more or less constant rate of stem wood production later in life accounted for the overall decline in height growth rate from midlife on. However, the links between height growth and mechanical constraints are inexact, as there is no strict requirement for a particular allometry, other than that the critical buckling diameter be exceeded – in free standing trees. Although breakage is a common cause of tree death (Putz et al. 1983), tradeoffs between height growth and mechanical safety are expected and will vary with environment, crowding by competitors and the ecological characteristics of trees (Givnish 1995).

The loose nature of mechanical constraints is nicely illustrated by the anomalously high height growth rate between ages of 57 and 67 years for the example noble fir tree (Fig. 6.5). During this interval diameter growth was lower over the lower bole and higher over the upper bole than in the preceding and following 10-year intervals – and the taper of the stem top shifted to a more slender form that was progressively reversed over the following decades (Fig. 6.4). However, continuation of the reduced lower bole thickening that contributed to this high extension rate would have put the tree at increasing risk of breakage.

Although support requirements eventually cause unavoidable declines in height growth rate, they do not necessitate the cessation of extension growth or directly limit tree height. Maximum height growth rates are typically attained when trees are less than 1/3 of their maximum heights (e.g. Curtis et al. 1974) and the factors affecting height growth may vary over the life of a tree. Trees of particularly long lived species may show more pronounced increases in the ratio of diameter to height growth than the tall noble fir considered here, as might be expected if other factors were limiting extension growth later in life. Recent studies indicate that the maximum heights of two of the tallest species, coast redwood and Douglas fir are limited by water transport (Koch et al. 2004; Domec et al. 2008).

On the other hand, old-growth trees may be more susceptible to breakage and tip up than trees of younger stands, despite their proportionately thicker stems – due to increases in wind speed with height and the prevalence of rot and other wood defects in old trees (King 1986; Frelich and Reich 1996; Van Pelt and Sillett 2008). The breaking strength of wood decreases with beam size, even for sound wood, due to increases in the probability of encountering stress-concentrating wood defects – though this effect is small (Bohannan 1966). Thus, continued diameter growth, as needed to overgrow rotten or damaged areas and maintain functional sapwood, likely increases long-term survival in trees not subject to overtopping by neighbors.

Many old trees cease height growth due to top breakage facilitated by fungal decay (Van Pelt and Sillett 2008) or dieback that may be due to hydraulic limits to water transport that are exacerbated during severe drought years. More than half of the 450 year-old Douglas-fir trees examined by Ishii et al. (2000) showed top dieback or broken tops. Species that are longer lived than noble fir, such as Douglas-fir, western redcedar (*Thuja plicata*) and coast redwood (*Sequoia sempervirens*) may attain maximum height in midlife preceding top breakage (common in Douglas-fir), pronounced top dieback and partial replacement (western redcedar) or localized top dieback and replacement (coast redwood) (Van Pelt 2001; Van Pelt and Nadkarni 2004; Koch et al. 2004). Stem diameter growth continues throughout life – independently of height increase or decrease in this late stage of growth.

Trees of old, all-aged forests exhibit high variation in growth and the growth rates of small juveniles are much reduced by the shade of taller neighbors (e.g. Clark and Clark 2001; Condit et al. 1999). Trees of such forests may also show substantial alterations in the ratio of height to diameter growth as they grow up towards the canopy, likely reflecting shifts in light environment and the positions of neighboring crowns (Sumida et al. 1997). Saplings of shade tolerant *Abies* species vary greatly in crown shape and branch vs. leader growth, with umbrella shaped crowns that provide efficient light interception in shade and conical crowns that achieve greater height growth per biomass increment in forest openings (Kohyama 1980; King 1997).

Thus, there is much circumstantial evidence that trees regulate apical and cambial growth in an adaptive fashion that depends on size and exposure to light and wind, as cited in this review. The physiological and genetic mechanisms that underlie this regulation are beyond the scope of this chapter, but are an area of ongoing research (e.g. Day and Greenwood 2011). However, the need for a general slowing of height growth to prevent undue risks of breakage or toppling suggests that declines in water potential and turgor pressure with increasing height (Woodruff et al. 2004; Woodruff and Meinzer 2011) could act as a positional signal to reduce extension growth, as well as directly limiting growth.

5.1 Future Research Needs

Expanded measurements of whole tree growth and allometry are needed to better define the links between height growth and allometry and integrate them with hypothesized physiological controls of growth, as listed below.

1. Direct measures of the allometric trajectories of individual trees are needed, as illustrated for the example noble fir. Many allometric analyses are based on samples of trees of varying size, drawn from forest plots, with the tacit assumption that individual trajectories will on average mirror the allometries derived from trees of differing size measured at one time. But most current forests are undergoing succession and even among all-aged forests in quasi-equilibrium, the juveniles that survive to reach the canopy may differ in proportions from current populations that include many suppressed trees that are destined to die.
2. Direct measures of the wood production of individuals are also needed, as trees of different crown classes show different efficiencies of wood production (Binkley et al. 2002), and trees may shift in crown position as stands age. Estimates of stem volume and growth are often inferred from dbh, based on empirical allometric equations fit to the stands of interest. But, the allometric trajectories of growing individuals may differ systematically from the within-stand allometries measured at a given time, as noted above. Furthermore, the thickness of annual growth rings at breast height vs. the average ring thickness along the whole bole varies considerably with degree of crowding and stand disturbance history (Kozlowski 1971). Thus, inferring past production from ring widths at breast height is problematical. Dbh is a poor predictor of stem volume for large trees due to butt swell (Van Pelt 2001), and diameter must be measured well above the progressively rising buttresses to insure accurate growth histories for large tropical trees.

   These direct measures of growth and proportions could be accomplished by stem analysis (Herman et al. 1975; Monserud 1985), although this method is limited to temperate and boreal trees and certain tropical species, mostly from forests with strong dry seasons, where annual growth rings can be reliably measured (Brienen and Zuidema 2005) – but see Fichtler et al. (2003) regarding growth rings in wet tropical forest trees. Climbing and coring trees from base to top is a less destructive method of gaining this information for those skilled in climbing trees (Sillett et al. 2010).

3. Further studies of the physiological controls of height growth that might integrate mechanisms related to allometry, carbon balance and water transport are also needed. One possible approach would be to alter allometry by cutting treetops and assessing the degree and rapidity of top replacement as a function of tree height and age – and any corresponding shifts in the distribution of diameter growth along the bole. Many trees experience stem breakage during the course of their lives and the capacity for rapid regrowth would be of particular advantage for trees of young stands, where height growth is still relatively rapid and the costs of failing to keep abreast of neighboring trees are greater than for more widely-spaced old-growth trees. Top cutting might also increase water potentials near the cut by reducing foliage area and hence water demand at this point. Hence, there may be links between allometric recovery and shifts in water potential.
4. Studies of allometric trajectories and the size dependence of height growth and wood production are also needed for regions outside of the Pacific Northwest and California, where different climatic and/or ecological conditions pertain, resulting in the evolution of trees with different crown proportions and height growth patterns than the tall, long-crowned conifers emphasized here (Falster and Westoby 2003). For example, lowland tropical forests contain a high diversity of understory and midstory species and the tallest trees tend to be widely spaced emergent species that expand their crowns out in all directions above the surrounding canopy (King 1996; Richards 1996). Certain disturbance-associated tropical trees, such as balsa (*Ochroma pyrimidale*), show extremely rapid height growth in youth, but slow greatly through adulthood. Palms show stem and crown allometries that differ substantially from those of dicotyledonous trees (Rich et al. 1986, 1995; Tomlinson 1990) and the implications of these differences for the course of height growth have yet to be explored.

Acknowledgements I thank Robert Monserud and an anonymous reviewer for helpful comments on the manuscript.

References

Aide TM (1987) Limbfalls: a major cause of sapling mortality for tropical forest plants. Biotropica 19:284–285

Alexander RR, Tackle D, Dahms WG (1967) Site indexes for lodgepole pine, with corrections for stand density: methodology. USDA Forest Service Research Paper RM – 29. Rocky Mountain Forest and Range Experiment Station, Fort Collins

Anton NPR, Schieving F (2010) The role of wood mass density and mechanical constraints in the economy of tree architecture. Am Nat 175:250–260

Beck DE, Della-Bianca L (1970) Yield of unthinned yellow-poplar. USDA Forest Service Research Paper SE-58. Southeast Forest Experiment Station, Ashville

Bertram JEA (1989) Size-dependent differential scaling in branches: the mechanical design of trees revisited. Trees 4:241–253

Binkley D, Stape JL, Ryan MG, Barnard HR, Fownes JH (2002) Age-related decline in forest ecosystem growth: an individual-tree, stand-structure hypothesis. Ecosyst 5:58–67
Bohannan B (1966) Effects of size on bending strength of wood members. U.S. Forest Service Research Paper FPL 56. USDA Forest Service, Forest Products Laboratory, Madison
Bond BJ, Czarnomski C, Cooper C, Day ME, Greenwood MS (2007) Developmental decline in height growth in Douglas-fir. Tree Physiol 27:441–453
Bormann FH (1965) Changes in the growth patterns of white pine trees undergoing suppression. Ecol 46:269–277
Brienen RJW, Zuidema PA (2005) Relating tree growth to rainfall in Bolivian rain forests: a test for six species using tree ring analysis. Oecologia 146:1–12
Büsgen M, Münch E (1929) The structure and life of forest trees, 3rd edn (trans: Thomson T). Chapman & Hall Ltd., London
Clark DA, Clark DB (2001) Getting to the canopy: tree height growth in a Neotropical rain forest. Ecol 82:1460–1472
Condit R, Ashton PS, Manokaran N, LaFrankie JV, Hubbell SP, Foster RB (1999) Dynamics of the forest communities at Pasoh and Barro Colorado: comparing two 50-ha plots. Philos Trans R Soc Lond B 334:1739–1748
Coutand C, Moulia B (2000) Biomechanical study of the effect of a controlled bending on tomato stem elongation: local strain sensing and spatial integration of the signal. J Exp Bot 51:1825–1842
Coutand C, Martin L, Leblanc-Fournier N, Decourteix M, Julien J-L, Moulia B (2009) Strain mechanosensing quantitatively controls diameter growth and PtaZFP2 gene expression in poplar. Plant Physiol 151:223–232
Curtis RO, Herman FR, DeMars DJ (1974) Height growth and site index for Douglas-fir in high-elevation forests of the Oregon-Washington Cascades. For Sci 20:307–316
Curtis RO, Clendenen GW, Reukema DL, DeMars DJ (1982) Yield tables for managed stands of coast Douglas-fir. Gen. Tech. Rep. PNW-135. US Dept. of Agriculture Forest Service, Portland
Day ME, Greenwood MS (2011) Regulation of ontogeny in temperate conifers. In: Meinzer FC, Lachenbruch B, Dawson TE (eds.) Size- and age-related changes in tree structure and function. Springer, Dordrecht, pp 91–119
Domec J-C, Lachenbruch B, Meinzer FC, Woodruff DR, Warren JM, McCulloh KA (2008) Maximum height in a conifer is associated with conflicting requirements for xylem design. Proc Natl Acad Sci USA 105:12069–12074
Ek AR (1974) Dimensional relationships of forest and open grown trees in Wisconsin. University of Wisconsin School of Natural Resources Forestry Research Note 181
Falster DS, Westoby M (2003) Plant height and evolutionary games. Trends Ecol Evol 18:337–343
Farnsworth KD, Niklas KJ (1995) Theories of optimization, form and function in branching architecture of plants. Funct Ecol 9:355–363
Fichtler E, Clark DA, Worbes M (2003) Age and long-term growth in an old-growth tropical rain forest, based on analyses of tree rings and ^{14}C. Biotropica 35:306–317
Flewelling J, Collier R, Bonyea B, Marshall D, Turnblom E (2001) Height-age curves for planted stands of Douglas-fir, with adjustments for density. Stand management Cooperative Working Paper Number 1. College of Forest Resources, Univ. Seattle, 34 p
Frelich LE, Reich PB (1996) Old growth in the Great Lakes region. In: Davis MB (ed.) Eastern old-growth forests: prospects for rediscovery and recovery. Island Press, Washington, DC, pp 144–160
Fujimori T (1977) Stem biomass and structure of a mature *Sequoia sempervirens* stand on the Pacific Coast of northern California. J Jpn For Soc 59:435–441
Givnish TJ (1995) Plant stems: biomechanical adaptation for energy capture and influence on species distributions. In: Gartner BL (ed.) Plant stems: physiology and functional morphology. Academic, San Diego, pp 3–49
Greene DF, Jones KF, Proulx OJ (2007) The effects of icing events on the death and regeneration of North American trees. In: Johnson EA, Miyanishi K (eds.) Plant disturbance ecology. Elsevier, Amsterdam, pp 181–213

Greenhill AG (1881) Determination of the greatest height consistent with stability that a vertical pole or mast can be made, and the greatest height to which a tree of given proportions can grow. Proc Cambridge Philos Soc 4:65–73

Herman FR, Curtis RO DeMars DJ (1978) Height growth and site index estimates for noble fir in high-elevation forests of the Oregon-Washington Cascades. USDA Forest Service Research Paper PNW-243, Pacific Northwest Forest and Range Experimental Station, Portland

Herman FR, DeMars DJ, Woollard RF (1975) Field and computer techniques for stem analysis of coniferous forest trees. USDA Forest Service Research Paper PNW-194. Pacific Northwest Forest and Range Experiment Station, Portland

Holbrook NM, Putz FE (1989) Influence of neighbors on tree form: effects of lateral shade and prevention of sway on the allometry of *Liquidambar styraciflua* (sweet gum). Am J Bot 76:1740–1749

Ickes K, DeWalt SJ, Thomas SC (2003) Resprouting of woody saplings following stem snap by wild pigs in a Malaysian rain forest. J Ecol 91:222–233

Ishii H (2011) How do changes in leaf/shoot morphology and crown architecture affect growth and physiological function of tall trees? In: Meinzer FC, Lachenbruch B, Dawson TE (eds.) Size- and age-related changes in tree structure and function. Springer, Dordrecht, pp 215–232

Ishii H, Reynolds JH, Ford ED, Shaw DC (2000) Height growth and vertical development of an old-growth *Pseudotsuga-Tsuga* forest in southwestern Washington State, USA. Can J For Res 30:17–24

Jacobs MR (1954) The effect of wind sway on the form and development of *Pinus radiata* D. Don. Aust J Bot 2:35–51

King D (1981) Tree dimensions: maximizing the rate of height growth in dense stands. Oecologia 51:351–356

King DA (1986) Tree form, height growth, and susceptibility to wind damage in *Acer saccharum*. Ecology 67:980–990

King DA (1987) Load bearing capacity of understory treelets of a tropical wet forest. Bull Torrey Bot Club 114:419–428

King DA (1990) The adaptive significance of tree height. Am Nat 135:809–828

King DA (1991) Tree allometry, leaf size and adult tree size in old-growth forests of western Oregon. Tree Physiol 9:369–381

King DA (1996) Allometry and life history of tropical trees. J Trop Ecol 12:25–44

King DA (1997) Branch growth and biomass allocation in *Abies amabilis* saplings in contrasting light environments. Tree Physiol 17:251–258

King DA (2005) Linking tree form, allocation and growth with an allometrically explicit model. Ecol Model 185:77–91

King DA, Davies SJ, Tan S, Nur Supardi MN (2006) The role of wood density and stem support costs in the growth and mortality of tropical trees. J Ecol 94:670–680

King DA, Davies SJ, Tan S, Nur Supardi MN (2009) Trees approach gravitational limits to height in tall lowland forests of Malaysia. Funct Ecol 23:284–291

Kira T (1978) Community architecture and organic matter dynamics in tropical lowland rain forests of Southeast Asia with special reference to Pasoh Forest, West Malaysia. In: Tomlinson PB, Zimmermann MH (eds.) Tropical trees as living systems. Cambridge University Press, Cambridge, pp 561–590

Koch GW, Sillett SC, Jennings GM, Davis SD (2004) The limits to tree height. Nature 428:851–854

Kohyama T (1980) Growth pattern of *Abies mariesii* saplings under conditions of open-growth and suppression. Bot Mag Tokyo 93:13–24

Kozlowski TT (1971) Growth and development of trees, vol II. Academic, New York

Lachenbruch B, Moore JR, Evans R (2011) Radial variation in wood structure and function in woody plants, and hypotheses for its occurrence. In: Meinzer FC, Lachenbruch B, Dawson TE (eds.) Size- and age-related changes in tree structure and function. Springer, Dordrecht, pp 121–164

Lawton RO (1982) Wind stress and elfin stature in a montane rain forest tree: an adaptive explanation. Am J Bot 69:1224–1230

Leiser AT, Kemper JT (1973) Analysis of stress distribution in the sapling tree trunk. J Am Soc Hort Sci 98:164–170

Long JN, Smith FW, Scott DRM (1981) The role of Douglas fir stem sapwood and heartwood in the mechanical and physiological support of crowns and development of stem form. Can J For Res 11:459–464

Maguire DA, Hann DW (1989) The relationship between gross crown dimensions and sapwood area at crown base in Douglas-fir. Can J For Res 29:557–565

Mäkelä A (2002) Derivation of stem taper from the pipe theory in a carbon balance framework. Tree Physiol 22:891–905

Mäkelä A, Valentine HT (2006) Crown ratio influences allometric scaling in trees. Ecology 87:2967–2972

Martinez-Vilalta J, Vanderklein D, Mencuccini M (2007) Tree height and age-related decline in growth in Scots pine (*Pinus sylvestris* L.). Oecologia 150:529–544

Mattheck C (1998) Design in nature: learning from trees. Springer, Berlin

Mattheck C, Kubler H (1995) Wood – the internal optimization of trees. Springer, Berlin

McDowell NG, Bond BJ, Hill LT, Ryan MG, Whitehead D (2011) Relationships between tree height and carbon isotope discrimination. In: Meinzer FC, Lachenbruch B, Dawson TE (eds.) Size- and age-related changes in tree structure and function. Springer, Dordrecht, pp 255–286

McMahon TA (1973) Size and shape in biology. Science 179:1201–1204

McMahon TA (1975) The mechanical design of trees. Sci Am 233:92–102

McMahon TA, Kronauer RE (1976) Tree structures: deducing the principle of mechanical design. J Theor Biol 59:443–466

Mencuccini M, Martinez-Vilalta J, Vanderklein D, Hamid HA, Korakaki E, Lee S, Michiels B (2005) Size-mediated aging reduces vigor in tall trees. Ecol Lett 8:1183–1190

Mencuccini M, Höltä T, Martinez-Vilalta J (2011) Comparative criteria for models of the vascular transport systems of tall trees. In: Meinzer FC, Lachenbruch B, Dawson TE (eds.) Size- and age-related changes in tree structure and function. Springer, Dordrecht, pp 309–339

Monserud RA (1984) Problems with site index: an opinionated review. In: Bockheim J (ed.) Forest Land Classification: Experiences, Problems, Perspectives. Symposium Proceedings, University of Wisconsin, Dept. of Soil Science, Madison, pp 167–180

Monserud RA (1985) Comparison of Douglas-fir site index and height growth curves in the Pacific Northwest. Can J For Res 15:673–679

Monserud RA (2003) Modeling stand growth and development. In: Monserud RA, Haynes RW, Johnson AC (eds.) Compatible forest management. Kluwer, Dordrecht, pp 145–173

Morgan J, Cannell MGR (1994) Shape of tree stems – a re-examination of the uniform stress hypothesis. Tree Physiol 14:49–62

Niklas KJ (1993) Influence of tissue-density-specific mechanical properties on the scaling of plant height. Ann Bot 72:165–172

Niklas KJ (1994) Interspecific allometries of critical buckling height and actual plant height. Am J Bot 8:1275–1279

Niklas KJ, Spatz H-C (2000) Wind induced stresses in cherry trees: evidence against the hypothesis of constant stress levels. Trees 14:230–237

Ossenbruggen PJ, Peters MA, Shigo AL (1986) Potential failure of a decayed tree under wind loading. Wood Fiber Sci 18:168–185

Osunkoya OO, Omar-Ali K, Amit N, Dayan J, Daud DS, Sheng TK (2007) Comparative height–crown allometry and mechanical design in 22 tree species of Kuala Belalong rainforest, Brunei, Borneo. Am J Bot 94:1951–1962

Poorter L, Bongers L, Bongers F (2006) Architecture of 54 moist-forest tree species: traits, tradeoffs, and functional groups. Ecology 87:1289–1301

Pruyn ML, Gartner BL, Harmon ME (2002) Respiratory potential in sapwood of old versus young ponderosa pine trees in the Pacific Northwest. Tree Physiol 22:105–116

Pruyn ML, Gartner BL, Harmon ME (2005) Storage versus substrate limitation to bole respiratory potential in two coniferous tree species of contrasting sapwood width. J Exp Bot 56:2637–2649

Putz FE, Coley PD, Lu K, Montalvo A, Aiello A (1983) Uprooting and snapping of trees: structural determinants and ecological consequences. Can J For Res 13:1011–1020
Quine CP, Gardiner BA (2007) Understanding how the interaction of wind and trees result in wind-throw, stem breakage, and canopy gap formation. In: Johnson EA, Miyanishi K (eds.) Plant disturbance ecology. Elsevier, Amsterdam, pp 103–155
Rich PM (1987a) Developmental anatomy of the stem of *Welfia georgii, Iriartea gigantea*, and other arborescent palms: implications for mechanical support. Am J Bot 74:792–802
Rich PM (1987b) Mechanical structure of the stem of arborescent palms. Bot Gaz 148:42–50
Rich PM, Helenurm K, Kearns D, Morse SR, Palmer MW, Short L (1986) Height and stem diameter relationships for dicotyledonous trees and arborescent palms of Costa Rican tropical wet forest. Bull Torrey Bot Club 113:241–246
Rich PM, Holbrook NM, Luttinger N (1995) Leaf development and crown geometry of two Iriartoid palms. Am J Bot 82:328–336
Richards PW (1996) The tropical rain forest: an ecological study, 2nd edn. Cambridge University Press, Cambridge
Ryan MG, Waring RH (1992) Maintenance respiration and stand development in a subalpine lodgepole pine forest. Ecology 73:2100–2108
Ryan MG, Yoder BJ (1997) Hydraulic limits to tree height and tree growth. Bioscience 47:235–242
Ryan MG, Binkley D, Fownes JH (1997) Age-related decline in forest productivity: pattern and process. Adv Ecol Res 27:213–262
Ryan MG, Phillips N, Bond BJ (2006) The hydraulic limitation hypothesis revisited. Plant Cell Environ 29:367–381
Sala A, Fouts W, Hoch G (2011) Carbon storage in trees: does relative carbon supply decrease with tree size? In: Meinzer FC, Lachenbruch B, Dawson TE (eds.) Size- and age-related changes in tree structure and function. Springer, Dordrecht, pp 287–306
Shinozaki K, Yoda K, Hozumi K, Kira T (1964) A quantitative analysis of plant form: the pipe model theory. I. Basic analyses. Jpn J Ecol 14:97–105
Shugart HH (1984) A theory of forest dynamics: the ecological implications of forest succession models. Springer, New York
Sillett SC, Van Pelt R, Cook GW, Ambrose AR, Carroll AL, Antoine ME, Mifsud BM (2010) Increasing wood production through old age in tall trees. For Ecol Manag 259:976–994
Sjolte-Jorgensen J (1967) The influence of spacing on the growth and development of coniferous plantations. Internatl Rev For Res 2:43–94
Spurr SH, Barnes BV (1980) Forest ecology, 3rd edn. Wiley, New York
Steppe K, Niinemets U, Teskey RO (2011) Tree size- and age-related changes in leaf physiology and their influence on carbon gain. In: Meinzer FC, Lachenbruch B, Dawson TE (eds.) Size- and age-related changes in tree structure and function. Springer, Dordrecht, pp 235–253
Sterck FJ, Bongers F (1998) Ontogenetic changes in size, allometry, and mechanical design of tropical rain forest trees. Am J Bot 85:266–272
Sterck FJ, van Gelder HA, Poorter L (2006) Mechanical branch constraints contribute to life-history variation across tree species in a Bolivian forest. J Ecol 94:1192–1200
Sumida A, Ito H, Isagi Y (1997) Trade-off between height growth and stem diameter growth for an evergreen Oak, *Quercus glauca*, in a mixed hardwood forest. Funct Ecol 11:300–309
Thomas SC (1996) Asymptotic height as a predictor of growth and allometric characteristics in Malaysian rain forest trees. Am J Bot 83:556–566
Thomas SC (2011) Age-related changes in tree growth and functional biology: the role of reproduction. In: Meinzer FC, Lachenbruch B, Dawson TE (eds.) Size- and age-related changes in tree structure and function. Springer, Dordrecht, pp 33–64
Tomlinson PB (1990) The structural biology of palms. Clarendon, Oxford
Tyree MT, Zimmermann MH (2002) Xylem structure and the ascent of sap. Springer, Berlin
USDA Forest Products Laboratory (1999) Wood handbook: wood as an engineering material. Gen. Tech. Rep. FPL-GTR-113. USDA Forest Service, Forest Products Laboratory, Madison
van Gelder HA, Poorter L, Sterck FJ (2006) Wood mechanics, allometry, and life history variation in a tropical tree community. New Phytol 171:367–378

Van Pelt R (2001) Forest giants of the Pacific Coast. Global Forest Society, Vancouver
Van Pelt R, Nadkarni NM (2004) Development of canopy structure in *Pseudotsuga menziesii* forests in the Southern Washington Cascades. For Sci 50:326–341
Van Pelt R, Sillett SC (2008) Crown development of coastal *Pseudotsuga menziesii*, including a conceptual model for tall conifers. Ecol Monogr 78:283–311
Vogel S (1984) Drag and flexibility in sessile organisms. Am Zool 24:37–44
Vogel S (1989) Drag and reconfiguration of broad leaves in high winds. J Exp Bot 40:941–948
Waterhouse JT, Quinn FLS, Quinn CJ (1978) Growth patterns in the stem of the palm *Archontophoenix cunninghamiana*. Bot J Lin Soc 77:73–93
Wiemann MC, Williamson GB (1989) Radial gradients in the specific gravity of wood in some tropical and temperate trees. For Sci 35:197–210
Wilson BF, Archer RR (1979) Tree design: some biological solutions to mechanical problems. Bioscience 29:293–298
Winter LE, Brubaker LB, Franklin JF, Miller EA, DeWitt DQ (2002) Initiation of an old-growth Douglas-fir stand in the Pacific Northwest: a reconstruction from tree-ring records. Can J For Res 32:1039–1056
Woodruff DR, Meinzer FC (2011) Size-dependent changes in biophysical control of tree growth: the role of turgor. In: Meinzer FC, Lachenbruch B, Dawson TE (eds.) Size- and age-related changes in tree structure and function. Springer, Dordrecht, pp 363–384
Woodruff DR, Bond BJ, Meinzer FC (2004) Does turgor limit growth in tall trees? Plant Cell Environ 27:229–236
Zobel BJ, van Buijtenen JP (1989) Wood variation: its causes and control. Springer, Berlin

Chapter 7
Influence of Tree Ontogeny on Plant-Herbivore Interactions

Karina Boege, Kasey E. Barton, and Rodolfo Dirzo

Abstract As trees develop, they undergo morphological and physiological changes that can influence not only their performance, but also their interactions with herbivores. The expression of their defenses is influenced by changes in the selective pressures exerted by herbivores and by the plant's tradeoffs in resource allocation, and can result in ontogenetic trajectories that show increases, decreases, or mixed trends in the expression of anti-herbivory traits. In some species, these trajectories occur as gradual transitions among ontogenetic stages, but in other species there are pronounced phase changes marked by heterophylly or by abrupt changes in chemical, physical, or biotic defenses. This chapter discusses the patterns of such trajectories and the multiple factors that can influence them, including the specific herbivores feeding on trees, the activities of herbivores' natural enemies, the switches among the different defensive mechanisms that trees can express to reduce herbivory, the particular ecosystem in which they grow, and the influence of phylogenetic constraints that restrict or allow the evolution of ontogenetic trajectories in plant defense. Studies that integrate the role of ontogeny into evolutionary ecology theory will advance our understanding of how natural selection can target the ontogenetic trajectories of plant defense. Such research will also have application for targeting pest control onto vulnerable ontogenetic stages, and for selection of lines

K. Boege (✉)
Instituto de Ecología, Universidad Nacional Autónoma de México, Apartado Postal 70–275, Ciudad Universitaria, UNAM, 04510 Mexico D.F., Mexico
e-mail: kboege@servidor.unam.mx

K.E. Barton
Botany Department, University of Hawai'i at Mānoa, 3190 Maile Way, St. John 101 Honolulu, HI, 96822, USA
e-mail: kbarton@hawaii.edu

R. Dirzo
Department of Biology, Stanford University, 371 Serra Mall, Stanford CA 94305, USA
e-mail: rdirzo@stanford.edu

F.C. Meinzer et al. (eds.), *Size- and Age-Related Changes in Tree Structure and Function*, Tree Physiology 4, DOI 10.1007/978-94-007-1242-3_7,

with improved defensive mechanisms to protect rare and endangered species as well as promote productivity in commercial stands.

Keywords Defenses • Herbivory • Ontogeny • Resistance • Tolerance

1 Why Is Plant Ontogeny Influential for Plant-Herbivore Interactions?

As plants develop, they undergo important morphological and physiological changes. These changes, defining their ontogenetic trajectories (Fusco 2001), can be particularly striking for trees, given that they develop from small seedlings to very large and long-lived individuals. Such morphological and physiological changes are likely to influence not only the performance of individual trees, but also their interactions with many species, for example, their herbivores. Overall, as trees develop, changes in the selective pressures exerted by herbivores, together with plant resource allocation tradeoffs, influence the expression of plant defense. Thus, the ontogenetic trajectories in anti-herbivory traits can be characterized by increases, decreases, or mixed trends in the expression of such traits. In some species, these ontogenetic trajectories occur with pronounced phase changes marked by heterophylly or abrupt changes in chemical, physical, or biotic defenses (e.g. *Eucalyptus globulus*, and many ant plants), while most other tree species show more gradual transitions among ontogenetic stages. As we will discuss in this chapter, the patterns of such trajectories can be influenced by multiple factors, including the specific herbivores feeding on trees, the activities of herbivores' natural enemies, the switches among the different defensive mechanisms that trees can express to reduce herbivory, the particular ecosystem in which they grow, and the influence of phylogenetic constraints that restrict or allow the evolution of ontogenetic trajectories in plant defense.

2 External and Internal Factors Driving Ontogeny-Related Changes in the Quality of Trees as Food for Herbivores

Herbivory has been identified as a biotic interaction that usually has negative impacts on plant fitness (Harper 1977; Dirzo 1984; Crawley 1997). Such impacts however, may depend on the ontogenetic stage at which trees are attacked (Ehrlén 1995; Weiner 2004). Seeds and seedlings, for example, are particularly vulnerable to herbivore attack and they commonly get killed, especially when eaten by mammalian herbivores (Dirzo et al. 2007). As trees develop and acquire more biomass, the impacts of herbivore damage are less drastic but can still cause important reductions in their growth rates, survival and reproduction (Marquis 1984; Domínguez and Dirzo 1994; Boege 2005b). These negative impacts commonly represent an important selective force that has favored the evolution of defensive mechanisms that allow plants to deal with their consumers. Defenses have been classified into (i) resistance traits that reduce the amount of damage received, which can be physical (e.g., spines, thorns or leaf toughness), chemical (secondary metabolites), or biotic (e.g., traits maintaining or enhancing the activity of natural enemies of herbivores); (ii) tolerance mechanisms that reduce the impact of herbivore damage once it has occurred, and (iii) escape strategies that reduce the probability of plants to be found by their consumers (Lubchenco and Gaines 1981; Dirzo 1984; Rosenthal and Kotanen 1994; Crawley 1997). Because the most dramatic impacts of herbivores on plant fitness occur during the youngest stages (Dirzo 1984; Marquis 1984), natural selection by herbivores is predicted to promote trees to have higher levels of any of these defensive mechanisms early than later during their development (Spiegel and Price 1996; Boege and Marquis 2005; Barton and Koricheva 2010).

However, despite the selective pressure of herbivores and the adaptive value of defensive mechanisms, plants are likely to optimize their levels of defense throughout ontogeny, depending on the priorities of other functions that also can influence plant fitness, such as growth, reproduction or maintenance, which commonly compete for the limited resources available for plants (Herms and Mattson 1992). The amount of resources allocated to each function, and in particular to defense against herbivores, depends on the physiological priorities at each ontogenetic stage and on resources available for plants at that particular moment of their lives. For example, in young plants it may be a priority to express high growth rates to establish and outcompete their neighbors; in contrast, when entering into their mature phase, reproduction may become a priority for resource use. In the specific case of trees, resource allocation to different functions during ontogeny also depends on the type of ecosystem and particular conditions in which they grow (e.g., open areas vs. forests, tree gaps vs. understory, and nutrient-poor vs. nutrient-rich soils). For instance, after producing the primary root system, seedlings growing in the understory of forests soon deplete the carbon stored in cotyledons, so further growth depends on the accumulation of new photosynthetic tissue (Swaine and Whitmore 1988; Martínez-Ramos et al. 1989; Bryant et al. 1991). Because in closed forests understory growth is constrained until a light gap is opened, saplings may remain in the same stage for many years, and allocate the few resources they can acquire to prevent herbivore attack on their limited amount of leaf area. As soon as light becomes available (after, for example, a branch or tree fall), growth can resume and is likely to have a priority over other functions, allowing plants to balance their root:shoot ratio and acquire enough above-ground biomass to reach the canopy (Blundell and Peart 2001). However, during such growth, as trees become larger and are able to accumulate resources, they can also start allocating them to other functions, such as defense or storage. Finally, when trees reach the canopy and enter into their mature stage, reproductive structures may compete for resources with defensive traits (Bryant et al. 1991; Boege and Marquis 2005). Thus, from the resource allocation perspective, two peaks of increased plant defense are likely to occur as plants develop: one at the seedling stage, when plants have enough maternal resources to defend their developing tissues, and a second one after they have reached a balance in their root:shoot ratio and before reproduction commences

(Bryant et al. 1991, and see Boege and Marquis 2005, Fig. 1). In contrast, in open areas or treefall gaps within forests, tradeoffs between growth and defense are likely to be defined by nutrient or water availability rather than by light limitation, and tolerance is likely to have a priority over the production of resistance traits (Coley 1983). For example, pioneer species in tree fall gaps of forests have a priority to grow fast to outcompete their neighbors (Blundell and Peart 2001). Having enough light and resources, their rapid intrinsic growth rates allow them to tolerate herbivore damage, by efficiently replacing the tissues lost to herbivores with no need to produce high levels of resistant traits. Once trees have reached the canopy, they are likely to keep compensating but also to allocate some resources to resistance and reproduction. Nevertheless, there are no available studies assessing switches in defensive strategies of pioneer species and this warrants further investigation. In the particular case of environments where water availability is the limiting factor for growth, distinct ontogenetic patterns in defense expression may be absent or weak if the same traits function for both water conservation and defense, as is the case with sclerophyllous leaves (Turner 1994). However, the specific role of water availability on the ontogeny of defense has not received much attention and remains unclear.

Interestingly, if we consider the magnitude of the impacts of herbivores throughout ontogeny, one would expect to observe a decline in the quantity or quality of defensive traits as plants develop, given that the young stages are more vulnerable, and thus should be better defended than old stages. In contrast, when considering resource allocation tradeoffs between defense and growth or reproduction, and the resources available for plants to produce defenses, the opposite pattern is expected (Barton and Koricheva 2010). As we will describe in the rest of the chapter, empirical evidence suggests that a combination of both mechanisms drives the ontogenetic trajectories of defense observed in trees. Furthermore, ontogenetic patterns in defense vary among tree species depending on the particular guild and abundance of herbivores feeding on them, on the type of defenses expressed, as well as on the ecosystem in which the trees grow.

3 Changes in the Quality of Trees as Food for Herbivores: Ontogenetic Patterns in Resistance

Several studies have detected that the number of herbivores feeding on trees and the amount of damage they produce are strongly dependent on the age of trees. In some cases herbivore abundance, performance and/or damage has been reported to increase with tree age (Tahvanainen et al. 1985; Jachmann 1989; Jogia et al. 1989; Kearsley and Witham 1989; Swihart and Bryant 2001; Boege and Marquis 2006), but in many others the opposite pattern has been found (Zagory and Libby 1985; Price et al. 1987; Kearsley and Witham 1989; Karban 1990; Fritz et al. 2001; Van Bael et al. 2003). In fact, a recent meta-analysis has detected that the overall pattern is that mammalian herbivores prefer to consume more tissue from mature than from juvenile trees, but the patterns of abundance and preference of insect herbivores is rather more variable and presents no general trend (Barton and Koricheva 2010). Plant quality as food for herbivores is highly dependent on the type and amount of resistance traits they produce, and on the expression of other defensive strategies, such as tolerance or escape. Thus, changes in the defensive strategies during the development of trees are likely to be responsible for the observed patterns in herbivore preference and performance. Only two studies have examined developmental trajectories of defense throughout the entire ontogeny of trees, and these reveal differences among kinds of plant resistance. In *Ryparosa kurrangii* (Achariaceae), levels of cyanogenic glycosides decrease from the cotyledon to the autonomous seedling stage, and further decrease from the juvenile to mature tree stage (Webber and Woodrow 2009). Interestingly, the decrease in cyanogenic glycosides occurs simultaneously with an increase in leaf mass per area, measured as a proxy for leaf toughness, suggesting an ontogenetic switch from chemical resistance during early ontogeny to physical resistance in late ontogeny (Webber and Woodrow 2009). This study, however, did not explore other possible defensive traits that might also have changed during ontogeny. An ontogenetic switch has also been observed in *Populus tremuloides* (Salicaceae), although in this case it is a switch among classes of secondary compounds (Donaldson et al. 2006). As ramets age from 0 to 25+ years, concentrations of condensed tannins increase while concentrations of phenolic glycosides decrease significantly (Donaldson et al. 2006). Neither of these studies examined herbivore damage, and so the relationships between the ontogenetic switches observed and herbivore performance and damage remain unclear.

Although general conclusions about whole-ontogeny patterns cannot be drawn from only two studies, additional insights can be gained by considering studies focusing on patterns of resistance within a single or across two ontogenetic stages. For example, the few studies comparing tree seedlings of different ages in different species show a consistent increase in plant defense with developmental age, specifically for phenolics (Close et al. 2001; Fritz et al. 2001) and terpenoids (Bryant and Julkunen-Tiitto 1995). A single study examining the pattern in chemical resistance from the seedling to the juvenile stage reports a decrease in levels of leaf resin terpenoids (Feibert and Langenheim 1988). Within the juvenile stage, consistent increases have been observed for physical defenses (Nomura et al. 2001; del-Val and Dirzo 2003), while chemical resistance shows no clear pattern. Comparing saplings of different ages, it has been shown that concentrations of cyanogenic glycosides increase with plant age (Goodger et al. 2007), that alkaloids decrease (Liu et al. 1998), and that the levels of phenolics may increase or decrease, depending on the tree species (Bonell and Selander 1974; Basey et al. 1988; Laitinen et al. 2005). These patterns suggest that in trees, plant defense increases via secondary chemistry in seedlings and then via physical defenses in saplings. Developmental patterns in secondary chemistry during the sapling stage vary among species and among classes of secondary compounds. However, it should be noted that these patterns rely on data from relatively few species; additional studies are needed to ascertain their generality.

In contrast, many more studies have examined the expression of defense traits across the transition from the juvenile to the mature stages of trees, revealing again that the ontogenetic pattern varies among species and among defense traits.

As a measure of physical defense, leaf toughness tends to increase from the juvenile to the mature stage of trees (Kearsley and Witham 1989; Macedo and Langenheim 1989; Loney et al. 2006; but see Boege 2005b). Ontogenetic patterns in phenolics are variable: some species show increases from the juvenile to mature stage (Eck et al. 2001; Erwin et al. 2001; del-Val and Dirzo 2003; Gowda and Palo 2003; Boege 2005a, b; Boege and Marquis 2006; Goodger et al. 2006; Neilson et al. 2006), while other species show a decrease (Rousi et al. 1987; Basey et al. 1988; Loney et al. 2006; Neilson et al. 2006). Similarly, a general pattern fails to emerge for terpenoids, given the increases (Sinclair and Smith 1984; Goralka and Langenheim 1996; Loney et al. 2006) and decreases (Reichardt et al. 1984; Sinclair and Smith 1984; Langenheim et al. 1986; Macedo and Langenheim 1989) that have been reported. The single study we found comparing the induction of total phenolics in juveniles versus mature trees reports a greater inducibility in mature trees of the tropical deciduous *Casearia nitida* (Salicaceae; Boege 2005b). In addition, considering that cyanogenic glycosides are regarded as inducible defenses (Karban and Baldwin 1997), there is existing evidence of an increase of these compounds from the juvenile to the mature transition of several species of the genus *Eucalyptus*, (Gleadow and Woodrow 2000; Goodger et al. 2004; Goodger et al. 2006; Neilson et al. 2006), although *Eucalyptus cladocalyx* var. *nana* is a striking exception to this general pattern (Goodger et al. 2006). Considering the vast diversity of compounds within the broad classes of phenolics and terpenoids, it is perhaps not surprising that we do not detect general ontogenetic patterns for these classes (Barton and Koricheva 2010). Furthermore, investigations into the ontogeny of induced defense will need to consider how plant architecture changes as plants grow and develop. The transmission of signaling compounds within small juvenile plants is likely to result in a more synchronized induced response to herbivory, while the modularity and limited connectivity among branches within large mature trees may limit induction or even lead to induced susceptibility (Nykänen and Koricheva 2004). Additional studies on more species including several defensive traits may help clarify these ontogenetic trajectories. Moreover, variability among patterns can be explained by additional factors, such as the guild of herbivores feeding on trees, the particular habitat where tree species grow, life history strategies, and tradeoffs between types of defense (e.g., direct vs. indirect defense strategies), as we will discuss below.

4 Tree Ontogeny and the Impact of the Natural Enemies of Herbivores: Indirect Defenses

Traits facilitating the visitation or colonization of mutualistic animals that defend the plants against herbivores, such as plant food rewards, nesting space or chemical cues that attract herbivores' natural enemies (predators and parasitoids) are defined as indirect defenses. As we discuss below, these defensive strategies can include the interaction with mutualistic or non-mutualistic animals, and can change during the development of trees.

4.1 Indirect Defenses Involving Defensive Mutualisms

The most commonly documented indirect defense, myrmecophytism (Janzen 1966, 1969), involves plant-ant mutualisms, whereby the myrmecophytic plant provides nesting space in structures known as domatia, and sometimes offers food rewards directly, in the form of extrafloral nectar and/or nutritious food bodies (Fiala and Maschwitz 1992), or indirectly, via, for example, coccoid-produced honeydew (Gaume et al. 1998). In turn, ants protect host plants from herbivores (Janzen 1966; Schupp 1986; Fiala et al. 1994; Heil and McKey 2003). Additionally, in a few documented cases, such defensive mutualisms occur between wasps and plants, and in these cases the plant provides a reward in the form of floral nectar (Domínguez et al. 1989; Narbona and Dirzo 2010).

The fundamental driver of the ontogenetic shifts observed in plants with defensive mutualistic relationships is an ontogenetic constraint imposed by the fact that mutualistic plants do not start their lives associated with their defending insects. Instead, they have to acquire their defenders in the course of their ontogeny. Typically, a mutualistic plant starts with a size and morphology that cannot maintain a colony of ants, given that young (uncolonized) plants have not developed functional domatia, nor structures for the production of food bodies or nectar (extrafloral nectar in the case of myrmecophytic plants, or floral nectar in the case of wasp-mutualistic plants). This, coupled with the fact that young stages of the myrmecophytic plant are particularly vulnerable to the impacts of herbivory, leads to the hypothesis that in the absence of defenders during this critical stage of their ontogeny, uncolonized plants may be more dependent upon direct defenses than older (colonized) plants (del-Val and Dirzo 2003; Heil and McKey 2003). Assuming that colonized plants are sufficiently defended against herbivores, we might then expect to see a reduction in direct defenses upon colonization by ants or wasps. This would lead to an ontogenetic switch from direct defenses in young uncolonized plants to indirect defenses in older colonized plants.

Only a handful of studies are available to assess these expectations. Table 7.1 summarizes the results derived from seven independent studies involving seven different tree (with one exception, *Croton suberosus*, a shrubby plant) species. In six of these species, biotic, indirect defense has been found to be present in the older (colonized) plants, while younger stages lack such biotic defense. In most of these studies, plant palatability has been measured as a proxy of direct (chemical or physical) defense (eight comparisons, Table 7.1), while four comparisons examined natural herbivory in the field. Only one study (del-Val and Dirzo 2003) examined, in addition to herbivory and palatability, direct defenses in terms of plant phenolics, tannin concentration, and pubescence in colonized and uncolonized plants of the myrmecophyte *Cecropia peltata*. One study also compared herbivore survival in young and old plants (also a proxy of direct defenses) of the myrmecophyte *Macaranga* under controlled conditions with captive animals (Nomura et al. 2001).

The specific hypotheses for the comparisons of Table 7.1 are that, in contrast with old plants, young plants should be: (i) less palatable, (ii) have higher levels of direct defenses (phenolics and tannin concentrations, and pubescence), and (iii) should cause

Table 7.1 Studies that have examined ontogenetic changes in defense strategy among tree species with indirect defenses involving mutualistic animals (ants, or wasps) present in the old but not young ontogenetic stages

Plant species	Indirect defense	Variable measured	Comparison young vs. old plants	Source
Macaranga myrmecophytes	Ants	Palatability	Young<old	Nomura et al. (2001)
		Herbivore survival	Young<old	
Cordia alliodora	Ants	Palatability	Young~old	Trager and Bruna (2006)
Cecropia peltata	Ants	Palatability	Young>old	del-Val and Dirzo (2003)
		Herbivory	Young>old	
		Chemical defense	Young<old	
		Physical defense	Young<old	
Omphalea oleifera	Ants	Palatability	Young~old	R. Dirzo, unpublished
		Herbivory	Young~old	
Cordia alliodora	Ants	Palatability	Young<old	Llandres et al. (2010);
Croton suberosus	Wasps	Palatability	Young<old	Narbona and Dirzo (2010)
		Herbivory	Young~old	
Acacia cornigera	Ants	Palatability	Young<old	R. Dirzo, unpublished
		Herbivory	Young~old	
Populus tremuloides	Ants	Extrafloral Nectaries	Young>old	Wooley et al. (2007)
		Chemical defense	Young>old	

lower survival of captive generalist herbivores. Only five out of eight tests found that uncolonized young plants were less palatable than older plants. The other three studies found the opposite pattern or no difference between young and old plants. Survival of generalist herbivores was lower on young plants than in older plants, supporting the prediction that direct defense decreases with plant age. Finally, patterns of herbivore damage were generally similar across ontogenetic stages, suggesting that direct and indirect defense provide similar levels of protection against herbivores. A striking exception to these results was observed in *Cecropia peltata*, in which young plants lack direct defenses (chemical and physical) as well as indirect defenses (ants). This resulted in higher palatability and levels of herbivore damage of young plants compared to older plants (del-Val and Dirzo 2003). Clearly, young *C. peltata* plants lack resistance against herbivory, and the authors suggest that perhaps they rely instead on tolerance (del-Val and Dirzo 2003). Although tolerance has not been tested in *C. peltata*, a study with a related species, *C. obtusifolia*, found that artificial defoliation during the uncolonized stage had no effect on plant survival or growth compared to undamaged control plants (Frías 1996), suggesting that tolerance may be indeed important in *Cecropia* juveniles.

As mentioned above, our predictions about ontogenetic switches in direct and indirect defense were only partially supported by the evidence presented in Table 7.1. In some cases, the inconsistency of the results can be explained on the grounds of methodological differences. For example, it has been found that young

Cordia alliodora plants are either less palatable (Llandres et al. 2010) or similarly palatable (Trager and Bruna 2006), when compared to older plants. Such discrepancy could be due to the facts that in Trager and Bruna's (2006) study, their young (1 year-old) plants included saplings that had domatia already occupied by ants, while in the other study all young plants were uncolonized. The young but occupied saplings of the former study might have already foregone direct defenses (see Llandres et al. 2010). In addition, the palatability trials in Trager and Bruna's study were performed using the *Cordia*-specialized beetle *Coptocyla leprosa*, while the other study used a generalist herbivore, the larvae of the lepidopteran *Hypercombe* sp. Conceivably, the specialized beetle might be more capable of dealing with the secondary compounds of *C. alliodora* than the larvae of the generalist herbivore, thus rendering the palatability of plants from the two stages indistinguishable. A similar interpretation can be applicable in the case of the Neotropical tree *Omphalea oleifera* (Table 7.1) in which only the old trees produce extrafloral nectar that attracts a variety of ants, while the young plants do not produce such a reward. In contrast, in the case of *Populous tremuloides,* the density of extrafloral nectaries is actually greater in young trees (1 year old) than in 4- and 10-year-old trees (Wooley et al. 2007), and there is no detectable trade-off between this indirect defense and direct chemical defenses or tolerance (Sugiura et al. 2006; Wooley et al. 2007).

Limited as it is, this set of studies collectively suggests that young uncolonized trees may invest in direct defenses to reduce herbivory, and then relax these defenses upon colonization by their animal defenders. Future studies measuring chemical, physical, and tolerance traits in young uncolonized and old colonized plants are needed and will provide key new insights into ontogenetic switches and trade-offs between direct and indirect defense. Furthermore, explicit comparisons of generalist and specialist herbivores will reveal whether direct and indirect defenses are in fact redundant, and thus involved in an ontogenetic switch, or whether they may act in synergy to defend plants against a wide range of herbivores.

4.2 Indirect, But Non-mututalistic Defense and Ontogeny

In some cases, plant species are defended by animals that are not engaged in mutualistic interactions with the plant. Such indirect defense is known in herbaceous plants and some trees, but knowledge is extremely scarce as to whether this exhibits any kind of ontogenetic trajectory. One relevant study (Van Bael et al. 2003) investigated how predators affected herbivore abundance and levels of herbivory in saplings and adult trees in three tropical tree species. Using cages to prevent access to bird predators, they found that caging dramatically increased herbivore abundance and levels of herbivory on trees but had no effect on saplings. Although a variety of factors could potentially account for these results, a proximate interpretation is that the third trophic level has a greater impact on the plant's herbivores at the older but not younger ontogenetic stages and that the levels of herbivory on adult trees would be significantly higher in the absence of predatory, top-down controls.

The consequences of such patterns on ontogenetic trajectories of plant defense and on plant fitness remain to be investigated.

Further evidence for the importance of tritrophic interactions on the ontogeny of plant defense and herbivory comes from studies on the dry forest deciduous tree species, *Casearia nitida*. Comparisons of caged and non-caged saplings and mature trees found that the foraging intensity of bird predators was significantly higher in mature trees (Boege and Marquis 2006). Moreover, foraging of parasitoid wasps was restricted to the canopies of mature trees, and they did not forage in saplings (Boege 2005a). Together with ontogenetic increases in total leaf phenolics, these patterns of indirect and direct defense led to significantly higher herbivore densities and levels of damage in saplings compared to mature trees (Boege 2005a). Determining whether the ontogenetic increases in predation and parasitism observed in these studies reflect general patterns requires further tests in additional species. Particular focus on the potential ontogenetic change in expression of plant volatile organic compounds (VOC's) may reveal a mechanism by which tritrophic interactions change during plant development. Although VOC release has been shown to decrease with plant age in cultivated herbs (Cole 1980), information about their ontogenetic patterns in trees remains unknown.

5 Ontogenetic Patterns in Escape and the Influence of Biogeography

Developmental patterns in herbivory may occur for reasons other than top-down (i.e. indirect) or bottom-up (i.e. defensive chemistry) forces. Specifically, apparency to herbivores (Feeny 1976) may change as plants grow and develop, leading either to ontogenetic increases or decreases in escape from herbivores. Several different mechanisms can drive developmental patterns in apparency to and escape from herbivores, including associational resistance, distance from conspecific trees, lag time in the herbivore colonization of young trees, phenology, and limited access to trees as they grow. Furthermore, these developmental patterns often show biogeographical variation.

There is a strong spatial component to developmental patterns in apparency and escape, evidenced by the importance of plant neighbors and distance from conspecific trees. Associational resistance occurs when susceptible plants escape from herbivory by growing in close proximity to well-defended plants. Typically, it is small and/or young plants that gain associational resistance from neighbors, sometimes called nurse plants (Niering et al. 1963). Associational resistance is often provided by nurse plants with physical defenses that effectively protect seedlings from mammalian herbivores (Rousset and Lepart 2000; Baraza et al. 2006; Vandenberghe et al. 2009), but can also be mediated by secondary chemistry (Smit et al. 2006). Although it has rarely been compared among species, associational resistance seems to be especially important for seedlings that lack their own resistance or tolerance to herbivory (Vandenberghe et al. 2009).

Spatial relationships of individuals within species are also important for escape from herbivory. In this case, it is distance from, rather than proximity to, an adult tree that provides a benefit to young and small plants. As proposed by the Janzen-Connell hypothesis (Janzen 1970; Connell 1971), seedlings and juveniles growing near conspecific adults receive high loads of specialist herbivores from nearby adult trees, leading to relatively high mortality near adults and a decrease in herbivory and mortality with distance from conspecific adults. A meta-analysis of 23 studies testing the Janzen-Connell hypothesis in seedlings confirmed that distance from conspecific adults increases seedling survival (Hyatt et al. 2003), but that this pattern is more common in tropical forests. In contrast, temperate species tend to show higher survival near conspecific adults, perhaps indicating that suitable establishment sites are limited (and associated with conspecific adults) and that the distribution of individuals from the same species is more homogeneous in temperate forests, in contrast with tropical forests. Like escape by associational resistance, distance from conspecific adults provides only an ephemeral defense against herbivory. As trees grow, herbivores will undoubtedly locate and colonize them, but by that time, the plants are likely to have started expressing their own resistance and tolerance traits.

To our knowledge, only two studies have explicitly linked ontogenetic patterns in defense traits with Janzen-Connell patterns of herbivory and mortality (Langenheim and Stubblebine 1983; Sánchez-Hidalgo et al. 1999). Both studies examined tropical tree species showing relatively high seedling survival beneath parental trees, notable exceptions to the Janzen-Connell hypothesis. Furthermore, in both studies, it was demonstrated that the surviving seedlings had distinct secondary chemical profiles compared to the adult trees. It was concluded that parent trees produced genetically variable offspring and that herbivores consumed only those seedlings with profiles similar to the parent trees, leaving chemically differentiated seedlings to survive. By consuming a specific subset of the seedling pool, these herbivores would be selecting for ontogenetic patterns in plant secondary chemistry, thereby creating a moving-target (*sensu* Adler and Karban 1994) and allowing seedlings to escape from herbivory.

Associational resistance and the Janzen-Connell hypothesis illustrate how space affects herbivore-seeking behavior, thereby driving developmental patterns of escape. Time is similarly important for escape from herbivory, and can lead to ontogenetic differences in herbivory resulting in the simple observation that herbivore loads increase over time, as trees develop, and more herbivores find and colonize them. Temporal increases are most commonly reported for insect herbivores (Guedes et al. 2000; Fonseca and Benson 2003; Campos et al. 2006) and may also apply to other invertebrate herbivores, such as slugs and snails, but have not been explicitly tested. In addition, phenological changes are also involved in the ability of plants to escape from herbivory. For example, young saplings of *Acer mono* flush their leaves earlier than their adult relatives, increasing their probabilities of escaping from predators and pathogens (Seiwa 1999).

In contrast to the previous examples showing how seedlings and juvenile plants escape from herbivory, some species escape from herbivory as adults. In particular, trees are able to escape from ground-dwelling mammalian herbivores only when they outgrow the reach of these browsers. Due to the high densities of ground-dwelling mammalian herbivores, this phenomenon is especially common and well-documented in boreal forests (Swihart and Bryant 2001). Intense herbivory on juvenile, but not on adult plants, results in an ontogenetic decrease in the selection pressure on resistance traits. Indeed, feeding trials with the relevant ground-dwelling browsers (hares, deer, voles) reveal that juvenile plant tissue is much less palatable than adult plant tissue (reviewed in Swihart and Bryant 2001), and the limited evidence from studies examining plant resistance traits indicates that levels of secondary compounds can be as much as 96% higher in juvenile plants than in conspecific adults (Reichardt et al. 1984). For these species, escape from ground-dwelling herbivores occurs with a concomitant ontogenetic shift away from chemical defense in adult trees.

Interestingly, developmental patterns in escape and associated ontogenetic switches in defense vary biogeographically. In boreal forests, seedlings and juveniles experience herbivory by mammals and express high levels of chemical defense, while adults show lower levels of chemical defense and escape from mammalian herbivores. Unlike boreal trees, temperate and tropical species show weak ontogenetic patterns in herbivory and tend to increase chemical defense across ontogeny (Barton and Koricheva 2010). Biogeographical trends in key herbivores may explain this difference. In temperate and tropical forests, insects (Coley and Barone 1996) and mollusks are the most important herbivores, and these taxa are not limited by tree height in the same way that boreal mammals are. Ontogenetic increases in chemical defense of temperate and tropical species (Barton and Koricheva 2010) may reflect allocation constraints during development that limit production of secondary chemicals and physical defenses in seedlings (Herms and Mattson 1992). Future research explicitly measuring ontogenetic patterns in herbivore selection pressure and growth-defense trade-offs in temperate and tropical tree species would shed light on this striking biogeographical pattern.

Although still under-studied, these examples illustrate how escape from herbivory can change as trees grow and develop. Future research should further explore the mechanisms driving developmental patterns in escape with particular focus on biogeography, the role of herbivore feeding guild (mammalian browsers vs. insect defoliators, etc.), and how escape is involved in ontogenetic switches among defense traits. Studies of non-model tree species, particularly in the temperate and tropical forests, are needed to ascertain the generality of these patterns.

6 Ontogenetic Patterns in Tolerance to Herbivory

For many plants, resistant traits reduce levels of herbivory but are unable to completely eliminate damage. In these cases, tolerance can become an important plant defense strategy. Tolerance occurs when herbivore damage induces changes in plant traits that allow for compensation, or replacement of lost tissues, and the maintenance of fitness levels comparable to undamaged conspecifics. Although it remains unclear which mechanisms are most important, commonly studied tolerance traits include increases in photosynthesis rates, activation of dormant meristems, utilization of stored reserves, shifts in allocation patterns, and changes in phenology (Tiffin 2000). Considering the small size of seedlings with their limited photosynthetic area, minimal stored reserves, and nearly non-existent bud bank, we would predict that tolerance increases dramatically during plant ontogeny. Interestingly, the relatively few studies examining the ontogeny of tolerance in trees suggest that the pattern is actually more complex than this simple prediction.

Only four studies have explicitly tested how tolerance changes with plant age or ontogeny in woody plants (Weltzin et al. 1998; Warner and Cushman 2002; Boege 2005b; Hódar et al. 2008), but our synthesis can be expanded if we include a recent meta-analysis that compared studies that separately measured compensation in seedlings, saplings, and mature trees (Nykänen and Koricheva 2004). In general, seedlings experience significant reductions in growth following herbivory and are unable to compensate for lost tissues (Weltzin et al. 1998). Demonstrating the importance of even small differences in age (and, probably more importantly, size) during early ontogeny is the observation that 5-week-old honey mesquite (*Prosopis glandulosa*, Fabaceae) seedlings compensate for defoliation significantly better than 3-week-old seedlings. However, 3-week-old seedlings show higher survival after ten successive defoliations than 5 week old seedlings (Weltzin et al. 1998), demonstrating that the fitness currency used (survival vs. growth) can influence interpretations of tolerance.

As juveniles develop from seedlings to saplings, they show significant increases in compensation for herbivory (Nykänen and Koricheva 2004; Hódar et al. 2008), presumably due to increases in photosynthetic area, stored reserves, and bud bank. Following a peak at the sapling stage, compensation appears to decrease as plants mature (Warner and Cushman 2002; Nykänen and Koricheva 2004; Boege 2005b). This is likely due to a combination of factors, including shifts in allocation priorities, increased sectoriality of large trees, and senescence. During the vegetative juvenile stage, resources for growth are allocated entirely to vegetative tissues. After the onset of reproduction, however, allocation patterns shift, and reproductive tissues may be produced at the expense of vegetative growth. Following herbivory, these resource allocation shifts can limit compensation in mature plants. Thus, under-compensating mature plants may actually have high tolerance if they are preferentially allocating resources to reproduction at the expense of growth.

Compensation may decrease from the sapling to mature tree stage because sectoriality increases with plant size, so that large trees have few physiological connections among branches. As a result, herbivore-induced tolerance traits are localized on the damaged branch, and whole-plant measures of compensation are low (Haukioja and Koricheva 2000). Moreover, gradual activation of dormant buds due to release from apical dominance may reduce the bud bank in mature trees compared to saplings, further limiting compensatory growth following herbivory (Lehtilä 2000).

Finally, mature plants may experience lower compensation for herbivory than saplings because they have relatively more senescent leaves. As Nykänen and Koricheva (2004) demonstrated, trees are more responsive to herbivory early in the season when leaves are more "vigorous" (sensu Price 1991). As the season progresses, leaves begin to senesce and show limited responses to herbivory. This is consistent with the general observation that actively growing leaf tissue can better detect and respond to herbivory than older, inactive leaves (Herms and Mattson 1992; Karban and Baldwin 1997).

Determining whether the observed non-linearity in the ontogeny of compensation and tolerance is the result of shifts in allocation priorities, sectoriality, senescence, or some combination of these factors, requires further research. Additional questions that will shed light on the ontogeny of compensation and tolerance include the following: (1) Does tolerance to invertebrate vs. vertebrate herbivory differ in seedlings vs. mature trees? Mammals and mollusks account for most of seedling herbivory, while insects cause most of the damage experienced by mature trees. Because tolerance mechanisms are likely to differ for vertebrates versus invertebrates (Haukioja and Koricheva 2000), it is important to test these responses separately at each stage. (2) Do deciduous and evergreen trees differ in their ontogenetic patterns? Herbivory tends to reduce growth in evergreens more than in deciduous trees (Nykänen and Koricheva 2004), but it remains unclear whether this response is consistent across ontogenetic stages. (3) Do the mechanisms of tolerance differ among ontogenetic stages? Despite the general assumption that compensatory growth increases fitness, there are very few tests of the relationship between compensation and reproductive output (Tiffin 2000). Thus, it remains unclear whether compensatory growth actually acts as a mechanism by which plants tolerate herbivory. Furthermore, plant responses to herbivory can be confusing. In their meta-analysis, Nykänen and Koricheva (2004) found that herbivory increases the rate of photosynthesis, concentrations of carbon and phenolics, but decreases concentrations of carbohydrates, terpenes, and plant growth. Apparently, damaged plants were unable to achieve growth rates comparable to undamaged control trees despite increases in photosynthesis. Examination of other putative tolerance traits in seedlings, saplings, and mature plants would provide new insights into how trees compensate for defoliation and would elucidate how tolerance changes across ontogeny.

7 Mixed Strategies: Switching Between Defensive Mechanisms During Ontogeny

During their prolonged life span trees are usually attacked by several species of herbivores, which have different impacts on their performance. On the other hand, resource allocation constraints and costs of defense are also likely to change during plant development. Hence, because the efficiency and the benefit:cost ratios of the different defensive mechanisms are likely to vary throughout tree development, it is not surprising to find that plants can switch from one defensive strategy to another as they develop. However, few studies have examined suites of defense traits across plant ontogeny (Boege et al. 2007), and only one has formally assessed such switches for the particular case of trees. Webber and Woodrow (2009) demonstrated that cyanogenesis decreases during ontogeny while a proxy of physical defense (leaf weight per unit area) increases in the tropical understory tree *Ryparosa kurrangii* (Achariaceae). Additional data from independent studies and observations suggest that ontogenetic switches in plant defense are an important aspect of plant-herbivore interactions. For example, as explained above, it is well known that trees can be defended through associational resistance as seedlings, and then switch from escape to secondary chemistry, physical defenses or tolerance later in development. In addition, when seedlings then become juveniles and mature trees, switches between tolerance and chemical defense are also likely to occur, driven by the risk of attack and resource allocation trade-offs between growth and defense (Boege 2005b). For example, saplings of the tropical tree *Casearia nitida* show greater ability to compensate for intense levels of herbivory (75% of leaf damage) but present lower concentrations of secondary metabolites (phenolic compounds) than mature trees, which show a lower ability to compensate damage but have higher concentration of phenolics (Boege 2005a, b). In addition, as discussed in Sect. 4, trees may shift from direct to indirect (biotic) defense as they age, particularly in the case of tropical trees where defensive mutualisms are particularly prominent.

An additional case of switches in defensive mechanisms occurs in systems where mammalian herbivores cannot climb or reach the canopy of trees (e.g., rabbits in boreal zones, deer in temperate and tropical zones). In those systems, it is common that physical defenses such as spines and thorns, which are effective largely against mammalian herbivores (Hanley et al. 2007), are expressed only early in ontogeny, when these herbivores have access to the branches of young trees. Once trees outgrow the reach of these herbivores, the expression of such defensive traits is likely to decrease, as they do not provide further benefits to plants (Gowda and Palo 2003). Hence, a switch in defensive strategies is likely if damage in older ontogenetic stages is performed by insects that can reach high canopies. In this case, the expression of tolerance or other resistance traits (e.g., trichomes, secondary metabolites, etc.) or indirect, biotic defense could be expected. Nevertheless, such expected switches have not been assessed and deserve further research.

8 Ecological and Evolutionary Implications of Ontogenetic Changes in Plant Defense

As mentioned in previous sections, ontogenetic changes in plant defense are likely to be promoted by resource allocation adjustments to physiological priorities of trees during their development, but also by the relative impact and intensity of herbivore attack as trees grow and mature. Overall, ontogenetic changes in defense should allow plants to reduce the costs of defense by restricting their production to only those stages in which (a) defenses have reduced physiological or ecological costs and/or (b) fitness is severely reduced by herbivore damage. Because reducing the costs of defense should be translated into greater plant fitness, natural selection should favor genotypes whose ontogenetic patterns of defense minimize the cost:benefit ratio of defense production. This adds a level of complexity to the traditional approach to study the evolution of plant defense, which has implicitly focused on investigating how natural selection targets different defensive traits at a given ontogenetic stage of trees (Latta and Linhart 1997; Andrew et al. 2007; Bailey et al. 2007, Donaldson and Lindroth 2007). In contrast, here we propose that natural selection could target directly the ontogenetic trajectories in plant defense. However, for this to be possible, the variation in ontogenetic trajectories in a population should have a genetic component. That is, some plant genotypes should present greater degree of ontogenetic change than others, and this variation should have relevant impacts on plant fitness. Genetic variation in ontogenetic trajectories has been detected for some herbs and shrubs (Bowers and Stamp 1993; Schappert and Shore 2000; Barton 2007), and only for two tree species (Donaldson et al. 2006, Rehill et al. 2006). Thus, exploring the genetic component of the ontogenetic trajectories in trees warrants further research, as a starting point to understand the microevolutionary patterns of the ontogenetic trajectories of defense in plants.

9 Phylogenetic Signal

Studies documenting genetic variation in and natural selection on developmental trajectories of defense traits would inform us about the microevolution of the ontogeny of defense. To understand macroevolutionary patterns in the ontogeny of defense, we must adopt a phylogenetic context for these studies. Phylogenetics is typically applied to plant defense research in the following ways: (i) to avoid pseudoreplication in tests of trade-offs among defense traits across species (Rudgers 2004; Agrawal and Fishbein 2008), (ii) as phylogenetic contrasts for testing a specific hypothesis, such as the growth rate hypothesis (Van Zandt 2007), and (iii) to examine within-lineage patterns of evolution, such as escalation (Agrawal et al. 2008; Agrawal et al. 2009). Of these three areas, the use of phylogenetic contrasts for testing hypotheses is likely to be the most useful for understanding the ontogeny of plant defense.

A recent meta-analysis on the ontogeny of plant defense and herbivory revealed several key questions that could benefit from a phylogenetic contrast approach (Barton and Koricheva 2010). First, as discussed in Sect. 5 of this chapter, developmental patterns in herbivory suggest that the ontogeny of defense varies biogeographically. Herbivores in boreal forests show a strong preference for adult plant foliage compared to juvenile foliage, while herbivores in temperate and tropical forests tend to show the opposite pattern, preferring juvenile foliage (Barton and Koricheva 2010). However, this result is confounded by herbivore type: studies in boreal forests focused on mammalian herbivores, but studies in the temperate and tropics were limited to insect and mollusk herbivores. Phylogenetic contrasts could be used to separately test how latitude and herbivore type influence the ontogeny of plant defense. In the first case, a single replicate would consist of a phylogenetic trio (related species from each of a tropical, temperate, and boreal forests), and ontogenetic patterns would be compared among the three geographical regions for as many phylogenetic replicates as possible. Similarly, phylogenetic replicates could be used to compare ontogenetic patterns in defense traits in species experiencing herbivory by different herbivores (mammals vs. mollusks vs. insects). Typically, phylogenetic replicates consist of congeners (e.g. Rudgers et al. 2004; Van Zandt 2007), but in the case of the ontogeny of defense, it will be important to consider phylogenetic patterns at the plant family level in addition to the generic level, considering the evidence that the ontogeny of defense varies significantly among plant families (Barton and Koricheva 2010).

A key insight from previous research on the macroevolution of plant defense that is relevant to the study of the ontogeny is the general failure to detect trade-offs among defense traits, as evidenced by a lack of negative correlations among them (reviewed in Agrawal 2007). However, some trade-offs become apparent when multivariate analyses are conducted on suites of traits or "defense syndromes" (*sensu* Agrawal and Fishbein 2006). Because plants encounter different herbivores and express different defense traits across ontogeny, multivariate analysis provides a method for simultaneously examining these factors.

10 Conclusions

Ontogenetic patterns in plant defense and herbivory are complex, involving shifting suites of herbivores, switches among defense traits, and changing resource allocation priorities. Yet, many studies continue to overlook this key aspect of plant ecology. Integrating the role of ontogeny into evolutionary ecology theory provides new perspectives to our understanding of the interaction of trees with their herbivores, and in particular on how natural selection can target the ontogenetic trajectories of plant defense. Furthermore, this perspective can have important applications for forest management. For example, identifying when trees are most susceptible to be attacked by herbivores, and when they are likely to express particular defensive traits could help to concentrate pest control efforts on the ontogenetic stages that are not able to resist or tolerate herbivore damage, or on those stages with greater risk of attack. If genetic variation in ontogenetic trajectories in plant defense is found, selection of lines which best optimize the expression of defensive mechanisms throughout ontogeny could enhance productivity of commercial stands and help conserve rare and endangered species suffering from stage-specific herbivory. In the near future, it is imperative as well to incorporate the insights of emerging fields such as functional genomics, to understand the epigenetic regulation of developmental changes in trees (Brunner et al. 2004), in particular, to identify the genes responsible for the ontogenetic trajectories of plant defense traits. Within this framework, evo-devo research offers new avenues to determine how environmental factors affect developmental processes at a mechanistic level and how they influence the expression of phenotypes during the life history of plants. In particular, the understanding of the genetic bases and transduction patterns of ontogenetic changes will help to understand their ecological and evolutionary consequences (Sultan 2005).

References

Adler FR, Karban R (1994) Defended fortresses or moving targets? Another model of inducible defenses inspired by military metaphors. Am Nat 144:813–832

Agrawal AA, Fishbein M (2006) Plant defense syndromes. Ecology 87:S132–S149

Agrawal AA (2007) Macroevolution of plant defense strategies. Trends Ecol Evol 22:103–109

Agrawal AA, Fishbein M (2008) Phylogenetic escalation and decline of plant defense strategies. P Indian Acad Sci 105:10057–10060

Agrawal AA, Lajeunesse MJ, Fishbein M (2008) Evolution of latex and its constituent defensive chemistry in milkweeds (*Asclepias*): a phylogenetic test of plant defense escalation. Entomol Exp Appl 128:126–138

Agrawal AA, Salminen JP, Fishbein M (2009) Phylogenetic trends in phenolic metabolism of milkweeds (*Asclepias*): evidence for escalation. Evolution 63:663–673

Andrew RL, Wallis IR, Harwood CE, Henson M, Foley WJ (2007) Heritable variation in the foliar secondary metabolite sideroxylonal in *Eucalyptus* confers cross-resistance to herbivores. Oecologia 153:891–901

Bailey JK, Schweitzer JA, Rehill BJ, Irschick DJ, Whitham TG, Lindroth RL (2007) Rapid shifts in the chemical composition of aspen forests: an introduced herbivore as an agent of natural selection. Biol Invasions 9:715–722

Baraza E, Zamora R, Hódar JA (2006) Conditional outcomes in plant-herbivore interaction: neighbors matter. Oikos 113:148–156

Barton EK (2007) Early ontogenetic patterns in chemical defense in *Plantago* (Plantaginaceae): genetic variation and trade-offs. Am J Bot 94:56–66

Barton EK, Koricheva J (2010) The ontogeny of plant defense and herbivory: characterizing general patterns using meta-analysis. Am Nat 175:481–493

Basey JM, Jenkings SW, Busher PE (1988) Optimal central place foraging by beavers: tree size selection in relation to defensive chemicals of quaking aspen. Oecologia 76:278–282

Blundell AG, Peart DR (2001) Growth strategies of a shade-tolerant tropical tree: the interactive effects of canopy gaps and simulated herbivory. J Ecol 89:608–615

Boege K (2005a) Herbivore attack in *Casearia nitida* influenced by plant ontogenetic variation in foliage quality and plant architecture. Oecologia 143:117–125

Boege K (2005b) Influence of plant ontogeny on compensation to leaf damage. Am J Bot 92:1632–1640

Boege K, Marquis RJ (2005) Facing herbivory as you grow up: the ontogeny of resistance in plants. Trends Ecol Evol 20:441–448

Boege K, Marquis RJ (2006) Plant quality and predation risk mediated by plant ontogeny: consequences for herbivores and plants. Oikos 115:559–572

Boege K, Dirzo R, Siemens DH, Brown P (2007) Ontogenetic switches from plant resistance to tolerance: minimizing costs with age? Ecol Lett 10:177–187

Bonell ML, Selander RK (1974) Elephant seals: genetic variation and near extinction. Science 184:908–909

Bowers MD, Stamp NE (1993) Effects of plant age, genotype and herbivory on *Plantago* performance and chemistry. Ecology 76:1778–1791

Brunner AM, Busov VB, Strauss SH (2004) Poplar genome sequence: functional genomics in an ecologically dominant plant species. Trends Plant Sci 9:49–56

Bryant JP, Julkunen-Tiitto R (1995) Ontogenetic development of chemical defense by seedling resin birch: energy cost of defense production. J Chem Ecol 21:883–896

Bryant JP, Kuropat PJ, Reichardt PB, Clausen TP (1991) Controls over the allocation of resources by woody plants to chemical antiherbivore defense. In: Palo RT, Robbins CT (eds) Plant defenses against mammalian herbivory. CRC Press, Boca Raton, pp 83–103

Campos RI, Vasconcelos HL, Ribeiro SP, Neves FS, Soares JP (2006) Relationship between tree size and insect assemblages associated with *Anadenanthera macrocarpa*. Ecography 29:442–450

Close DC, Davies NW, Beadle CL (2001) Temporal variation of tannins (galloylglucoses), flavonols and anthocyanins in leaves of *Eucalyptus nitens* seedlings: implications for light attenuation and antioxidant activities. Aust J Plant Physiol 28:269–278

Cole RA (1980) Volatile components produced during ontogeny of cultivated crucifers. J Sci Food Agr 31:549–557

Coley PD (1983) Herbivory and defensive characteristics of tree species in a lowland tropical forest. Ecol Monogr 53:209–233

Coley PD, Barone JA (1996) Herbivory and plant defenses in tropical forests. Annu Rev Ecol Syst 27:305–335

Connell JH (1971) On the role of natural enemies in preventing competitive exclusion in some marine animals andin rain forest trees. In: den Boer PJ, Gradwell GR (eds) Dynamics of populations. Wageningen, Pudoc, pp 298–312

Crawley MJ (1997) Plant herbivore dynamics. In: Crawley MJ (ed) Plant ecology. Blackwell Science, Cambridge, pp 401–475

del-Val E, Dirzo R (2003) Does ontogeny cause changes in the defensive strategies of the myrmecophyte *Cecropia peltata*? Plant Ecol 169:35–41

Dirzo R (1984) Herbivory: a phytocentric overview. In: Dirzo R, Sarukhan J (eds) Perspectives on plant ecology. Sinauer, Sunderland, pp 141–165

Dirzo R, Mendoza E, Ortíz P (2007) Size-related differential seed predation in a heavily defaunated Neotropical rain forest. Biotropica 39:355–362

Domínguez CA, Dirzo R (1994) Effects of defoliation on *Erythroxylum havanense*, a tropical proleptic species. Ecology 75:1896–1902

Domínguez CA, Dirzo R, Bullock SH (1989) On the function of floral nectar in *Croton suberosus* Euphorbiaceae. Oikos 56:109–114

Donaldson JR, Stevens MT, Barnhill HR, Lindroth RL (2006) Age-related shifts in leaf chemistry of clonal aspen (*Populus tremuloides*). J Chem Ecol 32:1415–1429

Donaldson JR, Lindroth RL (2007) Genetics, environment, and their interaction determine efficacy of chemical defense in trembling Aspen. Ecology 88:729–739

Eck G, Fiala B, Linsenmair KE, Bin Hashim R, Proksch P (2001) Trade-off between chemical and biotic antiherbivore defense in the southeast Asian plant genus *Macaranga*. J Chem Ecol 27:1979–1996

Ehrlén J (1995) Demography of the perennial herb *Lathyrus vernus*. I. Herbivory and individual performance. J Ecol 83:287–295

Erwin EA, Turner MG, Lindroth RL, Romme WH (2001) Secondary plant compounds in seedling and mature aspen (*Populus tremuloides*) in Yellowstone National Park, Wyoming. Am Midl Nat 145:299–308

Feeny PP (1976) Plant apparency and chemical defense. In: Wallace JW, Mansell RL (eds) Biochemical interactions between plants and insects. Plenum, New York, pp 1–40

Feibert EB, Langenheim JH (1988) Leaf resin variation in *Copaifera langsdorfii*: relation to irradiance and herbivory. Phytochemistry 27:2527–2532

Fiala B, Maschwitz U (1992) Food bodies and their significance for obligate ant-association in the tree genus *Macaranga* (Euphorbiaceae). Bot J Linn Soc 110:61–75

Fiala B, Grunsky H, Mashwitz U, Linsenmair KE (1994) Diversity of ant-plant interactions: protective efficacy in *Macaranga* species with different degrees of ant association. Oecologia 97:186–192

Fonseca F, Benson WW (2003) Ontogenetic succession in Amazonian ant trees. Oikos 202:407–412

Frías R (1996) Ecología del mutualismo entre *Cecropia obtusifolia* y hormigas *Azteca* en la selva de Los Tuxtlas. Universidad Nacional Autónoma de México, Mexico City

Fritz RS, Hochwender CG, Lewkiewicz DA, Bothwell S, Orians CM (2001) Seedling herbivory by slugs in a willow hybrid system: developmental changes in damage, chemical defense, and plant performance. Oecologia 129:87–97

Fusco G (2001) How many processes are responsible for phenotypic evolution? Evol Dev 3:279–286

Gaume L, McKay D, Terrin S (1998) Ant-planr momopteran mutualism: how the third partner affects the interaction between a plant-specialist ant and its myrmecophyte host. Proc R Entomol Soc B 265:569–575
Gleadow RM, Woodrow IE (2000) Temporal and spatial variation in cyanogenic glycosides in *Eucalyptus cladocalyx*. Tree Physiol 20:591–598
Goodger JQD, Ades PK, Woodrow IE (2004) Cyanogenesis in *Eucalyptus plynthemos* seedlings: heritability, ontogeny and effect of soil nitrogen. Tree Physiol 24:681–688
Goodger JQD, Gleadow RM, Woodrow IE (2006) Growth cost and ontogenetic expression patterns of defence in cyanogenic *Eucalyptus* spp. Trees 20:757–765
Goodger JQD, Choo TYS, Woodrow IE (2007) Ontogenetic and temporal trajectories of chemical defence in a cyanogenic eucalypt. Oecologia 153:799–808
Goralka RJL, Langenheim JH (1996) Implications of foliar monoterpenoid variation among ontogenetic stages of California bay tree (*Umbellularia californica*) for deer herbivory. Biochem Syst Ecol 24:13–23
Gowda JH, Palo RT (2003) Age-related changes in defensive traits of *Acacia tortilis* Hayne. Afr J Ecol 41:218–223
Guedes RNC, Zanuncio TV, Medeiros GB (2000) Species richness and fluctuation of defoliator Lepidotpera populations in Brazilian plantations of *Eucalyptus grandis* as affected by plant age and weather factors. For Ecol Manag 137:179–184
Hanley ME, Lamont BB, Fairbanks MM, Rafferty CM (2007) Plant structural traits and their role in anti-herbivore defence. Perspect Plant Ecol 8:157–178
Harper JL (1977) Population biology of plants. Academic, New York
Haukioja E, Koricheva J (2000) Tolerance to herbivory in woody vs. herbaceous plants. Evol Ecol 14:551–562
Heil M, McKey D (2003) Protective ant-plant interactions as model systems in ecological and evokutionary research. Annu Rev Ecol Syst 34:425–453
Herms DA, Mattson WJ (1992) The dilemma of plants: to grow or to defend. Quart Rev Biol 62:186–202
Hódar JA, Zamora R, Castro J, Gómez JM, García D (2008) Biomass allocation and growth responses of Scots pine saplings to simulated herbivory depend on plant age and light availability. Plant Ecol 197:229–238
Hyatt LA, Rosenberg MS, Howard TG, Bole G, Fang W, Anastasia J, Brown K, Grella R, Hinman K, Kurdziel JP, Gurevitch J (2003) The distance dependence prediction of the Janzen-Connell hypothesis: a meta-analysis. Oikos 103:590–602
Jachmann H (1989) Food selection by elephants in the Miobbo biome, in relation to leaf chemistry. Biochem Syst Ecol 17:15–24
Janzen DH (1966) Coevolution of mutualism between antes and *Acacias* in Central America. Evolution 20:249–275
Janzen DH (1969) Allelopathy by myrmecophytes: the ant *Azteca* as an allelopathic agent of *Cecropia*. Ecology 50:147–153
Janzen DH (1970) Herbivores and the number of tree species in tropical forests. Am Nat 104:501–527
Jogia MK, Sinclair ARE, Andersen RJ RJ (1989) An antifeedant in balsam poplar inhibits browsing by snowshoe hares. Oecologia 79:189–192
Karban R (1990) Herbivore outbreaks on only trees: testing hypotheses about aging and induced resistance. Oikos 59:27–32
Karban R, Baldwin IT (1997) Induced responses to herbivory. University of Chicago Press, Chicago
Kearsley MJ, Witham TG (1989) Developmental changes in resistance to herbivory: implication for individuals and populations. Ecology 70:422–434
Laitinen J, Julkunen-Tiitto R, Rousi M, Heinonen J, Tahvanainen J (2005) Ontogeny and environment as determinants of the secondary chemistry of three species of white birch. J Chem Ecol 31:2243–2262
Langenheim JH, Stubblebine WH (1983) Variation in leaf resin composition between parent tree and progeny in *Hymenea:* implications for herbivory in the humid tropics. Biochem Syst Ecol 11:97–106

Langenheim JH, Convis CL, Macedo CA, Stubblebine WH (1986) *Hymenaea* and *Copaifera* leaf sesquiterpenes in relation to lepidopteran herbivory in southeastern Brazil. Biochem Syst Ecol 14:41–49

Latta RG, Linhart YB (1997) Path analysis of natural selection on plant chemistry: the xylem resin of ponderosa pine. Oecologia 109:251–258

Lehtilä K (2000) Modeling compensatory regrowth with bud dormancy and gradual activation of buds. Evol Ecol 14:315–330

Liu Z, Stanley BC, Wayne JB, Ying Y, Roysell JC, Matthew JF, John CA (1998) Variations in the secondary metabolite camptothecin in relation to tissue age and season in Camptotheca acuminata. Tree Physio 18:265–270

Llandres LA, Rodríguez-Gironés MA, Dirzo R (2010) Tropical dry forest plants and plant stages with bodyguards are more palatable and suffer less herbivory than their undefended counterparts. Biol J Linn Soc 101:536–543

Loney PE, McArthur C, Potts BM, Jordan GJ (2006) How does ontogeny in a *Eucalyptus* species affect patterns of herbivory by Brushtail Possums? Funct Ecol 20:982–988

Lubchenco J, Gaines SD (1981) A unified approach to marine plant-herbivore interactions. I. Populations and communities. Ann Rev Ecol Syst 12:405–437

Macedo CA, Langenheim JH (1989) Microlepidopeteran herbivory in relation to leaf sesquiterpens in *Copaifera langsdorfii* adult trees and their seedling progeny in a brazilian woodland. Biochem Syst Ecol 17:217–224

Marquis RJ (1984) Leaf herbivores decrease fitness of a tropical plant. Science 226:537–539

Martínez-Ramos M, Álvarez-Buylla E, Sarukhán J (1989) Tree demography and gap dynamics in a tropical rain forest. Ecology 70:555–558

Narbona E, Dirzo R (2010) A reassessment of the function of floral nectar in *Croton suberosus* (Euphorbiaceae): a reward for plant defenders and pollinators. Am J Bot 97:672–679

Neilson EH, Goodger JQD, Woodrow IE (2006) Novel aspects of cyanogenesis in *Eucalyptus camphora* subsp. *humeana*. Funct Plant Biol 33:487–496

Niering WA, Whittaker RH, Lowe CH (1963) The saguaro: a population in relation to environment. Science 142:15–23

Nomura M, Itioka T, Murase K (2001) Non-ant antiherbivore defenses before plant-ant colonization in *Macaranga* myrmecophytes. Pop Ecol 43:207–212

Nykänen H, Koricheva J (2004) Damage-induced changes in woody plants and their effects on insect herbivore performance: a meta-analysis. Oikos 104:247–268

Price PW (1991) The plant vigor hypothesis and herbivore attack. Oikos 62:244–251

Price PW, Roinen H, Tahvanainen J (1987) Plant age and attack by the bud galler, *Euura mucronata*. Oecologia 73:334–337

Rehill BJ, Whitham TG, Martinsen GD, Schweitzer JA, Bailey JK, Lindroth RL (2006) Developmental trajectories in cottonwood phytochemistry. J Chem Ecol 32:2269–2285

Reichardt PB, Bryant JP, Clausen TP, Wieland GD (1984) Defense of winter-dormant Alaska paper birch against snowshoe hares. Oecologia 65:58–69

Rosenthal JP, Kotanen PM (1994) Terrestrial plant tolerance to herbivory. Trends Ecol Evol 9:145–148

Rousi M, Häggman J, Bryant JP (1987) The effect of bark phenols upon mountain hare barking of winter-dormant Scots pine. Holarctic Ecol 10:60–64

Rousset O, Lepart J (2000) Positive and negative interactions at different life stages of a colonizing species (*Quercus humilis*). J Ecol 88:407–412

Rudgers JA (2004) Enemies of herbivores can shape plant traits: selection in a facultative ant-plant mutualism. Ecology 85:192–205

Rudgers JA, Strauss SY, Wendel JE (2004) Trade-offs among anti-herbivore resistance traits: insights from Gossypieae (Malvaceae). Am J Bot 91:871–880

Sánchez-Hidalgo ME, Martínez-Ramos M, Espinoza-García F (1999) Chemical differentiation between leaves of seedlings and spatially close adult trees from the tropical rainforest species *Nectandra ambigens* (Lauraceae): an alternative test of the Janzen-Connell model. Funct Ecol 13:725–732

Schappert PJ, Shore JS (2000) Cyanogenesis in *Turnera ulmifolia* L. (Turneraceae): II. Developmental expression, heritability and cost of cyanogenesis. Evol Ecol Res 2:337–352
Schupp EW (1986) *Azteca* protection of *Cecropia:* ant occupation benefits juvenile trees. Oecologia 70:379–385
Seiwa K (1999) Changes in leaf phenology are dependent on tree height in *Acer mono*, a deciduous broad-leaved tree. Ann Bot 83:355–631
Sinclair ARE, Smith JNM (1984) Do plant secondary compounds determine feeding preferences of snowshoe hares? Oecologia 61:403–410
Smit C, Den Ouden J, Müller-Schärer H (2006) Unpalatable plants facilitate tree sapling survival in wooded pastures. J Appl Ecol 43:305–312
Spiegel LH, Price PW (1996) Plant aging and the distribution of *Rhyaciona neomexicana* (Lepidoptera: Tortricidae). Environ Entomol 25:359–365
Sugiura S, Abe T, Makino S (2006) Loss of extrafl oral nectary on an oceanic island plant and its consequences for herbivory. Am J Bot 93:491–495
Sultan SE (2005) An emerging focus on plant ecological development. New Phytol 166:1–5
Swaine MD, Whitmore TC (1988) On the definition of ecological species groups in tropical rain forests. Vegetatio 75:81–86
Swihart RK, Bryant JP (2001) Importance of biogeography and ontogeny of woody plants in winter herbivory by mammals. J Mammal 82:1–21
Tahvanainen J, Helle E, Julkunen-Tiitto R, Lavola A (1985) Phenolic compounds of willow bark as deterrents against feeding by mountain hare. Oecologia 65:319–323
Tiffin P (2000) Mechanisms of tolerance to herbivore damage: what do we know? Evol Ecol 14:523–536
Trager MD, Bruna EM (2006) Effects of plant age, experimental nutrient addition and ant occupancy on herbivory in a neotrpical myrmecophyte. J Ecol 94:1156–1163
Turner IM (1994) Sclerophylly: primarily protective? Funct Ecol 8:669–675
Van Bael SA, Brawn JD, Robinson SK (2003) Birds defend trees from herbivores in a Neotropical forest canopy. P Natl Acad Sci USA 100:8304–8307
Van Zandt P (2007) Plant defense, growth, and habitat: a comparative assessment of constitutive and induced resistance. Ecology 88:1984–1993
Vandenberghe C, Smit C, Pohl M, Buttler A, Freléchoux F (2009) Does the strength of facilitation by nurse shrubs depend on grazing resistance of tree saplings? Basic Appl Ecol 10:427–436
Warner PJ, Cushman JH (2002) Influence of herbivores on a perennial plant: variation with life history stage and herbivore species. Oecologia 132:77–85
Webber BL, Woodrow IE (2009) Chemical and physical plant defence across multiple ontogenetic stages in a tropical rainforest understorey tree. J Ecol 97:761–771
Weiner J (2004) Allocation, plasticity and allometry in plants. Perspect Plant Ecol Evol Syst 6:207–215
Weltzin JF, Archer SR, Heischmidt RK (1998) Defoliation and woody plant (*Prosopis glandulosai*) seedling regeneration: potential *vs* realized herbivory tolerance. Plant Ecol 138:127–135
Wooley SC, Donaldson JR, Gusse AC, Lindroth RL, Stevens ML (2007) Extrafloral nectaries in aspen (Populus tremuloides): heritable genetic variation and herbivore-induced expression. Ann Bot 100:1337–1346
Zagory D, Libby WJ (1985) Maturation-related resistance of *Pinus radiata* to western gall rust. Phytopathology 75:1443–1447

Chapter 8
How Do Changes in Leaf/Shoot Morphology and Crown Architecture Affect Growth and Physiological Function of Tall Trees?

Hiroaki Ishii

Abstract With increasing height within the crowns of tall trees, leaves tend to become smaller and thicker and shoots shorter. In tall trees, the vertical variation in leaf/shoot morphology is largely driven by water status. Morphological changes associated with increasing height in the crown present static constraints on photosynthesis, such as decreasing light intercepting area relative to leaf mass and decreasing CO_2 diffusion rate inside the leaf. Despite high light availability, leaf-area-based photosynthetic rates at the tops of tall trees tend to be low and this may limit height growth. However, the observed changes in leaf/shoot morphology as well as xylem/leaf anatomy, and crown architecture may compensate for various physiological constraints associated with increasing tree height. Continuous renewal of branches and foliage through epicormic shoot production and the change from hierarchic to polyarchic crown architecture may allow large trees to maintain physiological function and continue to grow.

1 Introduction

Leaf and shoot morphology change with height in the crown of a tree. With increasing height in the crown and increasing light intensity, leaves tend to become smaller and thicker and shoot length tends to increase (Kozlowski and Pallardy 1996). Shade leaves are found in the lower crown and show morphological and physiological adaptations to low-light conditions, whereas sun leaves in the upper crown are adapted to high-light conditions. Shade leaves are arranged with minimum

H. Ishii (✉)
Department of Plant Science, Graduate School of Agricultural Science,
Kobe University, Kobe 657-8501, Japan
e-mail: hishii@swallow.kobe-u.ac.jp

F.C. Meinzer et al. (eds.), *Size- and Age-Related Changes in Tree Structure and Function*, Tree Physiology 4, DOI 10.1007/978-94-007-1242-3_8,

Fig. 8.1 Morphological characteristics of upper-, middle- and lower-crown (**a**) (*left to right*) current-year shoots of a 15-m-tall *Picea jezoensis* tree. In the upper crown, the shoot is longer and the leaves are thicker, shorter and densely packed along the shoot axis, whereas in the lower crown the shoot is shorter and the leaves are thinner, longer and arranged with less overlap. The upper-crown leaves (**b**) have more than one layer of palisade cells (*brackets*), while lower-crown leaves (**c**) have only one layer

overlap so that they do not cast shadow on each other (self-shading, Leverenz and Hinckley 1990) and usually have only one layer of palisade cells (Fig. 8.1). The amount of chlorophyll b, the light harvesting chlorophyll, increases relative to chlorophyll a in shade leaves. Shade leaves have lower maximum photosynthetic rate than sun leaves, but their photosynthetic rate at low light levels is higher and they are able to maintain positive net photosynthetic rates in deep shade (lower light compensation point). Sun leaves have multiple layers of palisade cells and more chloroplasts per leaf area, which allows them to utilize the strong light of the upper crown to realize higher maximum photosynthetic rate. In conifers, sun leaves are densely packed along the shoot axis. Strong light penetrates into the dense foliage, but when the light is weak, such as on a cloudy day, leaf overlap results in self-shading and the photosynthetic rate declines. Sun leaves have higher photosynthetic productivity and this results in greater shoot elongation in the upper crown. In trees over 50 m tall, however, the maximum shoot elongation occurs in the middle of the crown (Ishii et al. 2007) and shoots at the tops of tall trees are short and stunted (Greenwood et al. 2010). This suggests that despite high light intensity, shoot growth in the upper crown of tall trees is constrained by some limiting factor.

As trees grow taller and larger, the height and distance over which water must travel to reach transpiring leaves increases. With increasing height in the crown, water must be transported against gravitational force. The negative pressure required to transport water upward decreases at a rate of approximately 0.01 MPa per vertical meter (Zimmermann 1983; Bauerle et al. 1999). This is known as the *hydrostatic gradient* and indicates that, in order to transport water upward, 0.1 MPa of negative pressure is needed for each additional 10 m of tree height. Furthermore, as the branches of the tree extend, frictional resistances associated with long-distance water transport increase (Mencuccini 2003). And with each year of extension growth, the number of internodes, which form constrictions in the hydraulic pathway, increases. As a result of these compounding effects on water transport, the terminal leaves of large, tall trees are believed to be under constant water stress caused by negative xylem pressure.

The hydraulic limitation hypothesis postulated that reduced photosynthesis resulting from water stress sets the upper limit for tree growth (Ryan and Yoder 1997). In tall trees, photosynthetic rate declines on sunny afternoons because as the air dries, more and more water is lost through transpiration from stomata. As a result, the tree closes its stomata in order to conserve water at the cost of not being able to take in carbon dioxide needed for photosynthesis. Although the decline in photosynthesis due to decreasing stomatal conductance partially explains the decline in growth rate with increasing tree height, it may not be the only cause (Barnard and Ryan 2003; Martinez-Vilalta et al. 2007). In Norway spruce (*Picea abies*) and Scots pine (*Pinus sylvestris*), both photosynthetic rate and stomatal conductance decreases with increasing tree height, but the decrease in maximum photosynthetic rate relative to intercellular CO_2 concentration is greater than the decrease in stomatal conductance (Niinemets 2002). In Douglas-fir (*Pseudotsuga menziesii*), the decrease in photosynthetic rate with increasing tree height is associated with leaf anatomic characteristics such as increasing leaf thickness, which results in decreasing mesophyll conductance (Woodruff et al. 2009). These observations suggest that in addition to the dynamic constraints on leaf photosynthesis related to stomatal conductance, morphological changes associated with increasing tree height pose fixed constraints on photosynthetic rate.

The name sun-/shade-leaf implies that light is the main determinant of leaf morphological and physiological properties and that photosynthetic rate should be higher at the tops of tall trees. However, as explained above, in tall trees, the water status of leaves changes greatly with height in the crown. Thus morphological and physiological acclimation to the vertical light gradient is simultaneously affected by the hydrostatic gradient. For example, in Scots pine, light intensity increases exponentially with increasing height in the crown, but the rate of change in leaf morphology decreases, resulting in a non-linear relationship between morphology and light intensity (Stenberg et al. 2001). Leaf morphology also tends to be different between the top-most leaves of tall and short trees even when both are growing in similar light conditions (Woodruff et al. 2008; Ambrose et al. 2009; Mullin et al. 2009). The leaves of an epiphytic sapling growing atop a 30-m tall Japanese cedar (*Cryptomeria japonica*) tree tend to be more expanded than the leaves of its mother tree, even though both leaves occur adjacent to each other in the same light environment, suggesting that light is not the only factor influencing leaf morphology in tall trees (Fig. 8.2).

Leaf morphology and the arrangement of leaves on shoots determine the photosynthetic light-intercepting area (Stenberg et al. 2001). Therefore, leaf/shoot morphology is an important factor affecting tree growth. In this chapter, I will explore how constraints on leaf and shoot morphology limit photosynthesis at the tops of very tall trees. I will also discuss how morphological changes at various scales (from leaf and xylem anatomy to crown architecture) might compensate for the physiological constraints associated with increasing tree height.

Fig. 8.2 The leaves of an epiphytic sapling of Japanese cedar (*arrow*) are more expanded than those on the adjacent branch of its mother tree (*background*), even though both leaves developed under similar light conditions

2 Hydrostatic Constraints on Morphological Development and Exploitation of Light

The water status of the leaf is especially important during morphogenesis because turgor pressure must be sufficient for cell elongation (Boyer 1968; Hsiao 1973; Woodruff and Meinzer 2011) . In 60-m-tall Douglas-fir (*Pseudotsuga menziesii*) trees, insufficient turgor pressure due to hydrostatic limitation results in reduced leaf and shoot elongation in the upper crown (Woodruff et al. 2004; Meinzer et al. 2008). In addition to such direct effects of water status on leaf morphology, there are also indirect effects involving biochemical properties and carbon allocation. Under constant water stress, more carbon is allocated to non-structural, soluble carbohydrates. By increasing the concentration of solutes inside cells, the leaf is able to take up more water via osmotic adjustment (Morgan 1984). This change in carbon allocation may occur during leaf morphogenesis in order to maintain turgor pressure at the cost of allocation to leaf structure.

Increasing leaf thickness also prevents water loss from the leaf by decreasing the ratio of leaf surface area to mass. The opposite ratio, leaf mass per area (LMA), is a measure of dry mass allocation to light intercepting area. LMA increases with increasing height in the crown and increasing light intensity, indicating that more mass is allocated per light intercepting area in the upper crown where light intensity

Fig. 8.3 Vertical change in leaf and shoot morphology in a 112 m-tall coast redwood (*Sequoia sempervirens*) tree in northern California. Numbers indicate the height in the crown (m) where the branches were sampled

is high. In the upper crown of very tall trees, however, the increase in LMA is not correlated with the increase in light intensity (Ishii et al. 2008). In coast redwood (*Sequoia sempervirens*), the tallest living tree species, which can reach heights over 115 m, leaf and shoot morphology change markedly with height in the crown (Fig. 8.3). Leaf/shoot morphology and leaf chemistry of the tallest redwood trees change continuously with height in the crown, whereas light intensity increases exponentially (Fig. 8.4). In the lower-crown, where light intensity is low, leaf/shoot morphology and leaf chemistry are linearly correlated with the light environment, but in the upper crown, where light intensity is high, the changes in leaf/shoot morphology and leaf chemistry do not correspond to the marked increase in light availability. This indicates that light is not the primary factor determining leaf morphology in the upper crown of the tallest coast redwood trees.

In the tallest redwoods, leaf water status changes linearly with height in the crown and bulk leaf water potential of treetop leaves regularly approaches the cavitation threshold for this species (–2 MPa) during the day (Koch et al. 2004; Ishii et al. 2008), indicating that leaf water supply is indeed constrained by gravity near the tops of the tallest trees. Such hydrostatic constraints appear to limit photosynthetic exploitation of light in the upper crowns of coast redwoods (Ishii et al. 2008). Although light intensity is highest in the upper crown, tall trees may not be able to realize maximum photosynthetic rates if they are unable to produce leaves with structural and biochemical properties capable of fully exploiting the available light. In coast redwood, photosynthetic rate per unit leaf mass decreases linearly with

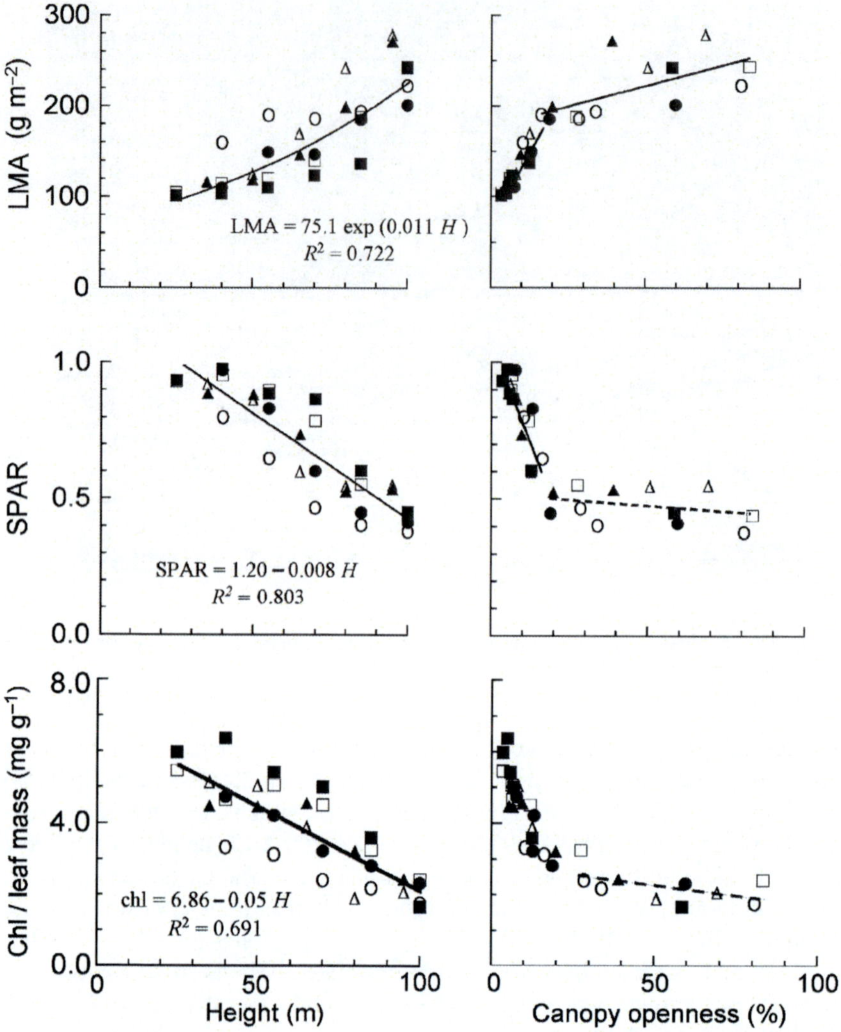

Fig. 8.4 Leaf mass per area (*LMA*), silhouette to projected area ratio (*SPAR*) and chlorophyll content (*Chl*) all change continuously with height in the crown of the world's tallest coast redwood (*Sequoia sempervirens*) trees. On the other hand, light intensity (*measured by canopy openness*) increases exponentially with increasing height in the crown. All three variables are correlated with light in the lower crown where canopy openness values are low, but change very little despite the marked increase in light intensity in the upper canopy. *Solid lines* indicate statistically significant relationships ($P<0.05$), while *broken lines* indicate insignificant relationships (after Ishii et al. 2008)

increasing height in the crown as more mass is allocated to photosynthetic tissue where light intensity is highest (Fig. 8.5). With increasing height in the crown, LMA increases and mesoporosity (proportion of leaf cross section devoted to air space) decreases (Oldham et al. 2010), which leads to slower CO_2 diffusion inside the leaf

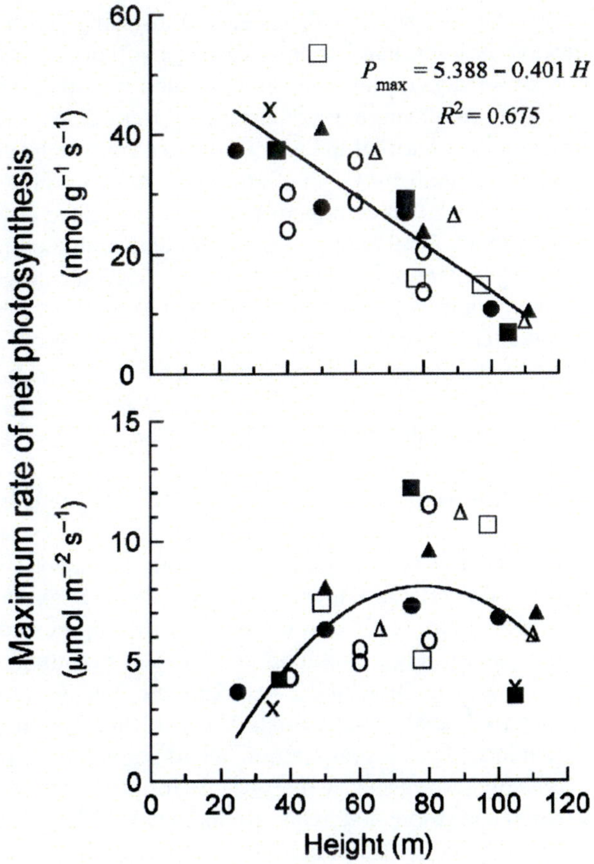

Fig. 8.5 Photosynthetic rate per leaf mass decreases linearly with increasing height in the crown of the world's tallest coast redwood (*Sequoia sempervirens*) trees, whereas photosynthetic rate per leaf area (*light intercepting area*) reaches maximum around 70–90 m (after Ishii et al. 2008)

(Parkhurst 1994; Hanba et al. 1999; Mullin et al. 2009; Steppe et al. 2011) and reduced photosynthetic rate per leaf area near the tree top (Ishii et al. 2008; Ambrose et al. 2009). In contrast to the linear decrease in mass-based photosynthetic rate, the highest photosynthetic rate per leaf area is observed around 70–90 m above the ground. In tall coast redwoods, the 70–90 m zone may be the threshold height where water stress does not limit shoot morphological development and leaf performance too much so that photosynthetic rate per light intercepting area is maximized.

3 How Morphological Changes Might Compensate for Constraints Associated with Increasing Tree Height

Crown architecture is the cumulative result of branch production and elongation, the main objective of which is the acquisition of space for light interception (Barthelemy and Caraglio 2007). Crown architecture also represents the hydraulic pathway through which water travels to reach transpiring leaves and optimization of water transport is important for maintaining hydraulic functional status, especially in large trees (Mencuccini 2003; Buckley and Roberts 2005; Mencuccini et al. 2011). The crown architecture of young trees is *hierarchic* such that architectural units are sequentially connected in order of decreasing size from the main stem to primary branches to small twigs. In contrast, the crown architecture of large, old trees is *polyarchic* such that architectural units are repeated at various positions within the tree (Millet et al. 1999, Fig. 8.6). Hierarchic architecture is the result of sequential branching, where new growth units are produced by regular elongation of terminal and lateral buds, whereas polyarchic architecture is the result of *reiteration*, where new architectural units are produced by irregular sprouting from dormant or adventitious buds on large-diameter axes such as the main stem and large branches (Briand et al. 1992; Remphrey and Davidson 1992; Begin and Filion 1999; Ishii et al. 2002). Reiteration from dormant buds is also known as epicormic shoot production.

Although the physiological mechanism controlling epicormic shoot production is not clear, auxin produced in the shoot apex and leaves is believed to inhibit the release of basal buds, which then become dormant (Bowersox and Ward 1968; Vogt and Cox 1970; Wignall and Browning 1988). Decapitation of the apex by pruning or injury leads to the release of the suppressed, dormant buds in various woody species (e.g., Guimond et al. 1998; Bellingham 2000; Kerr and Harmer 2001; Millet and Bouchard 2003). Epicormic shoot production is also associated with reduced cambial activity (Bachelard 1969; Nicolini et al. 2001). In large, old trees, epicormic shoot production occurs regularly in the absence of any injury or trauma, suggesting that it may be the result of internal factors, such as reduced meristem and cambial activity (Kauppi et al. 1991; Nicolini et al. 2003; Ishii et al. 2002). In large, old trees, epicormic shoot production is an important mechanism that maintains physiological function by mitigating the various factors that may lead to reduced growth, such as increasing hydraulic resistance, increasing respiration/photosynthesis ratio, and meristem aging (Ishii et al. 2007). Other chapters in this book provide details on the effects of these factors on physiological function. Here, we will consider how morphological changes, including epicormic shoot production and the change from hierarchic to polyarchic crown architecture might compensate for the various stresses associated with increasing tree size to maintain physiological function.

3.1 Improvement of Hydraulic Architecture

Epicormic shoot production shortens the hydraulic pathway to transpiring leaves and improves the hydraulic functional status of large, old trees by compensating for increasing hydraulic resistance. In mature *Cinnamomum camphora* trees, leaf-specific conductivity of epicormic shoots is higher than that of terminal shoots on the same branch (Otoda and Ishii 2009). The leaves on epicormic shoots of *Eucalyptus globulus*

Fig. 8.6 The crown architecture of large, old trees is polyarchic, where architectural units are repeated within the crown. Sprouting from epicormic buds produces repeating architectural units (*reiterations*) such as seedling-like structures on the branches of coast redwood (*upper*) and *Cinnamomum camphora* (*lower*)

are less resistant to water stress than terminal leaves suggesting they developed under more favorable water status. (Ito and Suzaki 1990). The improved water status of epicormic shoots may be the result of their direct connectivity to large-diameter branch axes, which have high conductivity. This is similar to stump sprouts, which

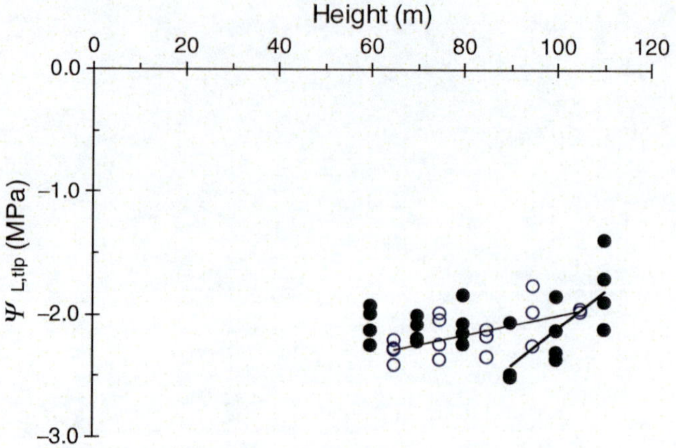

Fig. 8.7 Bulk leaf water potential at turgor loss ($\Psi_{L,tlp}$) in relation to height in the crown for two of the world's tallest coast redwood (*Sequoia sempervirens*) trees. The $\Psi_{L,tlp}$ increases with height in the crown of the shorter tree (*open symbols, thin line*, $P<0.05$), while in the taller tree, it is constant from 60 to 80 m and then increases above 90 m (*filled symbols, thick line*, $P<0.05$)

have higher hydraulic conductance than seedlings because of their connectivity to well-developed root systems (Kauppi et al. 1990; Ito et al. 1995).

In *C. camphora*, however, hydraulic conductance is similar between epicormic and terminal shoots when light conditions are more favorable at the branch terminal (Otoda and Ishii 2009). This suggests that some compensating mechanism may work to improve water transport to terminal shoots where greater photosynthetic gain can be obtained (Sellin and Kupper 2005). In branches of tall coast redwood trees, leaf-specific hydraulic conductivity, sapwood to leaf area ratio, and resistance to xylem cavitation all increase with increasing tree height (Burgess et al. 2006; Ambrose et al. 2009). This suggests increasing water transport efficiency along with increasing hydraulic safety at the tree top. The sapwood to leaf area ratio of the trunk, however, decreases with increasing tree height, suggesting that hydraulic compensation occurs at the branch level, but not in the trunk of the tallest coast redwoods (Ambrose et al. 2010). Measurements of leaf water relations characteristics indicate that bulk leaf water potential at turgor loss may increase with height in the crown, suggesting that the tree-top leaves may not be so water-stressed (Fig. 8.7). In redwood, the number of transfusion tracheids, which transport water and solutes between the leaf vein and mesophyll, increases with increasing height in the crown. In addition, they become deformed, suggesting they may collapse under water stress and act as a hydraulic buffer that decreases the probability of xylem dysfunction (Oldham et al. 2010). Morphological and anatomical traits that contribute to improved water status may compensate for hydraulic limitation to promote photosynthesis at the tree top and branch terminal where light availability is highest.

3.2 Reduction of Maintenance Respiration

Epicormic shoot production may also compensate for increasing respiration/photosynthesis ratio with increasing tree size. A classic explanation for decreasing growth in large, old trees is that, with increasing tree size, the proportion of non-productive tissue (e.g., branches and the trunk) increases relative to productive tissue (leaves) (the respiration hypothesis *sensu* Ryan and Yoder 1997). This hypothesis is based on the notion that leaf area culminates with increasing size, while secondary tissue continues to accumulate because the surface area over which the tree must put on a new layer of xylem increases with each year of radial growth. As a result, increasing maintenance respiration relative to photosynthetic production might lead to reduced net photosynthesis and growth decline (Clark 1983). However, recent studies indicate that large trees may not necessarily be carbon limited (Sala et al. 2011).

There are several processes that could potentially prevent increase in maintenance respiration with increasing tree size. In mature trees, shaded branches in the lower crown die. This process is called "self-pruning" because it appears that the tree is getting rid of unnecessary, non-productive branches (Makinen 1999). To take the place of the self-pruned branches, both conifers and broadleaved trees produce epicormic branches originating from suppressed buds in the branch axil (Jemison and Schumacher 1948; Krajicek 1959; Lanner 1996; Ishii and Wilson 2001; Nicolini et al. 2001; Burrows et al. 2003). The newly produced branches have low C/F ratio (dry mass ratio of non-productive to productive tissue) suggesting that epicormic branching may contribute to reducing maintenance respiration (Remphrey and Davidson 1992). In Douglas-fir, epicormic branches from the trunk begin to appear after ca 100 years and their number and proportion continue to increase with increasing tree age, such that in trees ca 650 years old, nearly 90% of the live branches are either epicormic or from reiterated trunks (Van Pelt and Sillett 2008). In large, old trees of Douglas-fir and coast redwood, epicormic shoot production also occurs regularly on small-diameter branch axes (Fig. 8.8). The newly produced epicormic shoots eventually grow on top of the parent branch and this process is repeated so that, when viewed from above, the entire length of the branch is covered with waves of newly produced foliage. The overtopped, parent branch eventually dies back, a form of self-pruning within branches. Epicormic shoot production results in constant renewal of foliage while preventing increase in the C/F ratio. It appears as if the tree is anticipating the eventual decline of existing branches by producing the next generation of epicormic branches to eventually take their place.

3.3 Prevention of Meristem Aging

Meristem aging has also been suggested as a factor leading to growth decline in large, old trees (Day et al. 2002). Early experiments using scions taken from young

Fig. 8.8 (**a**) Epicormic shoot production from the main stem of old-growth Douglas-fir (*arrow*). Note that the epicormic branch has produced epicormic shoots of its own near the base (*brackets*). (**b**) Epicormic shoot production from the branch axis in old-growth coast redwood (*arrows*). The distal end of the branch is to the left. (**c**) Top-view of a branch of old-growth Douglas-fir covered with waves, or layers of newly produced epicormic shoots

and mature trees grafted onto common root stock suggested the existence of a genetically programmed cellular senescence of the apical meristem that is irreversible (Bender 2004). However, more recent experiments have shown that growth rates are a function of tree size, not meristem age (e.g., Matsuzaki et al. 2005; Bond et al. 2007; Mencuccini et al. 2007) and that cellular senescence may be reversible, possibly through DNA demethylation (Fraga et al. 2002). Reciprocal grafting experiments using scions of red spruce (*Picea rubens*) from young, mature and old growth trees showed that leaf/shoot morphology is affected by both meristem age and tree size, while shoot elongation and apical control are determined by tree size only (Greenwood et al. 2010). These results suggest that intrinsic changes due to meristem aging interact with extrinsic factors, such as water status and hormone concentration, which are affected by tree size (Day and Greenwood 2011).

Epicormic shoot production may also compensate for the negative effects of increasing size on meristem vigor. In large, old trees of Douglas-fir, epicormic shoot production occurs from branch axes nearly 60-years old (Ishii et al. 2002). Theoretical explorations using a simulation model of branch growth (Ford et al. 1990), indicate that epicormic shoot production increases branch longevity by successful reiteration of productive architectural units, as well as compensate for multiple stresses associated with increasing tree age (Kennedy et al. 2010). For example, the maximum age of simulated branches in Douglas-fir was nearly 350 years (Kennedy et al. 2004), similar to actual maximum branch ages observed on old trees (Ishii et al. 2002; Winter et al. 2002). In contrast, that of grand fir (*Abies grandis*), which does not produce epicormic shoots regularly, was less than 200 years. Suppressed buds, which originate in the leaf axils, continue to grow with the expanding cambium for several years (Kozlowski 1971). In arboriculture, it is empirically known that the success rate of scions is higher for epicormic sprouts than for terminal branches. This suggests that relative to sequential buds, which continue to elongate, the physiological age of epicormic buds, which have undergone less cell division, may be younger.

4 Conclusions

The hydraulic limitation hypothesis proposed direct effects of water stress on photosynthetic rate through reduction in stomatal conductance (dynamic constraints *sensu* Woodruff et al. 2009). However, as discussed in this chapter, the mechanisms underlying reduced photosynthetic rates in tall trees may be more complex, involving interactions among water transport, biochemical properties, leaf/shoot morphology, dry matter allocation, and crown architecture. There may also be compensating mechanisms that increase tolerance to the multiple stresses associated with increasing tree height. Researchers are now attempting to separate the confounding effects of these factors on photosynthetic and growth rates of tall trees. In order to fully understand the physiological changes associated with tree growth, we must develop a further understanding of the causes and consequences of multiple stresses associated with increasing size and how trees respond to these stresses over the course their lifetime (Ford 1992; Ishii et al. 2007).

The arrangement of leaves on the shoot axis is an important morphological property determining photosynthetic gain of the tree. Leaf and shoot morphology affect photosynthetic productivity by influencing how trees utilize light within the crown. The ratio of shoot silhouette area to projected leaf area (SPAR) is an index of the degree of leaf overlap within conifer shoots. Tree species with higher maximum SPAR ($SPAR_{max}$) tend to have higher leaf area index (LAI, the amount of leaf area per unit land area) (Leverenz and Hinckley 1990). This indicates that species with high $SPAR_{max}$ are able to fully utilize the vertical gradient in light intensity from the strong light in the upper canopy to the weak light in the lower canopy. The $SPAR_{max}$ of coast redwood is nearly 1.0 (Ishii et al. 2008), indicating that there is very little overlap among shade leaves arranged along the shoot axis. The high morphological plasticity of its foliage may allow this species to utilize light across a wide gradient of light intensities within crowns up to 98 m deep (S.C. Sillett personal communication). Redwood forests have the highest observed LAI among forests of the world and greatest aboveground biomass (Westman and Whittaker 1975; Sillett and Van Pelt 2007).

In coast redwood forests, tree-level wood production, derived from direct measurement of branches in the canopy, increases with age up to and including trees over 112 m tall and 1850 years old (Sillett et al. 2010). At the whole tree level, the ratio of respiratory demand to photosynthetic capacity seems to be independent of tree size. Although diameter growth at breast height decreases with increasing tree size, wood production of the whole trunk and crown does not. Indeed the notion that wood production reaches an early peak followed by a decline with increasing tree size and age may not be true of coast redwoods. (Sillett et al. 2010). With increasing size and age, single trees may become as complex as the whole forest canopy, comprising reiterated stems and branches of various sizes that interact with each other as the tree continues to grow (Sillett and Van Pelt 2000). Ground-level measurements may indicate decreasing growth of the lower trunk with increasing size and age, but the crown may comprise vigorously growing young branches and reiterated trunks.

In order to understand how age/size-related changes in leaf/shoot morphology and crown architecture affect whole-tree growth rate, we must elucidate functional architectural units within large, old trees. For example, what is the unit of turnover resulting from reiteration by epicormic shoot production? What is the rate of turnover and is there a limit to their dynamics that may lead to eventual growth decline? How does crown disturbance affect these dynamics? Answering such questions would provide further insight toward answering the age-old question "Why do trees grow so big and live so long?"

Acknowledgements I thank Dr. S.C. Sillett for constructive comments on earlier versions of this chapter.

References

Ambrose AR, Sillett SC, Dawson TE (2009) Effects of tree height on branch hydraulics, leaf structure and gas exchange in California redwoods. Plant Cell Environ 32:743–757

Ambrose AR, Sillett SC, Koch GW, Van Pelt R, Antoine ME, Dawson TE (2010) Effects of height on treetop transpiration and stomatal conductance in coast redwood (*Sequoia sempervirens*). Tree Physiol 30:1260–1272

Bachelard EP (1969) Studies on the formation of epicormic shoots on eucalypt stem segments. Austr J Biol Sci 22:1291–1296

Barnard HR, Ryan MG (2003) A test of the hydraulic limitation hypothesis in fast-growing *Eucalyptus saligna*. Plant Cell Environ 26:1235–1245

Barthelemy D, Caraglio Y (2007) Plant architecture: a dynamic, multilevel and comprehensive approach to plant form, structure and ontogeny. Ann Bot 99:375–407

Bauerle WL, Hinckley TM, Cermak J, Kucera J, Bible K (1999) The canopy water relations of old-growth Douglas-fir trees. Trees 13:211–217

Begin C, Filion L (1999) Black spruce (*Picea mariana*) architecture. Can J Bot 77:664–672

Bellingham PJ (2000) Resprouting as a life history strategy in woody plant communities. Oikos 89:409–416

Bender J (2004) DNA methylation and epigenetics. Ann Rev Plant Biol 55:31–68

Bond BJ, Czarnomski NM, Cooper C, Day ME, Greenwood MS (2007) Developmental decline in height growth in Douglas-fir. Tree Physiol 27:441–453

Bowersox TW, Ward WW (1968) Auxin inhibition of epicormic shoots in white oak. For Sci 14:192–195

Boyer JS (1968) Relationship of water potential to growth of leaves. Plant Physiol 43:1056–1062

Briand CH, Posluszny U, Larson DW (1992) Differential axis architecture in *Thuja occidentalis* (eastern white cedar). Can J Bot 70:340–348

Buckley TN, Roberts DW (2005) How should leaf area, sapwood area and stomatal conductance vary with tree height to maximize growth? Tree Physiol 26:145–157

Burgess SSO, Pittermann J, Dawson TE (2006) Hydraulic efficiency and safety of branch xylem increases with height in *Sequoia sempervirens* (D. Don) crowns. Plant Cell Environ 29:229–239

Burrows GE, Offord CA, Meagher PF, Ashton K (2003) Axillary meristems and the development of epicormic buds in Wollemi pine (*Wollemia nobilis*). Ann Bot 92:835–844

Clark JR (1983) Age-related changes in trees. J Arboric 9:201–205

Day ME, Greenwood MS (2011) Regulation of ontogeny in temperate conifers. In: Meinzer FC, Lachenbruch B, Dawson TE (eds) Size- and age-related changes in tree structure and function. Springer, Dordrecht

Day ME, Greenwood MS, Diaz-Sala C (2002) Age- and size-related trends in woody plant shoot development: regulatory pathways and evidence for genetic control. Tree Physiol 22:507–513

Ford ED (1992) The control of tree structure and productivity through the interaction of morphological development and physiological processes. Int J Plant Sci 153:S142–S162

Ford ED, Avery A, Ford R (1990) Simulation of branch growth in the *Pinaceae*: interactions of morphology, phenology, foliage productivity, and the requirement for structural support, on the export of carbon. J Theor Biol 146:15–36

Fraga MF, Rodriguez R, Canal MJ (2002) Genomic DNA methylation – demethylation during aging and reinvigoration of *Pinus radiata*. Tree Physiol 22:813–816

Greenwood MS, Day ME, Schatz J (2010) Separating the effects of tree size and meristem maturation on shoot development of grafted scions of red spruce (*Picea rubens* Sarg.). Tree Physiol 30(4):459–468

Guimond CM, Lang GA, Andrews PK (1998) Timing and severity of summer pruning affects flower initiation and shoot regrowth in sweet cherry. HortSci 33:647–649

Hanba YT, Miyazawa S-I, Terashima I (1999) The influence of leaf thickness on the CO_2 transfer conductance and leaf stable isotope ratio for some evergreen tree species in Japanese warm temperate forests. Funct Ecol 13:632–639

Hsiao TC (1973) Plant responses to water stress. Ann Rev Plant Physiol 24:519–570

Ishii H, Wilson ME (2001) Crown structure of old-growth Douglas-fir in the western Cascade Range, Washington. Can J For Res 39:1250–1261

Ishii H, Ford ED, Dinnie CE (2002) The role of epicormic shoot production in maintaining foliage in old *Pseudotsuga menziesii* (Douglas-fir) trees II. Basal reiteration from older branch axes. Can J Bot 80:916–926

Ishii HT, Ford ED, Kennedy MC (2007) Physiological and ecological implications of adaptive reiteration as a mechanism for crown maintenance and longevity. Tree Physiol 27:455–462

Ishii H, Jennings GM, Sillett SC, Koch GW (2008) Hydrostatic constraints on morphological exploitation of light in tall *Sequoia sempervirens* trees. Oecologia 156:751–763

Ito S, Suzaki T (1990) Morphology and water relations of leaves of *Eucalyptus globulus* sprouts. Bull Kyushu Univ For 63:37–53

Ito S, Sakuta K, Gyokusen K (1995) Distribution of hydraulic resistance in seedlings, sprouts and an adult tree of *Pasania edulis* Makino. Ecol Res 10:143–149

Jemison GM, Schumacher FX (1948) Epicormic branching in old-growth Appalachian hardwoods. J For 46:252–253

Kauppi A, Kiviniitty M, Ferm A (1990) Leaf morphology and photosynthetic rate in birch seedlings and stump sprouts. Can J For Res 20:952–960

Kauppi A, Paukkonen K, Rinne P (1991) Sprouting ability of aerial and underground dormant basal buds of *Betula pendula*. Can J For Res 21:528–533

Kennedy MC, Ford ED, Ishii H (2004) Model analysis of the importance of reiteration for branch longevity in *Pseudotsuga menziesii* compared with *Abies amabilis*. Can J Bot 82:892–909

Kennedy MC, Ford ED, Hinckley TM (2010) Defining how aging *Pseudotsuga* and *Abies* compensate for multiple stresses through multi-criteria assessment of a functional-structural model. Tree Physiol 30:3–22

Kerr G, Harmer R (2001) Production of epicormic shoots on oak (*Quercus robur*): effects of frequency and time of pruning. Forestry 74:467–477

Koch GW, Sillett SC, Jennings GM, Davis SD (2004) The limits to tree height. Nature 428:851–854

Kozlowski TT (1971) Growth and development of trees. Academic, New York

Kozlowski TT, Pallardy SG (1996) Physiology of woody plants. Academic, San Diego

Krajicek JE (1959) Epicormic branching in even-aged, undisturbed white oak stands. J For 57:372–373

Lanner RM (1996) The role of epicormic branches in the life history of western larch. In: Schmidt WC, Wyman C (eds.) Ecology and management of *Larix* forests: a look ahead, vol. INT-319. USDA Forest Service General Technical Report, Logan, UT, pp 323–326

Leverenz JW, Hinckley TM (1990) Shoot structure, leaf area index, and productivity of evergreen conifer stands. Tree Physiol 6:135–149

Makinen H (1999) Growth, suppression, death, and self-pruning of branches of Scots pine in southern and central Finland. Can J For Res 29:585–594

Martinez-Vilalta J, Vanderklein D, Mencuccini M (2007) Tree height and age-related decline in growth in Scots pine (*Pinus sylvestris* L.). Oecologia 150:529–544

Matsuzaki J, Norisada M, Kodaira J, Suzuki M, Tange T (2005) Shoots grafted into the upper crowns of tall Japanese cedar (*Cryptomeria japonica*) show foliar gas exchange characteristics similar to those of intact shoots. Trees 19:198–203

Meinzer FC, Bond BJ, Karanian JA (2008) Biophysical constraints on leaf expansion in a tall conifer. Tree Physiol 28:197–206

Mencuccini M (2003) The ecological significance of long-distance water transport: short-term regulation, long-term acclimation and the hydraulic costs of stature across plant life forms. Plant Cell Environ 26:163–182

Mencuccini M, Hölttä T, Martinez-vilalta J (2011) Comparative criteria for models of the vascular transport system of tall trees. In: Meinzer FC, Lachenbruch B, Dawson TE (eds.) Size- and age-related changes in tree structure and function. Springer, Dordrecht

Mencuccini M, Martinez-Vilalta J, Hamid HA, Korakaki E, Vanderklein D (2007) Evidence for age- and size-mediated controls of tree growth from grafting studies. Tree Physiol 27:463–473

Millet J, Bouchard A (2003) Architecture of silver maple and its response to pruning near the power distribution network. Can J For Res 33:726–739

Millet J, Bouchard A, Edelin C (1999) Relationship between architecture and successional status of trees in the temperate deciduous forest. Ecoscience 6:187–203

Morgan JM (1984) Osmoregulation and water stress in higher plants. Ann Rev Plant Physiol 35:299–319

Mullin LP, Sillett SC, Koch GW, Tu KP, Antoine ME (2009) Physiological consequences of height-related morphological variation in *Sequoia sempervirens* foliage. Tree Physiol 29:999–1010

Nicolini E, Chanson B, Bonne F (2001) Stem growth and epicormic branch formation in understory beech trees (*Fagus sylvatica* L.). Ann Bot 87:737–750

Nicolini E, Caraglio E, Pelissier R, Leroy C, Roggy J-C (2003) Epicormic branches: a growth indicator for the tropical tree, *Dicorynia guianensis* Amshoff (Caesalpiniaceae). Ann Bot 92:97–105

Niinemets U (2002) Stomatal conductance alone does not explain the decline in foliar photosynthetic rates with increasing tree age and size in *Picea abies* and *Pinus sylvestris*. Tree Physiol 22:515–535

Oldham AR, Sillett SC, Tomescu AMF, Koch GW (2010) The hydrostatic gradient, not light availability, drives height-related variation in *Sequoia sempervirens* (Cupressaceae) leaf anatomy. Am J Bot 97:1–12

Otoda T, Ishii H (2009) Basal reiteration improves the hydraulic functional status of mature *Cinnamomum camphora* trees. Trees 23:317–323

Parkhurst DF (1994) Diffusion of CO_2 and other gases inside leaves. New Phytol 126:449–479

Remphrey WR, Davidson CG (1992) Spatiotemporal distribution of epicormic shoots and their architecture in branches of *Fraxinus pennsylvanica*. Can J For Res 22:336–340

Ryan MG, Yoder BJ (1997) Hydraulic limits to tree height and tree growth. Bioscience 47:235–242

Sala A, Fouts W, Hoch G (2011) Carbon storage in trees: does relative carbon supply decrease with tree size? In: Meinzer FC, Lachenbruch B, Dawson TE (eds.) Size- and age-related changes in tree structure and function. Springer, Dordrecht

Sellin A, Kupper P (2005) Effects of light availability versus hydraulic constraints on stomatal responses within a crown of silver birch. Oecologia 142:388–397

Sillett SC, Van Pelt R (2000) A redwood tree whose crown is a forest canopy. Northwest Sci 74:34–43

Sillett SC, Van Pelt R (2007) Structure of an old-growth redwood forest: trunk reiteration and limb formation promote epiphytes, soil development, and water storage in the canopy. Ecol Monogr 77:335–359

Sillett SC et al (2010) Increasing wood production through old age in tall trees. For Ecol Manage 259:976–994

Stenberg P, Palmroth S, Bond BJ, Sprugel DG, Smolander H (2001) Shoot structure and photosynthetic efficiency along the light gradient in a Scots pine canopy. Tree Physiol 21:805–814

Steppe K, Niinemets U, Teskey RO (2011) Tree size- and age-related changes in leaf physiology and their influence on carbon gain. In: Meinzer FC, Lachenbruch B, Dawson TE (eds.) Size- and age-related changes in tree structure and function. Springer, Dordrecht

Van Pelt R, Sillett SC (2008) Crown development of coastal *Pseudotsuga menziesii*, including a conceptual model for tall conifers. Ecol Monogr 78:283–311

Vogt AR, Cox GS (1970) Evidence for the hormonal control of stump sprouting by oak. For Sci 16:165–171

Westman WE, Whittaker RH (1975) The pygmy forest region of northern California: studies on biomass and primary productivity. J Ecol 63:493–520

Wignall TA, Browning G (1988) Epicormic bud development in *Quercus robur* L. Studies of endogenous IAA, ABA, IAA polar transport and water potential in cambial tissues and effects of exogenous hormones on bud outgrowth from stem explants. J Exp Bot 39:1667–1678

Winter LE, Brubaker LB, Franklin JF, Miller EA, DeWitt DQ (2002) Initiation of an old-growth Douglas-fir stand in the Pacific Northwest: a reconstruction from tree-ring records. Can J For Res 32:1039–1056

Woodruff DR, Meinzer FC (2011) Size-dependent changes in biophysical control of tree growth: the role of turgor. In: Meinzer FC, Lachenbruch B, Dawson TE (eds.) Size- and age-related changes in tree structure and function. Springer, Dordrecht

Woodruff DR, Bond BJ, Meinzer FC (2004) Does turgor limit growth in tall trees? Plant Cell Environ 27:229–236

Woodruff DR, Meinzer FC, Lachenbruch B (2008) Height-related trends in leaf xylem anatomy and hydraulic characteristics in a tall conifer: safety versus efficiency in foliar water transport. New Phytol 180:90–99

Woodruff DR, Meinzer FC, Lachenbruch B, Johnson DM (2009) Coordination of leaf structure and gas exchange along a height gradient in a tall conifer. Tree Physiol 29:261–272

Zimmermann MH (1983) Xylem structure and the ascent of sap. Springer, Berlin

Part IV
Photosynthetic Gas Exchange and Carbon Balance

Chapter 9
Tree Size- and Age-Related Changes in Leaf Physiology and Their Influence on Carbon Gain

Kathy Steppe, Ülo Niinemets, and Robert O. Teskey

Abstract Understanding how leaf-level physiology changes with tree size and age is important for scaling single leaf measurements to the whole plant and stand level and for quantifying carbon fluxes from forest ecosystems. This chapter reviews what is known about the influences of tree height and age on gas exchange and foliar structure in both gymnosperm and angiosperm trees. We address how the key physiological processes, photosynthesis, respiration and stomatal conductance vary with tree height and age. To help explain the observed patterns, the underlying factors that can be responsible for the changes in leaf physiology are assessed, including tree size- and age-related trends in foliar anatomy, morphology and chemistry. In addition to modifications in foliar morphology and chemistry, biochemical limitations to net assimilation rates associated with the diffusion of carbon dioxide from the atmosphere to the sites of carboxylation are examined. Our review emphasizes that a variety of factors collectively are responsible for tree height- and age-related decline in net photosynthetic rates, and that the importance of different limitations varies for different species and between gymnosperm and angiosperm trees. While there is still much to be learned, what is clear from our current understanding is that more integrated studies that consider the simultaneous roles of leaf structure, chemistry and stomatal and mesophyll factors are needed to disentangle and assign importance to the various factors responsible for decreases in carbon gain with tree age and size.

K. Steppe (✉)
Laboratory of Plant Ecology, Faculty of Bioscience Engineering, Ghent University, Coupure links 653, B-9000 Ghent, Belgium
e-mail: Kathy.Steppe@UGent.be

Ü. Niinemets
Institute of Agricultural and Environmental Sciences, Estonian University of Life Sciences, Kreutzwaldi 1, Tartu 51014, Estonia

R.O. Teskey
Daniel B. Warnell School of Forestry and Natural Resources, University of Georgia, Athens, GA 30602, USA

F.C. Meinzer et al. (eds.), *Size- and Age-Related Changes in Tree Structure and Function*, Tree Physiology 4, DOI 10.1007/978-94-007-1242-3_9,

1 Introduction

As trees grow older, they increase in size. Numerous structural and physiological characteristics change with tree size and age, including net assimilation rate and stem hydraulic conductance (Ryan and Yoder 1997; Bond 2000; Niinemets 2002; Mullin et al. 2009) as well as foliar dry mass per unit area (Niinemets and Kull 1995; Niinemets 2002). These profound adjustments in leaves have an impact on tree growth and forest net primary production due to their direct influence on photosynthetic carbon gain (Gower et al. 1996; Ryan et al. 1997). An understanding of changes in leaf physiology and leaf anatomy, morphology and chemistry that affect foliar photosynthesis with tree size and age is important for scaling single leaf measurements to the whole plant and stand level and thus, for quantifying carbon fluxes from forest ecosystems. It also helps to explain observed size- and/or age-dependent declines in forest net primary productivity.

In this chapter, we review the influences of tree height and age on gas exchange and foliage structure in gymnosperm and angiosperm tree species. We address how the key physiological processes, photosynthesis, respiration and stomatal conductance vary with tree height and age. To help explain the observed patterns, the underlying factors that can be responsible for the changes in leaf physiology are assessed, including tree size- and age-related trends in foliar anatomy, morphology and chemistry. In addition to modifications in foliar morphology and chemistry, biochemical limitations to net assimilation rates associated with the diffusion of carbon dioxide from the atmosphere to the sites of carboxylation are examined. The current review emphasizes that a variety of factors collectively are responsible for tree height- and age-related decline in net photosynthesis rates, and that the importance of different limitations varies for different species and between gymnosperm and angiosperm trees. We highlight both consistencies and inconsistencies in the changes that species undergo with age and size. We also examine the studies that have addressed the fascinating question whether these changes are attributable to genetic changes with age or to the physiological stresses imposed by increasing size.

2 Changes in Gas Exchange and Leaf Morphology with Tree Size and Age

2.1 Age- and Size-Related Modifications in Net Photosynthesis in Gymnosperm Trees

As trees age and concurrently grow larger, the rate of net photosynthesis can change dramatically. There is a generally consistent pattern among gymnosperm tree species in size- and age-related changes in light-saturated net photosynthesis (photosynthetic capacity) based on leaf mass (A_M). A_M is highest in the youngest and smallest trees, and declines precipitously from the seedling stage through the first decades of growth (Fig. 9.1). It then continues to decline linearly with progressive increases in size and

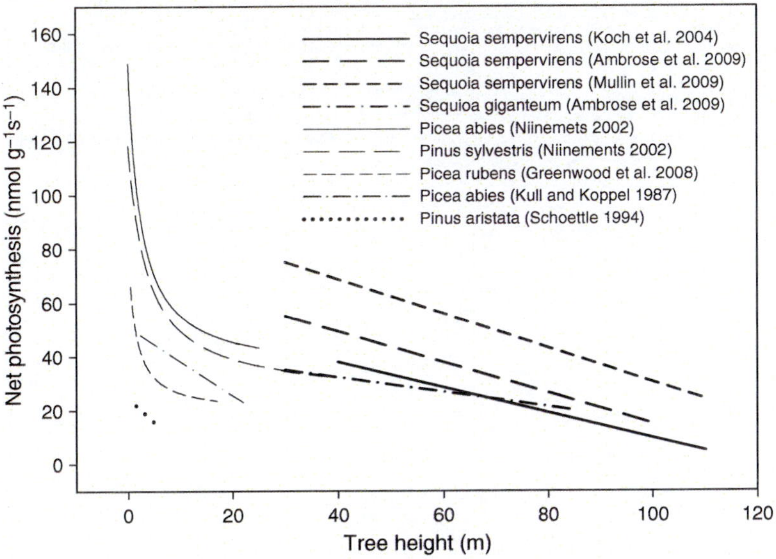

Fig. 9.1 Review of changes in net photosynthesis with tree height for several gymnosperm tree species

age. This declining pattern is often, but not always, similar when photosynthetic capacity is calculated on a leaf area basis (A_A). In *Pseudotsuga menziesii*, A_A declined in a manner consistent with those shown in Fig. 9.1, with A_A decreasing about 40% in trees between 5 and 55 m in height (Woodruff et al. 2009). This was also reported for *Picea rubens*, in which there was a 37% decline in A_A, and a 64% decline in A_M with increasing tree age from 3 year and height from 0.5 m to 127 year and 17 m (Greenwood et al. 2008). In *Pinus contorta* and *Pinus ponderosa* trees, A_A was significantly lower in older (275 year for *P. contorta* and 229 year for *P. ponderosa*) than in younger (45 year for *P. contorta* and 54 year for *P. ponderosa*) trees (Yoder et al. 1994). In *Picea abies* and *Pinus sylvestris*, the changes in rates of net photosynthesis with tree height and age were almost identical on either a mass or area basis (Kull and Koppel 1987; Niinemets 2002). However, in other species shown in Fig. 9.1, including *Pinus aristata* (Schoettle 1994) and *Sequoia sempervirens* (Mullin et al. 2009), and in other studies (*Pinus ponderosa*, Grulke and Retzlaff 2001; *Pinus pinaster*, Magnani et al. 2008; *Picea abies*, Merilo et al. 2009) it was reported that either A_A or the maximum rate of carboxylation on an area basis did not change significantly with tree age or size.

2.2 *Age- and Size-Related Patterns in Net Photosynthesis in Angiosperm Trees*

Although studied in only a few angiosperm tree species, net photosynthesis vs. tree age and size relationships are often different from the patterns in gymnosperm trees. For example, the change in A_M over the 0–20 m height range for angiosperm trees is

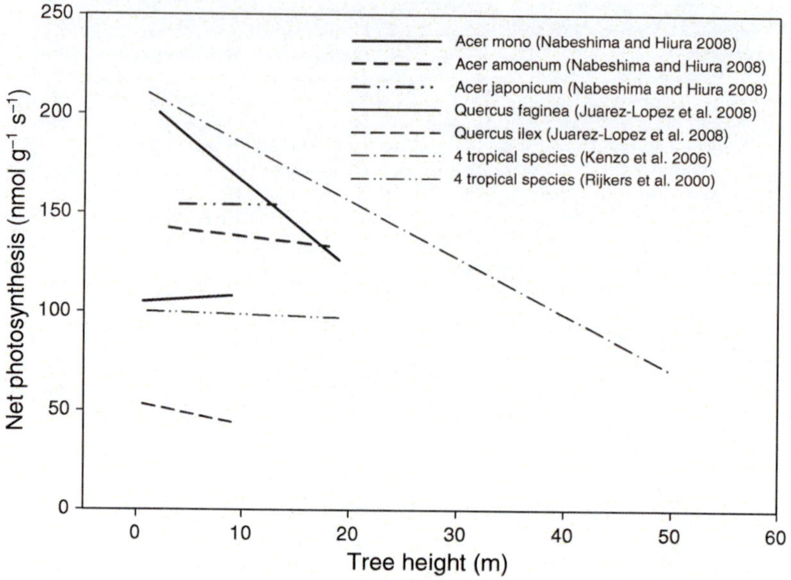

Fig. 9.2 Review of changes in net photosynthesis with tree height for several angiosperm tree species

generally small or nonexistent (Fig. 9.2), although for some species, patterns similar to gymnosperms can be observed (e.g. temperate deciduous trees *Acer amoenum* and *Acer mono* in Fig. 9.2).

When comparing A_A among angiosperm trees, the contrast with gymnosperms is even more distinct. In many instances, photosynthetic capacity on an area basis increases with tree size and age, e.g., in topical evergreens *Duguetia surinamensis*, *Vouacapoua americana*, *Dicorynia guianensis*, *Goupia glabra* (Rijkers et al. 2000), and *Dipterocarpus globosus*, *Dryobalanops aromatica*, *Shorea acuta*, *Shorea beccariana*, *Shorea macroptera* (Kenzo et al. 2006), in Mediterranean evergreens *Quercus ilex* (Juárez-López et al. 2008) and in temperate deciduous *Fagus sylvatica* (Herbinger et al. 2005), *Betula maximowicziana* and *Fraxinus mandschurica* (Koike 1984; Koike et al. 1986). In other instances, A_A does not change with tree size and age, e.g., in temperate deciduous *Acer mono* (Nabeshima and Hiura 2004; Nabeshima and Hiura 2008), *Acer amoenum* and *Acer japonicum* (Nabeshima and Hiura 2008) and in Mediterranean deciduous *Quercus faginea* (Juárez-López et al. 2008). A decrease in A_A with increasing tree size and age has been observed only in a very few cases, e.g., in temperate deciduous *Acer platanoides* and *Fraxinus excelsior* (Abdul-Hamid and Mencuccini 2009) and *Prunus serotina* (Fredericksen et al. 1996 for a comparison between seedlings and trees in late season). However, in some of the cases, especially in the studies in dense tropical canopies, light and tree height have changed simultaneously, making it difficult to disentangle the effects of environment from age and size per se (e.g., Kenzo et al. 2006; Rijkers et al. 2000 for confounding modifications of light with tree age and size).

2.3 Age- and Size-Related Changes in Mesophyll Diffusion, Stomatal Conductance and Respiration Rate

Along with photosynthetic capacity, many other changes occur in the leaves of trees as they age and increase in stature. Alterations in mesophyll CO_2 diffusion conductance (g_m) generally parallel the changes in A_A or A_M (Woodruff et al. 2009; Mullin et al. 2009). For example, in *Sequoia sempervirens* foliage, there was a steady decline in g_m per leaf mass across trees ranging in height from approximately 30–110 m, corresponding to an analogous pattern in A_M (Mullin et al. 2009). In the same trees, there was no change in either g_m per area or A_A with tree height or age. Likewise, in *Pseudotsuga menziesii*, g_m per area and A_A followed identical patterns of decline across trees ranging from 5 to 55 m in height (Woodruff et al. 2009).

Stomatal conductance to water vapor (g_s) typically declines with tree height and age. This observation is consistent over a range in age or height in both angiosperm and gymnosperm tree species that exhibit a decrease in A_M or A_A with age or height (Hubbard et al. 1999; Niinemets 2002; Nabeshima and Hiura 2004; Greenwood et al. 2008; Ambrose et al. 2009). Stomatal conductance also decreases with age or height in species that exhibit no change in A_M or A_A with age or height (Schoettle 1994; Juárez-López et al. 2008; Magnani et al. 2008; Nabeshima and Hiura 2008). Only a few studies have reported no change in g_s with age. In one study, constancy of g_s was associated with a decrease in A_M with tree age (*Picea rubens*, 60 vs. 120 years-old; Day et al. 2001) and in another with an increase in A_A (*Fagus sylvatica*, 2 vs. 60 years-old; Herbinger et al. 2005). Associated with changes in A_M, A_A and g_s with age and height is the relative enrichment of leaf carbon by the heavier isotope (^{13}C), resulting in increased leaf $\delta^{13}C$ values (McDowell et al. 2011), interpreted as an enhancement of the water use efficiency of leaves on taller and older trees (Koch et al. 2004; Greenwood et al. 2008; Magnani et al. 2008). However, enrichment of leaf carbon by the heavier isotope can also partly be associated with reductions in mesophyll diffusion conductance, thereby reflecting the overall greater limitation of photosynthesis by CO_2 availability (Woodruff et al. 2009).

Surprisingly, there is very little information available about changes in leaf respiration rate with increasing tree age and size and no broad conclusions can be drawn. In *Pinus ponderosa*, Grulke and Retzlaff (2001) reported no significant differences in leaf dark respiration across five tree age classes ranging from 0–10 to 41–60 years-old. In *Sequoia sempervirens*, Mullin et al. (2009) observed that leaf dark respiration increased substantially with increasing tree height, on an area basis from 1.6 μmol m^{-2} s^{-1} in 30 m tall trees to 5.9 μmol m^{-2} s^{-1} in 110 m tall trees, and on a mass basis from 6.8 nmol g^{-1} s^{-1} in 30 m tall trees to 10.3 nmol g^{-1} s^{-1} in 110 m tall trees.

2.4 Changes in Photosynthetic Attributes: Age or Size?

An obvious but hard to answer question is whether the changes observed in gas exchange and leaf morphology are due to age-dependent genetic changes (accumulation of mutations and/or altered gene expression), or altered physiological and environmental stresses due to tree height. As tree age and size are commonly correlated, it is difficult to separate the age- and size-dependent controls by field studies. Correlation analyses in large datasets using both height and age as independent variables have suggested that tree size correlates more strongly with the observed foliar modifications than tree age (Niinemets 2002; Ambrose et al. 2009). However, correlation analyses cannot provide conclusive insight into the relative importance of size and age. Several studies have attempted to address this question by grafting shoots of different ages onto rootstocks of common age. The foliage produced by the scions of different ages showed physiological characteristics related to the age of the original cutting in experiments with *Pinus taeda* (Greenwood 1984) and *Picea rubens* (Rebbeck et al. 1992). There were significant differences in rates of net photosynthesis observed between juvenile and mature *Larix laricina* scions consistent with the age of the original cutting (Hutchison et al. 1990). However, in a similar study with cuttings from *Pseudotsuga menziesii* trees the opposite was observed: within 2 years after grafting the scions of various aged shoots all exhibited photosynthetic and growth characteristics of the rootstock (Bond et al. 2007). Another study using *Pinus sylvestris* cuttings from trees ranging in age from 36 to 269 years-old found a comparable result with no effect of scion age on foliage photosynthetic characteristics 4–5 years after grafting (Vanderklein et al. 2007).

In comparison to the work in conifers just described, much less experimental work has been done on age- vs. size-determinants of photosynthetic changes in angiosperms. In *Acer pseudoplatanus* and *Fraxinus excelsior*, large old trees had lower photosynthesis rates than young trees, but scion age altered foliage photosynthesis to a minor extent, suggesting that tree size was dominating the reduction in photosynthesis in older and larger trees (Abdul-Hamid and Mencuccini 2009). Mencuccini et al. (2007) reviewed the available evidence and suggested that the discrepancy among the studies may have resulted from study-to-study differences in scion and rootstock size and age. It has further been concluded that the age factor affecting the photosynthetic characteristics is only significant during the first years of tree development, while further the size effects dominate (Mencuccini et al. 2007; Ambrose et al. 2009). Thus, both age and environmental effects can play a role in the observed decline in photosynthesis in mature relative to young trees, but clearly more work with different-sized and aged rootstocks and scions is needed to gain conclusive insight into the relative significance of tree age and size in determining the observed decline in assimilation with tree growth and maturation.

3 Underlying Factors Responsible for Size- and Age-Related Changes in Leaf Physiology

The decline in photosynthesis rates with tree size and age was initially only studied with respect to reductions in stomatal conductance due to longer (or more complex) hydraulic pathways and reduced allocation to roots as a tree increases in size (Yoder et al. 1994; Bond 2000; Day et al. 2001; Greenwood et al. 2008; Niinemets et al. 2009; Ishii 2011). However, as manipulations of photosynthetic rate in younger trees by decreasing hydraulic conductivity or in older trees by reducing transpiration often yield ambiguous results, the universal applicability of the hydraulic limitation model has been questioned (Day et al. 2001). If hydraulic limitation is the main driving variable, a steady decrease would be expected with increasing age and size, while this is not always the case (e.g., Greenwood et al. 2008). These observations suggest that size- and/or age-related hydraulic limitations do not act on photosynthesis as the sole driving variable and that other factors such as foliar anatomy, morphology or chemistry might also play a key role in the size- and age-related decline.

3.1 Trends in Foliar Anatomy, Morphology and Chemistry

Foliar phenotypic modifications in anatomy and morphology in response to the plant's developmental stage is known as heteroblasty (Sinnott 1960; Kozlowski 1971). As a result of maturation or phase change within the apical meristem, leaf primordia with different morphologies are produced as the plant ages and grows (Greenwood et al. 2008), resulting in dramatic changes in leaf morphology and anatomy in taller and older trees. Such structural changes play a key role in determining age- and size-dependent changes in net photosynthesis rate per unit area ($A_A = A_M$ times foliar dry mass per unit area, M_A) and in internal diffusion limitations.

3.1.1 Modifications in Foliar Dry Mass per Unit Area (M_A)

In virtually all instances, M_A increases substantially with tree height and age. This has been described for several gymnosperm trees including *Picea abies* (Kull and Koppel 1987; Niinemets and Kull 1995), *Picea rubens* (Greenwood et al. 2008), *Pinus taeda* (Greenwood 1984), *Pinus sylvestris* (Niinemets 2002), *Pseudotsuga menziesii* (Bond 2000), *Sequoia sempervirens* (Ambrose et al. 2009; Mullin et al. 2009) and *Sequoiadendron giganteum* (Ambrose et al. 2009) (Fig. 9.3a) as well as

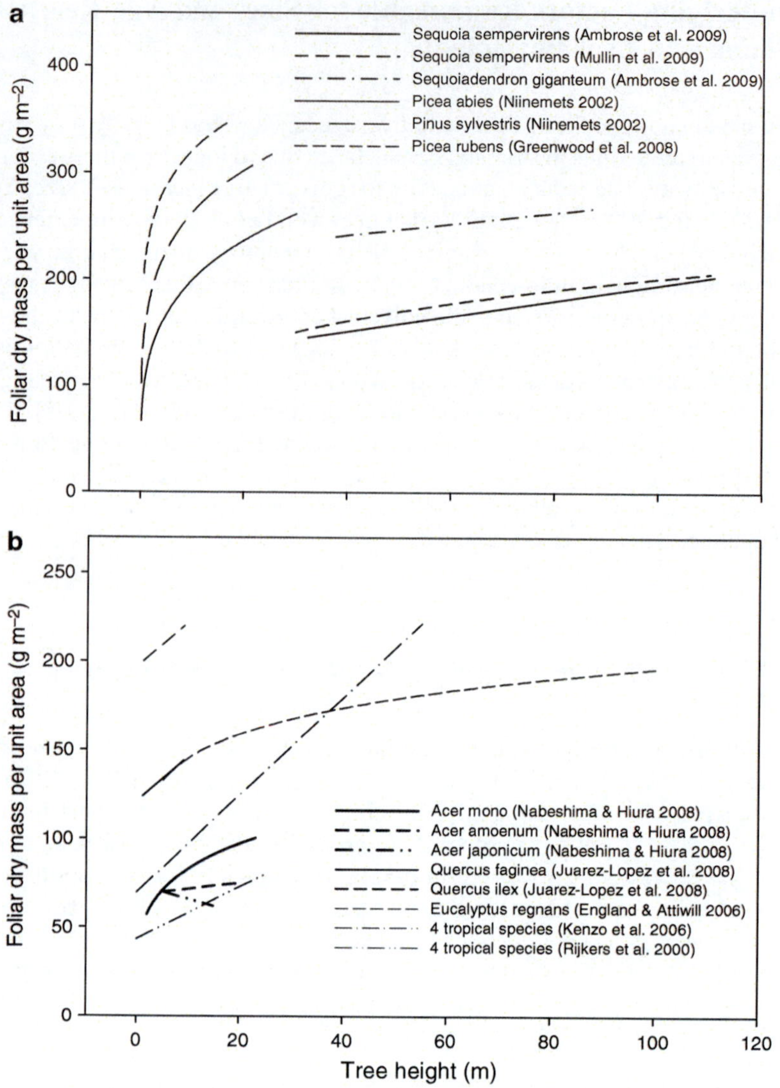

Fig. 9.3 Review of changes in foliar dry mass per unit area with tree height for several gymnosperm (**a**) and angiosperm (**b**) tree species

for angiosperm trees, including *Eucalyptus regnans* (England and Attiwill 2006), *Acer mono* (Nabeshima and Hiura 2004), *Quercus faginea* (Juárez-López et al. 2008), *Quercus ilex* (Juárez-López et al. 2008) and several tropical tree species (Rijkers et al. 2000; Kenzo et al. 2006) (see Nabeshima and Hiura 2008 for an exception) (Fig. 9.3b).

Although one might expect that age-related morphological trends are complicated by canopy position, it has been demonstrated that M_A increases with age in both sun and shade-adapted foliage (Niinemets and Kull 1995; Day et al. 2001).

The most pronounced increases in M_A were across age classes<40 years, with the rate of change lessening across the mature age classes. This relatively more rapid change of needle morphological characteristics in younger and shorter plants than in taller and older plants has also been observed in other conifers (Niinemets 2002). Niinemets (2002) reviewed the influence of both height and age on M_A in a data set consisting of 114 published studies and found a hyperbolic increase with increasing size (Fig. 9.3a) and age, and this increase was attributed to a greater foliar dry mass in larger and older trees. A similar hyperbolic relationship was observed for *Eucalyptus regnans* (England and Attiwill 2006) (Fig. 9.3b). Ambrose et al. (2009) studied the increase in M_A with increasing height in trees>20 m and found strong linear relationship in California redwoods, with different slopes for *Sequoia sempervirens* and *Sequoiadendron giganteum*. This linear part of the relationship in trees with a height>20 m is observed in all instances shown in Fig. 9.3.

The age- and size-dependent increases in M_A are still not fully understood, but they are possibly caused by the greater water stress in leaves of large trees (Niinemets and Kull 1995; Woodruff et al. 2004; Ambrose et al. 2009; Niinemets et al. 2009). Increased M_A is known to reduce water loss, because there is less leaf evaporative surface area per unit of foliar mass (Richardson et al. 2000). Photosynthetic capacity is also affected by increased M_A and this has been associated with changes in the ratio of internal air space and mesophyll area to total area (Sect. 3.1.2). In addition, greater M_A may also imply increased respiration rate per leaf surface area, reducing net carbon gain per unit leaf area (Koch et al. 2004).

3.1.2 Modifications in Leaf Size and Internal Architecture

An important factor responsible for larger M_A in taller and older trees is the decrease in foliar surface area. In line with the hypothesis of greater water stress in larger trees, the functional significance of smaller leaves is reduced area for transpiration, and improved efficiency of heat exchange due to a reduced boundary layer resistance (England and Attiwill 2006). Smaller leaves can result from greater water limitation and lower leaf turgor in crowns of taller trees (Woodruff et al. 2004; Ambrose et al. 2009). According to Lockhart's equation (1965), above the wall yielding threshold and if plastic extensibility is constant, cell expansion depends directly on turgor pressure and is sensitive to drought stress (Niinemets 2002; Steppe et al. 2006; Saveyn et al. 2007). The decline in leaf turgor with increasing height in response to decreasing xylem water potentials causes reduced cell expansion and slower cell division (Niinemets 2002; England and Attiwill 2006; Woodruff et al. 2004; Ambrose et al. 2009; Woodruff and Meinzer 2011).

All these changes not only reduce leaf size, but also result in increased M_A (Niinemets 2002; England and Attiwill 2006; Ambrose et al. 2009). England and Attiwill (2006) suggested that leaves of *Eucalyptus regnans* become more xeromorphic with increasing tree age and height. In their study, leaf area decreased about 50% from 6–7 years-old to 240 years-old, while M_A increased by about 27% from 6–7 years-old to 240 years-old trees (England and Attiwill 2006).

Apart from the surface area, leaf cross-sectional area can also vary with tree size. Greenwood et al. (2008) found that not only the total cross-sectional area of needle mesophyll tissue in *Picea rubens* increased with increasing tree size, but it increased relatively as a percentage of the cross-sectional area (from 61% to 66% across tree age classes from 3 to 127 years-old). These more massive leaves of taller trees have to deal with a longer pathway from the stomata to the innermost mesophyll cells, increasing the resistance to CO_2 diffusion within the needle. On the other hand, increased mesophyll thickness provides a greater cell wall area available for CO_2 diffusion that may contribute to decreased CO_2 liquid-phase resistance in mature trees (Terashima et al. 2005; Evans et al. 2009).

Different from *Picea rubens*, Apple et al. (2002) found a reduction in relative mesophyll cross-sectional area with increasing tree size in *Pseudotsuga menziesii*. Such limited investment of biomass in mesophyll can reduce photosynthesis per unit leaf mass. In addition, in *Picea rubens*, the internal air space between cells steadily declined with increasing age from 29% to 21% across tree age classes from 3 to 127 years-old (Greenwood et al. 2008). Decreased air space also decreases the surface area of mesophyll diffusion for gas exchange. The final effect of alterations in internal architecture on photosynthesis depends on modifications in cell wall thickness, in exposed to total chloroplast surface area ratio, in diffusion pathway lengths and in the fraction of mesophyll airspace (Terashima et al. 2005; Evans et al. 2009). Unfortunately, all these parameters together have not yet been determined for trees of varying size and age, calling for additional detailed anatomical studies to gain insight into the role of anatomy in limiting photosynthesis in various-sized and -aged trees.

3.1.3 Modifications in Leaf Thickness and Density

To further understand the role of variations in M_A in leaf photosynthesis, it is useful to distinguish between its components, leaf density (D) and thickness (T, $M_A = D \times T$). For species with complex three-dimensional leaf cross-section, M_A is the product of D and leaf volume to area ratio (Niinemets 1999; Poorter et al. 2009). In conifers, the size- and age-dependent increase in M_A is primarily associated with enhanced foliage density, while both thickness and density increase in broad-leaved species (Niinemets 1997a; Niinemets et al. 1999; England and Attiwill 2006; Juárez-López et al. 2008; Niinemets et al. 2009).

Thickness and density can have a different effect on photosynthesis. Leaves with a greater density generally have smaller and more tightly packed cells as well as thicker and more lignified cell walls than leaves of lower density, resulting in a lower volume of intercellular air space and longer diffusion path lengths from intercellular air space to the outer surfaces of cell walls as well as through the cell walls. Leaves with a greater density are also less elastic (higher leaf bulk elastic modulus) (Niinemets 2001), allowing them to develop a larger water potential gradient between soil and needles with lower tissue water loss. Their greater capacity for extracting water from the soil associated with greater density and, thus, thicker cell walls and more structural non-photosynthetic biomass, can lead to a lower photosynthetic capacity. It is noteworthy that an important difference exists when larger M_A arises from greater thickness because increases in foliar thickness allow accumulation of photosynthetic biomass per unit area, but should not necessarily increase mesophyll diffusion limitations (see Niinemets 1999 for a review). In *Eucalyptus regnans* the increase in M_A was associated with a significant increase in leaf thickness of about 17% from age 6–7 years-old to 240 years-old (England and Attiwill 2006), and similar modifications have been observed in other broad-leaved trees (Malkina 1983; Niinemets et al. 1999).

3.1.4 Modifications in Cuticle Thickness

Increases in cuticle thickness with tree age and size have been found in both conifers and broad-leaved species. Richardson et al. (2000) reported a 40% increase from age 15 to 145 years-old in a spruce hybrid. Mean cuticle thickness increased linearly with tree height by 0.044 $\mu m\ m^{-1}$ for *Pseudotsuga menziesii* along a height gradient of 53 m (from 7.25 μm for the 5 m height class to 9.5 μm for the 58 m height class) (Woodruff et al. 2010). In *Eucalyptus regnans* a clear increase in cuticle thickness was found with age and height from 7.7 μm at 6–7 years-old to 10.3 μm at 58 years-old, with no subsequent increase in cuticle thickness in older and larger trees (England and Attiwill 2006). Although there is no general correlation between cuticle thickness and the rate of cuticular transpiration across species (Burghardt and Riederer 2006), within species, the rate of cuticular transpiration has been shown to be negatively associated with cuticle thickness (e.g., DeLucia and Berlyn 1984). Thus, such increases in cuticle thickness with tree age and height may improve canopy water relations. In addition, a thicker cuticle may reflect a greater fraction of incident solar radiation and result in lower leaf temperatures (Slaton et al. 2001).

3.1.5 Modifications in Foliar Nitrogen Content

Reduction in nutrient availability with stand age has been suggested to play an important role in age-dependent decreases in growth rate and photosynthesis (Gower et al. 1996). Worldwide, there is a negative correlation between nitrogen content per mass (N_M) and M_A (e.g., Wright et al. 2004). Provided age-dependent species-specific changes in M_A fit the global trends, N_M is expected to decrease with age- and size-dependent increases in M_A. There is experimental evidence of reductions in foliage nitrogen content with increasing tree size in several studies (Niinemets 1997a; Merilo et al. 2009; Woodruff et al. 2009). Other studies, however, suggest that foliar nitrogen contents do not necessarily change with increases in tree size and age (Yoder et al. 1994; Day et al. 2001; Niinemets 2002; Nabeshima and Hiura 2008) or may even increase with increasing tree size (Nabeshima and Hiura 2004). In fact, variations in nutrient availability across stands on various soils and at different geographical locations seems to play a more important role in foliage nitrogen content than do tree height and age-related adjustments in foliar chemistry (Niinemets 2002; Juárez-López et al. 2008).

3.1.6 Modifications in Lignin Content

There is evidence that the leaves of older and taller trees are more lignified (Niinemets 1997b). Higher foliar lignin contents in larger trees further suggest that the diffusion resistance between leaf intercellular air space and carboxylation sites in the chloroplasts is greater in taller trees (Niinemets 2002), because lignification dramatically decreases the permeability of cell walls to both water and CO_2 (Brett and Waldron 1996).

3.2 Trends in Biochemical Limitations

In addition to modifications in foliar anatomy, morphology and chemistry just described, an understanding of the biochemical limitations to foliage net assimilation rates associated with the diffusion of carbon dioxide from the atmosphere to the sites of carboxylation is important to be able to explain the observed size- and age-related photosynthetic decline.

3.2.1 Limitations Due to Stomata

To separate between the various limitations of photosynthesis, foliage net assimilation rate (A) is conventionally expressed as (e.g. Nobel 1991):

$$A = g_s (C_a - C_i) = g_m (C_i - C_c) \tag{9.1}$$

where g_s is the stomatal conductance to CO_2, g_m is the mesophyll diffusion conductance to CO_2, C_a is the ambient CO_2 concentration, C_i is the CO_2 concentration in sub-stomatal cavities, and C_c is the CO_2 concentration in chloroplasts. As noted above, many studies have demonstrated that g_s decreases with increasing tree size, reflecting increases in leaf area to sapwood area ratio (Fischer et al. 2002; Köstner et al. 2002) as well as reductions in stem and leaf hydraulic conductance (Fischer et al. 2002; Mencuccini 2002; Martínez-Vilalta et al. 2007; Greenwood et al. 2008) with increases in tree age and size. However, changes in g_s alone do not necessarily mean that the reduction in A is due to stomata. A and g_s are often strongly correlated (Wong et al. 1979; Ball and Berry 1982; Wong et al. 1985), and may also decrease in parallel throughout tree ontogeny. In fact, the degree to which stomata limit photosynthesis is characterized by the CO_2 drawdown ($A / g_s = C_a - C_i$) implying that the magnitude of g_s relative to assimilation rate (photosynthetic capacity) determines the degree to which stomata constrain photosynthesis.

Several studies do demonstrate that C_i is reduced in taller trees (Nabeshima and Hiura 2004; Herbinger et al. 2005; McDowell et al. 2005; Juárez-López et al. 2008) and this has also been inferred based on carbon isotope composition of leaf material (Yoder et al. 1994; Martínez-Vilalta et al. 2007; Greenwood et al. 2008; Woodruff et al. 2009; McDowell et al. 2011). On the other hand, contrasting observations have been made in other studies. Ambrose et al. (2009) showed that A/g_s did not vary with tree height in *Sequoia sempervirens*, indicating that A and g_s changed in parallel, while A/g_s increased in *Sequoiadendron giganteum*, demonstrating that g_s was reduced more than A in taller trees. In *Pseudotsuga menziesii*, A/g_s also did not correlate significantly with tree height, which was different from what had been predicted on the basis of carbon isotope discrimination (Woodruff et al. 2009). Analogous patterns were observed in *Picea rubens* (Greenwood et al. 2008). These data collectively demonstrate that it is very important to examine the CO_2 drawdown (A/g_s) when making inferences on the control of stomata on photosynthesis in different-sized trees.

Lack of a clear indication of stronger stomatal limitation in larger trees is paralleled by contrasting trends in bulk canopy hydraulic characteristics. In some studies, leaf area to sapwood area ratio was negatively associated with tree size (Fischer et al. 2002; Martínez-Vilalta et al. 2007; Abdul-Hamid and Mencuccini 2009), and sapwood to basal area ratio and leaf hydraulic conductance were either independent (*Sequoiadendron giganteum*) or positively (*Sequoia sempervirens*) associated with tree size (Ambrose et al. 2009).

Most of the work on control of photosynthesis by g_s in different-size trees has focused on steady-state g_s values under light-saturated conditions. However, g_s is a highly variable quantity, changing during the day, between days and during the season. An important aspect in stomatal control of photosynthesis in different-sized trees is how rapidly stomatal conductance changes in response to environmental perturbations such as light, temperature and humidity alterations during the day as well as in response to soil drought. Not much work on the sensitivity of g_s to environmental drivers in different-sized trees has been conducted, but there is evidence that foliage in taller trees has greater cavitation resistance at the expense of reduced water transport capacity and efficiency (Woodruff et al. 2008; Ambrose et al. 2009) and is able to maintain higher leaf average g_s and photosynthesis at lower leaf water potentials than that in smaller trees (Woodruff et al. 2007). This may mean that under environmental stress, stomatal controls on carbon gain are less in larger old than in smaller young trees. In some species, this has been supported by carbon isotope discrimination data (Donovan and Ehleringer 1991, 1992).

3.2.2 Limitations Due to Mesophyll Conductance

Mesophyll diffusion conductance from sub-stomatal cavities to chloroplasts (g_m) constitutes a large part of the total diffusion conductance from the ambient air to the chloroplasts, often more than 50% (Niinemets et al. 2009). As with g_s, the importance of mesophyll diffusion on photosynthesis scales with the CO_2 drawdown

($A/g_m = C_i - C_c$) rather than with g_m alone (Eq. 9.1). Worldwide, there is a strong positive correlation between A/g_m and M_A, suggesting that mesophyll diffusion limits photosynthesis more in leaves with more robust structure (Niinemets and Sack 2006; Niinemets et al. 2009). Given this strong relationship, it is reasonable to expect that mesophyll diffusion conductance also limits photosynthesis more in taller trees having foliage with greater M_A. However, many fewer studies have investigated the variation in g_m than in g_s in different-sized trees. To our knowledge, g_m variation in different-sized trees has only been studied in Mullin et al. (2009) and Woodruff et al. (2009). In Mullin et al. (2009), A and g_m on leaf mass basis, and in Woodruff et al. (2009) A and g_m per unit area decreased with increasing tree height. However, the ratio A/g_m was independent of tree height in both studies (Mullin et al. 2009; Woodruff et al. 2009), suggesting that g_m limited photosynthesis similarly independent of tree height. Clearly more work on g_m limits in different-sized trees is needed to confirm whether the tree-size dependent variations in g_m limitations of photosynthesis within given species indeed depart from world-wide trends described across multiple species.

3.2.3 Limitations Due to Biochemical Capacities

Foliage biochemical capacities, the maximum carboxylase activity of Rubisco per leaf mass ($V_{cmax/mass}$) and the capacity for photosynthetic electron transport per mass ($J_{max/mass}$) can be expressed as the products of independent foliage chemical and structural characteristics. $V_{cmax/mass}$ can be expressed as follows:

$$V_{cmax/mass} = 6.25 V_{cr} F_R N_M \qquad (9.2)$$

where V_{cr} is the specific activity of Rubisco, i.e. the maximum rate of ribulose-1,5-bisphosphate carboxylation per unit Rubisco protein, F_R is the fraction of leaf nitrogen in Rubisco, N_M is leaf nitrogen content per dry mass and 6.25 is the nitrogen content of the Rubisco protein (Niinemets and Tenhunen 1997). $J_{max/mass}$ can be expressed analogously (Niinemets and Tenhunen 1997). Area-based characteristics can be further found by multiplying by M_A. As demonstrated above, the bulk of the studies report a reduction in A_M with tree height and size. On the other hand, fewer studies have found a negative correlation between A_A and tree height (Sect. 1) and some even a positive correlation, reflecting the confounding positive effect of tree height on M_A.

Equation 9.2 demonstrates that the reduction in mass-based foliage photosynthetic potentials can result from lower bulk foliage nitrogen content (N_M) as well as from reductions in foliage nitrogen allocation to photosynthetic machinery. Although there is no general trend in N_M across the studies (Sect. 3.1.5), reduced N_M clearly has limited the formation of high-capacity photosynthetic apparatus in several studies and is partly responsible for size-dependent reductions in mass-based foliage photosynthetic potentials in these studies (Niinemets 1997a; Merilo et al. 2009; Woodruff et al. 2009). Reduced nitrogen in such studies is also compatible with overall reductions in photosynthetic mesophyll and increases in support biomass (Apple et al. 2002).

It has been suggested that in higher M_A leaves, a greater proportion of leaf nitrogen is invested in defense and structural functions, which implies less nitrogen available for photosynthetic processes (Juárez-López et al. 2008), i.e., lower F_R (Eq. 9.2). Although very few studies have currently looked at nitrogen partitioning in different-sized trees, there is evidence of a lower fraction of nitrogen in the photosynthetic apparatus in older and larger trees (Niinemets 2002; Juárez-López et al. 2008; Merilo et al. 2009).

4 Conclusions and Directions for Future Research

Although the information at the mechanistic level is highly incomplete, evidence in this chapter indicates that there is no single factor that is universally responsible for the reductions in foliage photosynthesis rates in different-sized and -aged trees observed across the studies. Different factors explain the reduction in different studies, and more commonly, multiple factors are responsible for the observed reductions. Thus, integrated studies considering leaf structure, chemistry and stomatal and mesophyll factors are needed to disentangle the various factors responsible for decreases in carbon gain with tree age and size. Also, more experimental work is needed to gain insight into the role of the reduced physiological activity on plant competitive potential and tree responsiveness to stress, given that the severity of different stresses changes with changing tree size (Niinemets 2010). Additional emphasis needs to be placed on studies in angiosperms of size- and age-related effects on leaf morphology, physiology and biochemistry. There also is a critical need for an improved understanding of the basic causes of the changes that have been observed over the lifetime of trees. Creative approaches that allow the contribution of age (caused by changes in genetic expression) to be separated from that due to the physical limitations imposed by size, are especially needed in order to better understand the fundamental changes that occur over time in trees, which represent some of the largest and oldest organisms on earth.

References

Abdul-Hamid H, Mencuccini M (2009) Age- and size-related changes in physiological characteristics and chemical composition of *Acer pseudoplatanus* and *Fraxinus excelsior* trees. Tree Physiol 29:27–38

Ambrose AR, Sillett SC, Dawson TE (2009) Effects of tree height on branch hydraulics, leaf structure and gas exchange in California redwoods. Plant Cell Environ 32:743–757

Apple M, Tiekotter K, Snow M, Young J, Soeldner A, Phillips D, Tingey D, Bond BJ (2002) Needle anatomy changes with increasing tree age in Douglas-fir. Tree Physiol 22:129–136

Ball JT, Berry JA (1982) The C_i/C_s ratio: a basis for predicting stomatal control of photosynthesis. Carnegie Inst Wash Yearb 81:88–92

Bond BJ (2000) Age-related changes in photosynthesis of woody plants. Trends Plant Sci 5:349–353

Bond BJ, Czarnomski NM, Cooper C, Day ME, Greenwood MS (2007) Developmental decline in height growth in Douglas-fir. Tree Physiol 27:441–453

Brett CT, Waldron KW (1996) Physiology and biochemistry of plant cell walls, vol 1, 2nd edn, Topics in plant functional biology. Chapman and Hall, London, 225 pp

Burghardt M, Riederer M (2006) Cuticular transpiration. In: Riederer M, Müller C (eds) Biology of the plant cuticle. Blackwell, Oxford, pp 292–311

Day ME, Greenwood MS, White AS (2001) Age-related changes in foliar morphology and physiology in red spruce and their influence on declining photosynthetic rates and productivity with tree age. Tree Physiol 21:1195–1204

DeLucia EH, Berlyn GP (1984) The effect of increasing elevation on leaf cuticle thickness and cuticular transpiration in balsam fir. Can J Bot 62:2423–2431

Donovan LA, Ehleringer JR (1991) Ecophysiological differences among juvenile and reproductive plants of several woody species. Oecologia 86:594–597

Donovan LA, Ehleringer JR (1992) Contrasting water-use patterns among size and life-history classes of a semi-arid shrub. Funct Ecol 6:482–488

England JR, Attiwill PM (2006) Changes in leaf morphology and anatomy with tree age and height in the broadleaved evergreen species, *Eucalyptus regnans* F. Muell. Trees 20:79–90

Evans JR, Kaldenhoff R, Terashima I (2009) Resistances along the CO_2 diffusion pathway inside leaves. J Exp Bot 60:2235–2248

Fischer DG, Kolb TE, De Wald LE (2002) Changes in whole-tree water relations during ontogeny of *Pinus flexilis* and *Pinus ponderosa* in a high-elevation meadow. Tree Physiol 22:675–685

Fredericksen TS, Steiner KC, Skelly JM, Joyce BJ, Kolb TE, Kouterick KB, Ferdinand JA (1996) Diel and seasonal patterns of leaf gas exchange and xylem water potentials of different-sized *Prunus serotina* Ehrh. trees. For Sci 42:359–365

Gower ST, McMurtrie RE, Murty D (1996) Aboveground net primary production decline with stand age: potential causes. Trends Ecol Evol 11:378–382

Greenwood MS (1984) Phase change in loblolly pine: shoot development as a function of age. Physiol Plant 61:518–522

Greenwood MS, Ward MH, Day ME, Adams SL, Bond BJ (2008) Age-related trends in red spruce foliar plasticity in relation to declining productivity. Tree Physiol 28:225–232

Grulke NE, Retzlaff WA (2001) Changes in physiological attributes of ponderosa pine from seedling to mature tree. Tree Physiol 21:275–286

Herbinger K, Then C, Löw M, Haberer K, Alexous M, Koch N, Remele K, Heerdt C, Grill D, Rennenberg H, Häberle K-H, Matyssek R, Tausz M, Wieser G (2005) Tree age dependence and within-canopy variation of leaf gas exchange and antioxidative defence in *Fagus sylvatica* under experimental free-air ozone exposure. Environ Pollut 137:476–482

Hubbard RM, Bond BJ, Ryan MG (1999) Evidence that hydraulic conductance limits photosynthesis in old *Pinus ponderosa* trees. Tree Physiol 19:165–172

Hutchison KW, Sherman CD, Weber J, Smith SS, Singer PB, Greenwood MS (1990) Maturation in larch. II. Effects of age on photosynthesis and gene expression in developing foliage. Plant Physiol 94:1308–1315

Ishii H (2011) How do changes in leaf/shoot morphology and crown architecture affect growth and physiological function of large, old trees? In: Meinzer FC, Lachenbruch B, Dawson TE (eds.) Size- and age-related changes in tree structure and function. Springer, Dordrecht

Juárez-López FJ, Escudero A, Mediavilla S (2008) Ontogenetic changes in stomatal and biochemical limitations to photosynthesis of two co-occurring Mediterranean oaks differing in leaf life span. Tree Physiol 28:367–374

Kenzo T, Ichie T, Watanabe Y, Yoneda R, Ninomiya I, Koike T (2006) Changes in photosynthesis and leaf characteristics with tree height in five dipterocarp species in a tropical rain forest. Tree Physiol 26:865–873

Koch GW, Sillett SC, Jennings GM, Davis SD (2004) The limits to tree height. Nature 428:851–854

Koike T (1984) Comparison of photosynthetic response to habitat factors between seedlings and saplings of *Fraxinus mandshurica* var. japonica. Environ Control Biol 22:33–38

Koike T, Sakagami Y, Fujimura Y (1986) Characteristics of the leaf dynamics and the photosynthesis of the seedlings and saplings of *Betula maximowicziana* and *Fraxinus mandshurica* var. *japonica* in Hokkaido, Japan. In: Fujimori T, Whitehead D (eds) Crown and canopy structure in relation to productivity. Forestry and Forest Products Research Institute, Ibraki, pp 396–408

Köstner B, Falge E, Tenhunen JD (2002) Age-related effects on leaf area/sapwood area relationships, canopy transpiration and carbon gain of Norway spruce stands (*Picea abies*) in the Fichtelgebirge, Germany. Tree Physiol 22:567–574

Kozlowski TT (1971) Seed germination, ontogeny and shoot growth. In: Growth and development of trees, Vol 1. Academic, New York, 443 pp

Kull O, Koppel A (1987) Net photosynthetic response to light intensity of shoots from different crown positions and age in *Picea abies* (L.) Karst. Scand J Forest Res 2:157–166

Lockhart JA (1965) An analysis of irreversible plant cell elongation. J Theor Biol 8:264–275

Magnani F, Bensada A, Cinnirella S, Ripullone F, Borghetti M (2008) Hydraulic limitations and water-use efficiency in *Pinus pinaster* along a chronosequence. Can J For Res 38:73–81

Malkina IS (1983) Relationships between photosynthesis of oak leaves and their structure and age of trees. Lesovedeniye 4:68–71

Martínez-Vilalta J, Vanderklein D, Mencuccini M (2007) Tree height and age-related decline in growth in Scots pine (*Pinus sylvestris* L.). Oecologia 150:529–544

McDowell NG, Licata J, Bond BJ (2005) Environmental sensitivity of gas exchange in different-sized trees. Oecologia 145:9–20

McDowell NG, Bond BJ, Hill LT, Ryan MG, Whitehead D (2011) Relationships between tree height and carbon isotope discrimination. In: Meinzer FC, Lachenbruch B, Dawson TE (eds.) Size- and age-related changes in tree structure and function. Springer, Dordrecht

Mencuccini M (2002) Hydraulic constraints in the functional scaling of trees. Tree Physiol 22:553–565

Mencuccini M, Martínez-Vilalta J, Hamid HA, Korakaki E, Vanderklein D (2007) Evidence for age- and size-mediated controls of tree growth from grafting studies. Tree Physiol 27:463–473

Merilo E, Tulva I, Räim O, Kükit A, Sellin A, Kull O (2009) Changes in needle nitrogen partitioning and photosynthesis during 80 years of tree ontogeny in *Picea abies*. Trees 23:951–958

Mullin LP, Sillett SC, Koch GW, Tu KP, Antoine ME (2009) Physiological consequences of height-related morphological variation in *Sequoia sempervirens* foliage. Tree Physiol 29:999–1010

Nabeshima E, Hiura T (2004) Size dependency of photosynthetic water- and nitrogen-use efficiency and hydraulic limitation in *Acer mono*. Tree Physiol 24:745–752

Nabeshima E, Hiura T (2008) Size-dependency in hydraulic and photosynthetic properties of three *Acer* species having different maximum sizes. Ecol Res 23:281–288

Niinemets Ü (1997a) Distribution patterns of foliar carbon and nitrogen as affected by tree dimensions and relative light conditions in the canopy of *Picea abies*. Trees 11:144–154

Niinemets Ü (1997b) Energy requirement for foliage construction depends on tree size in young *Picea abies* trees. Trees 11:420–431

Niinemets Ü (1999) Research review. Components of leaf dry mass per area, thickness and density alter leaf photosynthetic capacity in reverse directions in woody plants. New Phytol 144:35–47

Niinemets Ü (2001) Climatic controls of leaf dry mass per area, density and thickness in trees and shrubs at the global scale. Ecology 82:453–469

Niinemets Ü (2002) Stomatal conductance alone does not explain the decline in foliar photosynthetic rates with increasing tree age and size in *Picea abies* and *Pinus sylvestris*. Tree Physiol 22:515–535

Niinemets Ü (2010) Responses of forest trees to single and multiple environmental stresses from seedlings to mature plants: past stress history, stress interactions, tolerance and acclimation. For Ecol Manag 260:1623–1639

Niinemets Ü, Kull O (1995) Effects of light availability and tree size on the architecture of assimilative surface in the canopy of *Picea abies*: variation in needle morphology. Tree Physiol 15:307–315

Niinemets Ü, Sack L (2006) Structural determinants of leaf light-harvesting capacity and photosynthetic potentials. In: Esser K, Lüttge UE, Beyschlag W, Murata J (eds) Progress in botany, vol 67. Springer Verlag, Berlin, pp 385–419

Niinemets Ü, Tenhunen JD (1997) A model separating leaf structural and physiological effects on carbon gain along light gradients for the shade-tolerant species *Acer saccharum*. Plant Cell Environ 20:845–866

Niinemets Ü, Kull O, Tenhunen JD (1999) Variability in leaf morphology and chemical composition as a function of canopy light environment in co-existing trees. Int J Plant Sci 160:837–848

Niinemets Ü, Díaz-Espejo A, Flexas J, Galmés J, Warren CR (2009) Role of mesophyll diffusion conductance in constraining potential photosynthetic productivity in the field. J Exp Bot 60:2249–2270

Nobel PS (1991) Physicochemical and environmental plant physiology, 4th edn. Academic, New York

Poorter H, Niinemets Ü, Poorter L, Wright IJ, Villar R (2009) Causes and consequences of variation in leaf mass per area (LMA): a meta-analysis. New Phytol 182:565–588

Rebbeck J, Jensen KF, Greenwood MS (1992) Ozone effects on the growth of grafted mature and juvenile red spruce. Can J For Res 22:756–760

Richardson AD, Berlyn GP, Ashton PMS, Thadani R, Cameron IR (2000) Foliar plasticity of hybrid spruce in relation to crown position and stand age. Can J Bot 78:305–317

Rijkers T, Pons TL, Bongers F (2000) The effect of tree height and light availability on photosynthetic leaf traits of four neotropical species differing in shade tolerance. Funct Ecol 14:77–86

Ryan MG, Yoder BJ (1997) Hydraulic limits to tree height and tree growth. What keeps trees from growing beyond a certain height? Bioscience 47:235–242

Ryan MG, Binkley D, Fownes JH (1997) Age-related decline in forest productivity: pattern and process. Adv Ecol Res 27:213–262

Saveyn A, Steppe K, Lemeur R (2007) Daytime depression in tree stem CO_2 efflux rates: is it caused by low stem turgor pressure? Ann Bot 99:477–485

Schoettle AW (1994) Influence of tree size on shoot structure and physiology of *Pinus contorta* and *Pinus aristata*. Tree Physiol 14:1055–1068

Sinnott EW (1960) Plant morphogenesis. McGraw-Hill, New York, 550 pp

Slaton MR, Hunt ER, Smith WK (2001) Estimating near-infrared leaf reflectance from leaf structural characteristics. Am J Bot 88:278–284

Steppe K, De Pauw DJW, Lemeur R, Vanrolleghem PA (2006) A mathematical model linking tree sap flow dynamics to daily stem diameter fluctuations and radial stem growth. Tree Physiol 26:257–273

Terashima I, Araya T, Miyazawa S-I, Sone K, Yano S (2005) Construction and maintenance of the optimal photosynthetic systems of the leaf, herbaceous plant and tree: an eco-developmental treatise. Ann Bot 95:507–519

Vanderklein D, Marinez-Vilalta J, Lee S, Mencuccini M (2007) Plant size, not age, regulates growth and gas exchange in grafted Scots pine trees. Tree Physiol 27:71–79

Wong SC, Cowan IR, Farquhar GD (1979) Stomatal conductance correlates with photosynthetic capacity. Nature 282:424–426

Wong S-C, Cowan IR, Farquhar GD (1985) Leaf conductance in relation to rate of CO_2 assimilation. III. Influences of water stress and photoinhibition. Plant Physiol 78:830–834

Woodruff DR, Meinzer FC (2011) Size-dependent changes in biophysical control of tree growth: the role of turgor. In: Meinzer FC, Lachenbruch B, Dawson TE (eds.) Size- and age-related changes in tree structure and function. Springer, Dordrecht

Woodruff DR, Bond BJ, Meinzer FC (2004) Does turgor limit growth in tall trees? Plant Cell Environ 27:229–236

Woodruff DR, McCulloh KA, Warren JM, Meinzer FC, Lachenbruch B (2007) Impacts of tree height on leaf hydraulic architecture and stomatal control in Douglas-fir. Plant Cell Environ 30:559–569

Woodruff DR, Meinzer FC, Lachenbruch B (2008) Height-related trends in leaf xylem anatomy and shoot hydraulic characteristics in a tall conifer: safety versus efficiency in water transport. New Phytol 180:90–99

Woodruff DR, Meinzer FC, Lachenbruch B, Johnson DM (2009) Coordination of leaf structure and gas exchange along a height gradient in a tall conifer. Tree Physiol 29:261–272

Woodruff DR, Meinzer FC, McCulloh KA (2010) Height-related trends in stomatal sensitivity to leaf-to-air vapour pressure deficit in a tall conifer. J Exp Bot 61:203–210

Wright IJ, Reich PB, Westoby M, Ackerly DD, Baruch Z, Bongers F, Cavender-Bares J, Chapin T, Cornelissen JHC, Diemer M, Flexas J, Garnier E, Groom PK, Gulias J, Hikosaka K, Lamont BB, Lee T, Lee W, Lusk C, Midgley JJ, Navas ML, Niinemets Ü, Oleksyn J, Osada N, Poorter H, Poot P, Prior L, Pyankov VI, Roumet C, Thomas SC, Tjoelker MG, Veneklaas E, Villar R (2004) The world-wide leaf economics spectrum. Nature 428:821–827

Yoder BJ, Ryan MG, Waring RH, Schoettle AW, Kaufmann MR (1994) Evidence of reduced photosynthetic rates in old trees. For Sci 40:513–527

Chapter 10
Relationships Between Tree Height and Carbon Isotope Discrimination

Nate G. McDowell, Barbara J. Bond, Lee T. Dickman, Michael G. Ryan, and David Whitehead

Abstract Understanding how tree size impacts leaf- and crown-level gas exchange is essential to predicting forest yields and carbon and water budgets. The stable carbon isotope ratio ($\delta^{13}C$) of organic matter has been used to examine the relationship of gas exchange to tree size for a host of species because it carries a temporally integrated signature of foliar photosynthesis and stomatal conductance. The carbon isotope composition of leaves reflects discrimination against ^{13}C relative to ^{12}C during photosynthesis and is the net result of the balance of change in CO_2 supply and demand at the sites of photosynthesis within the leaf mesophyll. Interpreting the patterns of changes in $\delta^{13}C$ with tree size are not always clear, however, because multiple factors that regulate gas exchange and carbon isotope discrimination (Δ) co-vary with height, such as solar irradiance and hydraulic conductance. Here we review 36 carbon isotope datasets from 38 tree species and conclude that there is a consistent, linear decline of Δ with height. The most parsimonious explanation of

N.G. McDowell (✉) • L.T. Dickman
Earth and Environmental Sciences Division, Los Alamos National Lab, MS-J495, Los Alamos, NM 87544, USA
e-mail: mcdowell@lanl.gov

B.J. Bond
Department of Forest Ecosystems and Society, Oregon State University, 330 Richardson Hall, Corvallis, OR 97331, USA

M.G. Ryan
Laboratory for Studies of the Forest Carbon Cycle, USDA Forest Service, Rocky Mountain Research Station, 240 West Prospect Road, Fort Collins, CO 80526, USA

Colorado State University, Fort Collins, CO 80523, USA

D. Whitehead
Landcare Research, P.O. Box 40, Lincoln 7640, New Zealand

F.C. Meinzer et al. (eds.), *Size- and Age-Related Changes in Tree Structure and Function*, Tree Physiology 4, DOI 10.1007/978-94-007-1242-3_10,

this result is that gravitational constraints on maximum leaf water potential set an ultimate boundary on the shape and sign of the relationship. These hydraulic constraints are manifest both over the long term through impacts on leaf structure, and over diel periods via impacts on stomatal conductance, photosynthesis and leaf hydraulic conductance. Shading induces a positive offset to the linear decline, consistent with light limitations reducing carbon fixation and increasing partial pressures of CO_2 inside of the leaf, p_c at a given height. Biome differences between tropical and temperate forests were more important in predicting Δ and its relationship to height than wood type associated with being an angiosperm or gymnosperm. It is not yet clear how leaf internal conductance varies with leaf mass area, but some data in particularly tall, temperate conifers suggest that photosynthetic capacity may not vary dramatically with height when compared between tree-tops, while stomatal and leaf internal conductances do decline in unison with height within canopy gradients. It is also clear that light is a critical variable low in the canopy, whereas hydrostatic constraints dominate the relationship between Δ and height in the upper canopy. The trend of increasing maximum height with decreasing minimum Δ suggests that trees that become particularly tall may be adapted to tolerate particularly low values of p_c.

1 Introduction

Understanding how tree size impacts gas exchange is essential to predicting forest yields, carbon budgets, hydrology, and regional climate, each of which has scientific, economic, and policy impacts throughout the globe (IPCC 2007; Bonan 2008). The stable carbon isotope ratio ($\delta^{13}C$) of organic matter has been frequently used to examine the relationship of gas exchange to tree size (Ryan et al. 2006) because it carries a temporally integrated signature of foliar photosynthesis and stomatal conductance (Ehleringer 1993; Seibt et al. 2008). This tool works because carbon isotope discrimination (Δ) against ^{13}C relative to ^{12}C during photosynthesis is driven by the net effect of CO_2 supply and demand on the CO_2 mole fraction at the sites of photosynthesis within the mesophyll (Farquhar et al. 1982). Interpretation of the patterns of Δ with tree size is not always clear, however, because multiple factors that regulate gas exchange and Δ co-vary with height, such as solar irradiance (Farquhar et al. 1982; Ehleringer et al. 1986) and hydraulic conductance (Waring and Silvester 1994). In the last two decades a relatively large volume of data regarding the relationship between Δ and tree height has been published; however, no synthetic review has yet been undertaken. Such a review may allow improved interpretation of size–Δ relationships and hence better utilization of $\delta^{13}C$ as a tool to understand adjustments in gas exchange as trees grow larger. Our objective was to review published datasets along with five additional, unpublished datasets on Δ–size patterns in relation to environmental and physiological factors that change as trees become larger. In addition, we examined a case study of the gas exchange characteristics of short and tall mountain beech (*Nothofagus solandrii*) trees to examine potential mechanisms driving Δ patterns with height.

A brief review of the dependence of tissue $\delta^{13}C$ on gas exchange is required to provide a hypothesis framework for the size-related factors that may influence Δ. The two stable isotopologues (molecules that differ only in their isotopic composition) of CO_2, $^{13}CO_2$ and $^{12}CO_2$ represent approximately 98.9% and 1.1% of atmospheric CO_2, respectively. Discrimination against $^{13}CO_2$ during photosynthesis arises mainly from gaseous diffusion from the atmosphere through the stomatal pores to the outer mesophyll cell walls (4.4‰, *a* in Eq. 10.1), and carboxylation by the enzyme Rubisco (~27‰, *b*). Discrimination associated with liquid diffusion across membranes to the sites of carboxylation in the chloroplasts is typically lumped into parameter *b* (Brugnoli and Farquhar 2000; Bickford et al. 2010). These processes are captured in Eq. 10.1 (Farquhar et al. 1982):

$$\Delta \approx a + (b - a) \cdot \frac{p_c}{p_a} \tag{10.1}$$

where p_c and p_a are the chloroplast and atmospheric partial pressures of CO_2, respectively (Hultine and Marshall 2000; Seibt et al. 2008). Thus, anything that affects p_c affects Δ. This is a simplified representation of the processes that influence Δ but it provides an initial hypothesis framework.

Gas exchange characteristics may change as trees become larger due to size-dependent variation in environmental and structural factors and their interactions, such as changes in light availability, hydraulic conductance, and carbon allocation (e.g., Yoder et al. 1994; Gower et al. 1996; Ryan and Yoder 1997). Some traits, such as leaf mass per area and sexual maturation, vary with meristem age independently of size (Bond 2000; Bond et al. 2007; Greenwood et al. 2008; Thomas 2011); however, developmental changes in gas exchange are driven by tree size rather than age (Day et al. 2002; Mencuccini et al. 2005; Bond et al. 2007; Steppe et al. 2011). Thus, we use the term size rather than age throughout this paper (McDowell et al. 2005). We know from Fick's law that photosynthetic assimilation (*A*) and conductance to CO_2 from the atmosphere to chloroplast (g_c, which includes stomatal g_s, and internal g_i) control p_c when p_a is relatively stable:

$$p_c \approx p_a - \frac{A}{g_c} \tag{10.2}$$

The interaction of tree size, physiology (Eqs. 10.1 and 10.2), and some environmental variables was originally conceptualized in a seminal paper by Francey and Farquhar (1982).

One important size-related environmental change is increased shading within tree crowns (Francey et al. 1985; Ehleringer et al. 1986; Hanba et al. 1997; Carswell et al. 2000; Parker et al. 2002; Poorter et al. 2005; Lloyd et al. 2009) which raises p_c because *A* declines via reduced light interception. This change in the light environment occurs within the height profile of individual trees within closed-canopy forests, but does not occur at the tops of canopy-dominant trees or in open-grown trees. Other environmental factors influencing p_c and Δ that may change during forest growth include atmospheric CO_2 concentration (Medina and Minchin 1980; Sternberg et al. 1989), humidity, temperature and wind speed (Baldocchi et al. 2002), and water and nutrient availability (Yoder et al. 1994; Gower et al. 1996; Magnani et al. 2000; Niinemets 2002; Martínez-Vilalta et al. 2007a).

There are numerous structural factors that may impact Δ as trees grow. Plant size directly impacts gas exchange through the relationship of hydraulic supply and demand (e.g. Meinzer and Grantz 1990; Mencuccini and Comstock 1999) including impacts on hydraulic conductance associated with increasing soil-to-leaf pathlength (above and belowground) (Yoder et al. 1994; Mencuccini and Grace 1996a; Ryan and Yoder 1997; Phillips et al. 2002; Mencuccini 2003). This relationship can be conceptualized via Darcy's law (Whitehead 1998):

$$g_c \approx \frac{k_s A_s (\psi_l - \psi_s)}{h A_l \eta D} \tag{10.3}$$

in which k_s is specific conductivity of the xylem, A_s and A_l are sapwood and leaf area, respectively, Ψ_l and Ψ_s are leaf and soil water potential, respectively, η is the viscosity of water, and D is vapor pressure deficit. We note that the original formulations of this model used g_s to represent stomatal conductance to water vapor, which is equivalent to g_c after correction for the 1.6 higher diffusivity of water vapor than CO_2. Height (h) is the most easily and commonly measured component of the pathlength, although it is not the only component. The use of height as a surrogate for pathlength can mislead people to think height alone is a driver of reduced gas exchange in trees, when in fact much of the increase in pathlength and hydraulic resistance is located in roots and branches (Magnani et al. 2000; Sperry et al. 2002; Martínez-Vilalta et al. 2007b). Nonetheless, if pathlength increases and all other components stay relatively constant, then Eq. 10.3 reduces to $g_c = c*1/h$ (where c is a constant), which takes the shape of a non-linear decline in g_c with pathlength. Both hydraulic and stomatal conductances follow this relationship (Mencuccini and Grace 1996b; McDowell et al. 2002a; Niinemets 2002; Schäfer et al. 2000). Thus far it has been shown that many structural factors vary with increasing h to minimize its impact on g_c as trees grow larger, but that these homeostatic adjustments fail to completely offset the negative impact of h on g_c (McDowell et al. 2002a, b; Mencuccini 2003), in part due to requirements to protect xylem from irreversible embolism (Domec et al. 2008). Height itself may be restricted in a homeostatic manner, i.e. reduced height growth facilitates maintenance of g_c above critically low thresholds, even for plants with short maximum heights (Addington et al. 2006; Ryan et al. 2006). Equation 10.3 is based on the premise that irreversible embolism is avoided in part through constraining g_s during periods of water limitation, and does not include variation in light or leaf properties, for example, so it provides only an approximate hypothesis about how gas exchange varies with size.

The interactive impacts of irradiance and pathlength on Δ were demonstrated by Waring and Silvester (1994), who sampled foliage from a range of branch lengths

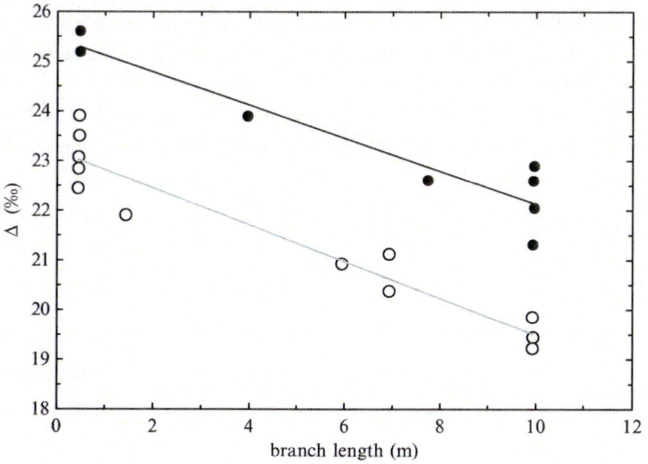

Fig. 10.1 The relationship of Δ calculated from foliar $\delta^{13}C$ to branch length on shaded (*closed circles*) and sunlit (*open circles*) sides of *Pinus radiata* trees. The regression lines are sunlit Δ = –0.37 • length (m) + 23.2 ($r^2 = 0.92$, $p < 0.001$) and shade Δ = –0.33 • length (m) + 25.4 ($r^2 = 0.89$, $p < 0.001$) (Recreated from Waring and Silvester 1994)

on sunny and shady sides of open-grown *Pinus radiata* trees (Fig. 10.1). Consistent with Eqs. (10.1–10.3), Δ declined with decreasing p_c due to increased *A* at higher irradiance, and decreased g_c with longer branch lengths. Other structural and physiological responses to increased tree size that can impact Δ include increases in leaf thickness (Vitousek et al. 1990; Bond et al. 1999; Hanba et al. 1997; Koch et al. 2004; Ishii et al. 2008; Ambrose et al. 2009; Ishii 2011) and hence decreasing g_i (Warren and Adams 2006), although correlations between g_i and leaf thickness are not always observed (Terashima et al. 2005). This morphological response is commonly thought to be adaptive to the light environment, although it also driven by the gravitational constraint on leaf water potential (0.01 MPa m^{-1}) and subsequent impacts on turgor during leaf expansion (e.g. Marshall and Monserud 2003; Woodruff et al. 2004; Cavaleri et al. 2010; Woodruff and Meinzer 2011). Foliar nitrogen content may increase with light availability within crowns (Bond et al. 1999; Duursma and Marshall 2006; Lloyd et al. 2009), which could increase *A* and hence reduce Δ via increased photosynthetic capacity. This is not observed between fully sunlit trees of different heights, however (e.g. Mencuccini and Grace 1996b; Ryan et al. 1997; McDowell et al. 2002a).

It is no surprise that mixed interpretations of the variation in Δ with tree height arise given the large number of environmental and structural factors that change with tree growth. Therefore, we reviewed all available data sets that provided both Δ and *h*, with consideration of these other factors when possible, in hopes of better understanding how Δ, and hence gas exchange, varies with tree size. This review includes woody trees and shrubs from any regions that we could find in the literature.

Based on Fig. 10.1, Eqs. 10.1–10.3, and the assumption that plants allocate their carbon and nutrient capital optimally to maximize whole-plant A while avoiding irreversible embolism (Bloom et al. 1985; Mencuccini 2003; Buckley and Roberts 2006; Lloyd et al. 2009), we hypothesized that a decline in Δ with h, as h approaches its site- and species-specific maximum, is universal regardless of climate or species. In such an observational analysis it is impossible to mechanistically partition the effects of each of the above-mentioned drivers on Δ. Open- and closed canopy trees were sampled together in some of the studies, however, allowing us to test the hypothesis that shade causes a positive offset of the foliar Δ/h relationship relative to sunlit foliage. To further examine mechanisms, we included a detailed case study on short and tall (maximum local height) mountain beech (*Nothofagus solandri*) trees as a test of the role of gas exchange characteristics in Δ/h patterns.

2 Methods

2.1 Case Study

We measured gas exchange (g_s, and A), Ψ_l and foliar and wood $\delta^{13}C$ for two tree sizes (average heights 4.3 and 11.2 m) in a *Nothofagus solandri* stand at 1,300 m elevation at Craigieburn Forest Park in New Zealand. To examine if patterns found in the stand at 1,300 m varied with climate, we also sampled foliar $\delta^{13}C$ from canopy tops of *Nothofagus solandri* trees at five additional sites distributed along an elevation gradient from 800 to 1,250 m above sea level. Cragieburn Forest Park is located in central South Island, New Zealand (43°15′S, 171°35′E). Mountain beech forest occurs as a mono-specific forest from valley bottoms at 650 m to tree line at 1,400 m elevation. At 914 m elevation, mean annual temperature is 8.0°C, mean annual precipitation is 1,447 mm, and mean annual radiation is 4,745 MJ/m^2 (McCracken 1980). At 1,240 m elevation, annual precipitation was estimated to be 1,700 mm and mean annual temperature 6°C (Tate et al. 2000). The soil was formed from mixed greywacke loess and colluvium over shattered rock, and was a fine sandy, mixed, frigid, Andic Dystrochrept called Bealey silt loam (Tate et al. 2000). Characteristics of forest stands similar to those sampled in this study are described by Tate et al. (2000), Davis et al (2003), and Allen et al. (1997).

The response of A and g_s to D was measured at the tops of the *Nothofagus* trees using a portable gas exchange system (LI-COR 6400, LI-COR Inc., Lincoln, NE, USA). Measurements were taken with C_a = 385 ppm, photosynthetically active radiation >1,000 μmol m^{-2} s^{-1}, and ambient humidity to allow for D to increase as temperature increased throughout the day. Because tree stomata are slow to respond, and humidity in the cuvette reflects an increase over ambient from transpiration, we calculated D from leaf temperature (LI-COR's energy balance calculation) and ambient humidity measured with a sling psychrometer or from a weather station 1 km distant. Gas exchange was measured on four trees of each size class and three to seven shoots per tree. Where possible the same shoots were measured periodically throughout the day. Measurements were made over a 2-week period, and a hydraulic lift was used to reach the canopy top for sampling. Leaf water potential was measured periodically on adjacent shoots with a Scholander pressure chamber (PMS Instrument Company, Albany, OR, USA). Leaf-specific hydraulic conductance (K_l) was calculated as transpiration divided by $\Psi_l - \Psi_s$, where Ψ_s was pre-dawn Ψ_l (Phillips et al. 2002).

Foliage for analysis of $\delta^{13}C$ and nitrogen was collected from canopy tops at the 1,300 m site, and along the elevation gradient. For the elevation gradient, we also sampled from two height classes: trees about 4 m in height, and trees >11 m in height. Maximum tree height increased as elevation decreased, from ~11 m at 1,300 m elevation to ~18 m at 800 m elevation. Wood was sampled for $\delta^{13}C$ either from disks from harvested trees (short trees) or increment cores (tall trees). Foliage and wood samples were dried at 65°C, ground, and analyzed for $\delta^{13}C$ composition using a an isotope ratio mass spectrometer (Europa Scientific 20/20) interfaced to a Dumas elemental analyser (Europa Scientific ANCA-SL, Europa Scientific Ltd., Crewe, UK). Nitrous oxide was removed by gas chromatography and corrections for ^{17}O were done for all runs (Craig 1957). Isotope ratios were presented relative to Vienna PeeDee belemnite carbonate standard (*VPDB*), calibrated against a certified secondary source from the Australian Commonwealth Scientific and Research Organization, Canberra, Australia.

We corrected the isotopic results for elevation impacts on CO_2 partial pressure (Hultine and Marshall 2000; Warren et al. 2001; Körner 2007; McDowell et al. 2010). Though this effect is minor, it is relatively straightforward to correct for when elevation and temperature are known. We used local mean temperature data for each site to calculate the drawdown of CO_2 partial pressure between the atmosphere and chloroplast ($p_a - p_c$, Pa, Hultine and Marshall 2000).

2.2 *Global Patterns*

We examined the relationship of Δ to h at the global scale through a literature review along with an additional five unpublished data sets of trees, shrubs, angiosperms, gymnosperms, and across all possible regions and latitudes. We searched the literature for papers that included figures or tables with both height and either $\delta^{13}C$ or Δ data from foliage or stemwood to maximize the breadth of our investigation. A few publications were excluded because height was provided qualitatively (i.e. low and high) rather than with quantifiable units (i.e. meters), and therefore could not be analyzed quantitatively. All stemwood samples were collected from breast height and most used data from the most recent year of growth. For graphical data we digitized the figures to generate numerical datasets (ImageDIG *Version 2.0.7*, SciCepts Engineering, Chiangmai, Thailand). Comparison of the digitized numerical output to known values using a somewhat fuzzy pdf copy of data from Yoder et al. (1994) showed strong correspondence (digitized $\delta^{13}C = 0.99 \cdot$ true $\delta^{13}C + 0.03$, $r^2 = 0.99$, $p < 0.001$). A slightly weaker relationship was observed using a clear pdf of Barnard and Ryan (2003) to actual data (digitized $\delta^{13}C = 0.91 \cdot$ true $\delta^{13}C - 3.2$, $r^2 = 0.99$, $p < 0.001$), with the error attributed to the loss of eight data points due to overlapping points that could not be discerned. Thus, large datasets with overlapping clouds of points may have lower accuracy than data sets with fewer points.

Our primary form of analysis was regression of Δ versus *h*. The slope of this relationship was used to compare the sensitivity of Δ to *h* across studies (e.g. McDowell et al. 2002b). The theoretical regression should be non-linear according to the simplest version of Darcy's law ($\Delta = 1/h$), however, we saw no evidence of deviations from linear patterns. Comparison of fit statistics of linear to various non-linear regressions confirmed this observation. Multi-parameter polynomials have been used for predicting maximum heights (Burgess and Dawson 2007) but we excluded them from this comparison because of our inability to interpret such relationships mechanistically. The five new (previously unpublished) datasets all had foliage harvested from the top 10% of short and tall tree crowns that were sunlit, canopy dominants, i.e. short trees were sampled in open or short statured forests, not in the understory of taller forests. These new datasets are (source location and contributor in parentheses): *Pinus teocote* (Mexico, Cornejo Oviedo), *Pinus ponderosa* (Oregon, C. Rose), *Pinus sylvestris* (Scotland, M. Mencuccini), *Artemisia tridentata* (California, W. Oechel), *Pseudotsuga menziesii* (Washington, N. McDowell, B. Bond, J. Janisch, A. Schauer, the $\delta^{13}C$ component of this dataset was published by Bond et al. 2008), and the case study *Nothofagus solandrii* (New Zealand, M. Ryan and D. Whitehead). The samples were analyzed on an isotope ratio mass spectrometer (Delta Plus, Finnigan, Bremen, Germany) interfaced with an elemental analyzer (ESC 4010, Costech, Valencia, CA) located at the Integrated Stable Isotope Research Facility at the Western Ecology Division of the US Environmental Protection Agency, Corvallis, Oregon. Similar to the New Zealand analyses, nitrous oxide was removed by gas chromatography, corrections for ^{17}O were made, and results are presented on the *VPDB* scale.

For published datasets, where possible, we partitioned the data collected in sun-exposed locations from data collected along light gradients within the forest canopy. When possible we also collected ancillary data associated with foliar, environmental or climatic characteristics. A list of the datasets used in this analysis is provided in Table 10.1.

$\delta^{13}C$ data were converted to Δ (Farquhar et al. 1982):

$$\Delta \approx \frac{\delta^{13}C - \delta^{13}C_a}{1 + \delta^{13}C/1000} \tag{10.4}$$

All $\delta^{13}C$ results were expressed relative to *VPDB*. We obtained annual atmospheric stable carbon isotope ratios ($\delta^{13}C_a$) from the Institute for Arctic and Alpine Research at the University of Colorado and the National Oceanic and Atmospheric Administration, Earth System Research Laboratory. Calculating Δ using Eq. 10.4 allows physiological interpretation of $\delta^{13}C$ by removing the impact of temporally-changing $\delta^{13}C_a$ due to fossil fuel emissions of CO_2 depleted in $\delta^{13}C$ (Leavitt and Long 1988; Keeling et al. 1989). This correction is necessary because a change in the $\delta^{13}C$ of CO_2 used as the substrate for photosynthesis causes a physiologically independent change in the $\delta^{13}C$ of plant material, and because the literature review includes data from >40 year period. Additional sources of analytical error arise from variation in chemical composition of tissues, cellulose extraction and elevation. Lipid, lignin, and cellulose vary in their $\delta^{13}C$ composition and their content within foliage and wood may vary across species and with height. The $\delta^{13}C$ offset for cellulose to whole-tissue ranges from <0.9‰ to >2.5‰ (Schleser 1990; Leavitt and Long 1991; Macfarlane et al. 1999; Loader et al. 2003; Harlow et al. 2006) due in part to variation with species and extraction methods. Only four of our compiled studies (12%) provided cellulose alone, three of which provided only wood cellulose (Grulke and Retzlaff 2001; Monserud and Marshall 2001; Delzon et al. 2004), and one of which provided only foliage cellulose (Phillips et al. 2003). The elevation effect was described above in the *Nothofagus* case study methods section. This effect was difficult to address across the literature review because only 24% of the studies we compiled provided site elevation. The impact may be relatively small; the approximate $\delta^{13}C$ error from assuming constant elevation is ~0.5‰, derived from the standard deviation of 513 m for studies that provided elevation. We highlighted only results in which we were confident the impacts of cellulose extraction and elevation were negligible. In addition, though we choose to analyze species and site specific slopes of Δ versus h for theoretical reasons, there is methodological value as well in that this slope is not affected by the $\delta^{13}C$ variation due to these errors because any error is constant *within* the datasets used to generate the slope.

3 Results

3.1 Case Study

Ψ_l declined more rapidly in the morning in the 11-m *Nothofagus* trees than the 4 m trees (Fig. 10.2a). The average Ψ_l at vapor pressure deficits (D) < 0.8 kPa was –0.83 MPa (se 0.21) and –1.29 MPa (se 0.18) for the 4- and 11-m trees, respectively (unequal variance t-test $p = 0.059$). Ψ_l was similar for 4 and 11-m trees at $D > 0.8$ kPa (–1.87 and –1.93 MPa, respectively, $p = 0.18$). For the same sample period, g_s was similar across tree sizes at $D < 0.8$ kPa ($p = 0.31$), but g_s was higher for the 4-m trees at $D > 0.8$ kPa (121.0 vs 91.5 mmol m^{-2} s^{-1} for 4 and 11-m trees, respectively, $p < 0.001$, Fig. 10.2b). K_l followed the same pattern as g_s, with non-significant differences during the morning at $D < 0.8$ kPa ($p = 0.29$) and significantly higher values for the 4-m trees than the 11-m trees at $D > 0.8$ kPa ($p < 0.001$, data not shown). Photosynthesis differed between 4 and 11-m trees (Fig. 10.2c, $p < 0.001$). Leaf [N] was significantly higher in the 4-m trees than the 11-m trees (1.54% and 1.19%, respectively, $p < 0.001$). Leaf $\delta^{13}C$ for the 4- and 11-m trees at this site averaged –25.8 and –24.3 ‰, respectively, equivalent to Δ of 18.3‰ and 16.8‰ ($p < 0.001$, Table 10.1). Wood $\delta^{13}C$ for the 4- and 11-m trees at this site averaged –25.7 and –24.4‰,

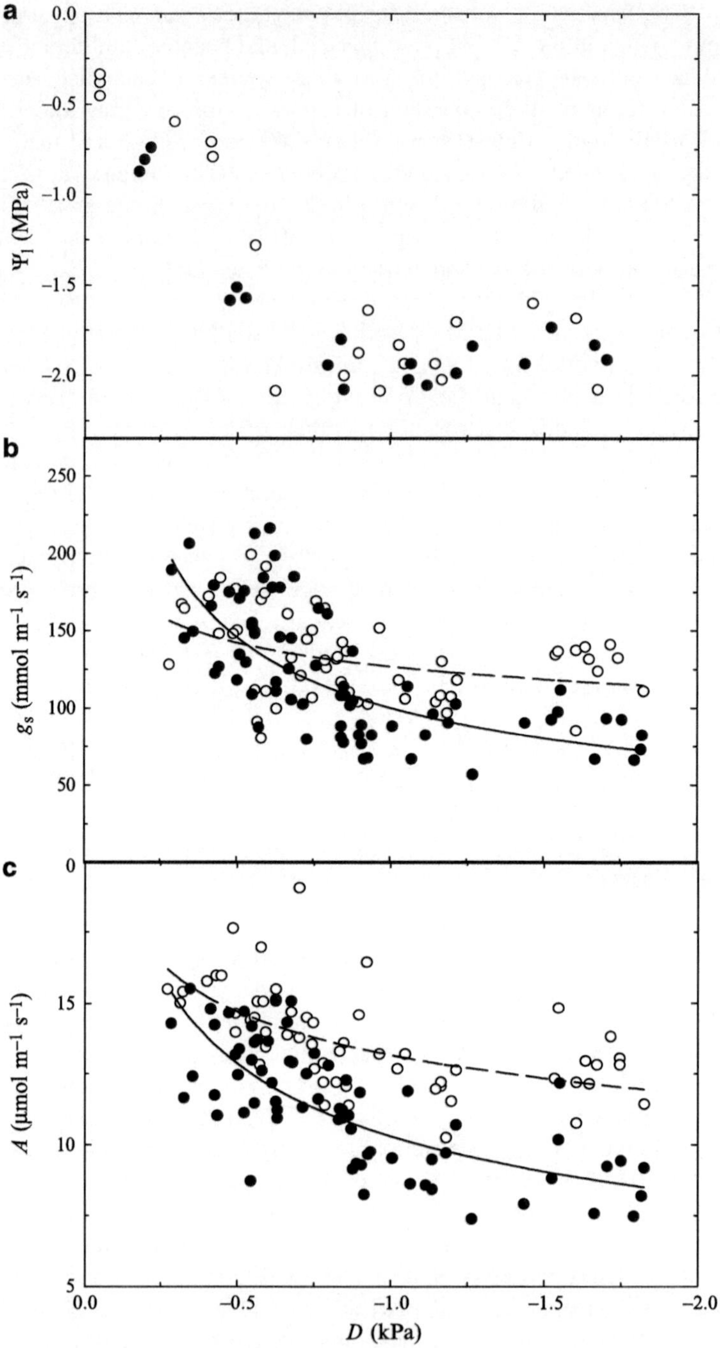

Fig. 10.2 The relationship of D to (**a**) Ψ_l, (**b**) g_s and (**c**) A for 4- (*open circles*) and 11-m (*closed circles*) tall *Nothofagus solandri* trees in New Zealand. The relationships in (**b**) are 4 m: $g_s = 126.8 \bullet D^{-0.166}$ ($r^2 = 0.14$); 11 m: $g_s = 100.4 \bullet D^{-0.541}$ ($r^2 = 0.50$). The relationships in (**c**) are 4 m: $A = 13.2 \bullet D^{-0.16}$ ($r^2 = 0.38$); 11 m: $A = 10.2 \bullet D^{-.321}$ ($r^2 = 0.54$)

Table 10.1 Summary of published and unpublished datasets used in analyses of the Δ–tree height relationship

Species	Height (m)	Δ (‰)	N	Slope (Δ/*h*)	Intercept	R^2	P	Citation
All species, foliar[1]	0.05–113.2	13.6–29.9	820	−0.19 (0.06)	23.4	0.66	0.07	This study
All species, wood[1]	1.01–52.8	14.6–27.2	285	−0.043 (0.01)	18.1	0.38	0.14	This study
Gymnosperms, foliar[1]	**0.16–113**	**13.6–24.9**	**390**	**−0.07 (0.008)**	**24.7**	**0.59**	**0.11**	This study
Lagarostrobos franklinii[2]	0.92–17.0	19.9–24.9	35	−0.13	22.8	0.41	<0.001	Francey et al. (1985)
Lagarostrobos franklinii[3]	0.97–17.0	19.9–21.3	13	−0.01	20.7	0.02	0.6	Francey et al. (1985)
Lagarostrobos franklinii[4]	0.95–14.3	22.2–24.9	6	−0.07	22.6	0.35	0.045	Francey et al. (1985)
Larix occidentalis Nutt[a]	3.85–5.68	19.0–21.4	5	−0.88	24.4	0.56	0.15	Zhang et al. (1994)
Picea rubens/mariana hybrid[a]	0.16–6.47	19.1–23.0	40	−0.48	22.4	0.94	<0.001	Major et al. (2007)
Picea sitchensis[a]	0.63–7.40	19.8–23.6	20	−0.38	22.4	0.42	0.002	Heaton and Crossley (1995)
Pinus elliottii/caribaea hybrid[a]	8.5–13.3	18.7–21.4	40	−0.39	24.6	0.44	<0.001	Xu et al. (2000)
Pinus pinaster	8.46–28.4	17.6–19.2	8	−0.03	19.2	0.22	0.24	Delzon et al. (2004)
Pinus ponderosa[9]	1.35–30.0	15.4–18.8	14	−0.091	19.4	0.63	<0.001	This study
Pinus ponderosa	2.10–37.9	15.5–20.5	27	−0.055	20.3	0.47	0.01	Yoder et al. (1994)
Pinus ponderosa Laws.[a]	0.41–6.83	16.3–18.0	8	−0.09	17.4	0.23	0.23	Grulke and Retzlaff (2001)
Pinus ponderosa[5]	5.60–37.7	18.0–19.7	4	−0.05	19.6	0.97	0.015	Sala and Hoch (2009)
Pinus ponderosa[6]	6.30–34.5	18.2–18.8	4	−0.02	18.9	0.76	0.13	Sala and Hoch (2009)
Pinus sylvestris[9]	0.90–25.0	18.8–21.3	7	−0.027	21.3	0.67	0.025	This study
Pinus sylvestris L.[7]	4.88–21.6	21.3–24.4	8	−0.11	24.1	0.49	0.055	Martínez-Vilalta et al. (2007a)
Pinus sylvestris L.[8]	4.08–21.4	20.2–24.4	23	−0.12	23.7	0.35	0.003	Martínez-Vilalta et al. (2007a)
Pinus teocote[9]	6.00–19.0	16.8–19.1	7	−0.001	18.0	0	0.99	This study
Pseudotstuga menzeseii	2.50–56.1	15.7–18.9	13	−0.03	18.5	0.43	<0.001	This study
Pseudotstuga menzeseii	24.0–53.0	16.0–18.1	2	−0.07	19.8	1	N/A	Winner et al. (2004)
Pseudotstuga menzeseii[9]	15.0–55.9	15.7–18.6	3	−0.071	19.7	0.99	0.033	McDowell et al. (2002a)
Sequoia sempervirens	28.7–112	14.8–19.4	15	−0.033	19.9	0.66	<0.001	Ambrose et al. (2009)
Sequoia sempervirens[2]	14.4–66.8	16.1–22.8	38	−0.092	24.1	0.68	<0.001	Burgess and Dawson (2007)
Sequoia sempervirens[2]	44.8–113	13.6–22.8	59	−0.11	28.0	0.71	<0.001	Koch et al. (2004)
Sequoia sempervirens[3]	14.4–53.8	16.7–22.3	15	−0.12	24.6	0.82	<0.001	Burgess and Dawson (2007)

(continued)

Table 10.1 (continued)

Species	Height (m)	Δ (‰)	N	Slope (Δ/*h*)	Intercept	R^2	P	Citation
Sequoia sempervirens[3]	45.0–100	16.5–23.4	11	−0.09	27.1	0.83	<0.001	Ishii et al. (2008)
Sequoia sempervirens[4]	16.5–66.8	16.1–22.8	11	−0.097	24.7	0.78	<0.001	Burgess and Dawson (2007)
Sequoia sempervirens[4]	40.0–80.0	18.6–24.1	31	−0.08	26.7	0.43	<0.001	Ishii et al. (2008)
Sequoia sempervirens[9]	85.5–112	14.4–19.2	17	−0.087	24.8	0.35	0.017	Koch et al. (2004)
Sequoiadendron giganteum	31.8–90.0	14.1–18.2	15	−0.041	19.1	0.66	<0.001	Ambrose et al. (2009)
Thuja plicata	24.0–53.0	13.6–18.2	2	−0.16	22.0	1	N/A	Winner et al. (2004)
Tsuga heterophila	24.0–53.0	16.2–20.0	2	−0.131	23.1	1	N/A	Winner et al. (2004)
Angiosperms, foliar[1]	**0.05–62.95**	**14.8–29.9**	**430**	**−0.33 (0.12)**	**22.1**	**0.76**	**0.026**	This study
Temperate[1]	**0.05–62.95**	**14.8–25.8**	**129**	**−0.56 (0.30)**	**23.2**	**0.76**	**0.06**	This study
Tropical[1]	**0.15–42.9**	**17.2–29.9**	**350**	**−0.18 (0.03)**	**23.0**	**0.74**	**0.01**	This study
Acer grandidentatum[a]	4.50–6.00	18.2–22.9	6	−2.4	33.1	0.91	0.004	Buchmann et al. (1997a)
All (*Alnus, Fraxinums, Ulmus*)	0.00–20.0	18.2–26.4	79	−0.2	23.6	0.53	<0.001	Hanba et al. (1997)
Alnus hirsute Turcz.	0.92–19.9	19.9–23.6	2	−0.19	23.8	1	N/A	Hanba et al. (1997)
Amazon Ji-Parana	0.72–32.9	19.8–27.2	10	−0.14	26.7	0.56	0.012	Ometto et al. (2002)
Amazon Manaus	0.33–31.8	22.0–29.9	7	−0.25	29.1	0.87	0.002	Ometto et al. (2002)
Amazon Saantarém	0.33–26.9	21.9–28.9	8	−0.25	29.5	0.92	<0.001	Ometto et al. (2002)
Amazon spp.	2.57–52.8	17.9–29.0	204	−0.066	26.1	0.1	<0.001	Martinelli et al. (1998)
Artemisia tridentate[9]	0.50–1.50	14.8–18.0	9	−0.81	17.3	0.16	0.29	This study
Castanopsis chinensis	2.95–23.9	21.2–23.4	3	−0.1	23.7	0.99	0.021	Ehleringer et al. (1986)
Chrysothamnus nauseosus	0.05–1.65	18.1–23.4	49	−2.57	22.3	0.74	<0.001	Donovan and Ehleringer (1994)
Cryptocarya chinensis	2.00–20.0	24.7–27.3	2	−0.24	27.5	1	N/A	Ehleringer et al. (1986)
Eucalyptus diversicolor	13.8–63.0	18.8–23.0	36	−0.064	23.7	0.56	<0.001	Burgess and Dawson (2007)
Eucalyptus saligna[10a]	1.43–7.90	21.1–23.4	23	−0.22	23.3	0.52	<0.001	Barnard and Ryan (2003)
Eucalyptus saligna[11a]	5.43–26.3	22.5–26.2	34	−0.18	27.5	0.76	<0.001	Barnard and Ryan (2003)
Fagus sylvatica L.	1.97–26.8	17.4–23.3	8	−0.12	23.7	0.98	<0.001	Schleser (1992)
Fraxinus mandshurica Rupr.	0.16–19.9	19.3–25.3	16	−0.27	24.2	0.9	<0.001	Hanba et al. (1997)
Juglans regia L.[a]	0.50–4.98	18.4–22.0	10	−0.76	22.7	0.91	<0.001	Le Roux et al. (2001)

Machilus velutina	1.00–14.9	21.3–27.1	2	−0.41	27.5	1	N/A	Ehleringer et al. (1986)
Nothofagus solandri[9]	4.00–11.0	16.8–19.1	2	−0.22	19.2	1	N/A	This study
Parashorea chinensis	35.0–49.9	20.1–21.6	4	−0.099	25.1	0.96	0.019	He et al. (2008)
Populus tremuloides[a]	2.00–6.00	17.7–20.4	6	−0.38	20.6	0.51	0.11	Buchmann et al. (1997a)
Quercus garryana[9]	10.0–25.0	15.6–16.9	2	−0.07	18.1	1	N/A	Phillips et al. (2003)
Schima superba	1.89–23.9	21.6–24.7	4	−0.13	24.9	0.94	0.03	Ehleringer et al. (1986)
Tropical[3]	0.15–25.0	17.2–25.4	26	−0.18	22.2	0.51	<0.001	Broadmeadow et al. (1992)
Tropical[4]	0.15–30.0	19.0–29.2	23	−0.11	24.8	0.25	0.015	Broadmeadow et al. (1992)
Ulmus davidiana Planch.	0.20–20.0	19.1–25.8	13	−0.28	24.0	0.69	< 0.001	Hanba et al. (1997)
Gymnosperms, wood[1]	**1.01–37.9**	**14.6–20.5**	**265**	**−0.029 (0.01)**	**16.8**	**0.31**	**0.16**	This study
Pinus monticola	1.01–34.5	14.9–16.4	77	−0.014	15.8	0.18	<0.001	Monserud and Marshall (2001)
Pinus ponderosa	1.14–32.5	14.6–16.8	80	−0.039	16.4	0.66	<0.001	Monserud and Marshall (2001)
Pinus ponderosa	1.35–30.0	15.4–18.8	12	−0.03	16.7	0.21	0.14	This study
Pinus ponderosa	2.10–37.9	15.5–20.5	12	−0.06	19.0	0.49	0.005	Yoder et al. (1994)
Pseudotsuga menzeseii	2.69–28.9	15.4–17.0	84	−0.002	16.1	0	0.68	Monserud and Marshall (2001)
Angiosperms, wood[1]	**1.97–52.8**	**16.8–29.0**	**81**	**−0.067 (0.03)**	**20.2**	**0.49**	**0.07**	This study
Amazon	2.57–52.8	17.9–29.0	71	−0.070	22.8	0.14	0.004	Martinelli et al. (1998)
Fagus sylvatica L.	1.97–26.8	17.4–23.3	8	−0.020	18.5	0.33	0.14	Schleser (1992)
Nothofagus solandri	4.00–11.0	16.8–19.1	2	−0.110	19.4	1.00	N/A	This study

Values include range of tree heights (m), range of Δ (‰), sample size (n), slopes (‰ m^{-1}) and regression statistics. Data are grouped by tissue type (foliage, wood), and vegetation type (gymnosperm and angiosperm), with averages (and total n) provided in the first row for each group (*standard errors in parentheses*). The averages include only datasets in which the trees approached the maximum potential height for that species at that site. Samples were collected within individual trees at different heights unless indicated otherwise.

Key: 1: average for the group, 2: sunny and shaded samples combined, 3: sunny, 4: shaded, 5: moist, 6: dry, 7: site 2, 8: sites 1 and 3 combined, 9: tree tops, 10: young, 11: old

[a]Trees sampled did not approach maximum potential height of that species at that site

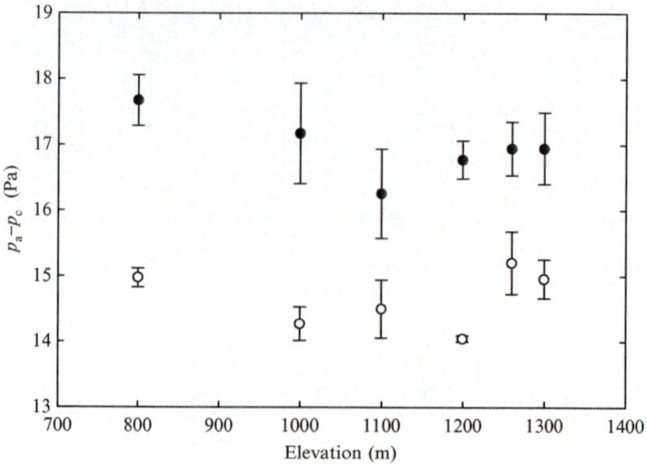

Fig. 10.3 The partial pressure difference between chloroplast and atmospheric CO_2 ($p_a - p_c$) calculated from foliar $\delta^{13}C$, across the *Nothofagus solandri* elevation transect in New Zealand. *Open circles* are short trees and *closed circles* are tall trees. *Bars* are standard errors

respectively, equivalent to Δ of 19.0 and 18.3‰ ($p = 0.06$, Table 10.1). For the elevation transect, leaf [N] was significantly higher for the short trees compared to the tall trees (1.27% and 1.04%, respectively, $p = 0.002$). $p_a - p_c$ calculated from foliage $\delta^{13}C$ averaged 14.7 and 17.0 Pa for the short and tall trees, respectively ($p < 0.001$, see Fig. 10.3). There was no trend in the $p_a - p_c$ difference between height classes with elevation ($p = 0.22$).

3.2 *Global Patterns*

We found 31 published datasets for inclusion with our five new datasets. This included a total of 15 gymnosperm and 23 angiosperm species, although we note that three publications from the Amazon sampled across a large numbers of species (Broadmeadow et al. 1992; Martinelli et al. 1998; Ometto et al. 2002), which we combined for analyses. After separation of all the individual datasets within publications (i.e. by species or sunny vs shady), we had a total of 61 regressions. These included 13, 48 and 0 from tropical, temperate and boreal regions, respectively. All gymnosperm datasets were from the temperate zone, whereas the angiosperm datasets were evenly divided between temperate and tropical zones. There was a total of 59 datasets of foliage $\delta^{13}C$, 9 of wood $\delta^{13}C$, and five studies that included both. The datasets spanned a range of maximum heights of 1.5–113 m and a Δ range of 13.6–29.9‰ (equivalent to $\delta^{13}C$ of −21.4 to −36.8‰, Table 10.1, Fig. 10.4a). Mean annual precipitation, for the 16 studies that provided precipitation data, ranged from 374 to 4,000 mm.

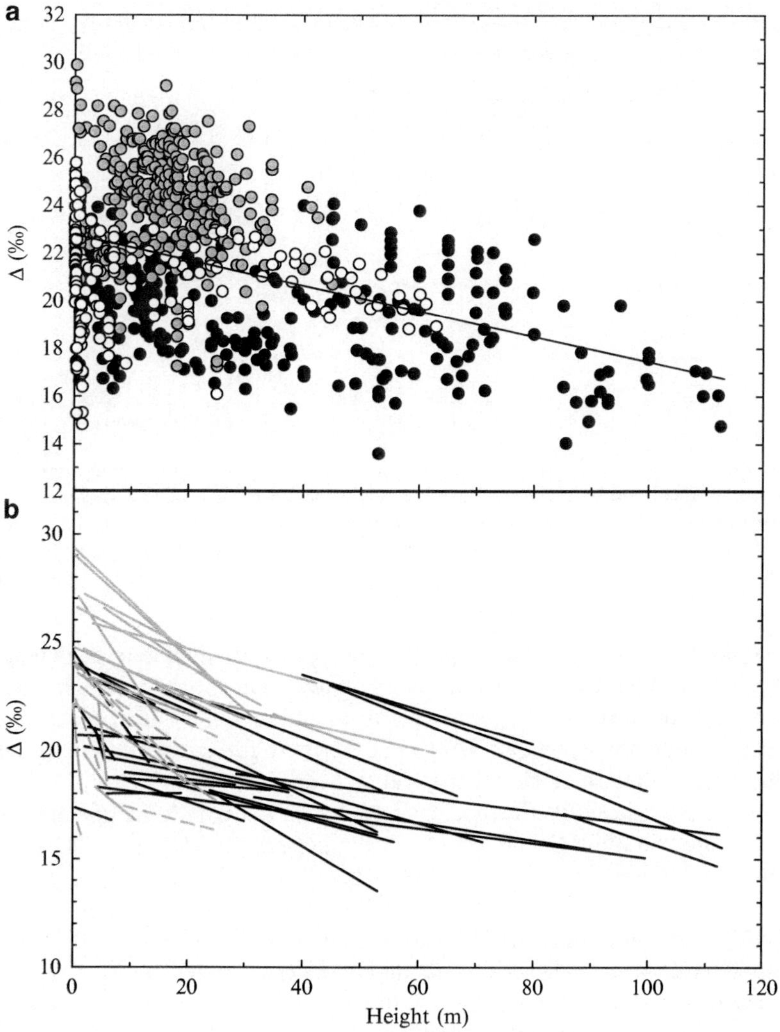

Fig. 10.4 (**a**) The relationship between Δ calculated from foliar $\delta^{13}C$ and height above the ground for temperate gymnosperms (*black circles*) and angiosperms (*open circles*), and tropical angiosperms (*gray circles*). Data include samples collected at different heights within trees as well as at the tops of individual trees of different heights. The relationship for all foliar datasets combined was: $\Delta = -0.053 \bullet \text{height} + 22.77$ ($r^2 = 0.25$, $p < 0.001$). (**b**) Individual regressions for each of the 59 datasets. Gymnosperms, temperate angiosperms, and tropical angiosperms are represented by *solid black*, *dashed gray*, and *solid gray lines*, respectively

Δ declined with increasing height for all foliar datasets (a negative Δ/h, Table 10.1) with 83% significant at $P < 0.10$. The consistent linear declines can be seen graphically in Fig. 10.4b, in which the regression line for each data set from Table 10.1 is shown, with the different lines denoting tropical angiosperms, temperate angiosperms,

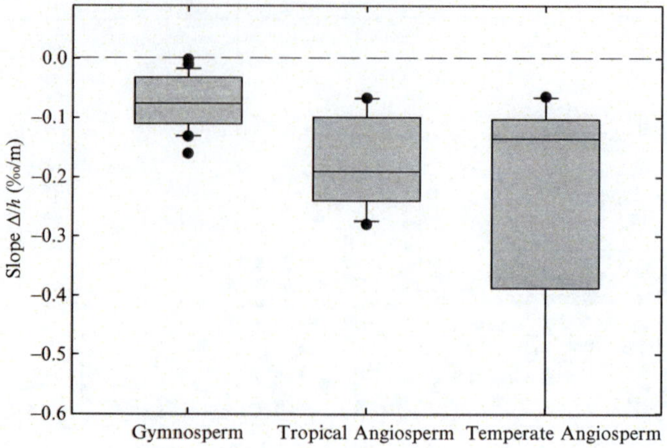

Fig. 10.5 Box plots summarizing the slopes of Δ to *h* for each dataset within gymnosperms and angiosperms (tropical and temperate). Box plots show the lower quartile, the median, and the upper quartile (*the box*), the upper and lower ten percentile values (*the thin lines*, and the range of the observations (*the closed circles*). Note the scale was held to a slope of −0.6‰ m^{-1} for clarity, which omits only the *Chrysothamnus* slope of −2.57‰ m^{-1}. All observations had a negative slope

and temperate gymnosperms. For all species that approached their maximum heights, the mean slope of Δ with *h* was −0.19‰ m^{-1}. Tropical angiosperms had higher Δ than either temperate angiosperms or gymnosperms (4, Table 10.1). The mean slope was shallowest for gymnosperms (−0.07‰ m^{-1}), steepest for temperate angiosperms (−0.56‰ m^{-1}), and intermediate for tropical angiosperms (−0.18‰ m^{-1}, Fig. 10.5). The mean values are heavily influenced by outliers, particularly for temperate angiosperms (e.g., the arid shrub *Chrysothamnus* had a slope of −2.57 ‰ m^{-1}). The median values for these three groups were −0.08, −0.24, and −0.20‰ m^{-1}, respectively (Fig. 10.5). The single foliar dataset which did not exhibit a negative slope was from *Eucalyptus saligna* trees in which the maximum height included in the dataset was only 40% of the site- and species-specific maximum (D. Binkley 2010). The within-canopy declines in Δ with increasing *h* in this dataset were consistent with expectations, but in contrast to the global trend, canopy-top Δ was higher at the tops of tall trees than small trees. It is important to note that studies in which the trees did not approach their maximum height had relatively high slopes (mean −0.59‰ m^{-1}) and were thus excluded from averages calculated across biomes, vegetation type, tissue type, etc. (Table 10.1, Zhang et al. 1994; Heaton and Crossley 1995; Buchmann et al. 1997a; Xu et al. 2000; Le Roux et al. 2001; Grulke and Retzlaff 2001; Barnard and Ryan 2003; Major et al. 2007).

Δ declined with increasing height for all wood datasets (Table 10.1, Fig. 10.6), with 50% of those studies significant at $P < 0.10$. The slope of Δ with increasing *h* was shallower than for foliage (Table 10.1). Similar to the foliage data, the wood slope was steeper and Δ was higher for angiosperms than gymnosperms (Table 10.1). Wood calculated Δ was lower than foliar Δ, as indicated by the lower intercepts (Table 10.1, Fig. 10.6).

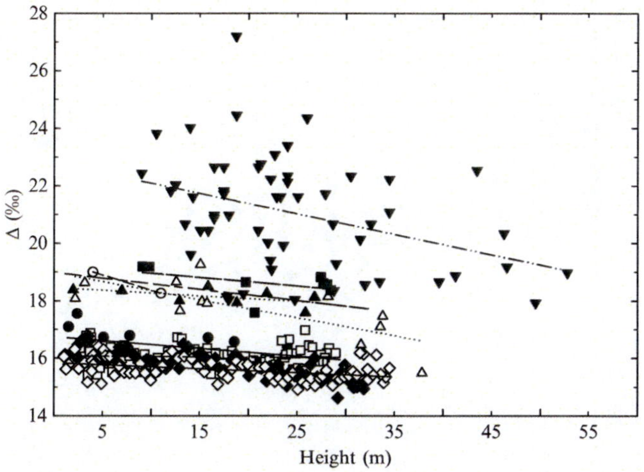

Fig. 10.6 The relationship between stemwood Δ and height for the nine stemwood datasets in this study, including extracted cellulose and whole-tissue data. The data include *Pinus ponderosa* (C. Rose, 2010 – *filled circles*, Yoder et al. 1994 – *upward open triangles*, and Monserud and Marshall 2001 – *filled diamonds*), *Nothofagus* (this study – *open circles*), Amazonian species (Martinelli et al. 1998 – *downward filled triangles*), *Pinus pinaster* (Delzon et al. 2004 – *filled squares*), *Fagus sylvatica* (Schleser 1992 – *upward filled triangles*), and *Pseudotsgua menziesii* (*open squares*) and *Pinus monticola* (*open diamonds*) from Monserud and Marshall (2001)

In five publications the authors sampled both sunlit and shaded crowns of trees. Additionally, sunlit and shaded crowns were sampled at the same *Pseudotsuga menziesii* sites by different groups. These datasets allowed us to test the hypothesis that the Δ response to *h* is exacerbated by light. The compiled sunlit and shaded data are shown in Fig. 10.7. The sunlit and shaded regression lines were offset from each other, with a steeper slope for the shaded than the sunlit trees (–0.065 vs –0.050‰ m^{-1}, respectively) and the shaded intercept more positive than the sunlit intercept (24.0 vs 24.2‰, respectively). To examine the consistency of the sunlit versus shaded Δ/*h* relationship within species, we plotted the sunlit and shaded *Sequoia sempervirens* data from three different publications (Fig. 10.8). Each study yielded results generally similar to the overall comparison, with a clear offset between the shaded and sunlit trees such that Δ was higher for any given *h* for shaded trees compared with sunlit trees.

Studies in which plants had maximum heights less than 20 m tended to have slopes steeper than –0.2‰ m^{-1}, however, most of the tall trees in this range were not near their maximum height (or local maximum height was not quantified), so interpretation of this result is ambiguous. There was also no relationship between the Δ/*h* slope and mean annual precipitation for angiosperms or gymnosperms (data not shown). There was, however, a negative relationship between minimum Δ (canopy top) and maximum height across all datasets (Fig. 10.9). We examined the relationship of Δ and Δ/*h* to structural and nutrient parameters (e.g. Eq. 10.3), however, no clear patterns emerged due in part to the limited availability of these ancillary measurements in the literature.

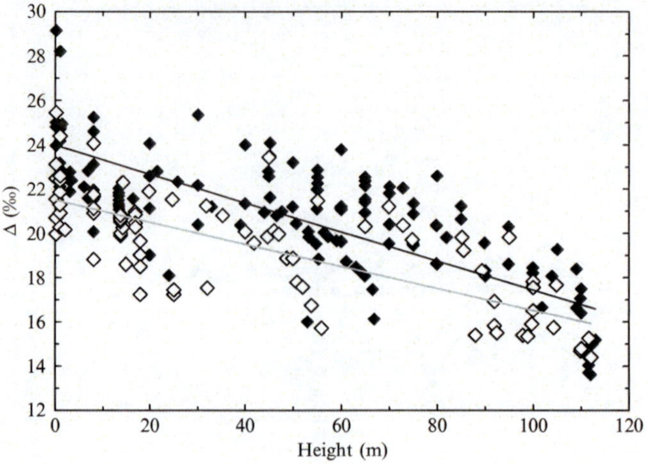

Fig. 10.7 The relationship between Δ calculated from foliage tissue and height for sunlit (*open diamonds*) and paired shaded (*closed diamonds*) trees. Data sets included are *Sequoia sempervirens* (Koch et al. 2004; Burgess and Dawson 2007; Ishii et al. 2008), Amazonian species (Broadmeadow et al. 1992), *Lagarostrobos franklinii* (Francey et al. 1985), and *Pseudotsuga menziesii* (McDowell et al. 2002a; Winner et al. 2004, this study). The regression lines are sunlit: Δ = –0.050 • height + 21.5, r^2 = 0.58, p = <0.001 (*dark line*), and shaded: Δ = –0.065 • height + 24.0, r^2 = 0.64, p = <0.001 (*gray line*)

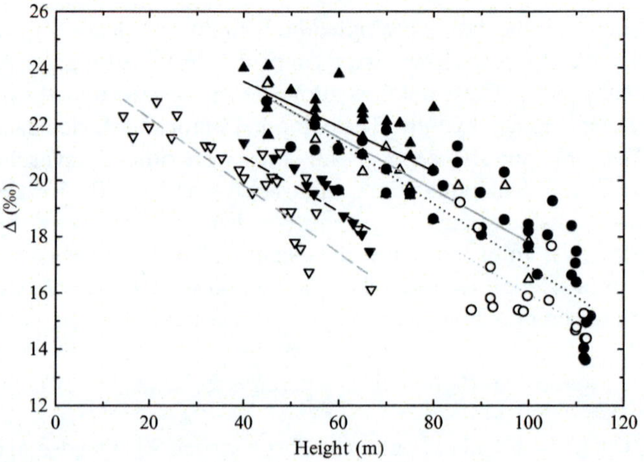

Fig. 10.8 The relationship between Δ (foliar) and height for sunlit and shaded samples for three *Sequoia sempervirens* studies. Shaded samples (*filled symbols*) were collected from trees in relatively closed canopies in contrast to sunlit samples (*open symbols*), which the authors indicated were collected from sunny or otherwise exposed locations. The regressions are: Koch et al. (2004, *circles*) shaded Δ = –0.11 • height (m) + 28.0, r^2 = 0.71, p = <0.001, sunlit Δ –0.10 · height (m) + 26.0, r^2 = 0.46, p = 0.017, Ishii et al. (2008, *up triangles*) shaded Δ = –0.08 • height (m) + 26.7, r^2 = 0.43, p = <0.001, sunlit Δ = –0.09 · height (m) + 27.1, r^2 = 0.84, p = <0.001, Burgess and Dawson (2007, *down triangles*) shaded Δ = –0.10 · height (m) + 24.7, r^2 = 0.78, p = <0.001, sunlit Δ = –0.12 · height (m) + 24.6, r^2 = 0.82, p = <0.001

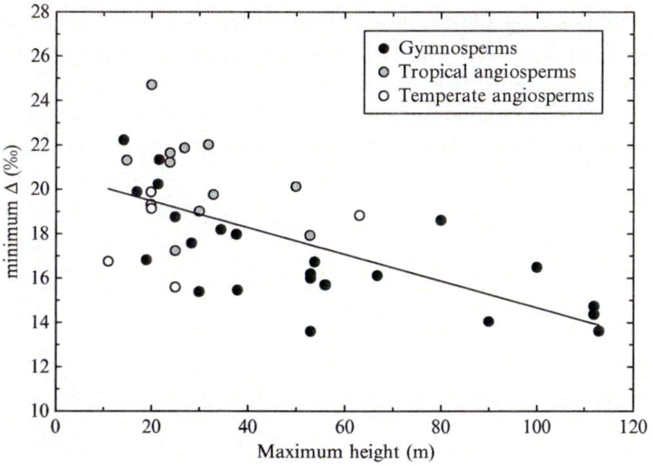

Fig. 10.9 The relationship between minimum Δ (canopy top) and maximum height. Shrubs (*Chrysothamnus* and *Artemesia*) and trees that did not approach their maximum heights were excluded. The regression line is for all data: minimum Δ = –0.06 · height + 20.7, $r^2 = 0.41$, p = <0.001

4 Discussion

Δ declined with increasing height in all 37 datasets and 61 regressions (Fig. 10.4). This decline results from multiple factors that change with increasing height such as increasing light availability and atmospheric $\delta^{13}C$, and decreasing water potential and hydraulic conductance. These changes drive concomitant changes in leaf mass per area, g_i, g_s, and leaf nutrient content. Despite this complexity, the constant decline in Δ with *h* found in our broad review demonstrates the existence of a consistent global relationship. The detailed case study results from *Nothofagus* demonstrate the combined importance of K_l, g_s and *A* on the dependence of Δ upon *h* (Fig. 10.2). We devote the discussion to examination of the driving mechanisms and how they are manifest in our observations.

4.1 The Response of Leaf-Level Gas Exchange and Δ to h in Nothofagus

Ψ_l, g_s, K_l and *A* all showed differences between tree size classes indicative of a hydraulic constraint on gas exchange (Fig. 10.2). These results are consistent with observations of other species (e.g. Yoder et al. 1994; Hubbard et al. 1999; Kolb and Stone 2000; McDowell et al. 2005; Martínez-Vilalta et al. 2007a) in support of the hydraulic limitation hypothesis (Ryan and Yoder 1997) that K_l is lower in taller trees, forcing greater stomatal closure and hence reduced *A*. There also appears to be a biochemical impact on *A* reflected in reduced internal conductance to CO_2 (g_i) with increasing height. In the *Nothofagus* dataset, *A* was always lower in taller trees, even when g_s was similar (compare Fig. 10.2b and c at $D < 0.8$ kPa). Because leaf [N] was significantly lower in the taller trees, it is likely that photosynthetic capacity was also lower (Field and Mooney 1986). This should drive Δ in the opposite direction of that observed (see below). Lower *A* at the same g_s for tall versus short trees could also result from reduced g_i. Thus, while hydraulics per se appear to play a direct role in driving diel gas exchange patterns, longer term impacts of hydraulics on g_i or leaf [N] appear to also be critical (see also Le Roux et al. 2001; Woodruff et al. 2004; Lloyd et al. 2009).

Foliar Δ and $p_a - p_c$ results for all seven *Nothofagus* sites showed evidence of lower p_c as *h* increased (Table 10.1, Fig. 10.3). This suggests that, for this species in this region, the dependence of Δ on *h* is regionally consistent despite variation in climate associated with elevation.

4.2 Why Does Δ Decline Linearly with Increasing *h*?

The clearest observation from this review is the linear relationship of Δ with *h* (Figs. 10.4b and 10.6–10.9). This is not consistent with the non-linear expectation according to Darcy's law (Eq. 10.3), confirming that other factors beyond hydraulics drive the response to height. Unfortunately, there was insufficient ancillary data on leaf nutrients, leaf mass per area, g_i, within-canopy CO_2 recycling, or the ratio of *h* to total hydraulic pathlength to determine if general relationships exist between these factors and height or Δ; a comprehensive study of these factors is needed in the future. Recycling of isotopically depleted canopy CO_2 should strengthen the gradient of Δ with *h*; however, field measurements have shown this effect to be relatively small (Lloyd et al. 1996, 1997; Brooks et al. 1997a, b; Buchmann et al. 1997a, b). Leaf nitrogen concentration per unit leaf area and photosynthetic capacity sometimes increase with height (and light availability) due to decreased leaf mass per area (e.g., Bond et al. 1999; Carswell et al. 2000; Barnard and Ryan 2003; Lloyd et al. 2009), which could further strengthen the observed relationship of Δ to *h*. This has been frequently observed in angiosperms, high leaf area forests, and trees that have not reached their maximum heights. In contrast, studies on particularly tall trees often find that leaf nitrogen per unit mass and photosynthetic capacity do not vary or even decline with *h* (McDowell et al. 2002a; Ambrose et al. 2009; Woodruff et al. 2009, *Nothofagus* data from this study). Internal conductance has been shown to increase with light or height (Le Roux et al. 2001; Warren et al. 2003). In two tall species, however, g_i and g_s declined linearly with increased *h* (Ambrose et al. 2009; Mullin et al. 2009; Woodruff et al. 2009). Patterns of g_i with *h* must be interpreted cautiously in light of the challenges to accurate measurement associated with g_i methodologies (Bickford et al. 2009; Pons et al. 2009). Recent evidence from both temperate conifer and tropical angiosperm forests suggests that light availability is a dominant driver of leaf morphology and physiology in the lower canopy while hydrostatic water potential gradients drive variation in the upper canopy (Ishii et al. 2007, 2008; Cavaleri et al. 2010; Ishii 2011).

We propose that the linear relationship between Δ and *h* observed in all 61 regressions results from a combination of factors, but is driven ultimately by the linear dependency of maximum leaf water potential on *h* (–0.01 MPa m^{-1}, Bauerle et al. 1999; Koch et al. 2004). This relationship sets boundaries on physiological properties that impact Δ. First, it sets an upper limit on foliar water potential during expansion, causing increasing leaf mass per unit area with increasing *h* (Marshall and Monserud 2003; Woodruff et al. 2004; Koch et al. 2004; Cavaleri et al. 2010), which may subsequently influence g_i and p_c via shifts in mesophyll surface area (Terashima et al. 2005; Oldham 2008). Second, as maximum xylem water potential declines with increasing *h*, the potential range of xylem water potential must also decline (unless the trees are anisohydric with *h*, which is rarely observed, allows only a few tenths MPa shift, and has consequences for xylem conductance and leaf area, Yoder et al. 1994; McDowell et al. 2002a). For example, if minimum g_s in both 10 and 100 m tall *Sequoia* trees occurred at leaf water potential of −2 MPa, then for the 10 m tall trees there is a Ψ range of 1.9 MPa over which water loss may occur prior to stomatal closure, whereas the Ψ range for the 100 m tree is only 1.0 MPa. This impacts Δ because taller trees spend more time with low g_s and hence low p_c than short trees. This was first demonstrated by Yoder et al. (1994) wherein tall ponderosa pines had the same maximum and minimum photosynthetic rates as neighboring short trees, but closed their stomata hours before the short trees because their minimum water potential was achieved earlier.

If the gravitational dependence of water potential on *h* drives the ultimate linearity of Δ with *h*, then how do the other variables in Eq. 10.3, plus light, nitrogen, and g_i, influence the relationship? At least four factors should enhance the non-linear relationship of Δ with increasing *h*: (1) hydraulic conductance declines non-linearly with *h* (Eq. 10.3, Mencuccini and Grace 1996a; McDowell et al. 2002a), which may be exacerbated by a non-linear increase in pathlength with increasing *h* (Magnani et al. 2000); (2) leaf nitrogen sometimes follows a Beer's law pattern with *h* (i.e. Lloyd et al. 2009); (3) occlusion of stomatal pores by cuticular wax may increase with *h* (England and Attiwill 2007); and (4) individual crown leaf area can increase with *h* to maximize whole-plant photosynthetic gain (Schoettle 1994; Buckley and Roberts 2006). In contrast, homeostatic shifts in plant structure, which operate to hold p_c constant within species and sites (Ehleringer 1993; McDowell et al. 2006), should push the Δ versus *h* relationship towards a linear pattern. Such shifts are well documented to occur with increasing *h*, such as changes in leaf area:sapwood area and root area:sapwood area ratios (Magnani et al. 2000; McDowell et al. 2002b; Mencuccini 2003) and sapwood permeability (England and Attiwill 2007). We propose that Δ moves away from the expected 1/*h* pattern at low heights due to *h*-driven homeostatic shifts (McDowell et al. 2002a, b; Mencuccini 2003) and light-driven leaf-photosynthetic shifts (Fig. 10.10). At some point, however, further shifts become infeasible and subsequent height growth forces Δ back towards the theoretical line from Eq. 10.3 (Fig. 10.10). This hypothesis can be tested by thorough

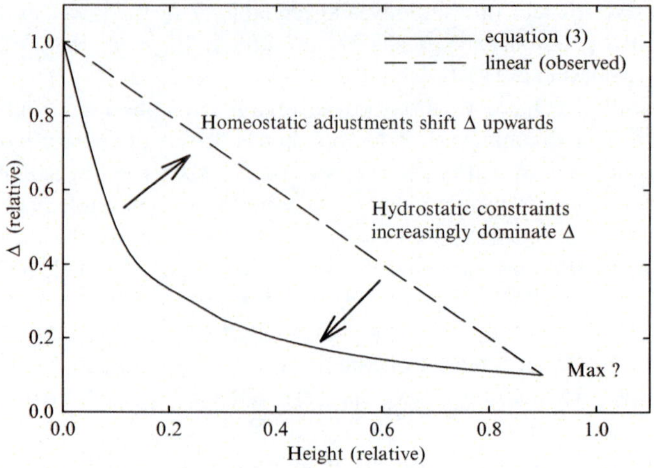

Fig. 10.10 A hypothesis regarding the interactive effects of environmental, physical and homeostatic factors on observed linear dependency of Δ with *h*. Many factors shift in the early phases of height growth to maximize p_c and hence Δ, thereby pushing Δ above the 1/*h* relationship. As height growth continues, the gravitational and hydraulic constraints dominate, forcing Δ to continue to a minimum

accounting of the vertical gradients of the above-mentioned key variables that influence Δ, both within the canopies of tall trees and across the sunlit tops of trees of different heights (e.g. Mullin et al. 2009).

A thorough accounting of all the factors that vary with height and that impact Δ would also aid our understanding of the trend of decreasing minimum Δ with increasing maximum *h* (Fig. 10.9). For the 43 datasets in which trees approached maximum height, there was a 0.06‰ decline in Δ for every meter increase in height. Though it is impossible with this dataset to conclusively interpret this pattern, it does suggest that the minimum Δ and maximum *h* are functionally linked. This may be due to a minimum sustainable p_c that allows a sufficient positive carbon balance at the leaf level. The minimum Δ values shown in Fig. 10.9 correspond to a p_c range of 320–147 ppm. The minimum p_c achievable by C3 plants is thought to be ~100–124 ppm (Tenhunen et al. 1984; Koch et al. 2004; Ward et al. 2005). This minimum Δ appears species- and site-specific (Fig. 10.9), perhaps due to variation in minimum CO_2-compensation points or thresholds for hydraulic failure due to prolonged embolism without refilling (Hacke et al. 2001; McDowell et al. 2002a; Ambrose et al. 2009). Understanding the drivers, consequences and homeostatic limitations of this minimum Δ is an important area for future research.

4.3 Irradiance and Height

In addition to the direct response of Δ to height, irradiance emerged as a clearly detectable environmental effect on Δ in our global review. Δ of sunlit foliage was

0.5‰–3.0‰ lower than that of shaded foliage (Figs. 10.7 and 10.8), consistent with prior observations (Fig. 10.1, Waring and Silvester 1994, also Livingston et al. 1998; Warren and Adams 2000; Le Roux et al. 2001; Geßler et al. 2001; Samuelson et al. 2003; Koch et al. 2004). There was no consistent difference in the Δ–*h* relationship between sunlit and shaded samples (Fig. 10.8). In this analysis, it appears that *h* dominates the trend in Δ while irradiance moderates the absolute values, presumably caused by the increased light interception and photosynthetic capacity with increased height and subsequent reduction in p_c (Ehleringer et al. 1986; Carswell et al. 2000; Lloyd et al. 2009). While *h* is clearly a global driver of Δ, the impact of irradiance on Δ should vary globally as a function of ecosystem leaf area and subsequent light interception profiles (Campbell and Norman 1998; Parker et al. 2002; Koch et al. 2004).

4.4 Patterns with Biomes and Vegetation Type

Clear differences emerged between biomes and vegetation type (gymnosperms and angiosperms). Tropical angiosperms had higher Δ than temperate gymnosperms or angiosperms (Table 10.1, Fig. 10.4). This Δ variation does not correspond with maximum height (data not shown), but does correspond with mean annual precipitation. For the studies that provided mean annual precipitation, it averaged 2,666 mm year^{-1} ($n = 6$) for tropical angiosperms, compared to 1,518 mm year^{-1} ($n = 19$) and 1,027 mm year^{-1} ($n = 7$) for temperate gymnosperms and angiosperms, respectively. Correlations between Δ and precipitation are routinely found in regional or global scale studies (e.g. Schulze et al. 1998; Warren et al. 2001; Bowling et al. 2002) due to the dominant effect of water availability on g_s and p_c. The lack of freezing temperatures in tropical regions may also allow higher g_s and p_c because xylem vessels can grow larger and have subsequently greater hydraulic conductance (Sperry et al. 2006).

The slope of Δ to *h* with biome and vegetation type differed from the absolute Δ values, being smallest for gymnosperms, largest for temperate angiosperms, and intermediate for tropical angiosperms (median values of −0.09, −0.24, and −0.20‰ m^{-1}, respectively, Table 10.1, Fig. 10.5). The greater slope of Δ to *h* indicates greater driving forces for a decline in Δ with *h* and/or greater flexibility of the trees to adjust Δ with height. The latter is consistent with the observation that species with higher sapling photosynthetic rates can achieve greater heights (Thomas and Bazzaz 1999); i.e. trees that start with high *A* may be able to grow relatively taller before *A* reaches threshold minimums. In our review, maximum Δ averaged 25.2‰ for angiosperms compared to 21.0‰ for gymnosperms, and the Δ range (maximum–minimum) was 4.0‰ and 5.8‰, respectively (31% greater for angiosperms). If the theoretical minimum Δ at maximum tree height is a fixed value at 13.6‰ ($\delta^{13}C$ of −20.0‰ estimated for *Sequoia sempervirens*, Koch et al. 2004; Burgess and Dawson 2007), then the maximum allowable Δ range is 36% greater for angiosperms than gymnosperms (11.6‰ vs 7.4‰, respectively). This can also be calculated using the minimum Δ observed for angiosperms (17.2‰) and gymnosperms (13.6‰, Fig. 10.9), yielding respective ranges of 7.4‰ and 13.0‰, or 43% greater for angiosperms.

The Δ patterns we observed between and within studies can also be influenced by post-assimilation isotopic effects. For example, differential day versus night respiratory fractionation, differential allocation of day versus night sucrose, and carbon fixation by PEP carboxylase, could all be manifest to influence $\delta^{13}C$ of leaf and stem tissues (Cernusak et al. 2009). Our understanding of these post-assimilation effects is rapidly progressing; however, it is premature for us to assess their magnitude without detailed isotopic flux and carbohydrate measurements (e.g. Badeck et al. 2007; Barbour et al. 2007; Brandes et al. 2006; Bickford et al. 2009). Stem tissues are particularly influenced by such post-photosynthetic metabolic effects, e.g. due to mixing of carbohydrate pools during the formation of wood (Helle and Schleser 2004; Ogée et al. 2009; Gessler et al. 2009). We observed a consistent offset between foliage and wood (similar to that summarized by Cernusak et al. 2009) and flatter Δ–*h* relationships in wood than foliage (Figs. 10.4 and 10.6), all consistent with post-photosynthetic effects. By comparing foliage and stem tissues separately we hoped to minimize these effects in comparisons of Δ to *h*. Inter-tree comparisons of different sized trees may be possible using either foliage or wood $\delta^{13}C$, however, we suggest that foliage $\delta^{13}C$ data may better integrate gas exchange when comparing Δ to *h* within trees.

4.5 Mixing Apples and Oranges

Five methodological constraints emerged from our global review that would benefit from careful research in the future. First, as noted earlier, foliage provides a better index of Δ in relation to *h* than wood, possibly due to the impacts of post-photosynthetic fractionation and mixing on the $\delta^{13}C$ composition of wood. Second, datasets from boreal forests were notably absent. Filling this biome-gap could yield new information since there was a clear pattern between tropical and temperate regions (Figs. 10.4 and 10.5) and since different climate drivers may dominate in boreal regions. Third, we need to separate the impacts of *h* on Δ between tops of differing size trees versus within crowns of tall trees. Comparing sunlit tops of trees of varying sizes excludes the confounding influence of light, allowing one to directly test the impacts of size-related hydraulics on Δ (e.g. McDowell et al. 2002a; Barnard and Ryan 2003; Koch et al. 2004, Figs. 10.7 and 10.8). Measuring within-canopy gradients avoids the assumptions of space-for-time substitution, however, and provides useful information on the combined light/hydraulic impacts on Δ, as well as information on within-tree homeostatic adjustments (Hubbard et al. 2002; Ishii et al. 2008). Since both approaches are useful, combining them in future studies may reveal more information than studying only tree tops or only within-tree gradients (Mullin et al. 2009). Fourth, as noted earlier, comprehensive measurements of environmental and physiological drivers of Δ with *h* are needed within studies, including light, humidity, gravitational potential, hydraulic conductance, gas exchange, leaf mass per area, leaf nutrients, woody anatomy, and leaf area:sapwood area ratios. Fifth, to understand the response of Δ throughout the height continuum, experimental designs should include trees near their site- and species-specific maximum heights. The slope of Δ to *h* tended to be steeper in studies of trees that were not at their maximum heights, suggesting that different responses are manifest during different phases of height growth. This may account for the outlier result from Barnard and Ryan (2003), in which the tallest measured trees had higher Δ rather than the expected lower Δ, but those trees were only 40% of maximum *h* for that site (D. Binkley 2010). Notably, factors such as leaf area:sapwood area follow a hyperbolic shape with increasing values to middle-heights followed by declining values to maximum heights in *Eucalyptus saligna* (N. McDowell 2010), thus Δ may also follow a similar pattern. Regardless, studies of height-related variation in Δ within trees that are not near their maximum heights may yield different insights than studies covering the entire height continuum.

5 Conclusions

Through our global review of 36 datasets with 38 species, we observed consistent, linear declines of Δ with *h*. The most parsimonious explanation of this result is that gravitational constraints on maximum leaf water potential set an ultimate boundary on the shape and sign of the relationship. These hydraulic constraints are manifest both over the long term through impacts on leaf structure, and over diel periods via impacts on K_l, g_s and *A*. Shading induces a positive offset to the linear decline, consistent with light limitations reducing *A* and increasing p_c at a given height (Fig. 10.1, Waring and Silvester 1994). Biome differences between tropical and temperate forests were more important in predicting Δ and its relationship to *h* than wood vasculature type (angiosperm vs gymnosperm). It is not yet clear how g_i varies with leaf mass per area (Terashima et al. 2005), but limited datasets in particularly tall, temperate conifers suggest that photosynthetic capacity does not vary dramatically with *h* when compared between tree-tops, while both g_s and g_i decline in unison with *h* within canopy gradients. We agree with previous conclusions that light is a critical variable low in the canopy whereas hydrostatic constraints dominate the Δ/*h* relationship in the upper canopy. A trend of increasing maximum *h* with decreasing minimum Δ suggests that trees that become particularly tall are adapted to tolerate particularly low values of p_c.

Acknowledgments We are grateful to the hundreds of scientists who have devoted long and challenging hours accessing tall trees to understand the fundamental constraints to tree size. We appreciate the generous provision of unpublished canopy-top samples from Maurizio Mencuccini, Walter Oechel, and Eladio Cornejo-Oviedo, and Renee Brooks for analyses of these samples. We are grateful for the shotgun assistance of Julian Licata and Andrew Schauer. Thoughtful edits were provided from three anonymous reviewers. Funding support during the writing of this manuscript was provided by DOE-BER and LANL-LDRD (N.G.M and L.T.H.), the National Science Foundation (DEB-0110142 and DEB-0823380, B.J.B.), a Hayward Scholarship from Landcare Research New Zealand (M.G.R.) and the Foundation for Research Science and Technology New Zealand (D.W.).

References

Addington RN, Donovan LA, Mitchell RJ, Vose JM, Pecot SD, Jack SB, Hacke UG, Sperry JS, Oren R (2006) Adjustments in hydraulic architecture of *Pinus palustris* maintain similar stomatal conductance in xeric and mesic habitats. Plant Cell Environ 29:535–545

Allen RB, Clinton PW, Davis MR (1997) Cation storage and availability along a *Nothofagus* forest development sequence in New Zealand. Can J For Res 27:323–330

Ambrose AR, Sillett SC, Dawson TE (2009) Effects of tree height on branch hydraulics, leaf structure and gas exchange in California redwoods. Plant Cell Environ 32:743–757

Badeck F-W, Tcherkez G, Nogues S, Piel C, Ghashghaie J (2007) Post-photosynthetic fractionation of stable carbon isotopes between plant organs – a widespread phenomenon. Rapid Commun Mass Spectrom 19:1381–1391

Baldocchi DD, Wilson KB, Gu L (2002) How the environment, canopy structure and canopy physiological functioning influence carbon, water and energy fluxes of a temperate broad-leaved deciduous forest – an assessment with the biophysical model CANOAK. Tree Physiol 22:1065–1077

Barbour MM, McDowell NG, Tcherkez G, Bickford C, Hanson D (2007) High frequency measurements reveal rapid post-illumination changes in the carbon isotope composition of leaf-respired CO_2. Plant Cell Environ 30:469–482

Barnard HR, Ryan MG (2003) A test of the hydraulic limitation hypothesis in fast-growing *Eucalyptus saligna*. Plant Cell Environ 26:1235–1245

Bauerle WL, Hinckley TM, Cermak J, Kucera J, Bible K (1999) The canopy water relations of old-growth Douglas-fir trees. Trees 13:211–217

Bickford CP, McDowell NG, Hanson DT (2009) High resolution field measurements of diurnal carbon isotope discrimination and internal conductance in a semi-arid species, *Juniperus monosperma*. Plant Cell Environ 32:796–810

Bickford CP, Hanson DT, McDowell NG (2010) Influence of diurnal variation in mesophyll conductance on modelled ^{13}C discrimination: results from a field study. J Exp Bot 61:3223–3233

Bloom AJ, Chapin FS III, Mooney HA (1985) Resource limitation in plants – an economic analogy. Annu Rev Ecol Syst 16:363–392

Bonan GB (2008) Forests and climate change: forcings, feedbacks, and the climate benefits of forests. Science 320:1444–1449

Bond BJ (2000) Age-related changes in photosynthesis of woody plants. Trends Plant Sci 5:349–353

Bond BJ, Farnsworth BT, Coulombe RA, Winner WE (1999) Foliage physiology and biochemistry in response to radiation gradients in conifers with varying shade tolerance. Oecologia 199:183–192

Bond BJ, Czarnomski NM, Cooper C, Day ME, Greenwood MS (2007) Developmental decline in height growth in Douglas-fir. Tree Physiol 27:441–453

Bond BJ, Meinzer FC, Brooks JR (2008) How trees influence the hydrological cycle in forest ecosystems. In: Wood PJ, Hannah DM, Sadler JP (eds.) Hydroecology and ecohydrology: past, present and future. Wiley, London, pp 7–28

Bowling D, McDowell NG, Bond BJ, Law BE, Ehleringer JR (2002) ^{13}C content of ecosystem respiration is linked to precipitation and vapor pressure deficit. Oecologia 131:113–124

Brandes E, Kodama N, Whittaker K, Weston C, Rennenberg H, Keitel C, Adams MA, Gessler A (2006) Short-term variation in the isotopic composition of organic matter allocated from the leaves to the stem of *Pinus sylvestris*: effects of photosynthetic and postphotosynthetic carbon isotope fractionation. Glob Change Biol 12:1922–1939

Broadmeadow MSJ, Griffiths H, Maxwell C, Borland AM (1992) The carbon isotope ratio of plant organic material reflects temporal and spatial variations in CO_2 within tropical forest formations in Trinidad. Oecologia 89:435–441

Brooks JR, Flanagan LB, Varney GT, Ehleringer JR (1997a) Vertical gradients in photosynthetic gas exchange characteristics and refixation of respired CO_2 within boreal forest canopies. Tree Physiol 17:1–12

Brooks JR, Flanagan LB, Buchmann N, Ehleringer JR (1997b) Carbon isotope composition of boreal plants: functional grouping of life forms. Oecologia 110:301–311
Brugnoli E, Farquhar GD (2000) Photosynthetic fractionation of carbon isotopes. In: Leegood RC, Sharkey TD, von Caemmerer S (eds.) Photosynthesis: physiology and metabolism. Kluwer, Dordrecht, pp 399–434
Buchmann N, Kao W-Y, Ehleringer J (1997a) Influence of stand structure on carbon-13 of vegetation, soils, and canopy air within deciduous and evergreen forests in Utah, United States. Oecologia 110:109–119
Buchmann N, Guehl JM, Barigah TS, Ehleringer JR (1997b) Interseasonal comparison of CO_2 concentrations, isotopic composition, and carbon cycling in an Amazonian rainforest (French Guiana). Oecologia 110:120–131
Buckley TN, Roberts DW (2006) How should leaf area, sapwood area and stomatal conductance vary with tree height to maximize growth? Tree Physiol 26:145–157
Burgess SSO, Dawson TE (2007) Predicting the limits to tree height using statistical regressions of leaf traits. New Phytol 174:626–636
Campbell GS, Norman JM (1998) An introduction to environmental biophysics, 2nd edn. Springer, New York, 286 pp
Carswell FE, Meir P, Wandelli EV, Bonates LCM, Kruijt B, Barbosa EM, Nobre AD, Grace J, Jarvis PG (2000) Photosynthetic capacity in a central Amazonian rain forest. Tree Physiol 20:179–186
Cavaleri MA, Oberbauer SF, Clark DB, Clark DA, Ryan MG (2010) Height is the primary determinant of leaf morphology in a tropical rain forest canopy. Ecology 91:1730–1739
Cernusak LA, Tcherkez G, Keitel C, Cornwell WK, Santiago LS, Knohl A, Barbour MM, Williams DG, Reich PB, Ellsworth DS, Dawson TE, Griffiths HG, Farquhar GD, Wright IJ (2009) Why are non-photosynthetic tissues generally ^{13}C enriched compared to leaves in C_3 plants? Review and synthesis of current hypotheses. Funct Plant Biol 36:199–213
Craig H (1957) Isotopic standards for carbon and oxygen and correction factors for mass spectrometric analysis of carbon dioxide. Geochim Cosmochim Acta 12:133–149
Davis MR, Allen RB, Clinton PW (2003) Carbon storage along a stand development sequence in a New Zealand *Nothofagus* forest. For Ecol Manag 177:313–321
Day ME, Greenwood MS, Diaz-Sala C (2002) Age- and size-related trends in woody plant shoot development: regulatory pathways and evidence for genetic control. Tree Physiol 22:507–513
Delzon S, Sartore M, Burlett R, Dewar R, Loustau D (2004) Hydraulic responses to height growth in maritime pine trees. Plant Cell Environ 27:1077–1087
Domec JC, Lachenbruch B, Meinzer FC, Woodruff DR, Warren JM, McCulloh KA (2008) Maximum height in a conifer is associated with conflicting requirements for xylem design. Proc Natl Acad Sci U S A 105:12069–12074
Donovan LA, Ehleringer JR (1994) Carbon isotope discrimination, water-use efficiency, growth, and mortality in a natural shrub population. Oecologia 100:347–354
Duursma RA, Marshall JD (2006) Vertical canopy gradients in $\delta^{13}C$ correspond with leaf nitrogen content in a mixed-species conifer forest. Trees 20:496–506
Ehleringer JR (1993) Carbon and water relations in desert plants: an isotopic perspective. In: Ehleringer JR, Hall AE, Farquhar GD (eds.) Stable isotopes and plant carbon/water relations. Academic, San Diego, pp 155–172
Ehleringer JR, Field CB, Lin Z-F, Kuo C-Y (1986) Leaf carbon isotope and mineral composition in subtropical plants along an irradiance cline. Oecologia 70:520–526
England JM, Attiwill PM (2007) Changes in sapwood permeability and anatomy with tree age and height in the broad-leaved evergreen species Eucalyptus regnans. Tree Physiol 27:1113–1124
Farquhar GD, O'Leary MH, Berry JA (1982) On the relationship between carbon isotope discrimination and the intercellular carbon dioxide concentration in leaves. Aust J Plant Physiol 9:121–137
Field C, Mooney HA (1986) The photosynthesis–nitrogen relationship in wild plants. In: Givnish TJ (ed.) On the economy of form and function. Cambridge University Press, Cambridge, pp 25–55
Francey RJ, Farquhar GD (1982) An explanation of $^{13}C/^{12}C$ variations in tree rings. Nature 297:2831

Francey RJ, Gifford RM, Sharkey TD, Weir B (1985) Physiological influences on carbon isotope discrimination in huon pine (*Lagarostrobos franklinii*). Oecologia 66:211–218

Geßler A, Schrempp S, Matzarakis A, Mayer H, Rennenberg H, Adams MA (2001) Radiation modifies the effect of water availability on the carbon isotope composition of beech (*Fagus sylvatica*). New Phytol 150:653–664

Gessler A, Brandes E, Buchmann N, Helle G, Rennenberg H, Barnard RL (2009) Tracing carbon and oxygen isotope signals from newly assimilated sugars in the leaves to the tree-ring archive. Plant Cell Environ 32:780–795

Gower ST, McMurtrie RE, Murty D (1996) Aboveground net primary production decline with stand age: potential causes. Trends Ecol Evol 11:378–382

Greenwood MS, Ward M, Day ME, Adams S, Bond BJ (2008) Age-related trends in red spruce foliar plasticity in relation to declining productivity. Tree Physiol 28:225–232

Grulke NE, Retzlaff WA (2001) Changes in physiological attributes of ponderosa pine from seedling to mature tree. Tree Physiol 21:275–286

Hacke UG, Stiller V, Sperry JS, Pittermann J, McCulloh K (2001) Cavitation fatigue: embolism and refilling cycles can weaken the cavitation resistance of xylem. Plant Physiol 125:779–786

Hanba YT, Mori S, Lei TT, Koike T, Wada E (1997) Variations in leaf $\delta^{13}C$ along a vertical profile of irradiance in a temperate Japanese forest. Oecologia 110:253–261

Harlow BA, Marshall JD, Robinson AP (2006) A multi-species comparison of $\delta^{13}C$ from whole wood, extractive-free wood and holocellulose. Tree Physiol 26:767–774

He C-X, Li J-Y, Zhou P, Guo M, Zheng Q-S (2008) Changes of leaf morphological, anatomical structure and carbon isotope ratio with the height of the wangtian tree (*Parashorea chinensis*) in Xishuangbanna, China. J Integr Plant Biol 50:168–173

Heaton THE, Crossley A (1995) Carbon isotope variations in a plantation of Sitka spruce, and the effect of acid mist. Oecologia 103:109–117

Helle G, Schleser GH (2004) Beyond CO_2-fixation by Rubisco – an interpretation of $^{13}C/^{12}C$ variations in tree rings from novel intra-seasonal studies on broad-leaf trees. Plant Cell Environ 27:367–380

Hubbard RM, Bond BJ, Ryan MG (1999) Evidence that hydraulic limitation explains differences in photosynthesis between young and old *Pinus ponderosa*. Tree Physiol 19:165–172

Hubbard RM, Bond BJ, Senock RS, Ryan MG (2002) Effects of branch height on leaf gas exchange, branch hydraulic conductance and branch sap flux in open-grown ponderosa pine. Tree Physiol 22:575–581

Hultine KR, Marshall JD (2000) Altitude trends in conifer leaf morphology and stable isotope composition. Oecologia 123:32–40

IPCC (2007) Climate change 2007: the physical science basis. Contribution of working group I to the fourth assessment report of the intergovernmental panel on climate change, IPCC Secretariat, Geneva, Switzerland

Ishii H (2011) How do changes in leaf/shoot morphology and crown architecture affect growth and physiological function of large, old trees? In: Meinzer FC, Lachenbruch B, Dawson TE (eds.) Size- and age-related changes in tree structure and function. Springer, Dordrecht

Ishii HT, Ford ED, Kennedy MC (2007) Physiological and ecological implications of adaptive reiteration as a mechanism for crown maintenance after trees reach maximum size. Tree Physiol 27:455–462

Ishii HT, Jennings GM, Sillett SC, Koch GW (2008) Hydrostatic constraints on morphological exploitation of light in tall *Sequoia sempervirens* trees. Oecologia 156:751–763

Keeling CD, Bacastow RB, Carter AF, Piper SC, Whorf TP, Heimann M, Mook WG, Roeloffzen H, Peterson H (1989) A three-dimensional model of atmospheric CO_2 transport based on observed winds:1. Analysis of observational data. In: Peterson DH (ed.) Aspects of climate variability in the Pacific and the western Americas, Geophysical monographs. American Geophysical Union, Washington, DC, pp 165–236

Koch GW, Sillett SC, Jennings GM, Davis SD (2004) The limits to tree height. Nature 428:851–854

Kolb TE, Stone JE (2000) Differences in leaf gas exchange and water relations among species and tree sizes in an Arizona pine-oak forest. Tree Physiol 20:1–12

Körner C (2007) The use of 'altitude' in ecological research. Trends Ecol Evol 22:569–574
Le Roux X, Bariac T, Sinoquet H, Genty B, Piel C, Mariotti A, Girardin C, Richard P (2001) Spatial distribution of leaf water-use efficiency and carbon isotope discrimination within an isolated tree crown. Plant Cell Environ 24:1021–1032
Leavitt SW, Long A (1988) Stable carbon isotope chronologies from trees in the southwestern United States. Glob Biogeochem Cycles 2:189–198
Leavitt SW, Long A (1991) Seasonal stable-carbon isotope variability in tree rings: possible paleoenvironmental signals. Chem Geol 87:59–70
Livingston NJ, Whitehead D, Kelliher FM, Wang Y-P, Grace JC, Walcroft AS, Byers JN, Mcseveny TM, Millard P (1998) Nitrogen allocation and carbon isotope fractionation in relation to intercepted radiation and position in a young *Pinus radiata* D. Don tree. Plant Cell Environ 21:795–803
Lloyd J, Kruijt B, Hollinger DY, Grace J, Francey RJ, Wong SC, Kelliher FM, Miranda AC, Farquhar GD, Gash JH, Vygodskaya NN, Wright IR, Miranda HS, Schulze ED (1996) Vegetation effects on the isotopic composition of atmospheric CO_2 at local and regional scales: theoretical aspects and a comparison between rain forest in Amazonia and boreal forest in Siberia. Aust J Plant Physiol 23:371–399
Lloyd J, Kruijt B, Hollinger DY, Grace J, Francey RJ, Wong SC, Kelliher FM, Miranda AC, Farquhar GD, Gash JH, Vygodskaya NN, Wright IR, Miranda HS, Schulze ED (1997) An alternative interpretation of the appropriateness and correct means for the evaluation of CO_2 recycling indices. Aust J Plant Physiol 24:399–405
Lloyd J, Patino S, Paiva RQ, Nardoto GB, Quesada CA, Santos AJB, Baker TR, Brand WA, Hilke I, Gielmann H, Raessler M, Luizao FJ, Martinelli LA, Mercado LM (2009) Variations in leaf physiological properties within Amazon forest canopies. Biogeosci Discuss 6:4639–4692
Loader NJ, Robertson I, McCarroll D (2003) Comparisons of stable carbon isotope ratios in the whole-wood, cellulose and lignin of oak tree-rings. Palaeogeog Palaeoclimatol Palaeoecol 196:394–407
Macfarlane C, Warren CR, White DA, Adams MA (1999) A rapid and simple method for processing wood to crude cellulose for analysis of stable carbon isotopes in tree rings. Tree Physiol 19:831–835
Magnani F, Mencuccini M, Grace J (2000) Age-related decline in stand productivity: the role of structural acclimation under hydraulic constraints. Plant Cell Environ 23:251–263
Major JE, Mosseler A, Barsi DC, Campbell M (2007) Comparative nutrient economy, stable isotopes, and related adaptive traits in *Picea rubens*, *Picea mariana*, and their hybrids. Trees 21:677–692
Marshall JD, Monserud RA (2003) Foliage height influences specific leaf area of three conifer species. Can J For Res 33:164–170
Martinelli LA, Almeida S, Brown IF, Moreira MZ, Victoria RL, Sternberg LSL, Ferreira CAC, Thomas WW (1998) Stable carbon isotope ratio of tree leaves, boles and fine litter in a tropical forest in Rondonia, Brazil. Oecologia 114:170–179
Martínez-Vilalta J, Vanderklein D, Mencuccini M (2007a) Tree height and age-related decline in growth in Scots pine (*Pinus sylvestris* L.). Oecologia 150:529–544
Martínez-Vilalta J, Korakaki E, Vanderklein D, Mencuccini M (2007b) Below-ground hydraulic conductance is a function of environmental conditions and tree size in Scots pine. Funct Ecol 21:1072–1083
McCracken IJ (1980) Mountain climate in the Craigieburn Range, New Zealand. In: Benecke U, Davis MR (eds.) Mountain environments and subalpine growth. New Zealand Forest Service Forestry Research Institute Technical Paper 70, pp 41–60
McDowell NG, Phillips N, Lunch CK, Bond BJ, Ryan MG (2002a) An investigation of hydraulic limitation and compensation in large, old Douglas-fir trees. Tree Physiol 22:763–774
McDowell NG, Barnard HR, Bond BJ, Hinckley T, Hubbard RM, Ishii H, Köstner B, Magnani F, Marshall JD, Meinzer FC, Phillips N, Ryan MG, Whitehead D (2002b) The relationship between tree height and leaf area: sapwood area ratio. Oecologia 132:12–20
McDowell NG, Licata J, Bond BJ (2005) Environmental sensitivity of gas exchange in different-sized trees. Oecologia 145:9–20

McDowell NG, Adams HA, Bailey JD, Hess M, Kolb TE (2006) Homeostatic maintenance of ponderosa pine gas exchange in response to stand density changes. Ecol Appl 16:1164–1182

McDowell NG, Allen C, Marshall L (2010) Growth, carbon isotope discrimination, and mortality across a ponderosa pine elevation transect. Glob Change Biol 16:399–415

Medina E, Minchin P (1980) Stratification of ^{13}C-values of leaves in Amazonian rain forests. Oecologia 45:377–378

Meinzer FC, Grantz DA (1990) Stomatal and hydraulic conductance in growing sugarcane: stomatal adjustment to water transport capacity. Plant Cell Environ 13:383–388

Mencuccini M (2003) The ecological significance of long distance water transport: short-term regulation and long-term acclimation across plant growth forms. Plant Cell Environ 26:163–182

Mencuccini M, Comstock J (1999) Variability in hydraulic architecture and gas exchange of common bean (*Phaseolus vulgaris*) cultivars under well-watered conditions: interactions with leaf size. Aust J Plant Physiol 26:115–124

Mencuccini M, Grace J (1996a) Hydraulic conductance, light interception and needle nutrient concentration in Scots pine stands and their relation with net primary productivity. Tree Physiol 16:459–468

Mencuccini M, Grace J (1996b) Developmental patterns of aboveground hydraulic conductance in a Scots pine (*Pinus sylvestris* L.) age sequence. Plant Cell Environ 19:939–948

Mencuccini M, Martinez-Vilalta J, Vanderklein D, Hamid HA, Korakaki E, Lee S, Michiels B (2005) Size-mediated ageing reduces vigour in trees. Ecol Lett 8:1183–1190

Monserud RA, Marshall JD (2001) Time-series analysis of δ^{13}C from tree rings. I. Time trends and autocorrelation. Tree Physiol 21:1087–1102

Mullin LP, Sillett SC, Koch GW, Tu KP, Antoine ME (2009) Physiological consequences of height-related morphological variation in *Sequoia sempervirens* foliage. Tree Physiol 29:999–1010

Niinemets U (2002) Stomatal conductance alone does not explain the decline in foliar photosynthetic rates with increasing tree age and size in *Picea abies* and *Pinus sylvestris*. Tree Physiol 22:515–535

Ogée J, Barbour MM, Wingate L, Bert D, Bosc A, Stievenard M, Lambrot C, Pierre M, Bariac T, Loustau D, Dewar RC (2009) A single-substrate model to interpret intra-annual stable isotope signals in tree-ring cellulose. Plant Cell Environ 32:1071–1090

Oldham AR (2008) Height-associated variation in leaf anatomy of tall redwoods: potential impacts on whole-tree carbon balance. M.A. thesis, Humboldt State University, 59 pp

Ometto JPHB, Flanagan LB, Martinelli LA, Moreira MZ, Higuchi N, Ehleringer JR (2002) Carbon isotope discrimination in forest and pasture ecosystems of the Amazon Basin, Brazil. Glob Biogeochem Cycles 16:1109–1119

Parker GG, Davis MM, Chapotin SM (2002) Canopy light transmittance in Douglas-fir-western hemlock stands. Tree Physiol 22:147–157

Phillips N, Bond BJ, McDowell NG, Ryan MG (2002) Canopy and hydraulic conductance in young, mature, and old Douglas-fir trees. Tree Physiol 22:205–211

Phillips N, Bond BJ, McDowell NG, Ryan MG, Schauer A (2003) Leaf area compounds height-related hydraulic costs of water transport in Oregon White Oak trees. Funct Ecol 17:832–840

Pons TL, Flexas J, von Caemmerer S, Evans JR, Genty B, Ribas-Carbo M, Brugnoli E (2009) Estimating mesophyll conductance to CO_2: methodology, potential errors, and recommendations. J Exp Bot 60:2217–2234

Poorter L, Bongers F, Sterck FJ, Wöll H (2005) Beyond the regeneration phase: differentiation of height-light trajectories among tropical tree species. J Ecol 93:256–267

Ryan MG, Yoder BJ (1997) Hydraulic limits to tree height and tree growth. Bioscience 47:235–242

Ryan MG, Binkley D, Fownes JH (1997) Age-related decline in forest productivity: pattern and process. Adv Ecol Res 27:213–262

Ryan MG, Phillips N, Bond BJ (2006) The hydraulic limitation hypothesis revisited. Plant Cell Environ 29:367–381

Sala A, Hoch G (2009) Height-related growth declines in ponderosa pine are not due to carbon limitation. Plant Cell Environ 32:22–30

Samuelson LJ, McLemore PC III, Somers GL (2003) Relationship between foliar $\delta^{13}C$ and hydraulic pathway length in *Pinus palustris*. For Sci 49(5):790–798
Schäfer KVR, Oren R, Tenhunen JD (2000) The effect of tree height on crown level stomatal conductance. Plant Cell Environ 23:365–375
Schleser GH (1990) Investigations of the $\delta^{13}C$ pattern in leaves of *Fagus sylvatica* L. J Exp Bot 41:565–572
Schleser GH (1992) $\delta^{13}C$ pattern in a forest tree as an indicator of carbon transfer in trees. Ecology 73:1922–1925
Schoettle AW (1994) Influence of tree size on shoot structure and physiology of *Pinus contorta* and *Pinus aristata*. Tree Physiol 14:1055–1068
Schulze E-D, Williams RJ, Farquhar GD, Schulze W, Langridge J, Miller JM, Walker BH (1998) Carbon and nitrogen isotope discrimination and nitrogen nutrition of trees along a rainfall gradient in northern Australia. Aust J Plant Physiol 25:413–425
Seibt U, Rajabi A, Griffiths H, Berry JA (2008) Carbon isotopes and water-use efficiency: sense and sensitivity. Oecologia 155:441–454
Sperry JS, Hacke U, Oren R, Comstock J (2002) Water deficits and hydraulic limits to leaf water supply. Plant Cell Environ 25:251–263
Sperry JS, Hacke UG, Pittermann J (2006) Size and function in conifer tracheids and angiosperm vessels. Am J Bot 93:1490–1500
Steppe K, Niinemets Ü, Teskey RO (2011) Tree size- and age-related changes in leaf physiology and their influence on carbon gain. In: Meinzer FC, Lachenbruch B, Dawson TE (eds.) Size- and age-related changes in tree structure and function. Springer, Dordrecht
Sternberg L da SL, Mulkey SS, Wright SJ (1989) Ecological interpretation of leaf carbon isotope ratios: influence of recycled carbon dioxide. Ecology 70:1317–1324
Tate KR, Scott NA, Ross DJ, Parshotam A, Claydon JJ (2000) Plant effects on soil carbon storage and turnover in a montane beech (*Nothofagus*) forest and adjacent tussock grassland in New Zealand. Aust J Soil Res 38:685–698
Tenhunen JD, Lange OL, Gebel J, Beyschlag W, Weber JA (1984) Changes in photosynthetic capacity, carboxylation efficiency, and CO_2 compensation point associated with midday stomatal closure and midday depression of net CO_2 exchange of leaves of *Quercus suber*. Planta 162:193–203
Terashima I, Araya T, Miyazawa S, Sone K, Yano S (2005) Construction and maintenance of the optimal photosynthetic systems of the leaf, herbaceous plant, and tree: an eco-developmental treatise. Ann Bot 95:507–519
Thomas SC (2011) Age-related changes in tree growth and functional biology: the role of reproduction. In: Meinzer FC, Lachenbruch B, Dawson TE (eds.) Size- and age-related changes in tree structure and function. Springer, Dordrecht
Thomas SC, Bazzaz FA (1999) Asymptotic height as a predictor of photosynthetic characteristics in Malaysian rain forest trees. Ecology 80:1607–1622
Vitousek PM, Field CB, Matson PA (1990) Variation in foliar $\delta^{13}C$ in Hawaiian *Metrosideros polymorpha*: a case of internal resistance? Oecologia 84:362–370
Ward JK, Harris JM, Cerling TE, Wiedenhoeft A, Lott MJ, Dearing M-D, Coltrain JB, Ehleringer JR (2005) Carbon starvation in glacial trees recovered from the La Brea tar pits, southern California. Proc Natl Acad Sci U S A 102:690–694
Waring RH, Silvester WB (1994) Variation in foliar ^{13}C values within the crowns of *Pinus radiata* trees. Tree Physiol 14:1203–1213
Warren CR, Adams MA (2000) Water availability and branch length determine δ13C in foliage of *Pinus pinaster*. Tree Physiol 20:637–643
Warren CR, Adams MA (2006) Internal conductance does not scale with photosynthetic capacity: implications for carbon isotope discrimination and the economics of water and nitrogen use in photosynthesis. Plant Cell Environ 29:192–201
Warren CR, McGrath JF, Adams MA (2001) Water availability and carbon isotope discrimination in conifers. Oecologia 127:476–486

Warren CR, Ethier GJ, Livingston NJ, Grant NJ, Turpin DH, Harrison DL, Black TA (2003) Transfer conductance in second growth Douglas-fir (*Pseudotsuga menziesii* (Mirb.) Franco) canopies. Plant Cell Environ 26:1215–1227

Whitehead D (1998) Regulation of stomatal conductance and transpiration in forest canopies. Tree Physiol 18:633–644

Winner WE, Thomas S, Berry J, Bond BJ, Cooper CE, Hinckley TM, Ehleringer JR, Fessenden JE, Lamb B, McCarthy S, McDowell NG, Phillips N, Williams M (2004) Canopy carbon gain and water use: analysis of old-growth conifers in the Pacific Northwest. Ecosystems 7:482–497

Woodruff DR, Meinzer FC (2011) Size-dependent changes in biophysical control of tree growth: the role of turgor. In: Meinzer FC, Lachenbruch B, Dawson TE (eds.) Size- and age-related changes in tree structure and function. Springer, Dordrecht

Woodruff DR, Bond BJ, Meinzer FC (2004) Does turgor limit growth in tall trees? Plant Cell Environ 27:229–236

Woodruff DR, Meinzer FC, Lachenbruch B, Johnson DM (2009) Coordination of leaf structure and gas exchange along a height gradient in a tall conifer. Tree Physiol 29:261–272

Xu ZH, Saffigna PG, Farquhar GD, Simpson JA, Haines RJ, Walker S, Osborne DO, Guinto D (2000) Carbon isotope discrimination and oxygen isotope composition in clones of the F_1 hybrid between slash pine and Caribbean pine in relation to tree growth, water-use efficiency and foliar nutrient concentration. Tree Physiol 20:1209–1218

Yoder BJ, Ryan MG, Waring RH, Schoettle AW, Kaufmann MR (1994) Evidence of reduced photosynthetic rates in old trees. For Sci 40(3):513–527

Zhang J, Fins L, Marshall JD (1994) Stable carbon isotope discrimination, photosynthetic gas exchange, and growth differences among western larch families. Tree Physiol 14:531–539

Chapter 11
Carbon Storage in Trees: Does Relative Carbon Supply Decrease with Tree Size?

Anna Sala, Willa Fouts, and Günter Hoch

Abstract Until very recently, age- and size-related declines in productivity of individual trees and stands have been attributed to reductions of carbon availability due either to progressive increases in carbon sinks other than growth (Respiration hypothesis) or to decreases carbon sources (Assimilation hypothesis). Although the validity of these hypotheses is now questioned and new alternative explanations have been proposed, actual data on size-dependent changes of stored mobile carbon in mature trees (used as indicators of carbon balance between source and sink activities) are surprisingly limited. Based on available data for mature trees and consistent with evidence that mature trees under current atmospheric CO_2 concentrations are not carbon limited, the relative carbon supply in trees does not become increasingly limited as they grow large. In spite of many uncertainties, research to date points to the need to question the historically carbon-centric mechanisms proposed to explain age-related growth declines in trees and forests. Future research should focus on whether and when alternative growth limiting factors (e.g. turgor, long distance transport of assimilates, nutrients) may contribute to growth limitations in tall trees.

1 Introduction

Height growth poses significant challenges in trees including increases in biomechanical risks associated with heavier loads (Niklas 2007; King 2011), increase of the proportion of woody, non-photosynthetic tissues (Kira and Shidei 1967;

A. Sala (✉) • W. Fouts
Division of Biological Sciences, The University of Montana, Missoula, MT 59812, USA
e-mail: sala@mso.umt.edu

G. Hoch
Institute of Botany, University of Basel, Schönbeinstrasse 6, CH-4056 Basel, Switzerland

F.C. Meinzer et al. (eds.), *Size- and Age-Related Changes in Tree Structure and Function*, Tree Physiology 4, DOI 10.1007/978-94-007-1242-3_11,

Whittaker and Woodwell 1967; Peet 1980; Kozlowski et al. 1991), increase of distance over which vascular tissues need to move water, nutrients and assimilated substances throughout the tree (Mencuccini et al. 2007; Mencuccini and Hölttä 2010), and an increase of the gravitational pull against the ascent of water (Koch et al. 2004; Woodruff and Meinzer 2011). Because trees (like any other plant) must ensure favorable water and carbon balance over time, it is not surprising that they undergo a variety of biochemical, physiological, anatomical and structural adjustments that minimize the negative effects of height growth, thereby sustaining them as biomechanically, biophysically and physiologically viable organisms (Ishii 2011; Scholz et al. 2011; Steppe et al. 2011; Woodruff and Meinzer 2011).

A very well-known and common phenomenon occurring as individual trees or mono-specific stands age and increase in size is a decline of growth rates and net primary productivity (Ryan et al. 1997). However, the underlying mechanisms for this general pattern are still under debate and remain unresolved. Several explanations have been proposed over the years to explain the so called age-related decrease of tree and forest productivity, including increased relative respiration losses, nutrient limitations and decreases of carbon assimilation, which in turn, are caused by a variety of factors such as hydraulic constraints, reductions of leaf area, and genetic factors (Gower et al. 1996; Ryan et al. 1997, 2004; Day and Greenwood 2011; Steppe et al. 2011). These classical explanations can be grouped into two main classes of competing hypotheses: 'Respiration hypotheses' (age-related declines in productivity are due to proportional increases of respiratory demand as trees increase in size) and 'Assimilation or Gross Primary Productivity hypotheses' (reduced growth rates result from declining carbon assimilation in response to a variety of causes). A common feature of these two competing groups of hypotheses is that they both ultimately predict that age-related declines in growth rates are due to changes in C-balance with tree size and age. In other words, they both predict reductions of carbon available for growth as trees age and grow tall. Although the validity of these hypotheses has now been questioned and alternative hypotheses have been proposed (see below), the carbon-centric view embodied by these two classes of hypotheses has never been properly tested. Surprisingly little research has examined patterns of stored non-structural carbon as trees age and grow large, which is the main focus of this chapter. We ask the following two questions: (a) Does the concentration of stored non-structural carbon compounds change as trees increase in size or age? (b) If so, do these changes suggest that trees become carbon limited as they increase in size? We first provide a very brief outline of the historical development of the two classical hypotheses proposed to explain height/age-related growth declines and outline current alternative hypotheses. We then examine the use of stored non-structural carbon compounds in trees as indicators of carbon balance and relative carbon source-sink activities, and briefly review evidence to date on whether mature trees under ambient CO_2 levels are carbon limited. We follow by presenting the limited available data on changes in stored carbon pools with tree height and their implications for the mechanisms responsible for the height/age-related decline of productivity in trees and forests. Finally, we provide suggestions and methodological considerations for future research.

2 Mechanisms for the Age-Related Decline in Productivity

Until recently, the mechanisms underlying the age-related decline in above ground net primary productivity of forests (Ryan et al. 1997) have been embodied in two main categories of competing hypotheses (Gower et al. 1996; Ryan et al. 2004). The 'Respiration Hypothesis', initially proposed by Kira and Shidei (1967), predicted that age/size related declines in growth rates result from a progressive imbalance between carbon assimilation and respiration as trees age and grow tall due to the increased respiratory load imposed by the greater proportion of woody, non-photosynthetic tissues. However, neither initial field observations and modeling approaches (Ryan and Waring 1992), nor recent experimental approaches (Ryan et al. 2004) provide support for the respiration hypothesis. The alternative "Carbon Assimilation hypothesis" in contrast, predicted that age/size-related decreases in carbon assimilation, rather than increases in respiration, were responsible for the age-related declines in growth rates. While a number of factors have been invoked to explain such age-related declines in gross primary production (Ryan et al. 1997), intense recent research has focused on the negative effect of tree size on water relations and gas exchange (Ryan et al. 2006). This is because as trees grow tall, the increase in gravitational pull against the ascent of water combined with the increase of the overall xylem hydraulic resistance from roots to leaves result in greater xylem tensions and a potentially higher risk of cavitation and hydraulic failure (Zimmermann 1983; Tyree and Sperry 1989). Because stomatal closure is the most effective mechanism to reduce water loss and minimize the risk of hydraulic failure, tall trees are expected to close stomata and reduce carbon assimilation to a greater extent than shorter trees (Friend 1993). Indeed, this prediction was first corroborated by Yoder et al. (1994) in the field. Subsequently, Ryan and Yoder (1997) proposed the Hydraulic Limitation Hypothesis (HLH) as a possible mechanism to explain the age-related decline of productivity in individual trees and stands. The HLH as initially stated proposed that hydraulic constraints on gas exchange as trees grow and increase in height were responsible for the age-related decline of productivity in trees. The HLH was extraordinarily well received because of the initial lack of support for the respiration hypothesis (Ryan and Waring 1992) and because compensatory mechanisms to minimize height-related hydraulic constraints (e.g. Coyea and Margolis 1992; Goldstein et al. 1998; Barnard and Ryan 2003; Zhang et al. 2009; Woodruff and Meinzer 2011) are generally not enough to offset the negative effect of height on gas exchange (Ryan et al. 2006).

About a decade after the formal publication of the HLH, a review of the literature by Ryan et al. (2006) concluded that in spite of compensatory mechanisms to offset path length effects and to cope with the negative effects of gravity, hydraulic constraints on gas exchange are very common in tall trees. Importantly, however, they found no evidence supporting that reductions in carbon assimilation in tall trees are responsible for growth reductions. In other words, hydraulic limitations on gas exchange and on growth are very common in trees, but the first is not the cause for the latter. This left the mechanisms responsible for the age-related decline in tree and forest productivity unresolved.

It is noteworthy that both the respiration and the carbon assimilation hypothesis ultimately predicted that trees become carbon limited as they grow tall and mature. Such a prediction is in stark contrast with increasing evidence that mature trees growing under ambient atmospheric CO_2 concentrations are not carbon limited (Körner 2003, see below). Surprisingly, up to very recently, this evidence was almost entirely ignored in the debate, perhaps because the implicit prediction that reduced available carbon would lead to C-limitation for growth was assumed to hold true.

There is now evidence that reduced turgor during cell expansion at the top of tall trees is responsible, at least in part, for reduced cell extension and growth (Koch et al. 2004; Woodruff et al. 2004; Ishii et al. 2008; Meinzer et al. 2008; Woodruff and Meinzer 2011). This and the lack of support for the respiration and the hydraulic limitation hypotheses as initially stated led to questioning of the critical assumption that growth in tall trees is limited by carbon availability (Ryan et al. 2006) and to suggestions that factors limiting cell growth and carbon demand may be ultimately responsible for the reduced growth rates in tall trees.

Although support for the carbon assimilation and respiration hypotheses is lacking and current research invokes mechanisms independent of carbon availability to explain the age- and size-related decrease in tree productivity, the notion that carbon availability limits growth in tall or old trees remains common. This is due in part to the scarcity of actual data on changes of non-structural carbon availability as trees increase in size and age.

3 Mobile C-Reserves in Trees

About half of all dry plant biomass consists of C, rendering C the most abundant plant nutrient. In addition to current photosynthesis, plants temporarily rely on stored C-reserves in order to maintain their metabolism. During periods of high photosynthetic activity, a fraction of recently assimilated C is generally directly fed into C storage pools. This process might be an active formation of C-stores, that otherwise would be used for C-sink activities (e.g. respiration, growth), or a passive accumulation of surplus photoassimilates during periods when photosynthate production exceeds the total C requirements of a plant (Chapin et al. 1990). Hereafter, we refer to all compounds that are orderly degraded and re-synthesized in response to changes in C source-sink-balances as 'mobile C compounds'. In principle, this definition includes all compounds that serve any cell function (in addition to storage), but which can be converted (recycled) to provide additional C-sources. However, the most important and ubiquitous mobile C compound is starch, which is exclusively synthesized for storage. Besides starch, other strict storage compounds are fructans and neutral lipids. While fructans are used alternatively to starch in some species, neutral lipids are considered to serve mainly as long-term C-stores, due to the relatively high energy demand for re-mobilization of fatty acids (Hoch et al. 2002). Low molecular weight sugars are highly mobile (readily 'available' for C-sink processes) and therefore are considered important C-reserves in plants.

However, their function as metabolic intermediates and osmotically active compounds (e.g. Arndt et al. 2008) clearly restricts a complete exhaustion of free sugars in living tissue. The frequency with which C-reserves are used and re-synthesized varies from diel (day-night-cycle of leaf-starch) to seasonal (e.g. winter dormancy in temperate climates) to episodic (e.g. re-flushing after leaf-loss to herbivores) cycles. Finally, the extent to which plants depend on mobile C compounds differs among life forms (e.g. annual vs. perennial plants) and environmental conditions (e.g. seasonal vs. non-seasonal climates).

Trees are distinguished from other functional plant types by a high proportion of heterotrophic tissue (i.e. wood), leading to higher C-demand per leaf area basis compared to other plant types (Stevens and Fox 1991), although a significant fraction of the wood consists of dead, non-respiring tissue (heartwood, Larcher 2003). In addition, hydraulic constraints in tall trees reduce leaf gas exchange and net-photosynthesis at the leaf level (see Ryan et al. 2006). Therefore, trees are generally considered to be more susceptible to C-limitation than most other plant types (Stevens and Fox 1991; Kozlowski and Pallardy 1997), emphasizing the importance of C-storage in this plant functional group, especially in mature, tall individuals. In seasonal climates where trees are periodically dormant (e.g. during winter or the dry season), stored C-reserves are needed to maintain cell function. Leaf flushing in deciduous trees depends on C stored during the previous season because of the lack of photosynthetically active tissue at bud break (Landhäusser and Lieffers 2003; Schädel et al. 2009). In evergreen trees of temperate climates, shoot growth is also partially supplied from C-reserves, which accumulate in mature leaves prior to bud break (Fischer and Höll 1991). Fruiting and especially mast fruiting has been traditionally thought to resemble a major sink for stored C-reserves in trees (Kozlowski and Pallardy 1997). However, current research suggests that the formation of fruits is largely supplied by current photoassimilates and thus independent of the import of stored C, even in masting tree species (Hoch 2005; Hoch and Keel 2006, and unpublished data by A. Sala and G. Hoch).

To the extent that mobile C-compounds resemble a longer term ‘buffer’ between C-source and -sink activities, their tissue concentrations mirror (at least to a certain degree) the net-balance between C-acquisition by photosynthesis and the sum of C used for biosynthesis and respiration (Fig. 11.1). Thus, deliberate comparative analyses of mobile carbon concentrations may be used as a proxy for the C-supply status of a plant. Clearly, the C-source and -sink activities are not independent processes, but are mutually influenced by complex and still not completely clarified feed-back and feed-forward mechanisms (Smith and Stitt 2007). Progressively decreasing C-reserve concentrations will ultimately induce negative effects on C-sink activities (i.e. C-limitation), while a sustained C surplus will increase the tissue concentration of mobile C compounds to a threshold at which negative feed-back mechanisms will lead to a down-regulation of photosynthesis (Fig. 11.1). Therefore, a plant’s mobile C pool will neither be completely exhausted at C limitation, nor increase infinitely in situations of C oversupply, but its size will respond within certain limits to changing C source-sink relations. Evidence of the responsiveness of mobile C-stores to C source-sink imbalances suggests the validity of this approach.

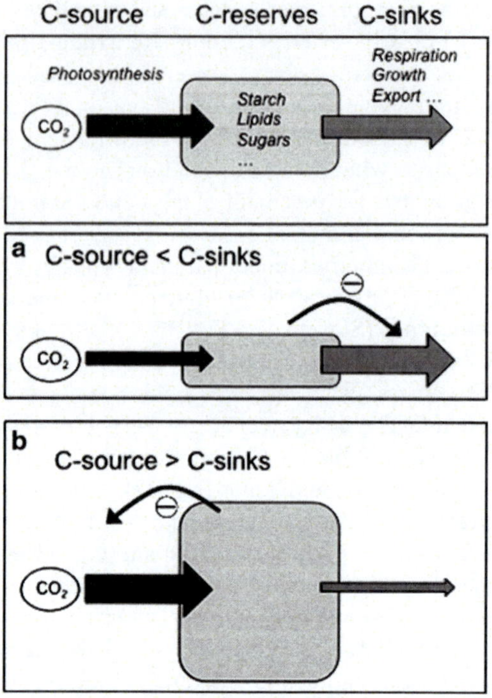

Fig. 11.1 Schematic illustration of the potential use of mobile carbon reserve concentration as indicators of carbon balance between sources and sinks. The thickness of the *arrows* is proportional to the amount of activity. Potential negative feedback and feedforward mechanisms are indicated by the *circled* minus sign. When sink activity is greater than source activity, the concentration of stored C pools is expected to decrease. Eventually, such decrease may limit sink activity and carbon demand (*panel a*). Conversely, when source activity is greater than sink activity, the concentration of stored C pools is expected to increase, which could down-regulate source activity (*panel b*)

For example, tissue concentrations of starch and low molecular weight carbohydrates generally increase in plants exposed to elevated atmospheric CO_2 (e.g. Teng et al. 2006) and decrease in plants growing at low CO_2 concentrations (e.g. Agüera et al. 2006). Similarly, scion-trunk girdling of 3-year-old *Prunus* trees induced higher concentrations of starch in leaves and shoots and decreased starch concentrations in roots (Jordan and Habib 1996). Finally, non-structural carbohydrate concentrations have been shown to decrease in plant tissues after experimental defoliation (Canham et al. 1999; Li et al. 2002) and shading (Piper et al. 2009).

So far, comparative analyses of C-reserve concentrations have been mainly used to investigate the extent to which insufficient photosynthetic C-supply explains the limits of tree growth in cold climates. A series of studies along altitudinal gradients from lower elevations to the alpine treeline found no indication of decreasing tissue concentrations of mobile C-reserves in treeline-forming trees (e.g. Hoch and Körner 2003, 2005; Hoch et al. 2002; Piper et al. 2006; Shi et al. 2006, 2008). Rather, there is a common trend of increasing concentrations of C-stores with elevation, which suggests C-sink limitation, rather than a limitation of C-source activities, as the physiological mechanism behind cold climate treeline formation. Increased concentrations of mobile C reserves at low, growth-limiting temperatures were also found in tree seedlings under controlled greenhouse-conditions (Domisch et al. 2001; Solfjeld and Johnsen 2006; Hoch and Körner 2009), in conifer seedlings transplanted to different altitudes across the treeline ecotone (Bansal and Germino 2008), as well as in dwarfed spruce trees growing on permafrost amidst tall forest trees (Hoch 2008). Similar to the above described studies along altitudinal gradients, comparative analyses of tissue concentrations of mobile C compounds in trees of different size classes can be employed to estimate the C supply status of trees and to elucidate the question whether insufficient C-balance explains the decreasing productivity of trees approaching their maximum height (Sala and Hoch 2009).

4 Existing Evidence for Sufficient C-Supply at Current Atmospheric CO_2 Concentrations

Independent of the comparative approach mentioned above, there is existing evidence for sufficient C supply of mature trees at current atmospheric CO_2 concentrations. For example, year-round high concentrations of mobile C reserve pools were described for mature trees in temperate (Hoch et al. 2003) and tropical (Würth et al. 2005) climates. Hoch et al. (2003) calculated that the mobile C reserves stored as starch and low molecular weight sugars in the aboveground wood of mature deciduous trees at a temperate mixed forest, would be sufficient to re-place the entire leaf canopy four times (disregarding the high degree of C-autonomy of developing leaves, which strongly decreases the effective need for stored C). Unlike mobile N- or P-compounds, mobile C-compounds are often not recovered from senescent leaves prior to leaf fall (but see Newell et al. 2002), and a considerable fraction of the starch pool stored in wood is likely never re-mobilized in trees, suggesting that parts of the starch synthesized by trees is sequestered rather than stored (Millard et al. 2007). Consequently, Körner (2003) and Millard et al. (2007) argued that the large pools of stored C-reserves in mature trees reflect an overall sufficient C-supply at ambient atmospheric CO_2 concentrations (which have increased by over 30% since the beginning of the industrial revolution in the middle of the eighteenth century, from c. 280 ppm CO_2 to over 385 ppm CO_2 in the present). If we consider the changes in atmospheric CO_2 over relatively recent geological times, the presumed overabundant C-supply of trees at present times seems logically consistent. For instance, Antarctic ice core analyses by Petit et al. (1999) and Siegenthaler et al. (2005) revealed that during the last 650 000 years the CO_2 concentration in the earth's atmosphere has never significantly exceeded 300 ppm (with minima as low as c. 180 ppm during the peaks of the ice ages). Further, the present CO_2 concentrations are likely the highest over the past 20–25 Ma (Pearson and Palmer 2000; Pagani et al. 2005). Thus, most modern plant species evolved at CO_2 concentrations markedly lower than today, and may have evolved to maximize the uptake and use of this 'scarce' nutrient (Körner 2006). Millard et al. (2007) even suggested that if CO_2 was limiting growth during plant evolution, the relatively inefficient carboxylation function of Rubisco would have been improved, which has not been the case. Instead, they suggested that Rubisco may serve as a form of nitrogen storage, which may lead to Rubisco concentrations in leaves even higher than actually needed to sustain a favorable carbon balance. In conclusion, the large C-reserve pools found in mature trees might reflect the current overabundance of atmospheric CO_2. In turn, this could also mitigate potential increases of carbon demands as trees grow old and tall.

5 Does Carbon Supply Become Limiting as Trees Increase in Size?

The approach used to asses C-balance at treeline mentioned above, was recently used by Sala and Hoch (2009) in ponderosa pine (*Pinus ponderosa*) to examine whether carbon supply limits growth as mature trees increase in size. To our knowledge, this and the recent work of Genet et al. (2010) remain the only published data to date on changes in stored C with size in mature trees. In contrast to the predictions of the respiration and carbon assimilation hypotheses, but consistent with the argument that mature trees are not carbon limited, Sala and Hoch (2009) found that size/height-related growth reductions were associated with significant increases of non-structural carbohydrates (NSC) in needles (not shown), branches and bole sapwood (Fig. 11.2) as well as of lipids in branch sapwood (Fig. 11.3) once active growth had ceased (mid-August). Although lipids in the bole decreased significantly with height and size, this did not necessarily indicate insufficient C-supply of tall trees (cf. Sala and Hoch 2009). These results not only suggest that carbon supply does not limit growth in ponderosa pine as trees increase in size, but that net C-supply increases with tree height. Further, we also found that increases of the concentration of mobile carbon compounds with tree height were stronger in a drier site relative to a moister site (Figs. 11.2 and 11.3), indicating that water availability might negatively impact C-sink activities (e.g. growth) more than net-photosynthesis.

In a study of relative carbon allocation to growth, storage and reproduction in European beech (*Fagus sylvatica*) and sessile oak (*Quercus petraea*) Genet et al. (2010) found that NSC concentrations in stems did not change in trees from stands between 50 and ca. 150–200 years old. In contrast, concentrations in beech roots decreased as stand age increased from 50 to 200 years old. Overall, however, estimated whole tree NSC concentrations and carbon allocation to storage was constant (oak) or slightly increased (in beech) with stand age. These results led the authors to interpret that carbon investment in storage with stand age was the cause for an observed decrease of carbon allocation to growth. However, we suggest (see above) that the age-related growth decline could result from hydraulic constraints, independently of carbon availability. If so, carbon investment into storage could be a consequence, rather than a cause of limited growth. Ultimately, both studies (Sala and Hoch 2009; Genet et al. 2010) show no age- or size-related declines of mobile C

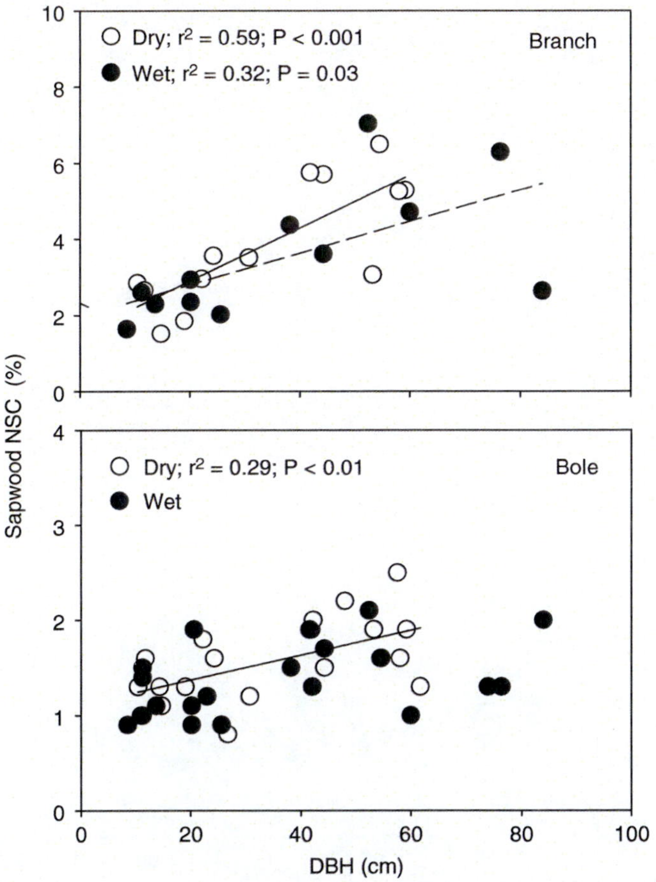

Fig. 11.2 Sapwood non structural carbohydrate (*NSC*) concentration in terminal, upper canopy branches (*upper panel*) and in the main trunk (bole; *lower panel*) of *Pinus ponderosa* as a function of tree diameter at breast height (*DBH*). Only the outer ten rings of sapwood were analyzed in the bole samples. Data were collected in August 2007 in a relatively moister site (wet, *black circles*) and drier site (dry; *open circles*) in Western Montana, USA (E-L Ranch site). *Lines* denote significant ($P<0.05$) linear relationships within each site (Redrawn from Sala and Hoch 2009)

reserves, which we interpret as evidence that carbon availability does not decrease with tree size or age.

In mature coast redwoods (*Sequoia sempervirens*; among the tallest trees on earth) Bentrup (2009) found increases, no change, or decreases of NSC with tree height depending on the tissue analyzed and the season of the year, with a tendency for an increase in NSC concentrations with tree height at the end of the season. Such mixed results led Bentrup (2009) to tentatively conclude that the tallest trees may temporarily be carbon limited early in the growing season. However, as we discuss below, seasonal trends of NSC concentrations need to be interpreted carefully due to potential confounding effects of phenology.

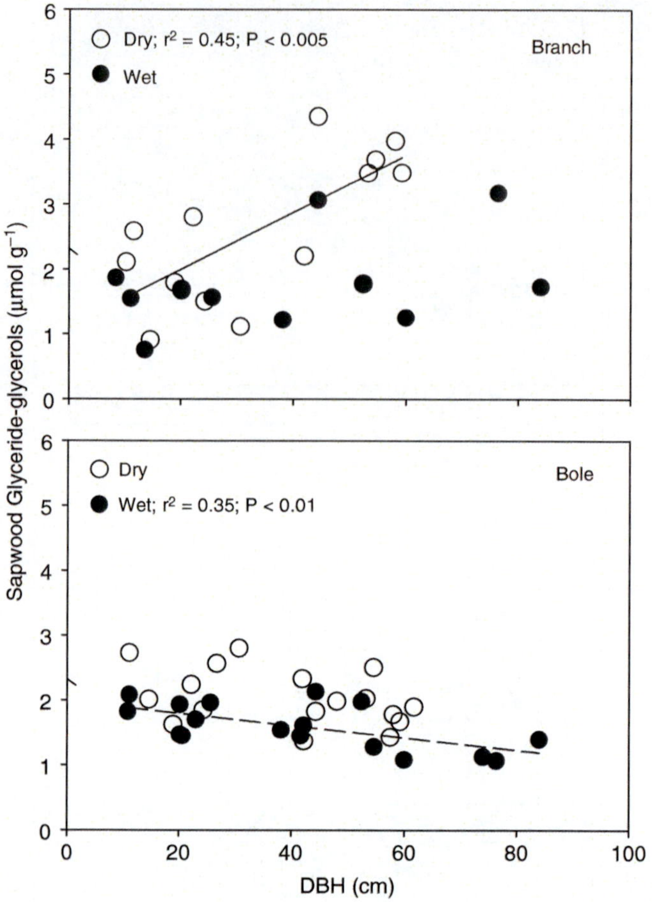

Fig. 11.3 Sapwood lipid concentration (measured as glyceride-glycerols) in terminal, upper canopy branches (*upper panel*) and in the main trunk (bole; *lower panel*) of *Pinus ponderosa* as a function of tree diameter at breast height (*DBH*). Only the outer ten rings of sapwood were analyzed in the bole samples. Data were collected in August 2007 in a relatively moister site (wet, *black circles*) and drier site (dry; *open circles*) in Western Montana, USA (E-L Ranch site). *Lines* denote significant ($P<0.05$) linear relationships within each site (Redrawn from Sala and Hoch 2009)

In addition to the published studies outlined above, here we present preliminary data for mature trees of two additional species: Douglas-fir (*Pseudotsuga menziesii*), a common and wide-spread species in western North America, and whitebark pine (*Pinus albicaulis*) a high elevation, five needle pine in the Rocky and Cascade Mountains and the Sierra Nevada. Along with Douglas-fir, we also collected supplementary data for ponderosa pine. Data for Douglas-fir and ponderosa pine were collected during the summer (mid-August) of 2009 on a dry, steep South-facing slope in the Rattlesnake Valley of western Montana (46.6°N and 113.2°W; 1,180 m elevation). The site has sparse trees which allowed us to sample isolated trees of

Table 11.1 Species sampled at each site, number of individuals sampled, DBH (cm), height (m) and age (yr) ranges

Species	Site	Number of trees	DBH range	Height range	Age range
Ponderosa pine	EL Ranch				
	Moist	12–20	8.5–84	3.9–40	15–126
	Dry	12–20	10.4–61.7	5.7–37.2	18–110
	Rattlesnake	10	10.8–67.2	5.5–32.7	n.m.
Douglas-fir	Rattlesnake	7	11.7–69	6.1–22.1	n.m.
Whitebark pine	Flint Creek	9	19–53.5	9.5–17.2	63–157

Tissues sampled were terminal branch sapwood (1- and 2-yr-old segments) at EL Ranch, bole sapwood at DBH (ten most recent rings) at EL Ranch and Rattlesnake, and 1-yr-old needles, terminal branch sapwood (6–8-yr-old segments) and bole sapwood at DBH (all sapwood; 9–90 rings) at Flint Creek. Data for Ponderosa pine at the EL Ranch site are from Sala and Hoch (2009) and are shown here for comparative purposes. n.m. = not measured

different sizes (Table 11.1). Mean annual air temperature in Missoula, Montana (ca. 5 km from the site) during the last 10 years is 8.1°C, with mean January and July temperatures of −2.4°C and 21.1°C, respectively. Mean annual precipitation is 417 mm.

Data for whitebark pine were collected at the end of the growing season (end of September) in 2007 at the Flint Creek range of western Montana (46.2°N and 113.3°W; 2,300 m elevation) in a mixed stand dominated by whitebark pine and subalpine fir (*Abies lasiocarpa*). Mean annual air temperature at the closest weather station (Anaconda, Montana, ca. 1,700 m elevation) is 6.1°C, with mean January and July temperatures of −4.5°C and 17.6°C, respectively. Mean total annual precipitation is 350 mm. Although the Flint Creek area is generally dry, the specific site is located on a local plateau, where the sampled stand surrounds a relatively moist meadow with an ephemeral spring dissecting it. For simplicity, we only show the data from samples collected in 2007, a low cone year, although data for other years shows similar patterns. At this site, we sampled nine trees with no or minimal shading effects by neighboring trees (Table 11.1).

Details on tree size, age and tissues sampled at each site are given in Table 11.1. All samples were immediately stored in a cooler with ice. After return to the laboratory (on the same day of sampling), all samples were microwaved (i.e. shock-heated) to interrupt all enzymatic activity. The bark and phloem was removed from bole and branch samples. For bole sapwood, the ten most recent rings were analyzed in all cases, except for whitebark pine, where all sapwood was analyzed (Table 11.1). All samples were oven dried for 2 days at 65°C, ground to a fine powder using a Wiley mill, and analyzed for NSC. For whitebark pine, samples were also analyzed for total acylglycerols (lipids). Chemical analyses for NSC and lipids followed the procedures described in Sala and Hoch (2009). In whitebark pine, which often has a unique growth form with multiple trunk individuals, DBH is a good indicator of overall tree size (biomass; Callaway et al. 2000). For instance, at our site trees are relatively short with rounded canopies and height is less meaningful than DBH. Because of this, and because DBH is correlated with age and height in ponderosa

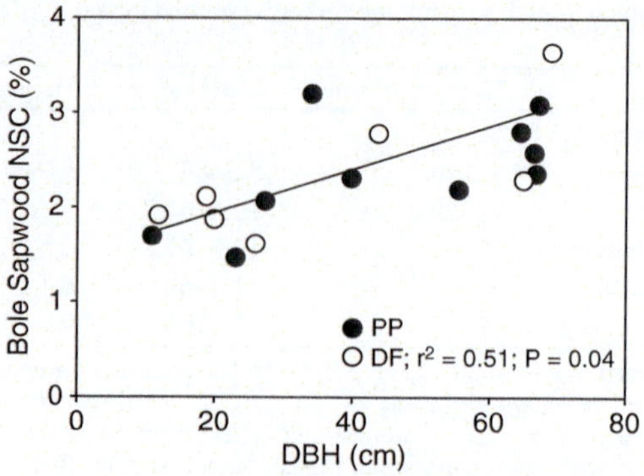

Fig. 11.4 Bole sapwood NSC concentrations in *P. ponderosa* (PP) and *Pseudotsuga menziesii* (DF) as a function of tree diameter at breast height (*DBH*). NSC concentrations were measured in the outer ten rings of sapwood. Trees were sampled in mid-August 2009 in the Rattlesnake Valley of western Montana. *Lines* denote significant ($P<0.05$) linear relationships within a species

pine and Douglas-fir, we opted to present all data as a function of DBH. However, results did not change when expressed as a function of height (not shown).

In ponderosa pine and Douglas-fir NSC concentrations in the outer sapwood tended to increase with DBH (Fig. 11.4), which corroborates our previous conclusion (Sala and Hoch 2009). However, in contrast to the drier site (E-L Ranch) sampled by Sala and Hoch (2009) during an exceptionally hot and dry year, the increase of NSC with tree size in ponderosa pine at the Rattlesnake site was not statistically significant (precipitation during the 2009 summer was above average). Apparently, both short- and long-term differences in water availability influence NSC concentrations and changes with tree size. In no case, however, did NSC decrease with tree size as expected based on predictions derived form the original hydraulic limitation hypothesis.

In whitebark pine neither NSC in bole and branch sapwood and in needles, nor lipids in branch and bole sapwood (Fig. 11.5) decreased with increasing tree diameter. On the contrary, there was a statistically significant increase of lipids in branch sapwood with tree age ($r^2=0.52$ $P=0.04$; not shown). Note, however, that these results are for the entire sapwood and we have not systematically measured changes of NSC concentrations with sapwood depth. However, bole sapwood NSC in our data did not change significantly as a function of sapwood depth or number of rings (not shown), suggesting little change in NSC concentrations with sapwood depth. Overall, results for whitebark pine are consistent with our previous conclusion that carbon supply does not become increasingly limited as trees increase in size.

Besides mature trees, changes in NSC concentrations with plant size have also been examined in saplings of *Abies balsamea* (a shade tolerant tree), *Acer rubrum* (intermediate shade tolerance) and *Pinus strobus* (a pioneer, shade-intolerant tree) in a moderately deeply shaded environment (up to 13% full sunlight) in Minnesota (Machado and Reich 2006). Saplings in this study were small (up to 1.3 m in height) and ranged from 6–8 to 21–24 years of age, depending on the species. The small

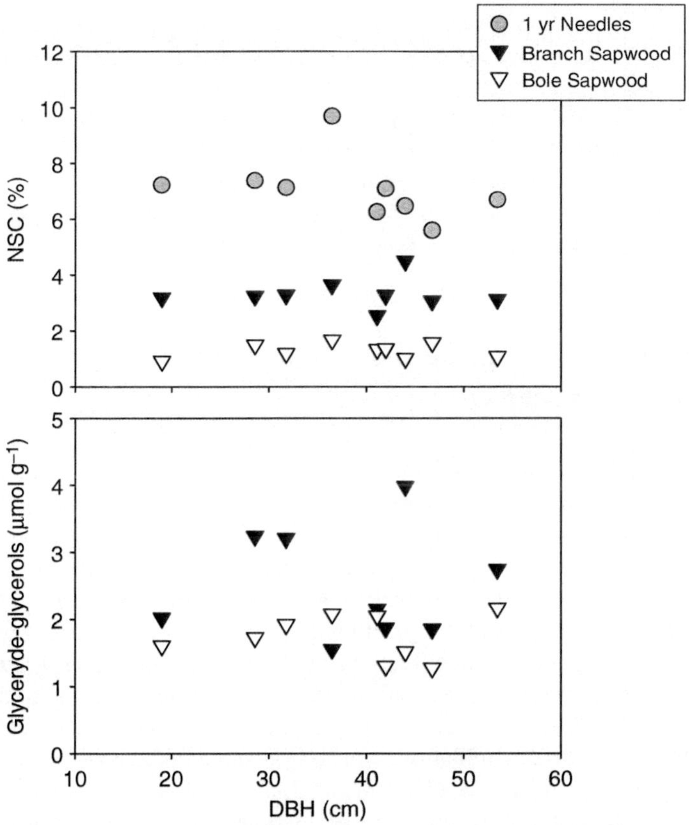

Fig. 11.5 NSC concentrations (*upper panel*) and total lipid concentration (measured as glyceryl-glycerols; *lower panel*) in whitebark pine (*Pinus albicaulis*) as a function of diameter at breast height (*DBH*). *Gray circles*: 1 year old needles; *black triangles*: terminal branch sapwood; *white triangles*: bole sapwood. All sapwood was analyzed in bole samples (see text). Lipids were not analyzed in needles. Trees were sampled in late September 2007 at the Flint Creek range of western Montana

stature of the saplings combined with the moist, shaded environment, suggests that hydraulic constraints on gas exchange were not an issue. However, carbon acquisition was likely constrained by low light, and changes in NSC with sapling size provide insight into size-related carbon source-sink dynamics. NSC concentrations in this study were measured in leaves, stems and roots at the end of the summer, which allowed estimation of whole plant NSC concentration as a function of plant size. While patterns of NSC concentration with sapling size within a single species were not necessarily consistent among different tissues, whole plant NSC concentrations did not change significantly with sapling size in *Abies* or *Acer* (with high and intermediate shade tolerance, respectively), but it significantly decreased in *Pinus* (shade-intolerant). To our knowledge this is the only published convincing evidence of increasingly limited carbon supply via photosynthesis with increases in plant size.

Such carbon limitation, however, is to be expected in this particular case involving a pioneer, light-demanding species growing in a deeply shaded environment. Notably, no such pattern was found in the more shade tolerant species equipped to function under low light environments. It remains to be seen whether the size-dependent decreases of NSC concentrations observed in *Pinus* saplings also occur once trees reach the canopy and the upper foliage becomes exposed to full light. Our previous (Sala and Hoch 2009) and current results in mostly isolated, mature trees suggest these patterns disappear or even reverse.

6 Methodological Considerations and Future Research Questions

Assuming that the concentration of stored carbon compounds at the end of the season reflects the end balance between carbon demand for earlier growth, and carbon supply by photosynthesis, data available so far suggest that the relative carbon supply in mature trees does not become increasingly limited as they grow larger. However, the limited available data preclude broad generalizations and highlight the need for further research to address pending questions. First, additional research is needed to elucidate the mechanisms responsible for the decline in their growth rates as trees become larger and older, including additional work on the role of turgor in cell expansion (Woodruff and Meinzer 2011), genetic and ageing factors (Day and Greenwood 2011), nutrient limitation, and long distance transport of assimilates. The latter is a major void in our understanding of tree physiology under environmental stress (Sala et al. 2010) and may be involved in growth declines as trees grow old and large (Woodruff and Meinzer 2011). With respect to tree carbon dynamics, there are also many open questions. For instance: (1) in what tissues and at what times during the year is the concentration of mobile carbon compounds most likely to reflect actual imbalance between sources and sinks (see Bansal and Germino 2008)?; (2) are height-related changes in stored carbon pools species-specific?; (3) how does environmental stress influence relative source and sink activities?; (4) when and to what extent does carbon allocation to reserve and other functions compete against growth as a carbon sink?; (5) does stored mobile carbon in trees become inaccessible? The latter question is very relevant as there is some evidence that plants fail to use stored C pools even under stress (Chapin et al. 1990; Li et al. 2002; Piper et al. 2009, but see Vargas et al. 2009), which suggests that these pools become unavailable over time (Chapin et al. 1990; Millard et al. 2007; Srichuwong and Jane 2007), and thus, by definition, are no longer mobile carbon pools. The question remains whether such inaccessibility varies over time and depends on stress (e.g. Vargas et al. 2009). If carbon reserve compounds become sequestered over time and there is a temporal mismatch between carbon supply by photosynthesis and carbon demand (i.e. times of maximum photosynthesis do not coincide with times of maximum carbon demand), then the issue of carbon limitation for certain functions still remains on the table. Therefore, it is crucial to understand the relative stress-sensitivity of carbon demand functions vs. carbon supply from photosynthesis and/or storage. However, current experimental evidence that changes in tissue concentrations of mobile carbon compounds reflect relative source-sink activities (see above) suggest that only surplus carbon is sequestered. Clearly, much research is needed to fully understand how tree size, environment, stress, and species-specific life history and ecology influence the balance between instantaneous carbon demand and carbon supply, active and passive allocation to storage, and accessibility of storage pools.

The concentration and seasonal dynamics of mobile carbon compounds varies depending on the organ and the species-specific phenological development (e.g. Barbaroux and Bréda 2002; Newell et al. 2002; Hoch et al. 2003; Sayer and Haywood 2006). Such variation suggest that an accurate picture of source-sink dynamics in trees requires detailed seasonal measurements of phenology in major organs (leaves, branches, roots and stems) and their corresponding mobile carbon compound concentrations. When such sampling intensity is unfeasible, the decision on when and what to sample becomes important. If access to stored carbon pools to meet instantaneous carbon demands is not an issue, once seasonal growth is completed, the residual mobile carbon compound concentration should reflect the end balance (deficit, surplus, or no change; see above). In this case, if we are interested in evaluating whether carbon supply becomes increasingly limiting for growth as trees increase in size, sampling tissues that significantly contribute to total stored carbon pools at the end of the active growing season is best. Further, because the concentrations of mobile C-reserves in plants are the net-outcome of multiple active (e.g. reserve formation and use) and passive (e.g. accumulation of overabundant photoassimilates) processes, patterns of change in NSC concentration with size are very sensitive to timing and phenology and need to be interpreted with caution. For instance, in deciduous and evergreen temperate tree species, transient, fast changes of C-reserve concentration have been shown to occur just prior and after bud break (Barbaroux and Bréda 2002; Hoch et al. 2003; Schädel et al. 2009). Therefore, comparative studies on changes of NSC concentrations in response to factors other than phenology need to rule out potential confounding effects due to phenology. Otherwise, even slight temporal phenology shifts (such as the time of leaf flushing from saplings to mature trees; Mediavilla and Escudero 2009), could induce differences in mobile C reserve concentrations which are unrelated to the integrated net C-balance.

In deciduous species, roots and main stems generally contain the largest fraction of the total pool of stored carbon (Kozlowski and Keller 1966; Barbaroux et al. 2003). In contrast, needles usually represent the largest C-storage pool in conifers (Hoch et al. 2002; Li et al. 2002), although roots and stems also contribute significantly. However, the strong and rapid fluctuations of NSC concentrations in needles of conifers over short-time scales, particularly around bud break (see above) may complicate the interpretation of age-related changes in the concentration of stored carbon pools. Despite the fact that trunks and main stems may constitute a very large fraction of the total stored carbon pools due to their large biomass, in most trees NSC concentrations in trunks and main stems are lower relative to branches, fine roots and foliage, and tend to exhibit the lowest seasonal variation (Hoch et al. 2003). This suggests that trunks and main stems are less dynamic storage organs, whose NSC concentration reflects longer term carbon source-sink balance. This and the fact that they represent a significant fraction of the total stored C pool, makes them good candidates to evaluate tree size effects on carbon source-sink dynamics. A complication when using stems, however, is that concentration of mobile carbon compounds often varies from the outer to the inner sapwood, and accessibility to these different compartments may vary. Therefore, the specific sapwood fraction sampled is also an important consideration for the interpretation of results, particularly in species with large amounts of sapwood. While in many species NSC concentration is highest in the outer sapwood and decreases with distance from the bark (Barbaroux and Bréda 2002; Hoch et al. 2003), in other species the reverse is true (Newell et al. 2002; Pruyn et al. 2005). In pines, lipid concentrations have also been show to increase from the outer to the inner sapwood (Saranpää and Nyberg 1987). In species with large amounts of sapwood, larger trees may have much more sapwood than smaller trees. In this case, if the concentration of NSC is greatest in the outer sapwood and declines appreciably as a function of distance from the bark, then the larger amount of sapwood in larger trees may cause a dilution effect leading to an apparent decrease of NSC with tree size. Yet results from the outer sapwood may not show any change or may even show an increase with tree size. Because of this, it may be best to standardize analysis to specific radial portions of the sapwood (i.e. a specific number of tree rings from the cambium) where accessibility is highest (e.g. next to the cambium), particularly in species with strong radial changes in sapwood NSC or lipid concentration and with large sapwood.

7 Conclusions

The two classical hypotheses to explain size/age- related declines of growth rates in trees and stands predicted that carbon availability for growth becomes increasingly limiting as trees increase in size (due either to progressive increases in carbon sinks other than growth or to decreases carbon sources, respectively). Although the validity of these hypotheses is now questioned and new alternative explanations have been proposed, actual data on size-dependent changes of stored mobile carbon in mature trees as indicators of carbon balance between source and sink activities is surprisingly limited. This has contributed to a common default assumption that carbon availability limits tree growth in large, old trees. This assumption conflicts with increasing evidence that mature trees under current atmospheric CO_2 concentrations have substantial amounts of stored carbon and are not carbon limited. To date, convincing evidence showing a decrease of the relative carbon supply with increasing tree size exists only for small saplings of shade-intolerant species in deeply shaded environments. In contrast, and assuming that concentration of stored carbon compounds at the end of the season reflects the end balance between carbon demand for earlier growth and carbon supply by photosynthesis, available data to date for four different evergreen temperate conifers (*Pinus ponderosa, P. albicaulis, Pseudotsuga menziesii* and *Sequoia sempervirens*) and two broadleaved trees (*Fagus sylvatica* and *Quercus petraea*) suggest that the relative carbon supply in trees does not become increasingly limited as they grow larger. On the contrary, in some cases, the data indicate an increasing oversupply of carbon with increasing size. Clearly, much research is still needed to understand how tree size, environment, stress, and species-specific life history and ecology influence the balance between instantaneous carbon demand and carbon supply, the allocation to storage, and the accessibility of storage pools.

In spite of many uncertainties, research to date points to the need to question the historically carbon-centric mechanisms proposed to explain age-related growth declines in trees and forests. Future research should focus on whether and when alternative growth limiting factors (e.g. turgor, long distance transport of assimilates, nutrients) may contribute to growth limitations in tall trees.

Acknowledgements Two anonymous reviewers provided very useful comments to improve this manuscript. Funding for this research was provided in part by NSF (DEB- 05–15756) and subsequent REU supplement to support WF.

References

Agüera E, Ruano D, Cabello P, de la Haba P (2006) Impact of atmospheric CO_2 on growth, photosynthesis and nitrogen metabolism in cucumber (*Cucumis sativus* L.) plants. J Plant Physiol 163:809–817

Arndt SK, Livesley SJ, Merchant A, Bleby TM, Grierson PF (2008) Quercitol and osmotic adaptation of field-grown *Eucalyptus* under seasonal drought stress. Plant Cell Environ 31:915–924

Bansal S, Germino MJ (2008) Carbon balance of conifer seedlings at timberline: relative changes in uptake, storage, and utilization. Oecologia 158:217–227

Barbaroux C, Bréda N (2002) Contrasting distribution and seasonal dynamics of carbohydrate reserves in stem wood of adult ring-porous sessile oak and diffuse-porous beech trees. Tree Physiol 22:1201–1210

Barbaroux C, Bréda N, Dufrêne E (2003) Distribution of above-ground and below-ground carbohydrate reserves of adult trees of two contrasting broad-leaved species (*Quercus petraea* and *Fagus sylvatica*). New Phytol 157:605–616

Barnard HR, Ryan MG (2003) A test of the hydraulic limitation hypothesis in fast-growing *Eucalyptus saligna*. Plant Cell Environ 26:1235–1245

Bentrup CR (2009) Evidence for carbon limitations to height growth in the tallest trees. M.S. thesis, Northern Arizona University, Flagstaff, 42 p

Callaway RM, Sala A, Keane RE (2000) Succession may maintain high leaf area: sapwood area ratios and productivity in old subalpine forests. Ecosystems 3:254–268

Canham CD, Kobe RK, Latty EF, Chazdon RL (1999) Interspecific and intraspecific variation in tree seedling survival: effects of allocation to roots versus carbohydrate reserves. Oecologia 121:1–11

Chapin FS, Schulze ED, Mooney HA (1990) The ecology and economics of storage in plants. Annu Rev Ecol Syst 21:423–447

Coyea MR, Margolis HA (1992) Factors affecting the relationship between sapwood area and leaf area of balsam fir. Can J For Res 22:1684–1693

Day ME, Greenwood MS (2011) Regulation of ontogeny in temperate conifers. In: Meinzer FC, Lachenbruch B, Dawson TE (eds) Size- and age-related changes in tree structure and function. Springer, Dordrecht

Domisch T, Finér L, Lehto T (2001) Effects of soil temperature on biomass and carbohydrate allocation in Scots pine (*Pinus sylvestris*) seedlings at the beginning of the growing season. Tree Physiol 21:465–472

Fischer C, Höll W (1991) Food reserves of Scots pine (*Pinus sylvestris* L.) I. Seasonal changes in the carbohydrate and fat reserves of pine needles. Trees 5:187–195

Friend AD (1993) The prediction and physiological significance of tree height. In: Solomon AM, Shugart HH (eds) Vegetation dynamics and global change. Chapman & Hall, New York, pp 101–115

Genet H, Bréda N, Dufrêne E (2010) Age-related variation in carbon allocation at tree and stand scales in beech (*Fagus sylvatica* L.) and sessile oak (*Quercus petraea* (Matt.) Liebl.) using a chronosequence approach. Tree Physiol 30:177–192

Goldstein G, Andrade JL, Meinzer FC, Holbrook NM, Cavelier J, Jackson P, Celis A (1998) Stem water storage and diurnal patterns of water use in tropical forest canopy trees. Plant Cell Environ 21:397–406

Gower ST, McMurtrie RE, Murty D (1996) Aboveground net primary production decline with stand age: potential causes. Trends Ecol Evol 11:378–382

Hoch G (2005) Fruit-bearing branchlets are carbon autonomous in mature broad-leaved temperate forest trees. Plant Cell Environ 28:651–659

Hoch G (2008) The carbon supply of *Picea abies* trees at a Swiss montane permafrost site. Plant Ecol Divers 1:13–20

Hoch G, Keel SG (2006) 13 C labelling reveals different contributions of photoassimilates from infructescences for fruiting in two temperate forest tree species. Plant Biol 8:606–614

Hoch G, Körner C (2003) The carbon charging of pines at the climatic treeline: a global comparison. Oecologia 135:10–21

Hoch G, Körner C (2005) Growth, demography and carbon relations of *Polylepis* trees at the world's highest treeline. Funct Ecol 19:941–951

Hoch G, Körner C (2009) Growth and carbon relations of tree line forming conifers at constant vs. variable low temperatures. J Ecol 97:57–66

Hoch G, Popp M, Körner C (2002) Altitudinal increase of mobile carbon pools in *Pinus cembra* suggests sink limitation of growth at the Swiss treeline. Oikos 98:361–374

Hoch G, Richter A, Körner C (2003) Non-structural carbon compounds in temperate forest trees. Plant Cell Environ 26:1067–1081

Ishii H (2011) How do changes in leaf/shoot morphology and crown architecture affect growth and physiological function of large, old trees? In: Meinzer FC, Lachenbruch B, Dawson TE (eds) Size- and age-related changes in tree structure and function. Springer, Dordrecht

Ishii HT, Jennings JM, Sillett SC, Koch GW (2008) Hydrostatic constraints on morphological exploitation of light in tall *Sequoia sempervirens* trees. Oecologia 156:751–763

Jordan MO, Habib R (1996) Mobilizable carbon reserves in young peach trees as evidenced by trunk girdling experiments. J Exp Bot 47:79–87

King DA (2011) Size-related changes in tree proportions and their potential influence on the course of height growth. In: Meinzer FC, Lachenbruch B, Dawson TE (eds) Size- and age-related changes in tree structure and function. Springer, Dordrecht

Kira T, Shidei T (1967) Primary production and turnover of organic matter in different forest ecosystems of the western Pacific. Jpn J Ecol 13:70–83

Koch GW, Sillett SC, Jennings GM, Davis SD (2004) The limits of tree height. Nature 428:851–854

Körner C (2003) Carbon limitation in trees. J Ecol 91:4–17

Körner C (2006) Plant CO_2 responses: an issue of definition, time and resource supply. New Phytol 172:393–411

Kozlowski TT, Keller T (1966) Food relations of woody plants. Bot Rev 32:293–383

Kozlowski TT, Pallardy SG (1997) Physiology of woody plants, 2nd edn. Academic, San Diego

Kozlowski TT, Kramer PJ, Pallardy SG (1991) The physiological ecology of woody plants. Academic, San Diego

Landhäusser SM, Lieffers VJ (2003) Seasonal changes in carbohydrate reserves in mature northern *Populus tremuloides* clones. Trees 17:471–476

Larcher W (2003) Physiological plant ecology. Springer, Berlin

Li MH, Hoch G, Körner C (2002) Source/sink removal affects mobile carbohydrates in *Pinus cembra* at the Swiss treeline. Trees 16:331–337

Machado J-L, Reich PB (2006) Dark respiration rate increases with plant size in saplings of three temperate tree species despite decreasing tissue nitrogen and nonstructural carbohydrates. Tree Physiol 26:915–923

Mediavilla S, Escudero A (2009) Ontogenetic changes in leaf phenology of two co-occurring Mediterranean oaks differing in leaf life span. Ecol Res 24:1083–1090

Meinzer FC, Bond BJ, Karanian JA (2008) Biophysical constraints on leaf expansion in a tall conifer. Tree Physiol 28:197–206

Mencuccini M, Hölttä T (2010) The significance of phloem transport for the speed of link between canopy photosynthesis and belowground respiration. New Phytol 185:189–203

Mencuccini M, Hölttä T, Petit G, Magnani F (2007) Sanio's Laws revisited. Size-dependent changes in the xylem architecture of trees. Ecol Lett 10:1084–1093

Millard P, Sommerkorn M, Grelet GA (2007) Environmental change and carbon limitation in trees: a biochemical, ecophysiological and ecosystem appraisal. New Phytol 175:11–28

Newell EA, Mulkey SS, Wright JS (2002) Seasonal patterns of carbohydrate storage in four tropical tree species. Oecologia 131:333–342

Niklas KJ (2007) Maximum plant height and the biophysical factors that limit it. Tree Physiol 27:433–440

Pagani M, Zachos JC, Freeman KH, Tipple B, Bohaty S (2005) Marked decline in atmospheric carbon dioxide concentrations during the Paleogene. Science 309:600–603

Pearson PN, Palmer MR (2000) Atmospheric carbon dioxide concentrations over the past 60 million years. Nature 406:695–699

Peet RK (1980) Changes in biomass and production during secondary forest succession. In: West DC, Shugart HH, Botkin DB (eds) Forest succession concepts and application. Springer, New York, pp 324–338

Petit JR, Jouzel J, Raynaud D, Barkov NI, Barnola JM, Basile I, Bender M, Chappellaz J, Davis M, Delaygue G, Delmotte M, Kotlyakov VM, Legrand M, Lipenkov VY, Lorius C, Pepin L, Ritz C, Saltzman E, Stievenard M (1999) Climate and atmospheric history of the past 420,000 years from the Vostok ice core, Antarctica. Nature 399:429–436

Piper FI, Cavieres LA, Reyes-Diaz M, Corcuera LJ (2006) Carbon sink limitation and frost tolerance control performance of the tree *Kageneckia angustifolia* D. Don (Rosaceae) at the treeline in central Chile. Plant Ecol 185:29–39

Piper FI, Reyes-Diaz M, Corcuera LJ, Lusk CH (2009) Carbohydrate storage, survival, and growth of two evergreen *Nothofagus* species in two contrasting light environments. Ecol Res 24:1233–1241

Pruyn ML, Gartner BL, Harmon ME (2005) Storage versus substrate limitation to bole respiratory potential in two coniferous tree species of contrasting sapwood width. J Exp Bot 56:2637–2649

Ryan MG, Waring RH (1992) Maintenance respiration and stand development in a subalpine lodgepole pine forest. Ecology 73:2100–2108

Ryan MG, Yoder BJ (1997) Hydraulic limits to tree height and tree growth. Bioscience 47:235–242

Ryan MG, Binkely D, Fownes JH (1997) Age-related decline in forest productivity: pattern and process. Adv Ecol Res 27:213–262

Ryan MG, Binkley D, Fownes JH, Giardina CP, Senock RS (2004) An experimental test of the causes of forest growth decline with stand age. Ecol Monogr 74:393–414

Ryan MG, Phillips N, Bond BJ (2006) The hydraulic limitation hypothesis revisited. Plant Cell Environ 29:367–381

Sala A, Hoch G (2009) Height-related growth declines in ponderosa pine are not due to carbon limitation. Plant Cell Environ 32:22–30

Sala A, Piper F, Hoch G (2010) Physiological mechanisms of drought-induced tree mortality are far from being resolved. New Phytol 186:274–281

Saranpää P, Nyberg H (1987) Lipids and sterols of *Pinus sylvestris* L. sapwood and heartwood. Trees 1:82–87

Sayer MAS, Haywood J (2006) Fine root production and carbohydrate concentrations of mature longleaf pine (*Pinus palustris* P. Mill.) as affected by season of prescribed fire and drought. Trees 20:165–175

Schädel C, Blöchl A, Richter A, Hoch G (2009) Short-term dynamics of nonstructural carbohydrates and hemicelluloses in young branches of temperate forest trees during bud break. Tree Physiol 29:901–911

Scholz F, Phillips NG, Bucci SJ, Meinzer FC, Goldstein G (2011) Hydraulic capacitance: biophysics and functional significance of internal water sources in relation to tree size. In: Meinzer FC, Lachenbruch B, Dawson TE (eds) Size- and age-related changes in tree structure and function. Springer, Dordrecht

Shi P, Körner C, Hoch G (2006) End of season carbon supply status of woody species near the treeline in western China. Basic Appl Ecol 7:370–377

Shi P, Körner C, Hoch G (2008) A test of the growth-limitation theory for alpine tree line formation in evergreen and deciduous taxa of the eastern Himalayas. Funct Ecol 22:213–220

Siegenthaler U et al (2005) Stable carbon cycle-climate relationship during the late Pleistocene. Science 310:1313–1317

Smith AM, Stitt M (2007) Coordination of carbon supply and plant growth. Plant Cell Environ 30:1126–1149

Solfjeld I, Johnsen O (2006) The influence of root-zone temperature on growth of *Betula pendula* Roth. Trees 20:320–328

Srichuwong S, Jane JL (2007) Physicochemical properties of starch affected by molecular composition and structures: a review. Food Sci Biotechnol 16:663–674

Steppe K, Niinemets U, Teskey RO (2011) Tree size- and age-related changes in leaf physiology and their influence on carbon gain. In: Meinzer FC, Lachenbruch B, Dawson TE (eds) Size- and age-related changes in tree structure and function. Springer, Dordrecht

Stevens GC, Fox JF (1991) The causes of treeline. Annu Rev Ecol Syst 22:177–191

Teng NJ, Wang J, Chen T, Wu XQ, Wang YH, Lin JX (2006) Elevated CO_2 induces physiological, biochemical and structural changes in leaves of Arabidopsis thaliana. New Phytol 172:92–103

Tyree MT, Sperry JS (1989) Vulnerability of xylem to cavitation and embolism. Annu Rev Plant Biol 40:19–38

Vargas R, Trumbore S, Allen MF (2009) Evidence of old carbon used to grow new fine roots in a tropical forest. New Phytol 182:710–718

Whittaker RH, Woodwell GM (1967) Surface area relations of woody plants and forest communities. Am J Bot 54:931–939

Woodruff DR, Meinzer FC (2011) Size-dependent changes in biophysical control of tree growth: the role of turgor. In: Meinzer FC, Lachenbruch B, Dawson TE (eds) Size- and age-related changes in tree structure and function. Springer, Dordrecht

Woodruff DR, Bond BJ, Meinzer FC (2004) Does turgor limit growth in tall trees? Plant Cell Environ 27:229–236

Würth MKR, Pelaez-Riedl S, Wright SJ, Körner C (2005) Non-structural carbohydrate pools in a tropical forest. Oecologia 143:11–24

Yoder BJ, Ryan MG, Waring RH, Schoettle AW, Kaufmann MR (1994) Evidence of reduced photosynthetic rates in old trees. For Sci 40:513–527

Zhang Y-I, Meinzer FC, Hao G-Y, Scholz FG, Bucci SJ, Takahashi FSC, Villalobos-Vega R, Giraldo JP, Cao K-F, Hoffmann WA, Goldstein G (2009) Size-dependent mortality in a Neotropical savanna tree: the role of height-related adjustments in hydraulic architecture and carbon allocation. Plant Cell Environ 32:1456–1466

Zimmermann MH (1983) Xylem structure and the ascent of sap. Springer, Berlin

Part V
Water Relations and Transport Processes

Chapter 12
Comparative Criteria for Models of the Vascular Transport Systems of Tall Trees

Maurizio Mencuccini, Teemu Hölttä, and Jordi Martinez-Vilalta

Abstract During the past decade, several detailed models of plant vascular transport systems (i.e. xylem and phloem) have been presented in the literature, and many of them are currently capable of accurately modelling the hydraulic characteristics of trees, including tall trees. This marks a departure from earlier modelling exercises in plant water relations, when models were intended primarily to promote an understanding of the biophysical and physiological mechanisms of transport, but whose performance was comparatively poor when they were required to predict the behaviour of organisms spanning the logarithmic ranges from small seedlings to gigantic old trees. In addition, many of these modelling efforts have assumed, more or less explicitly, that a principle of optimality operates in the design of the transport systems, i.e. they have assumed criteria by which various aspects of the transport system should be optimised. Moreover, these models are characterised by very different approaches, structures and objectives, and differ significantly with regard to several other important characteristics. Because models formally organise our knowledge, we review them here, in the hope of highlighting the theoretical progress achieved so far and the challenges remaining in our understanding of the vascular transport systems of trees.

M. Mencuccini (✉)
School of Geosciences, University of Edinburgh, Crew Building, West Mains Road, Edinburgh, EH9 3JN, UK

Centre for Ecology Research and Forestry Applications (CREAF), Universitat Autonoma de Barcelona, Bellaterra, Cerdanyola, Barcelona, Spain
e-mail: m.mencuccini@ed.ac.uk

T. Hölttä
Department of Forest Ecology, University of Helsinki, Helsinki, Finland

J. Martinez-Vilalta
Centre for Ecology Research and Forestry Applications (CREAF), Universitat Autonoma de Barcelona, Bellaterra, Cerdanyola, Barcelona, Spain

F.C. Meinzer et al. (eds.), *Size- and Age-Related Changes in Tree Structure and Function*, Tree Physiology 4, DOI 10.1007/978-94-007-1242-3_12,

1 Introduction

Plants in general, and trees in particular, undergo tremendous transformations in size during growth and development. The biomass of a mature tree can be ten or more orders of magnitude larger than that of a small seedling and plant hydraulic systems must similarly be designed in such a way that their performance is not impaired by these tremendous increases in size (Mencuccini 2003). How this is achieved without incurring significant costs and limitations in their physiological performance is still far from clear, but the literature of the last 10 years, including modelling attempts, has paid much more attention to these issues than earlier literature. Hence, some reflections can be made. In much of this modelling literature, the main insights into how plant transport systems are designed (especially in the case of large organisms such as trees) have been obtained using ideas that owe much to the use of a principle of optimality, i.e. the principle that the design of a transport system must satisfy some pre-defined criterion of optimal performance, which translates into a maximization or a minimization function. We will look at these in detail later on.

While much of this theoretical and modelling literature has focussed on the upward transport of water from roots to leaves via the xylem of plants, significant progress has also been made in the understanding of downward solute plus water transport via plant phloem. Plants are integrated units, with sub-systems that must be closely co-ordinated in order for the organism to work well, hence examining one sub-system in isolation without proper consideration for the interactions with other sub-systems is unlikely to be very productive in the long run (e.g. Zimmermann and Brown 1971; Holbrook and Zwieniecki 2005). We will therefore summarise theoretical and modelling literature that contains valuable information on both xylem and phloem transport sub-systems.

Because of the large quantity of resources necessary to construct water and solute transport systems in large organisms, these findings have central importance in the understanding of the whole-plant physiology of tall trees. A tight co-ordination between plant assimilation, carbohydrate allocation (including storage) and respiration must be invoked to explain the enormous evolutionary and ecological success of trees in colonising and retaining dominance in large swaths of the terrestrial biosphere. In other words, our capacity to predict the 'ecosystem physiology' of forests over long time periods seems limited without a careful consideration of how short-term processes such as gas exchange and long-distance transport are coupled to long-term processes such as growth and development. Understanding tree transport systems has a central role for two reasons, (a) because transport systems of tall organisms are expensive to build and maintain, and (b) because their performance in delivering resources is inextricably linked to the acquisition of other resources by the tree, either in the leaves by photosynthesis, or by the roots in the form of nutrients to be transported to leaves.

An expanding number of studies have been published recently on the physiology of tall trees and much of it is reflected in the chapters of this book (e.g. Ishii 2011; McDowell et al. 2011; Sala et al. 2011; Steppe et al. 2011; Woodruff and Meinzer 2011). This literature is precious and absolutely necessary to test the behaviour of models and to suggest further areas in which current models are deficient. Because of space limitations, we do not review it unless absolutely necessary, and focus instead on the literature directly relevant to the modelling of the physiology of tall trees.

We focus initially on presenting the main principles that we will use to summarize the modelling literature concerned with the 'optimal' design of transport systems of tall trees. We then present examples of such theoretical and modelling advances for xylem and phloem transport. Finally, we look at the frontiers of this research and point out what remains to be done and what can be achieved in the near future.

2 Comparative Criteria for the Construction of Transport Models of Tall Trees

There is a very large literature on the use of optimality criteria in ecology. It is not our intention to review it here, not even briefly. We will only highlight those principles which we believe are important for the construction of models of transport systems for tall trees and which have already been discussed in different contexts in the ecological literature on optimality.

The modular construction of their bodies allows plants, and particularly trees, to increase their sizes over several orders of magnitude (Ishii 2011; King 2011). Correspondingly, plant functions must scale accordingly (i.e. isometrically or allometrically). For example, Mencuccini (2003) reviewed the literature on plant hydraulics and reported that the published data on hydraulic conductance, G_h (amount of water moved from soil to leaf per unit of time and per unit of drop in water potential, i.e. mmol H_2O s^{-1} MPa^{-1}), showed a log normal distribution profile, with changes in the magnitude of this parameter of five orders of magnitude (from around 0.01 to more than 100 mmol H_2O s^{-1} MPa^{-1}). The reason for this strongly right-skewed frequency distribution was due to the logarithmic (and equally varying) distribution of leaf areas and biomasses across plants.

Does this general increase in size, and specifically the need to build long-distance transport systems, result in other physiological or ecological trade-offs? Is there a specific challenge that tall trees have to face, compared to other 'minor' organisms within the plant kingdom? Or does increased size provide unexpected benefits and possibilities for compensations?

Increased stature results in a number of different processes, i.e. increased costs of construction of a long transport system, an accentuated impact of the hydrostatic effects of gravity, plus the hydrodynamic effects due to water movement in conduits. In addition, the modular construction itself leads to expanding crowns for much of the life cycle of trees, with the consequent need to transport ever greater amounts of water and solutes. On the other hand, a bulky structure may provide substantial benefits in terms of accumulated water (Scholz et al. 2011), sugar and nutrient storage capabilities and the possibility of a more effective exploration of above- and below-ground biological spaces.

There probably exist challenges for plants (and especially trees) both in the sense of constructing systems that can work under very different biophysical regimes as well as in the sense that the use and allocation of carbohydrates produced by photosynthesis is crucial to their present and future performance. While many of these challenges are not unique to the plant kingdom (large animals such as elephants in terrestrial environments or whales in marine ones are likely to face analogous construction challenges), the unique feature of trees (and plants in general) is in the direct link between the efficacy in the distribution of resources by the transport systems (e.g. water from root tips to canopy leaves) and the capacity of other organs of acquiring and processing food e.g. poorly designed xylem would affect stomatal opening and carbon uptake). Similar examples can be derived with regard to the feedbacks between photosynthesis and root functioning. This an important topic to which we will return later.

In part as a consequence of what was said above, there also exist challenges to scientists in understanding the functionality of these transport systems. Typically, plant physiologists and ecologists have focussed on developing models that explain the 'working' of a particular system, i.e. the mechanisms necessary for that system to work properly. In the modern literature, starting from Dixon and Joly (1895) for xylem transport and Münch (1927) for phloem transport, physiological investigations have long focussed on understanding the biophysical and physiological mechanisms by which transport occurred, with models being constructed to summarise this sort of knowledge. Comparatively scant attention was paid to how these principles applied when faced with the challenge of modelling transport systems of large organisms.

However, it quickly becomes obvious that, once one starts to employ these models in the context of understanding and predicting performance of both small and tall plants, very little insight is provided by them. Most of the time, one has to resort to empirical parameter calibration to adjust model performance across these vast ranges of allometric scaling. It is then apparent that, in order for physiological models to be applicable across scales of logarithmic magnitudes, different principles must be incorporated, in addition to the known physiological mechanisms. We refer to models that incorporate these scaling (or allometric) principles as *scalable*. A scalable model is one in which attention is paid to the allometry of the system, in an empirical or a semi-mechanistic fashion, by incorporating components which scale with size, thereby ensuring that the system behaves logically across logarithmic scales. The level of detail of these representations in a scalable model can vary, from simple representations of the overall biomass cost of plant transport systems, to those in which allometric scaling is represented at the anatomical level. We will see a few examples later on.

In addition to scalability, theoretical models differ in the way they represent plant efficiency in constructing transport systems. Strictly speaking, *efficiency* can be defined as the ratio of energy exiting a system (seen as a black box) divided by energy required, as an input, to make that system work. If the two forms of inputs and outputs can be put on the same units, then efficiency is unitless and varies between 0 and 1. Hence a car engine has an energy conversion efficiency typically of 10% to over 40%, wind turbines of up to 60%, solar panels of around 50% and plant photosynthesis of up to 6%. Expressing efficiency on common units allows broad comparability across systems. Unfortunately, this practice is rare in the plant physiological literature, and particularly so in the design of plant transport systems. In some cases, the efficiency of a transport system is defined as the ratio of the water transportable per unit of time and water potential difference divided by the amount of photosynthate required to construct it (or more rarely, to construct and maintain it). This is still acceptable because it provides a measure of costs of the inputs (carbon required to construct and maintain the system) against which to assess the gains obtained by the system's outputs (water transported and expended in transpiration during photosynthesis). In many other cases, however, the concept of efficiency is used in a looser sense, to indicate the *efficacy* of a system in transporting water (e.g. as judged by the values of plant hydraulic conductances). While the difference is largely semantic, we will use it here to differentiate between models that explicitly calculate transport costs and those that do not.

In the first case (when costs and gains are compared), it is relatively easy to define a criterion for the construction of an optimal conducting system. As a consequence, optimal models exist that maximise transport efficiency per se (e.g. maximise hydraulic conductance gains given fixed construction costs). However, in the second case this is more difficult, because there does not exist an upper bound to how efficacious a transport system can be and, by definition, larger values of hydraulic conductance are always preferable. Hence for this class of models, criteria for optimal performance cannot be based on the maximisation of transport efficacy per se, but must be based on arguments of functional balance, e.g. the maximization of the carbon return by leaves given the hydraulic gains of the conducting system relative to other limiting factors or the efficacy of transport relative to its loss of safety because of, e.g. embolism.

A third criterion that we propose relates to the *choice of the optimality criterion* itself. How should a transport system for tall trees be designed? Should it maximise the total conductance per se, the organ-level efficiency of the resources used in the construction of such system, the total growth of the organism, the total net carbon gain (gross photosynthesis minus construction and maintenance costs) or the tree-level efficiency of resource use (e.g. the amount of carbon obtained by leaf photosynthesis divided by the amount required for organ construction, i.e. leaves with their associated cohorts of conducting pipes)? The answer given is frequently that the best optimality criterion is the one that relates more closely to evolutionary fitness, in the sense of promoting the long term survival of the species. In this sense, it may be difficult to decide which criterion is actually the 'best', because several of them are likely to be applicable under different circumstances. Practice shows that, in fact, optimality models incorporate a range of different criteria.

A fourth criterion by which modelling principles for plant transport systems can be assessed is, as alluded to before, the *level of physiological detail* which they incorporate. Some transport models have been developed primarily to allow scaling physiological processes against logarithmic axes of plant size, whereas their level of anatomical detail is limited. The cost of construction of a transport system may be assessed in terms of the total biomass invested in the organ (branch, stem, roots) carrying out the transport function (typically upward xylem water transport), without a detailed consideration of the other functions performed by that organ (e.g. mechanical support, carbohydrate storage, etc.). Hence, by definition, the costs of the transport system are overestimated and the corresponding efficiency under-estimated. Other models have instead incorporated the concept of efficiency by focussing on the actual carbon costs involved in constructing the individual units involved in water and solute transport, i.e., the conduits, in an attempt to separate hydraulic transport costs from other costs. There is an expanding empirical literature on the allometric scaling of various characteristics of individual conduits which is useful for this purpose (e.g. Wheeler et al. 2005; Sperry et al. 2006; Pittermann et al. 2006a, b). Because of the complexities involved in scaling up the results obtained at the anatomical level to the whole tree using this approach, empirical data on allometric scaling at the level of a whole plant are particularly important tests of the performance of these models.

Linked to the issue of the level of physiological detail embedded in the model, is our fifth and last criterion of the *scope of the model* and of its *transferability for the prediction at larger scales*. Models that attempt to represent many or all of the processes involved in plant transport systems are more likely to be employed to predict processes at larger spatial and temporal scales; on the other hand, they are also likely to provide more synthetic descriptions of the details of each process. *Vice versa*, more mechanistic models (which are constructed to describe only one or a few processes in great detail) frequently pay the price of being unusable for the description of processes at larger spatial and temporal scales because they cannot be easily 'contextualised'.

In the next section, we will review the plant transport models that have been employed in recent years to study the role of tree size on tree performance. Hence, our main focus will be on models centred at the organismal level while we will not attempt to review other models operating clearly at lower (conduit) or larger (ecosystem) scales. We will use these five criteria (i.e. (a) presence/absence of model scalability, (b) use of efficiency versus efficacy principles, (c) choice of the optimality criterion if one was present, (d) level of physiological detail and (e) transferability to other scales) to highlight similarities and differences among them and to explore how their differences have led to the advancement of our knowledge in the understanding of the physiology of tall trees. It is not our intention to rank these models according to our own preference of what constitutes the 'best' model. Our intention is rather to use them as examples of logical summaries of our knowledge to illustrate what we have learnt on the significance of transport systems in tall trees.

In dealing with the transport models appearing in the literature, we will refer to Table 12.1 (for the xylem) and Table 12.2 (for the phloem), which summarise the main features of each model relative to the five criteria previously introduced.

Table 12.1 Classification of xylem water transport models based on five criteria of model structure, philosophy and applicability

Models \ Criteria	Model scalability	Efficiency versus efficacy of transport system	Choice of optimality criterion	Level of physiological detail for transport system	Scope of theory/ transferability to other processes
Pipe model theory	No (at least originally). Subsequent applications by Mäkelä (see text).	Efficacy. Computes cross-sectional area required to transport water and give mechanical support to a given leaf.	Not an optimality model in original formulation	Limited to functional balance between leaves and stem. Somehow incorporates biomechanical constraints.	Yes: (1) C allocation
Hydrall	Yes	Efficiency. Computes total biomass costs of transport system.	Maximises net photosynthesis, while maintaining constant plant water potential.	Limited to computation of total biomass costs of transport system and total hydraulic conductance of roots and stems.	Yes: (1) C allocation; (2) response to environmental variables.
WBE	Yes, both ontogenetically and inter-specifically.	Efficacy, bounded by biomechanical and fractal constraints. Computes conductance of each tapered pipe.	Not an explicit optimality model.	Incorporates conduit tapering in successive hierarchical levels of quasi-fractal organism	Yes, very large range of applications
Murray's law	Yes, for conduit diameter. No, for other aspects.	Efficiency. Computes costs of conduit wall construction.	Maximises xylem hydraulic conductance while keeping wall investment constant.	Models conduit tapering and furcation in successive branches of furcated pipe.	Limited so far.
Net C gain model	Yes, with regard to conduit numbers, diameters and tapering.	Efficiency. Computes costs of conduit wall construction.	Maximises net whole-plant C gains (gross gains minus costs) or C use efficiency (ratio of C gross gains over C costs), while maintaining constant plant water potential.	Conduit tapering and number are included. Derives optimal size-dependent patterns of plant photosynthesis. Pits are not included.	Yes: (1) C allocation; (2) response to environmental variables. So far limited to heuristic solutions.

Table 12.2 Classification of phloem water transport models based on five criteria of model structure, philosophy and applicability

Models \ Criteria	Model scalability	Efficiency versus efficacy of transport system	Choice of optimality criterion	Level of physiological detail for transport system	Scope of theory/transferability to other processes
Information transmission model	No. Distance is incorporated but it fails to represent transport realistically in large organisms.	Efficacy in solute and water transport and information transmission in one single tube.	Not an optimality model in original formulation.	Highly detailed for the physiological processes occurring within a single tube; does not incorporate notions of conduit numbers, conduit tapering or conduit branching. Additional solutes or relays are not represented.	Understanding physiology. Little transferability.
Minimum investment model	Only if the ratio of number of conduits to leaf area supplied is specified.	Efficiency. Computes costs of conduit wall construction and number of pipes required to transport sugars.	Minimise number of pipes required, given the photosynthesis.	Conduit tapering and number of conduits are included. Incorporates relays and additional solutes.	Understanding physiology. No transferability.
Net C gain model	Yes, with regard to conduit numbers, diameters and tapering.	Efficiency. Computes costs of conduit wall construction and number of pipes required to transport sugars.	Maximises net whole-plant C gains (gross gains minus costs) or C use efficiency (ratio of C gross gains over C costs) based on combined xylem and phloem costs.	Conduit tapering, number and furcation are included. Phloem and xylem transport systems are jointly represented and interacting.	Yes: (1) C allocation; (2) response to environmental variables; (3) Estimates of turnover times of photosynthetic C.

3 Comparative Criteria for Models of Xylem Water Transport in Tall Trees: Empirical Evidence and Optimality Modelling

3.1 Pipe Model Theory

Since its inception, the pipe model theory (PMT) has been very successful at attracting the attention of foresters, ecologists and modellers for its intellectual simplicity, which allows understanding and modelling some important aspects of carbon allocation in plants. Hence, it is an obligate, albeit old, starting point. In its original formulation (Shinozaki et al. 1964a, b), the PMT assumed that, because of the need to support leaves both mechanically and hydraulically, simple linear relationships were to be expected between amount of leaf mass (or, by extension, area) on a plant and its branch, stem or root cross-sectional area (or, by extension, functional sapwood area), either within a single tree from top to bottom, or across trees of different sizes (Table 12.1). In some sense, it is the equivalent of the formulation of Corner's rule for individual leaves, which states that a constant ratio exists between petiole size and leaf area. The PMT has been used innumerable times as the theoretical justification for the derivation of empirical relationships between tree leaf area and stem cross-sectional or sapwood areas.

The original model was not scalable, in the sense that it did not account for size-related effects on the functional ratios between leaf mass and stem cross-sectional areas. In the original publication, Shinozaki et al. (1964a) noted that tapering at the bottom of trees caused systematic departures from the predicted linear relationship between stem cross-sectional area and leaf mass and attributed these departures to the accumulation of disused pipes after the lower branches had been shed. It is now well known that, for a range of other reasons including biomechanical effects, hydraulic compensation effects and sapwood ageing (Lachenbruch et al. 2011), large trees can frequently maintain larger values of both cross-sectional area and sapwood area per unit of leaf area (e.g. McDowell et al. 2002) than smaller trees (although the opposite pattern has also been reported). It is also obvious that, other things being equal, both construction and maintenance costs must increase more than proportionally with stem cross-sectional area because of the associated increase in height (King 2011). Mäkelä (1986, 1989) have subsequently proposed versions of the pipe model which scale more realistically with tree size, as construction and maintenance costs of longer pipes are explicitly accounted for. The concepts embedded in this class of models have proved to be far more useful than those of the original theory and have frequently been employed in whole forest carbon balance models (e.g. Mäkelä and Hari 1986).

Because the original PMT was primarily concerned with the functional balance between leaves and conducting stems, and without consideration for the actual costs involved, it was essentially based on a criterion of efficacy. In addition, without any additional constraints, the PMT was not an optimality model as such, but rather a proposal to interpret and understand the frequently observed linear relationships between leaf mass and stem cross-sectional areas. The level of physiological detail employed to represent the transport system was limited. Note however that the PMT did not make any assumption about size or shapes of individual pipes, which are better understood as strands of wood rather than individual conduits. Subsequent studies (e.g. West et al. 1997, 1999; McCulloh et al. 2003), however, have interpreted it as indicating that each pipe represented unbranched, untapered conduits. In any case, modelling of factors such as the vertical changes in stem hydraulic conductance, or of the age-related changes in anatomical properties, is not possible with the PMT. The biggest limitation of this theory was probably also its main strength: because hydraulic and bio-mechanical functions were jointly considered (albeit very empirically), its heuristic value was limited, but its practical application was both straightforward and robust within the size range for which the relationships were constructed.

Roderick and Berry (2001) published an extension of the PMT that linked the hydraulic properties of a set of parallel rigid pipes to a macroscopic variable, wood density. This linkage allowed the authors to make predictions of the expected direction of change of the hydraulic properties with the change in important environmental variables (water temperature, nitrogen availability, air CO_2 concentration) and to test these predictions against the observations of changes in wood density reported in the wood technology literature. By and large, the model predicted the qualitative direction of change in hydraulic properties correctly.

3.2 *Hydrall*

As the name suggests, Hydrall was designed to increase our understanding of the allocational constraints imposed on assimilation by the construction and maintenance of the xylem transport system (i.e. roots and stems) (Magnani et al. 2002). The model was designed from the beginning to scale with tree size, albeit not allometrically (Table 12.1). Also, because construction and maintenance costs were accounted for, a tree's carbon budget could be calculated. In Hydrall the transport costs were equated with the total biomass costs for a particular organ, without consideration for the various components within that organ. Hydrall was based on an explicit optimization criterion, i.e. maximising stand level assimilation while keeping leaf water potential just below the threshold for xylem embolism. Because of its flexibility and the incorporation of additional processes (mainly plant photosynthesis), it has been employed for the description and analysis of additional processes, namely the description of C allocation during growth, the age-related patterns of stand net primary productivity and the response of forest C accumulation to a range of environmental variables (Magnani et al. 2000, 2002). More recently, its central principle of allocational constraints based on plant water relations has been incorporated in a regional model of carbon fluxes across Europe (Zaehle et al. 2006). Similar hydraulic concepts have been incorporated into a wider model recently developed by Buckley and Roberts (2005).

3.3 WBE

The WBE model (sometimes, referred to as FBN, fractal branching network, cf. Weitz et al. 2006) has attracted enormous attention since its initial presentation in 1997. Its main strengths have been in providing a new language, that of power law relationships applied to the quasi-fractal representation of plants, to the old problem of description of plant form and function. It did so by formalising the description of plant form via a clever system of ratios which were assumed to scale constantly across the hierarchical order of segments inside the plant (e.g. the ratios of branch diameter and length, and conduit diameter of the mother branches against those of the daughter branches). This allowed considerable simplification in the calculation of their scaling as plant size increased. Volume filling by the branching network and scaling between diameters and lengths to maintain elastic similarity were employed to constrain two of these three allometric ratios. The third one, the allometric scaling of conduit diameter as a function of distance from the apex (called conduit tapering) was constrained empirically (West et al. 1997, 1999; Mäkelä and Valentine 2006) using the relationship between conduit diameter and number of segments from the apex and assuming complete independence of transport resistance from length, which resulted in universal scaling relationships. However, both higher and lower rates of tapering are observed than the conduit tapering function in the model, especially close to the apex and at the bottom of tree trunks, respectively (Mencuccini et al. 2007), and independence of resistance from length can only very approximately be achieved, even theoretically. Note that the use of the term 'tapering' is slightly confusing here, because in WBE the petiole conduit size acts as the fixed reference conduit size. Hence, high rates of tapering effectively indicate high rates of basipetal enlargement, rather than high rates of apical narrowing. Also, individual conduits do not normally taper, therefore the proposed tapering represents the diameter change occurring from one conduit to the next.

Although by definition, the WBE model scales with plant size (because it was designed to do precisely that), its success in doing so has been hotly debated, with questions raised about the comparability of inter-specific with intra-specific and especially ontogenetic analyses (e.g. Mencuccini 2002). Conduit tapering obviously helps to reduce the height-related increase in hydraulic resistance and several authors have reported rates of tapering consistent with the predictions from WBE (e.g. Mencuccini 2002; Anfodillo et al. 2006; Weitz et al. 2006; Mencuccini et al. 2007; Petit et al. 2008). However, notable exceptions have also been reported. For example, far from continually increasing their conduit size, tall gymnosperm and angiosperm trees frequently appear to keep size constant or even reduce the size in their trunks (Mencuccini et al. 2007; Petit et al. 2010), probably in order to minimise costs. Similarly, departures from model predictions have been reported in very small plants (Sack et al. 2002), in scaling nitrogen content and plant respiration across a range of sizes within a species (Reich et al. 2006), in scaling leaf hydraulic conductance (Coomes et al. 2008), in scaling tree size–density distributions (Coomes et al. 2003) and in scaling net primary productivity and water use with forest age (Magnani 1999).

The WBE model is based on an efficacy concept, since carbon costs specific to the transport systems are not quantified. In addition, WBE is not really an optimality model, since larger tapering exponents will always result in lower values of the total resistance of each pipe (albeit with lower returns and with larger cross-sections required to fit the conduits; cf. Mäkelä and Valentine 2006).

The fundamental contribution of the WBE model compared to earlier descriptions of plant water transport systems (e.g. the PMT) has been the introduction of conduit tapering. That conduits tapered had been known for a long time, but WBE were the first to realise that this process had important consequences that had not been explored. By assuming that a tapered pipe resulted in a transport system whose hydraulic resistance was independent of length (such that a small seedlings could be considered equivalent to a tall tree in terms of its transport capacities and photosynthetic performance per unit of leaf area), they were able to extend their model to a vast range of other processes and situations, up to ecosystem-scale analyses of gas fluxes and ecosystem metabolism. Finally, the authors of the WBE model claimed that their model could easily be extended to describe phloem transport in plants but so far, this extension has not been carried out.

Becker et al. (2003) examined the implications of including inter-conduit resistances in the WBE model. They examined three possible scenarios, depending on how pit resistance was assumed to scale with lumen resistance and found that the effect on the derived WBE exponents depended on the particular assumption made. If the pit resistance was assumed to represent a constant fraction of lumen resistance or if tracheid dimensions were scaled to minimize wood resistivity, then the derived exponents would be identical to the ones obtained by WBE. If instead, the proportion of resistance in pits increased with increasing conduit diameter, the derived exponents doubled in magnitude approximating those obtained by Murray's law (see later section). It is unclear on what bases should the choice be made given that each alternative has its own merit. Despite this empirical nature, this paper is important because it shows that theoretical scaling exponents cannot be obtained without a complete description of the hydraulic structure, which should incorporate pits.

3.4 *Murray's Law*

Murray (1926) examined the scaling of conduit diameters in animal arteries. He derived a cubic relationship between blood flow and vessel diameters at different positions along the vascular network, thereby explaining why diameters rapidly decrease from the central aorta to the peripheral arterioles. He also explained this relationship by showing that the cubic exponent resulted from the maximisation of the conductance of a branched system containing a fixed amount of blood (or, in his words, minimizing the total work carried out by the act of blood flowing creating friction and the energy invested in blood).

McCulloh et al. (2003) proposed that a similar relationship holds for plant xylem. They reasoned that water, compared to blood, was cheap but that, conversely, substantial costs were involved in building the walls to reinforce them against the potential threat of implosion caused by the tension present in plant xylem. They proved that under such circumstances an identical cubic scaling would result and they also showed that conduit furcation (i.e. the ratio of number of conduits between distal and proximal ranks) would increase hydraulic conductance. They also presented data confirming that, in peripheral leaflets and leaves of four species, the stems of two vines and in the initial stages of branch ramification of one species of tree (*Fraxinus*), conduit numbers did decrease from distal to basal positions (suggesting that the conduits were furcating, thereby violating the assumptions of WBE, although not necessarily of PMT), while their conduit diameters scaled following the proposed cubic relationship. Conversely, no furcation was seen for the stems of the diffuse porous *Acer* and the coniferous *Abies* (McCulloh et al. 2003, 2004; McCulloh and Sperry 2005). Murray's law as applied to plants by McCulloh et al. (2003) does show scalability with regard to conduit diameter. Murray's law is based on efficiency and the optimality criterion is to maximise hydraulic conductance for a fixed investment in conduit wall reinforcement. The level of description of the transport system is greater than in previous treatments, with a detailed representation of how conduits taper and of their furcation ratios, but the scope of the theory is more limited. Murray's law says nothing regarding the total number of conduits necessary to support a certain leaf area, nor regarding the relationship between tapering and distance from the apex. As such, it is not a scalable model. So far, the theory has not been applied beyond its original aim to predict other processes.

Strictly speaking, the predictions of Murray's law only apply to cases in which conduits do not carry out a support function. Applying it to lignified organisms creates problems. For example, in the absence of conduit furcation, Murray's law predicts that the optimal investment in hydraulic conductance occurs for totally untapered conduits. Hence, presence of conduit furcation is crucial to obtain vertical profiles of tapered conduits. Empirical data show that conduits generally do not furcate below the initial few bud scars in self-supporting trees, particularly diffuse porous and conifers (McCulloh et al. 2004; Petit et al. 2008). However, there is ample evidence that conduits continue to change their diameters all the way down the trunk and into the roots (sometimes with inverted taper; cf. Mencuccini et al. 2007). Consequently, this tapering cannot be explained by Murray's law.

3.5 Net C Gain Model

Mencuccini et al. (2007) presented a scaling model of plant xylem water transport which contained elements taken from many of the previous modelling approaches. The plant was represented following the quasi-fractal system of WBE, ensuring scalability, and the costs of conduit construction were estimated using the theory of conduit wall reinforcement outlined in McCulloh et al. (2003), which means that hydraulic efficiency was determined. However, Mencuccini et al. (2007) diverged in the choice of the optimality criterion. Rather than use the efficiency in the use of resources for the construction of the xylem system, this model used the combination of a constant leaf water potential plus the maximisation of the net C gains of the whole-plant, i.e. it linked the costs of xylem water transport with the gains obtained by photosynthesis in a manner resembling the approach of Hydrall. The rationale for choosing this optimisation criterion is that it would be adaptive for a plant to construct an inefficient transport system (with respect to the maximum set by Murray's law), provided that it resulted in proportionally more carbohydrates being photosynthesised or in proportionally lower transport costs. This differs from the analysis based on Murray's law, which predicts instead what the maximum hydraulic conductance will be for a *fixed*, not a *variable*, amount of resources spent on conduit walls. It is likely that the net carbon gains change considerably during ontogeny, since the xylem is cheap to construct when a plant is small and becomes increasingly costly as a plant grows in size, in which case resource conservation may became a predominant consideration relative to resource use efficiency. This approach provides an explanation for inverse tapering at the base of (especially coniferous) trees, an otherwise puzzling phenomenon. The published version of the model maximised net C gains (gross photosynthesis minus xylem construction costs) but essentially identical results in terms of the predictive patterns were obtained by employing alternative formulations for the costs and gains (e.g. incorporating xylem maintenance costs in addition to construction costs or incorporating a mesophyll conductance term in series with stomatal conductance), or by using alternative definitions of the optimality criterion to be maximised (i.e. by employing the ratio of total C gained by photosynthesis divided by the total C costs of the xylem transport system). The level of physiological detail of the model was relatively high compared to some of the other examples previously examined, with conduit tapering, conduit number and conduit furcation all considered (albeit the published version of the model was more restricted). Relative to Hydrall, the Mencuccini et al. (2007) model incorporated a more mechanistic calculation of costs, which are here based on the construction of conduit walls (whose thickness depends on conduit diameter) and not on the total biomass invested in wood. Plant photosynthesis was represented and made dependent on environmental variables. The high level of physiological detail resulted in a loss of simplicity of the model, which did not provide analytical solutions to the problem of optimal tapering, but required parameterization using site-specific and species-specific parameters.

4 Comparative Criteria for Models of Phloem Water Transport: Empirical Evidence and Optimality Modelling

Our knowledge of the scaling principles to be applied to the design of phloem transport systems for tall trees is considerably less advanced than the knowledge for the xylem. Thompson and Holbrook (2003a, b, 2004) have repeatedly attracted attention to the fundamental importance played by phloem anatomical parameters in affecting sugar and water transport properties in the phloem, and to the need to develop a sounder knowledge base of their variation within and across individuals and species, and yet these fundamental details of tree anatomy remain poorly known.

The paucity of data on tree phloem anatomy (particularly with regard to tall trees) is all the more important when one considers the importance of phloem transport in the context of the movement of carbohydrates from forest canopies to storage compartments, roots and mycorrhiza and the relevance of this movement for the carbon cycle. In addition, recent theories that attempt to explain age-related patterns in tree growth based on sink limitations (Sala and Hoch 2009; Sala et al. 2011) are directly concerned with the transport of sugars away from the crown and the presence of potentially negative feedbacks between photosynthesis, sink activities and efficacy of sugar export. As a consequence of the paucity of empirical information on phloem structure and function of tall trees, theories and models that are useful for understanding phloem transport systems of tall trees are limited. The most relevant ones are reviewed here (Table 12.2).

4.1 Information-Transmission Model

Thompson and Holbrook (2003a, 2003b, 2004) significantly expanded the analytical scope of phloem modelling relative to several previous theoretical exercises of phloem function (Tyree et al. 1974; Goeschl et al. 1976; Phillips and Dungan 1993). By using dimensionless quantities, they greatly simplified the analysis of phloem transport in several important respects (Thompson and Holbrook 2004). In the context of this review, perhaps the most important innovation was the analysis of information transmission along phloem axes. Thompson (2006) pointed out that, the smaller the turgor pressure differential between opposite ends of the phloem translocation pathway, the better the whole-plant integrative control of phloem transport, particularly in tall trees, where the turgor pressure drop would otherwise rapidly build up over long distances. With more-or-less uniform turgor along the axis, information on changes in water or sugar status would be transmitted rapidly from one end of the plant to the other, and the behaviour of the sieve cell–companion cell complex would not need to depend greatly on the knowledge of where these cells were located along the pathway. Information on changes in source and sink activity and water status could be transmitted via individual molecules moving at the solution bulk flow velocity or in the form of pressure and concentration waves (Phillips and Dungan 1993; Thompson and Holbrook 2004). Following Phillips and Dungan (1993), Thompson and Holbrook (2003b) used dimensionless ratios to show that, given realistic ranges of parameter values, tall plants (several metres of height) were likely to behave like 'reservoirs' rather than a 'rivers', i.e. their rates of both phloem transport and information transmission were likely to become sluggish and unresponsive to changes in prevailing conditions. In addition, Thompson (2006) showed that of all the factors affecting phloem transport, sieve tube length (potentially equal to total plant height) was by far the most important, but that data on this dimension are lacking.

The Thompson and Holbrook (2003a, 2003b) information-transmission model was not conceived to be scalable, as changes in sieve tube anatomical properties and their number with tree size were not considered, but it was intended as a tool to understand transport mechanisms per se. Indeed, when their model was parameterised for giant sequoia trees of height equal to 100 m, pressure-concentration waves were predicted to take at least a week to traverse the tree from canopy to roots, raising questions on how regulation of phloem transport occurs in tall trees. The model was based on an efficacy criterion, because construction costs were not calculated and, strictly speaking, it was not an optimality model because no equation related to the behaviour of the phloem tube was explicitly either maximised or minimised. In essence, the authors proposed a 'workable' regime of conditions, limited by the geometry of the system, under which flow of material and information occurs efficaciously and certain critical conditions associated with phloem transport (i.e. cell plasmolysis, water potential disequilibrium with the xylem and the occurrence of the 'reservoir' regime) are avoided. An additional characteristic of this model is that it describes in great detail the behaviour of a single conducting tube but it incorporates no notion of the number of tubes required to sustain a certain level of sugar export from a leaf, nor does it consider the possibility that tubes may be tapered, branched or contain solute relays. As such, its main application is in process understanding, whereas scalability and transferability are limited.

4.2 Minimum Investment Model

The Hölttä et al. (2009b) minimum investment model was designed to be scalable, by incorporating principles (such as tube tapering or variable number of phloem tubes per leaf) that allowed it to perform realistically across a range of tree sizes, although allometric theory was not employed. It also explicitly considered the risk that phloem transport under steady-state conditions was limited by excessive viscosity caused by high solute concentrations. The model contained an optimization function: given fixed photosynthetic and transpiration rates, it calculated the minimum number of combined xylem and phloem tubes required to obtain transport under steady-state conditions (Hölttä et al. 2006).

The rationale for analysing a coupled xylem / phloem transport system is that the structure and flux rates in the xylem control the water potential, which in turn affects phloem transport. Interestingly, because xylem water potential affects both xylem and phloem transport, the conductance of the two systems must be coupled, such that large structural investments in the xylem are predicted to reduce the need for investment in the phloem and *vice versa*.

Because the model incorporated the basic physiological mechanisms needed to describe transport as well as the components required to make it scalable, its scope was larger than the previous model. However, its transferability and applicability was limited by its increased complexity, the lack of a mechanistic description of photosynthesis and the lack of coupling to the final fate of the carbohydrates transported.

The system was designed to minimise the number of tubes required to achieve xylem and phloem transport (rather than optimize information transmission). This condition should make the predictions of this model very different from the predictions from the Thompson and Holbrook model. For example, it should result in extremely slow transfer times of pressure-concentration waves for tall plants (i.e. in the same order of magnitude as molecule transfer times or even slower), thereby negating the existence of a localised system of physiological control, as invoked by Thompson and Holbrook (2004). In reality, this delay in transfer times was greatly alleviated provided that the assumption was made that phloem conduits tapered too. Relatively small linear rates of tapering proved to be sufficient to keep the dimensionless ratios within the expected range, while wave propagation time and the minimum number of required phloem tubes declined dramatically (Hölttä et al. 2009b).

4.3 Speed-of-Link Model

Mencuccini and Hölttä (2010) built on the previous models developed by Hölttä et al. (2006, 2009b) to expand the range of processes to which they could be applied. As such, the new version did not contain any innovation compared to the previous models with regard to how the transport system was optimised. It is reviewed separately here to illustrate the importance of incorporating size-related processes to make a whole-plant phloem transport model scalable, such that its applicability to explain additional processes is enhanced.

The coupled xylem-phloem model of Hölttä et al. (2006) was expanded by linking it to a new routine that modelled root respiration and soil CO_2 gas diffusion as a function of soil temperature, moisture content and the amount of sugars transported by phloem and respired by roots. The coupled gas exchange/fluid transport/soil diffusion model was then employed to help interpret the observed temporal patterns of movement of isotopic tracers introduced in trees by photosynthetic labelling. Several studies have shown that, upon application of an isotopic C label over plant canopies, a spike in either depleted $^{13}CO_2$ (e.g. Steinmann et al. 2004), enriched $^{13}CO_2$ (e.g. Johnson et al. 2002; Högberg et al. 2008) or $^{14}CO_2$ (e.g. Howarth et al. 1994; Mikan et al. 2000) appeared in the soil respiratory flux after a delay of about 1–5 days, suggesting a close coupling between photosynthesis and soil/ecosystem respiration. Similar conclusions were obtained by several other authors (e.g. Ekblad et al. 2005; Gärdenäs 2000) using simultaneous measurements of photosynthetic and respiratory fluxes, but without employing isotopic tracers, although in this second case, the corresponding respiratory spikes tended to occur much earlier and more closely in time to the changes in photosynthesis.

Mencuccini and Hölttä (2010) tested the hypothesis that the 'speed-of-link' between canopy photosynthesis and soil respiration depended on whether the speed estimates were obtained by using the isotopically-based approach or the flux-based approach (cf., Mencuccini and Hölttä 2010). Fifty-four independent estimates of

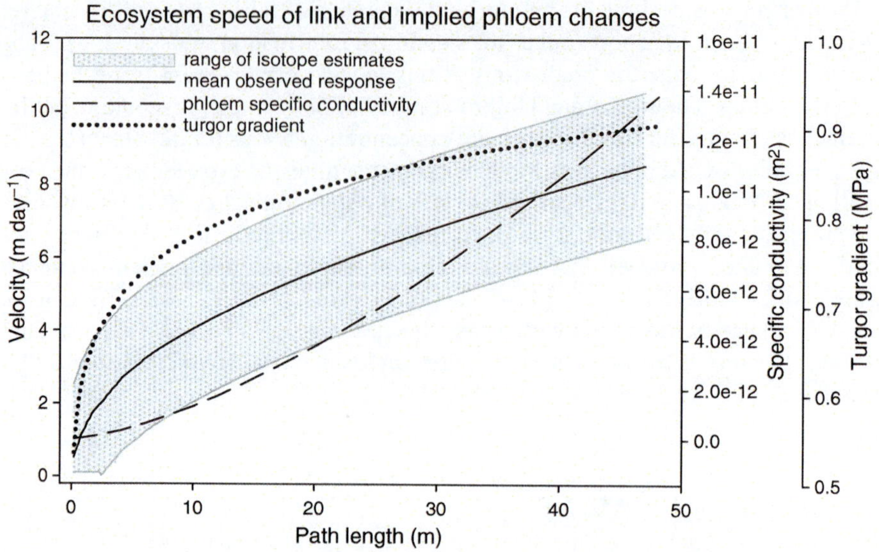

Fig. 12.1 Summary of the empirical and modelling results obtained by Mencuccini and Hölttä (2010) related to the effective transport velocities of isotopically labelled sugars in the pathway from canopy leaves to the soil surface (i.e. via phloem transport, root respiration and soil gaseous diffusion). The *dotted surface* represents the range of calculated transport velocities and the *continuous black line* the calculated mean velocity (in m/day, left axis), based on the observed times of appearance of the isotopic spikes out of the soil as gaseous CO_2 after the time of labelling of canopy photosynthesis, for a range of organisms ranging from small herbaceous plants to tall trees. The *dashed* and the *dotted lines* represent the inferred changes in phloem specific conductivity (m^2, inner right-hand side axis) and in phloem turgor pressure gradient (MPa, outer right-hand side axis), respectively, that must have taken place in the phloem transport systems of these organisms to account for the observed velocities. Sugar storage in the system and gaseous diffusion in the soil were also accounted for in the model but did not explain the observed size-related pattern

'speed-of-link' were available across a range of plant heights ranging from 0.1 to around 40 m, i.e. over three orders of magnitude (Fig. 12.1).

For the isotopically-based measurements, one would predict that the length of the phloem transport system would exert significant constraints on the speed of the transfer of the isotopically labelled molecules from the canopy, through the phloem and roots and out of the soil as gaseous CO_2, and particularly so in tall trees. However, the extent by which this constraint would operate is open to question. Without any specific size-related adjustments in phloem properties, the constraint imposed by the increase in height over three orders of magnitude is likely to be extremely large, as shown by the results reported by Thompson and Holbrook (2003b) for the model parameterised for giant sequoia trees.

The data did indeed show that the time delay between 1 and 5 days mentioned above was largely the result of differences in plant size (longer times in tall trees compared to herbaceous plants and tree seedlings). However, the measurements also showed an increase in the apparent velocity of the isotopic signal as a function of plant size (Fig. 12.1): the time required for the isotopic tracer to emerge from the soil surface when applied to 40-m tall trees was longer by only a few days compared to the time observed for much smaller plants (Fig. 12.1). Two models (the combined photosynthesis/phloem transport/gas diffusion model just mentioned and a simplified version based on average properties) were then employed to understand the nature of the mechanism(s) that could explain these observations. The authors found that they could only be explained by assuming that the phloem transport system operated differently depending on plant size, i.e. by assuming that both the turgor drop along the phloem pathway and the phloem specific conductivity increased as a function of plant size (Fig. 12.1, dotted and dashed curves and right axes). An alternative explanation could involve the presence of solute relays. If one assumes that individual sieve tubes are shorter than the length of the whole pathway and are connected in series with the solutes metabolically transferred from one sieve tube to the next (as hypothesized by Lang 1979), then a similar relationship would be found. Size-related changes in sugar storage within the plant or in soil gas/water transport properties could not explain the observations. Hence, the strong suggestion from these ecosystem-level observations was that significant size-related adjustments had taken place in the phloem transport system of tall trees and that size-related properties must be incorporated in phloem transport models to increase their applicability across a range of plant sizes, i.e. to make them scalable.

5 Empirical Evidence for Size- or Age-Related Phloem Sieve Cell Tapering

Do phloem tubes taper in a manner similar to what is seen in the xylem? The (limited) empirical evidence has never been systematically summarised and put into context. Because of the importance of this process, such a review is provided here, together with additional unpublished data (Table 12.3).

It appears that the large majority of the studies support the hypothesis that phloem sieve cells taper basipetally, similarly to what is found in the xylem, but a few exceptions are present (Table 12.3). Connor and Lanner (1990) studied xylem and phloem anatomy of Great Basin bristlecone pines (*Pinus longaeva* D. K. Bailey) of various ages in California and Utah. They found no age-related trend in phloem characteristics, i.e. annual increments and number of sieve cells produced per year along a radial file, but sieve cell diameters were not measured. For this report, it is unclear at which stem height the measurements refer to and, in addition, those trees were likely to have had fairly similar heights and maturational processes were likely to have ended, given the range of ages employed (site means ranging from 751 to 1,863 years). Quilhó et al. (2000) also found no evidence of vertical sieve tube tapering in 15-year-old *Eucalyptus globulus* Labill. trees sampled at six height levels (from 5% to 75% of tree height), possibly because, as in the xylem, much of the tapering may occur near the apex and a plateau is then rapidly reached. A report for a conifer species (*Picea abies* (L.) Karst., in Rosner et al. 2001) showed evidence of an inverse pattern of reductions in sieve cell diameters at the bottom of the stem

Table 12.3 Synthesis of published studies on age-related changes in phloem sieve cell diameters

Species	Type of measurement	Observed trend	Author(s)
Angiosperms			
Acacia nilotica var. *Telia*	Along the stem	Downward increase	Ghouse and Iqbal (1977)
Betula pendula	Along the stem	Downward increase	Trockenbrodt (1994)
Bauhinia aculeata (shrub)	Along the stem	Downward increase	Ewers and Fisher (1991)
Bauhinia galpinii (shrub)			
Bauhinia blakeana (tree)			
Bauhinia variegata (tree)			
Bauhinia fassoglensis (vine)	Along the stem	Downward increase	Ewers and Fisher (1991)
Bauhinia vahii (vine)			
Fagus sylvatica	Along the stem	Downward increase	Schulz and Behnke (1986)
Fagus sylvatica	Young versus old trees	More numerous and smaller sieve tubes in 10-m compared to 30-m-tall trees	Vollenweider et al. (1994)
Eucalyptus globulus	Along the stem	No trend	Quilhó et al. (2000)
Populus deltoides	Along the stem	Downward increase	MacDaniels (1918)
Populus euramericana cv. *Robusta*	Along the stem	Downward increase	Stahel (1968)
Populus tremula	Along the stem	Downward increase	Raskatov and Kosichenko (1968)
Populus tremula	Along the stem	Downward increase	Trockenbrodt (1994)
Prosopis spicigera	Along the stem	Downward increase	Ghouse and Iqbal (1977)
Quercus robar	Along the stem	Downward increase	Trockenbrodt (1994)
Ulmus Americana	Along the stem	Downward increase	MacDaniels (1918)
Ulmus glabra	Along the stem	Downward increase	Trockenbrodt (1994)
Conifers			
Picea abies	Along the stem	Smaller sieve cells at the stem bottom compared to base and inside of crown	Rosner et al. (2001)
Pinus longaeva	Stems of trees of age varying between 751 and 1,863 years	No change in phloem increment or in number of sieve cells	Connor and Lanner (1990)
Herbaceous plants			
Lolium temulentum	Along leaf veins	Diameter increase in major compared to minor veins	Lush (1976)
Panicum maximum	Along leaf veins	Diameter increase in major compared to minor veins	Lush (1976)
Zea mays	Along the primary unbranched root	Diameter increase and fewer cells towards the tip	Enns et al. (2006)

compared to the base and the inside of the crown. The report by Enns et al. (2006) on 21-days-old *Zea mays* seedlings is interesting because it showed clear evidence of furcation in sieve tube cells: their number drastically declined towards the root tips of unbranched primary roots. The decrease in sieve cell diameters in small leaf veins compared to major leaf veins (Lush 1976) is qualitatively similar to the vessel tapering normally found in tree leaves (Coomes et al. 2008).

Ewers and Fisher (1991) were probably the first to suggest that the concept of the trade-off between safety and efficiency could also be applied to phloem transport. They collected anatomical data along branches of different individuals of species of the genus *Bauhinia* (some of which were lianas, some shrubs and some trees). They showed clear increases in both xylem and phloem cross-sectional areas and in both vessel and sieve tube diameters with increases in the amount of leaf area distal to the stem segment. Visual inspection of their graphs shows that vertical sieve cell tapering was similar to vessel tapering in shrubs and trees, whereas phloem cell tapering was slower than vessel tapering in vines. As far as we know, this is the only study that simultaneously compared xylem and phloem anatomical properties along an axial gradient and put them in relation to the leaf area distal to those segments (cf., Gartner 2002 for a similar dataset comparing leaf-sapwood area ratios with leaf-inner bark area ratios along the stem of Douglas fir (*Pseudotsuga menziesii* (Mirbel) Franco), trees).

Because of the paucity of data on these trends, we recently collected a similar dataset, using the four chronosequences of *Pinus sylvestris* (at Guisachan and Selm Muir), *Acer pseudoplatanus* and *Fraxinus excelsior* described in Mencuccini et al. (2005), Martinez-Vilalta et al. (2007a, b) and Hamid and Mencuccini (2009), all from Scotland (UK). Xylem and phloem anatomical parameters were measured on cross-sections taken at breast height from trees of age varying between 3 and 250 (Figs. 12.2 and 12.3). Our results confirm the trends reported by Ewers and Fisher (1991) for the genus *Bauhinia*. A highly significant correlation was found between xylem sapwood area and functional phloem area (as estimated in regions of the cross-section where uncrushed phloem could be measured), with a relationship that was broadly similar across the three species studied at the four sites. The overall power-law scaling with a slope of $b = 0.78$ suggested a more rapid accumulation of functional tissue in the xylem than in the phloem in large plants and the individual slopes for the four cases were all below 1.00 (Fig. 12.2a). Changes in density and diameter of phloem sieve cells during ontogeny also broadly matched the corresponding changes in xylem conduit density and diameters (Fig. 12.2b, c).

Because functional areas of both xylem and phloem tissue increased during ontogeny, there were negative relationships in the ratios of leaf area-sapwood area and leaf area-functional phloem area, although for the pine such reductions were surprisingly very small. Because functional xylem tissue accumulates faster than functional phloem tissue (Fig. 12.3), the slopes observed for the leaf-functional phloem area ratios are less steep than those for the leaf area-sapwood area ratios for the angiosperms.

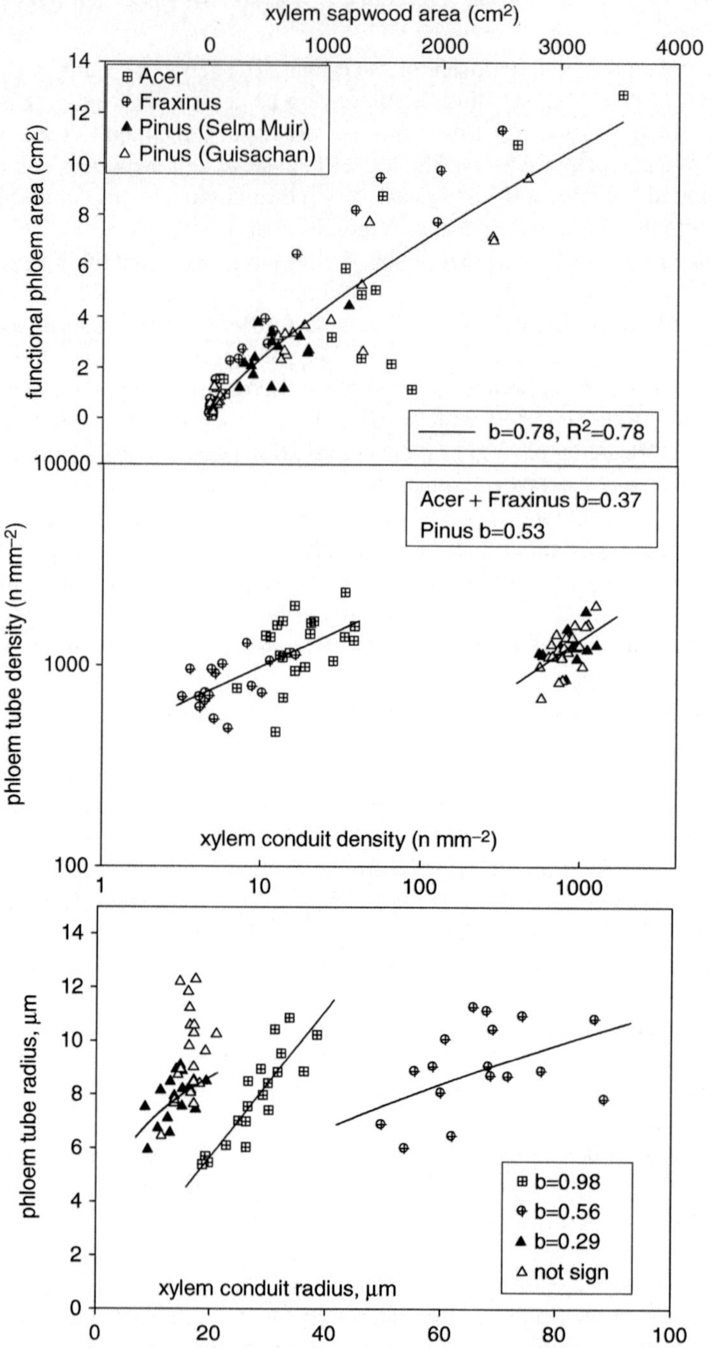

Fig. 12.2 Simultaneous changes in xylem and phloem properties during ontogeny in three different tree species, i.e. *Acer pseudoplatanus* (*crossed squares*), *Fraxinus excelsior* (*crossed circles*) *and Pinus sylvestris* (*white* and *black triangles*) at four sites across Scotland (UK). Xylem and phloem properties were obtained by sampling tree cores at breast height from a range of age and size classes. *Top panel*: Average xylem and phloem functional areas. *Middle Panel*: Average xylem and phloem conduit density. *Bottom Panel*: Average xylem and phloem conduit radius. Numbers in the panels refer to the slopes of the log–log scaling relationships between the various properties, when significant

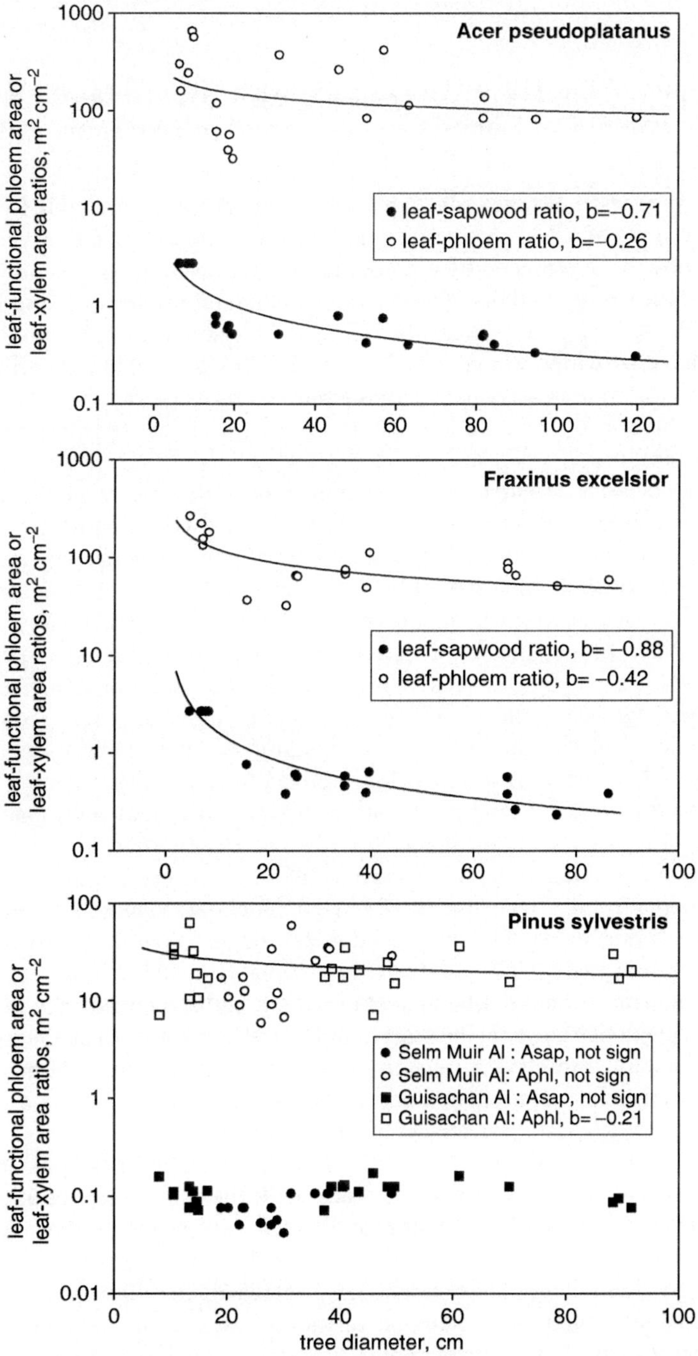

Fig. 12.3 Size-related changes in the ratios of leaf to functional sapwood area and leaf to functional phloem area across three species of trees studied in Scotland (UK). *Top Panel*: *Acer pseudoplatanus*. *Middle Panel*: *Fraxinus excelsior. Bottom Panel*: *Pinus sylvestris*. Note that the ratios are plotted on a log scale to accommodate the different values present for phloem versus xylem-based calculations. Figures in the panels refer to the slopes of the log–log scaling relationships between the various properties plotted against stem diameter, when these were significant

6 Critique: What Have We Learned with Regard to Modelling the Physiology of Long-Distance Vascular Transport in Trees?

Significant progress has taken place in understanding how to model the vascular transport systems of trees, particularly tall ones. For the xylem the original pipe model has now been replaced by a more detailed representation of a network of tapered and furcating conduits. These two fundamental properties, tapering and furcation, allow models to address the question of xylem transport in tall organisms and have provided some initial answers. Similar conduit tapering (and possibly furcation) appears to occur in the phloem as well, and the implications of the introduction of hypothetical solute relays for sugar transport have also begun to be explored.

It has also become apparent that an examination of the size-related changes in the hydraulic conductance of tall trees needs to consider construction and maintenance costs, again both for the xylem and for the phloem. It had been hypothesised for some time that such costs would be ecologically relevant (e.g. Magnani et al. 2000), but their significance has only become apparent when a much more detailed approach was adopted, i.e. when the wall reinforcement costs of individual conduits were considered, as opposed to generic biomass costs per unit of hydraulic conductance. A challenge here is in separating the conduit wall costs associated directly with hydraulic reinforcement versus those linked to mechanical reinforcement of the structure. The advancement of our knowledge of the allometric scaling of anatomical properties (conduit and pit diameter and length) should contribute to this objective.

Perhaps more strikingly, the last 10 years have seen the explicit introduction of optimality modelling of plant hydraulic architecture for both xylem and phloem transport systems. Optimality models have proved useful and have contributed to explaining otherwise puzzling phenomena. Most of these optimality models operate at the scale of entire organisms or above (Magnani et al. 2000); perhaps a challenge now is to employ their predictive power at the scale of individual conduits, to make sense of the emerging allometric datasets at the anatomical scale (e.g. Lancashire and Ennos 2002; Sperry and Hacke 2004; Hacke and Sperr Pittermann 2004).

Concomitantly to these advancements, new questions have arisen as a result of the application of a specific optimality criterion. Specifically for the phloem, how can rapid information transmission be obtained while the construction costs are also kept at relatively low levels? How can xylem and phloem properties be simultaneously coordinated with leaf-level transpiration and photosynthetic rates?

Collectively these modelling exercises have increased our understanding of the significance of the structural adjustments that allow tall trees to maintain levels of physiological activity identical or similar to the ones of small trees. In turn, by forcing us to examine and incorporate size-dependent properties into xylem and phloem transport models, these models have also achieved scalability, i.e. they can now be applied successfully on both small and large trees.

7 Optimal Transport Systems: Future Frontiers

It is remarkable that the models reviewed above give reasonably realistic answers to the questions of the optimal behaviour of tree vascular-transport systems, when many important properties are either not represented or represented only in a very crude fashion. For example, the predicted tapering that minimised the dependency of hydraulic resistance on path length in the WBE model depended crucially on the assumption of how pit resistance scaled with conduit lumen resistance (Becker et al. 2003). Similarly, none of the current models incorporate information on conduit packing or on conduit spatial arrangement, and only very crudely on conduit length, three components which are also likely to be important, because of their impact on the connectivity of the system (cf. Loepfe et al. 2007; Mencuccini et al. 2010). Without a mechanistic description of pit resistance, none of the existing optimality models can provide a picture of the expected profile of plant water potentials. The significance of this deficiency is that the trade-off between system efficiency (or efficacy) and safety from embolism cannot presently be incorporated in any of these optimality models.

It is interesting to note that much of the recent literature on the topic of tall trees has focussed on the description of the anatomical and physiological features occurring along the branches (e.g. Burgess et al. 2006; Ishii et al. 2008; Domec et al. 2008), in leaves (e.g. Koch et al. 2004; Woodruff et al. 2004, 2007, 2008) or in trunks (Domec et al. 2008) at various heights in a tree or at the top of trees of various heights, without linking these parameters to the rate of tapering occurring within individual axes. These two strands of investigation (one dealing with the effects of gravity, the other with the minimization of the hydraulic resistance caused by longitudinal growth, i.e. in other words, the effects of height versus those of length) should be reconciled and brought together.

Secondly, while progress has been made in avoiding empirical descriptions of the carbon costs of hydraulic conductance, it is still not clear how these costs should be represented mechanistically at the anatomical level. For example, it is very difficult to separate the costs due to the hydraulic versus the mechanical support function carried out by conduit walls. An optimality model that incorporated both hydraulic and mechanical properties at the anatomical level would probably be a significant step forward.

Thirdly, similar uncertainties remain with regard to how to incorporate hydraulic capacitance into optimality models, because we still do not understand to what extent the use of capacitance water by trees results from the emptying of embolized conduits (i.e. a trade-off with conductance, Hölttä et al. 2009a) or from processes independent from embolism (e.g. Meinzer et al. 2003, 2006; Scholz et al. 2007). Incidentally, if water capacitance resulted largely from the use of stored water in the inner bark, this would likely impact the transport properties of the phloem in ways currently not predicted by the existing models. The significance of our inability to model the optimal xylem structure that simultaneously maximises both hydraulic transport and storage is that, among other things, we cannot account for the fact that significant increases in xylem hydraulic resistances are both generally found in tall trees and theoretically predicted to occur (Mencuccini 2003; Mencuccini et al. 2007; Sperry et al. 2008), while major limitations to their photosynthesis and growth are generally not found (e.g. Ryan et al. 2006; Sillett et al. 2010). Many recent papers have highlighted the importance of capacitance in regulating the scaling of important physiological properties (e.g. Sperry et al. 2008).

Finally, the persisting lack of suitable empirical methods to estimate phloem physiological properties (water and solute fluxes, turgor pressure, hydraulic conductivity) still slows our progress in understanding the significance of phloem physiology at the whole plant level and slows our progress in obtaining a clearer picture of the integration and the interactions occurring between these two conducting systems.

8 Conclusions

Models of the transport systems of trees have progressed enormously in the past 10 years. We identified and used five criteria to compare models with a common set of terminology. Many of these models have paid attention to the peculiarities of transport in tall trees and the need to incorporate properties that allow their deployment across a range of plant sizes. In practice, this has involved the identification of properties that show size-dependent adjustment or, in other words, properties allowing the model to perform realistically for both large and small trees. A wide range of approaches have been employed to define these properties. Similarly, the criteria by which to assess the optimality of a particular function are also very varied, reflecting the particular questions posed by the researchers. In addition to the xylem, such models have also been applied to understand phloem behaviour and many similarities in the scaling of the two systems are apparent. The future development of models will continue to help direct empirical research that has the potential to advance our understanding of long-distance transport substantially.

Acknowledgements The authors thank the Catalan government (AGAUR – Generalitat de Catalunya) for a visiting Professorship in UAB to Maurizio Mencuccini. Thanks are due to two anonymous reviewers, for providing useful suggestions that improved the manuscript.

References

Anfodillo T, Carraro V, Carrer M, Fior C, Rossi S (2006) Convergent tapering of xylem conduits in different woody species. New Phytol 169:279–290

Becker P, Gribben RJ, Schulte PJ (2003) Incorporation of transfer resistance between tracheary elements into hydraulic resistance models for tapered conduits. Tree Physiol 23:1009–1019

Buckley TN, Roberts DW (2005) How should leaf area, sapwood area and stomatal conductance vary with tree height to maximize growth? Tree Physiol 26:145–157

Burgess SSO, Pittermann J, Dawson TE (2006) Hydraulic efficiency and safety of branch xylem increases with height in *Sequoia sempervirens* (D. Don) crowns. Plant Cell Environ 29:229–239

Connor KF, Lanner RM (1990) Effects of tree age on secondary xylem and phloem anatomy in stems of Great Basin bristlecone pine (*Pinus longaeva*). Am J Bot 77:1070–1077

Coomes DA, Duncan RP, Allen RB, Truscott J (2003) Disturbances prevent stem size-density distributions in natural forests from following scaling relationships. Ecol Lett 6:980–989

Coomes DA, Heathcote S, Godfrey ER, Shepherd JJ, Sack L (2008) Scaling of xylem vessels and veins within the leaves of oak species. Biol Lett 4:302–306

Dixon HH, Joly J (1895) On the ascent of sap. Philos Trans R Soc B (London) 186:563–576

Domec J-C, Lachenbruch B, Meinzer FC, Woodruff DR, Warren FM, McCulloh KA (2008) Maximum height in a conifer is associated with conflicting requirements for xylem design. Proc Natl Acad Sci USA 105:12069–12074

Ekblad A, Boström B, Holm A, Comstedt D (2005) Forest soil respiration rate and $\delta^{13}C$ is regulated by recent above ground weather conditions. Oecologia 143:136–142

Enns LC, McCully ME, Canny MJ (2006) Branch roots of young maize seedlings, their production, growth, and phloem supply from the primary root. Funct Plant Biol 33:391–399

Ewers FW, Fisher JB (1991) Why vines have narrow stems: histological trends in *Bauhinia* (Fabaceae). Oecologia 88:233–237

Gärdenäs AI (2000) Soil respiration fluxes measured along a hydrological gradient in a Norway spruce stand in south Sweden (Skogaby). Plant Soil 221:273–280

Gartner BL (2002) Sapwood and inner bark quantities in relation to leaf area and wood density in Douglas-fir. IAWA J 23:267–285

Ghouse AKM, Iqbal M (1977) Trends of size variation in phloem fibres and sieve-tube cells within the bark of some arid-zone trees. Flora 166:517–521

Goeschl JD, Magnuson CE, DeMichele DW, Sharpe PJH (1976) Concentration-dependent unloading as a necessary assumption for a closed form mathematical model of osmotically driven pressure flow in phloem. Plant Physiol 58:556–562

Hacke UG, Sperr Pittermann J (2004) Analysis of circular bordered pit function. II. Gymnosperm tracheids with torus-margo pit membranes. Am J Bot 91:386–400

Hamid HA, Mencuccini M (2009) Age- and size-related changes in physiological characteristics and chemical composition of *Acer pseudoplatanus* and *Fraxinus excelsior* trees. Tree Physiol 29:27–38

Högberg P, Högberg MN, Göttlicher SG, Betson NR, Keel SG, Metcalfe DB, Campbell C, Schindlbacher A, Hurry V, Lundmark T, Linder S, Näsholm T (2008) High temporal resolution tracing of photosynthate carbon from the tree canopy to forest soil microorganisms. New Phytol 177:220–228

Holbrook NM, Zwieniecki MA (2005) Vascular transport in plants, Physiological ecology series. Academic, New York, 592 pp

Hölttä T, Vesala T, Sevanto S, Perämäki M, Nikinmaa E (2006) Modeling xylem and phloem water flows in trees according to cohesion theory and Münch hypothesis. Trees-Struct Funct 20:67–78

Hölttä T, Cochard H, Mencuccini M, Nikinmaa E (2009a) Capacitive effect of cavitation in xylem conduits. Plant Cell Environ 32:10–21

Hölttä T, Nikinmaa E, Mencuccini M (2009b) Linking phloem function to structure: analysis with a coupled xylem-phloem transport model. J Theor Biol 259:325–337

Howarth WR, Pregitzer KS, Paul EA (1994) ^{14}C allocation in tree soil systems. Tree Physiol 14:1163–1176

Ishii H (2011) How do changes in leaf/shoot morphology and crown architecture affect growth and physiological function of tall trees? In: Meinzer FC, Lachenbruch B, Dawson TE (eds.) Size- and age-related changes in tree structure and function. Springer, Dordrecht

Ishii HT, Jennings GM, Sillett SC, Koch GW (2008) Hydrostatic constraints on morphological exploitation of light in tall *Sequoia sempervirens* trees. Oecologia 156:751–63

Johnson D, Leake JR, Ostle N, Ineson P, Read DJ (2002) In situ $^{13}CO_2$ pulse labelling of upland grassland demonstrates a rapid pathway of carbon flux from arbuscular mycorrhizal mycelia to soil. New Phytol 153:327–334

King DA (2011) Size-related changes in tree proportions and their potential influence on the course of height growth. In: Meinzer FC, Lachenbruch B, Dawson TE (eds.) Size- and age-related changes in tree structure and function. Springer, Dordrecht, pp 165–191

Koch GW, Sillett SC, Jennings GM, Davis SD (2004) The limits to tree height. Nature 428:851–854

Lachenbruch B, Moore JR, Evans R (2011) Radial variation in wood structure and function in woody plants, and hypotheses for its occurrence. In: Meinzer FC, Lachenbruch B, Dawson TE (eds.) Size- and age-related changes in tree structure and function. Springer, Dordrecht, pp 121–164

Lancashire JR, Ennos AR (2002) Modelling the hydrodynamic resistance of bordered pits. J Exp Bot 53:1485–1493

Lang A (1979) A relay mechanism for phloem translocation. Ann Bot 44:141–145

Loepfe L, Martinez-Vilalta J, Piñol J, Mencuccini M (2007) The relevance of xylem network structure for plant hydraulic efficiency and safety. J Theor Biol 247:788–803

Lush WM (1976) Leaf structure and translocation of dry matter in a C3 and a C4 grass. Planta 130:235–244

MacDaniels LH (1918) The histology of the phloem in certain woody angiosperms. Am J Bot 5:347–378

Magnani F (1999) Plant energetics and population density. Scientific correspondence. Nature 398:572

Magnani F, Mencuccini M, Grace J (2000) Age-related decline in stand productivity: the role of structural acclimation under hydraulic constraints. Plant Cell Environ 23:251–263

Magnani F, Grace J, Borghetti M (2002) Adjustment of tree structure in response to the environment under hydraulic constraints. Funct Ecol 16:385–393

Mäkelä A (1986) Implications of the pipe model theory on dry matter partitioning and height growth in trees. J Theor Biol 123:103–120

Mäkelä A (1989) Modeling structural-functional relationships in whole-tree growth: resource allocation. In: Dixon RH, Meldahl RS, Ruark GA, Warren WG (eds.) Process modelling of forest growth responses to environmental stresses. Timber Press, Portland, pp 81–95

Mäkelä A, Hari P (1986) Stand growth model based on carbon uptake and allocation in individual trees. Ecol Model 33:205–229

Mäkelä A, Valentine H (2006) The quarter-power scaling model does not imply size-invariant hydraulic resistance in plants. J Theor Biol 243:283–285

Martinez-Vilalta J, Korakaki E, Vanderklein D, Mencuccini M (2007a) Belowground hydraulic conductance is a function of environmental conditions and tree size in Scots pine. Funct Ecol 6:1072–1083

Martinez-Vilalta J, Vanderklein D, Lee S, Mencuccini M (2007b) Tree height and the age-related decline in growth in Scots pine (*Pinus sylvestris* L.). Oecologia 150:529–544

McCulloh KA, Sperry JS (2005) Patterns in hydraulic architecture and their implications for transport efficiency. Tree Physiol 25:257–267

McCulloh KA, Sperry JS, Adler FR (2003) Water transport in plants obeys Murray's law. Nature 421:939–942

McCulloh KA, Sperry JS, Adler FR (2004) Murray's law and the hydraulic vs. mechanic functioning of wood. Funct Ecol 18:931–938

McDowell NG, Barnard H, Bond BJ, Hinckley T, Hubbard RM, Ishii H, Köstner B, Magnani F, Marshall JD, Meinzer FC, Phillips N, Ryan MG, Whitehead D (2002) The relationship between tree height and leaf area: sapwood area ratio. Oecologia 132:12–20

McDowell NG, Bond BJ, Hill LT, Ryan MG, Whitehead D (2011) Relationships between tree height and carbon isotope discrimination. In: Meinzer FC, Lachenbruch B, Dawson TE (eds.) Size- and age-related changes in tree structure and function. Springer, Dordrecht, pp 255–286

Meinzer FC, James SA, Goldstein G, Woodruff D (2003) Whole-tree water transport scales with sapwood capacitance in tropical forest canopy trees. Plant Cell Environ 26:1147–1155

Meinzer FC, Brooks JR, Domec JC, Gartner BL, Warren JM, Woodruff DR, Bible K, Shawn DC (2006) Dynamics of water transport and storage in conifers studied with deuterium and heat tracing techniques. Plant Cell Environ 29:105–114

Mencuccini M (2002) Hydraulic constraints in the functional scaling of trees. Tree Physiol 22:553–565

Mencuccini M (2003) The ecological significance of long distance water transport: short-term regulation and long-term acclimation across plant growth forms. Plant Cell Environ 26:163–182

Mencuccini M, Hölttä T (2010) The significance of phloem transport for the speed of link between canopy photosynthesis and belowground respiration. New Phytol 185:189–203

Mencuccini M, Martinez-Vilalta J, Vanderklein D, Hamid HA, Korakaki E, Lee S, Michiels B (2005) Size-mediated ageing reduces vigour in tall trees. Ecol Lett 8:1183–1190

Mencuccini M, Hölttä T, Petit G, Magnani F (2007) Sanio's Laws revisited. Size-dependent changes in the xylem architecture of trees. Ecol Lett 10:1084–1093

Mencuccini M, Martinez-Vilalta J, Piñol J, Loepfe L, Burnat M, Alvarez X, Camacho J, Gil D (2010) A quantitative and statistically robust method for the determination of xylem conduit spatial distribution. Am J Bot 97:1247–1259

Mikan CJ, Zak DR, Kubiske ME, Pregitzer KS (2000) Combined effects of atmospheric CO_2 and N availability on the belowground carbon and nitrogen dynamics of aspen mesocosms. Oecologia 124:432–445

Münch E (1927) Versuche über den Saftkreislauf. Ber Dtsch Bot Gesel 45:340–356

Murray CD (1926) The physiological principle of minimum work. I. The vascular system and the cost of blood volume. Proc Natl Acad Sci USA 12:207–214

Petit G, Anfodillo T, Mencuccini M (2008) Tapering of xylem conduits and hydraulic limitations in sycamore (*Acer pseudoplatanus* L.) trees. New Phytol 177:653–664

Petit G, Pfautsch S, Anfodillo T, Adams MA (2010) The challenge of tree height in *Eucalyptus regnans*: when xylem tapering overcomes hydraulic resistance. New Phytol 187:1146–1153

Phillips RJ, Dungan SR (1993) Asymptotic analysis of flow in sieve tubes with semi-permeable walls. J Theor Biol 162:465–485

Pittermann J, Sperry JS, Hacke UG, Wheeler JK, Sikkema EH (2006a) Inter-tracheid pitting and the hydraulic efficiency of conifer wood: the role of tracheid allometry and cavitation protection. Am J Bot 93:1265–1273

Pittermann J, Sperry JS, Wheeler JK, Hacke UG, Sikkema EH (2006b) Mechanical reinforcement of tracheids compromises the hydraulic efficiency of conifer xylem. Plant Cell Environ 29:1618–1628

Quilhó T, Pereira H, Richter HG (2000) Within–tree variation in phloem cell dimensions and proportions in *Eucalyptus globulus*. IAWA J 21:31–40

Raskatov PB, Kosichenko NE (1968) On the characterization of the phloem elements in different sections of the aspen (*Populus tremula* L.) trunk. Bot Z SSSR 53:1612–16

Reich PB, Tjolker MG, Machado J-L, Oleksin J (2006) Universal scaling of respiratory metabolism, size and nitrogen in plants. Nature 439:457–461

Roderick ML, Berry SL (2001) Linking wood density with tree growth and environment: a theoretical analysis based on the motion of water. New Phytol 149:473–485

Rosner S, Baier P, Kikuta SB (2001) Osmotic potential of Norway spruce [*Picea abies* (L.) Karst.] secondary phloem in relation to anatomy. Trees 15:472–482

Ryan MG, Phillips N, Bond BJ (2006) The hydraulic limitation hypothesis revisited. Plant Cell Environ 29:367–381

Sack L, Marañón T, Grubb P (2002) Comment on: global allocation rules for patterns of biomass partitioning (Science, 295:1517–20). Science 296:1923

Sala A, Hoch G (2009) Height-related growth declines in ponderosa pine are not due to carbon limitation. Plant Cell Environ 32:22–30

Sala A, Fouts W, Hoch G (2011) Carbon storage in trees: does relative carbon supply decrease with tree size? In: Meinzer FC, Lachenbruch B, Dawson TE (eds.) Size- and age-related changes in tree structure and function. Springer, Dordrecht, pp 287–306

Scholz FG, Bucci SJ, Goldstein G, Meinzer FC, Franco AC, Miralles-Wilhelm F (2007) Biophysical properties and functional significance of stem water storage tissues in neo-tropical savanna trees. Plant Cell Environ 30:236–248

Scholz FG, Phillips NG, Bucci SJ, Meinzer FC, Goldstein G (2011) Hydraulic capacitance: biophysics and functional significance of internal water sources in relation to tree size. In: Meinzer FC, Lachenbruch B, Dawson TE (eds.) Size- and age-related changes in tree structure and function. Springer, Dordrecht, pp 341–361

Schulz A, Behnke H-D (1986) Fluorescence and electron microscopic investigations of the phloem of beech, Norway spruce and silver fir, showing different degrees of damage symptoms. Conference2 Statuskolloquium des PEF, Karlsruhe (Germany)

Shinozaki K, Yoda K, Hozumi K, Kira T (1964a) A quantitative analysis of plant form – the pipe model theory. I. Basic analyses. Jpn J Ecol 14:97–105

Shinozaki K, Yoda K, Hozumi K, Kira T (1964b) A quantitative analysis of plant form – the pipe model theory. II. Further evidence of the theory and its application in forest ecology. Jpn J Ecol 14:133–139

Sillett SC, Van Pelt R, Koch GW, Ambrose AR, Carroll AL, Antoine ME, Mifsud BM (2010) Increasing wood production through old age in tall trees. For Ecol Manag 259:976–994

Sperry JS, Hacke UG (2004) Analysis of circular bordered pit function. I. Angiosperm vessels with homogenous pit membranes. Am J Bot 91:369–385

Sperry JS, Hacke UG, Pittermann J (2006) Size and function in conifer tracheids and angiosperm vessels. Am J Bot 93:1490–1500

Sperry JS, Meinzer FC, McCulloh KA (2008) Safety and efficiency conflicts in hydraulic architecture: scaling from tissues to trees. Plant Cell Environ 31:632–645

Stahel J (1968) Quantitative und qualitative Alterungsphänomene in Pappeln (*Populus* × *euramericana* (Dode) Guinier cv. 'Robusta'). Wood Prod 26:418–427

Steinmann K, Siegwolf RTW, Saurer M, Körner C (2004) Carbon fluxes to the soil in a mature temperate forest assessed by ^{13}C isotope tracing. Oecologia 141:489–501

Steppe K, Niinemets Ü, Teskey RO (2011) Tree size- and age-related changes in leaf physiology and their influence on carbon gain. In: Meinzer FC, Lachenbruch B, Dawson TE (eds.) Size- and age-related changes in tree structure and function. Springer, Dordrecht, pp 235–253

Thompson M (2006) Phloem: the long and the short of it. Trends Plant Sci 11:26–32

Thompson MV, Holbrook NM (2003a) Application of a single-solute non-steady-state phloem model to the study of long-distance assimilate transport. J Theor Biol 220:419–455

Thompson MV, Holbrook NM (2003b) Scaling phloem transport: water potential equilibrium and osmoregulatory flow. Plant Cell Environ 26:1561–1577

Thompson MV, Holbrook NM (2004) Scaling phloem transport: information transmission. Plant Cell Environ 27:509–519

Trockenbrodt M (1994) Quantitative changes of some anatomical characters during bark development in *Quercus robur*, *Ulmus glabra*, *Populus tremula* and *Betula pendula*. IAWA J 15:387–398

Tyree MT, Christy AL, Ferrier JM (1974) A simpler iterative steady state solution of Münch pressure flow systems applied to long and short translocation paths. Plant Physiol 54:589–600

Vollenweider P, Dustin I, Hofer R, Vittoz P, Hainard P (1994) A study of the cambial zone and conductive phloem of common beech (*Fagus sylvatica* L.) using an image analysis method. Trees 9:106–112

Weitz JS, Ogle K, Horn HS (2006) Ontogenetically stable hydraulic design in woody plants. Funct Ecol 20:191–199

West GB, Brown JH, Enquist BJ (1997) A general model for the origin of allometric scaling laws in biology. Science 276:122–126

West GB, Brown JH, Enquist BJ (1999) A general model for the structure and allometry of plant vascular system. Nature 400:664–667

Wheeler JK, Sperry JS, Hacke UG, Hoang N (2005) Inter-vessel pitting and cavitation in woody Rosaceae and other vesselled plants: a basis for a safety versus efficiency trade-off in xylem transport. Plant Cell Environ 28:800–812

Woodruff DR, Meinzer FC (2011) Size-dependent changes in biophysical control of tree growth: the role of turgor. In: Meinzer FC, Lachenbruch B, Dawson TE (eds.) Size- and age-related changes in tree structure and function. Springer, Dordrecht, pp 363–384

Woodruff DR, Bond BJ, Meinzer FC (2004) Does turgor limit growth in tall trees? Plant Cell Environ 27:229–236

Woodruff DR, McCulloh KA, Warren JM, Meinzer FC, Lachenbruch B (2007) Impacts of tree height on leaf hydraulic architecture and stomatal control in Douglas-fir. Plant Cell Environ 30:559–569

Woodruff DR, Meinzer FC, Lachenbruch B (2008) Height-related trends in leaf xylem anatomy and shoot hydraulic characteristics in a tall conifer: safety versus efficiency in water transport. New Phytol 180:90–99

Zaehle S, Sitch S, Prentice IC, Liski J, Cramer W, Erhard M, Hickler T, Smith B (2006) The importance of age-related decline in forest NPP for modeling regional carbon balances. Ecol Appl 16:1555–1574

Zimmermann MH, Brown CL (1971) Trees, structure and function. Springer, Berlin

Chapter 13
Hydraulic Capacitance: Biophysics and Functional Significance of Internal Water Sources in Relation to Tree Size

Fabian G. Scholz, Nathan G. Phillips, Sandra J. Bucci, Frederick C. Meinzer, and Guillermo Goldstein

Abstract Components of the tree water transport pathway; roots, trunks, branches and leaves; can also serve as water storage compartments and therefore act transiently as intermediate sources of water for transpiring leaves. However, most previous work has focused on gradual depletion and recharge of tree internal water reserves as soil water availability varies over seasonal cycles. This chapter focuses on the underappreciated role that internal water storage plays in stabilizing the physiological function of trees under the dynamic conditions that prevail over the

F.G. Scholz (✉)
Consejo Ncional de Investigaciones Cientificas y Tecnicas (CONICET), Argentina

Departamento de Biología, Facultad de Ciencias Naturales,
Universidad Nacional de la Patagonia San Juan Bosco, Comodoro Rivadavia, Argentina
e-mail: fgscholz@unpata.edu.ar; fgscholz@yahoo.com

N.G. Phillips
Department of Geography and Environment, Boston University,
675 Commonwealth Avenue, Boston, MA 02215, USA

S.J. Bucci
Consejo Ncional de Investigaciones Cientificas y Tecnicas (CONICET), Argentina

Departamento de Biología, Facultad de Ciencias Naturales,
Universidad Nacional de la Patagonia San Juan Bosco, Comodoro Rivadavia, Argentina

F.C. Meinzer
USDA Forest Service, Pacific Northwest Research Station, 3200 SW Jefferson Way,
Corvallis, OR 97331, USA

G. Goldstein
Consejo Ncional de Investigaciones Cientificas y Tecnicas (CONICET), Argentina

Laboratorio de Ecología Funcional, Departamento de Ecología, Genética y Evolución,
Facultad de Ciencias Exactas y Naturales, Universidad de Buenos Aires,
Buenos Aires, Argentina

Department of Biology, University of Miami, P.O. Box 249118, Coral Gables, FL 33124, USA

F.C. Meinzer et al. (eds.), *Size- and Age-Related Changes in Tree Structure and Function*, Tree Physiology 4, DOI 10.1007/978-94-007-1242-3_13,

course of a day. Capacitive discharge of water into the transpiration stream can buffer daily fluctuations in xylem tension, thereby diminishing the risk of xylem embolism and hydraulic failure under dynamic conditions. Intrinsic sapwood capacitance and reliance on stored water increase with tree size. An inverse relationship between sapwood capacitance and resistance to embolism across diverse woody species suggests that above a minimum threshold value of capacitance, the tree survives by using capacitance to provide hydraulic safety by buffering fluctuations in tension, rather by relying on xylem structural features that directly reduce vulnerability to embolism. Progress in understanding the physiological role of capacitance in trees is impeded by non-uniformity in the way capacitance is measured and expressed, preventing much of the available information from being synthesized. To remedy this, standard protocols are described for defining and expressing capacitance and water storage capacity.

1 Introduction

Characterizing water flow along the soil–plant–atmosphere continuum with soil water as the sole source and the atmosphere as the sole sink with the plant's xylem linking the two is an oversimplification. It has been shown that components of the plant water transport pathway – roots, trunks, branches and leaves – can also serve as water storage compartments and therefore act transiently as intermediate sources of water for transpiring leaves (Waring and Running 1978; Waring et al. 1979; Goldstein et al. 1984; Holbrook and Sinclair 1992; Stratton et al. 2000; Meinzer et al. 2001). However, most previous work has focused on gradual depletion and recharge of plant internal water reserves as soil water availability varies over seasonal cycles. In this chapter we focus on the underappreciated role that internal water storage plays in stabilizing the physiological function of trees under the dynamic conditions that prevail over the course of a day.

True steady-state rates of water transport rarely, if ever, occur because both the properties of the water transport pathway and the environmental variables that drive transpiration change dynamically throughout the day (Hunt et al. 1991; Čermák and Nadezhdina 2011; Ewers et al. 2011). If instantaneous water fluxes through different plant compartments are not equal, then according to continuity laws a network of transient water sources and sinks must exist. These elements are capable of buffering transpiration-induced fluctuations in xylem tension that could result in embolism and hydraulic dysfunction. The transient buffering action of internal water reservoirs has been incorporated into models of whole-plant water transport that employ electric circuit analogues (Ohm's law) to specify a series of hydraulic capacitances and variable hydraulic resistances (e.g. Cowan 1972; Edwards et al. 1986; Tyree 1988; Williams et al. 1996; Phillips et al. 1997; Perämäki et al. 2001; Zweifel and Häsler 2001; Pettijohn et al. 2009). In terms of an electric circuit analogy, capacitive discharge of water into the transpiration stream in response to transpiration-induced increases in xylem tension should serve to dampen the rate and magnitude of fluctuations in xylem tension, thereby diminishing the risk of embolism and hydraulic failure under dynamic conditions.

"The rhythm of trees" (Zweifel 1999) is an apt term for capturing the role that capacitance plays in plant water transport dynamics, conferring elasticity or "buffering effects" on a system that otherwise would be highly inelastic and consequently susceptible to hydraulic failure. Steady-state treatments of water flow may or may not be appropriate for studies at a daily or longer time step, depending on the degree to which diurnal non-steady state behavior affects net daily fluxes of water or carbon (Phillips et al. 2003a), but non-steady state treatments must be used when the determination of diurnal dynamics of tree water uptake, xylem water flux, leaf water potential, and canopy transpiration are required. For example, isotopic methods for partitioning ecosystem transpiration from evaporation have found it necessary to account for diurnal non-steady state leaf and xylem water transport (e.g. Lai et al. 2006; Barnard et al. 2007).

Despite an increasing amount of information on plant water storage and its role in maintaining the integrity of the soil-to-leaf hydraulic continuum, it is difficult to compare results among studies and across species because of inconsistent use of terminology and associated units. For example, it is necessary to distinguish between the terms water storage capacity (WSC) and hydraulic capacitance (*C*), which, although related, have very different meanings but are sometimes used interchangeably. Water storage capacity (kg) is an extensive property and consequently is a function of the integrated volume of all the water storage compartments within a plant (Tyree and Yang 1990; Hunt et al. 1991; Tyree and Zimmermann 2002), whereas *C* reflects intrinsic differences in the biophysical properties of the storage tissues and their potential ability to release water to the xylem in response to variations in water status of different plant organs. *C* is generally defined as the ratio of change in tissue water volume to change in its water potential ($C = \Delta W/\Delta\Psi$, kg MPa^{-1}). Defined in this manner, it is clear that *C* as well as WSC will depend on plant size even though *C* is an intrinsic property of each type of storage tissue. Thus, for comparisons of intrinsic differences in *C* among species and individuals of different sizes, *C* should be normalized on a tissue volume basis, yielding intrinsic *C* in terms of the mass of water released per tissue volume and per change in water potential (kg m^{-3} MPa^{-1}, Meinzer et al. 2003; Scholz et al. 2007). Alternatively, relative water content-based capacitance (RWC MPa^{-1}) has been used to compare intrinsic *C* within and among species (Domec and Gartner 2001, 2002; Domec et al. 2005; Pratt et al. 2007). This measure shows the water content relative to the maximum that could be stored in a tissue, and its rate of change with changing Ψ. Although *C* varies in relation to tissue Ψ, it is typically greatest and nearly constant over the normal physiological operating range of Ψ under non-extreme conditions (Meinzer et al. 2003, 2008b; Scholz et al. 2007).

Although WSC and *C* are fundamentally different properties, both influence whole-tree transport properties and behavior. Tissues or organs with high WSC could have high or low *C*, and the relationship between WSC and *C* will depend on the morpho-anatomical characteristics of the tissues. On the other hand, the total amount of available water contained in a tissue (an estimate of WSC) should not be confused with the amount of water that is withdrawn and used from that tissue on a daily basis, which will depend on the tissue *C*, stomatal regulatory behavior and the environmental conditions experienced by the plant.

Information obtained using both experimental and theoretical approaches over the last few decades has increased our understanding of the functional significance of stored water and hydraulic capacitance for maintaining adequate water supply to the leaves. Accumulating experimental and theoretical information suggests convergent behavior among different plant life forms in relation to utilization of stored water and the potential physiological trade-offs involved. We have gained knowledge from direct and indirect methods to estimate tissue and whole plant water storage capacity and hydraulic capacitance, as well as from modeling approaches representing the hydraulic resistance (*R*) of different conductive tissues and the resistance pathway associated with discharge and recharge of tissue capacitance. Data are now available for several species from a variety of ecosystems covering a gradient of stem diameter or tree size and age in species with very different architectures, wood types and densities. In this context we analyze issues such as the role of hydraulic capacitance in regulating stomatal behavior and apparent stem and leaf hydraulic conductance under dynamic conditions. We also examine whether water released via embolism is the primary source of capacitance that buffers fluctuations in xylem tension, or whether this water comes from other storage compartments. We discuss how these relationships vary with tree size and wood anatomy.

2 Effects of Tree Size on Water Storage Capacity and Hydraulic Capacitance

Water flow in trees is subject to frictional resistance that potentially increases with path-length as the tree grows taller, hypothesized to ultimately lead to decreasing leaf-specific hydraulic conductance, stomatal conductance, photosynthesis and potentially further height growth (Ryan and Yoder 1997; Ryan et al. 2006; Woodruff and Meinzer 2011). Tall trees often exhibit morphological and physiological adjustments that mitigate hydraulic resistance limitations (Ishii 2011), including increased sapwood area to leaf area ratio (McDowell et al. 2002), greater sapwood hydraulic conductivity and embolism resistance (Burgess et al. 2006), lower minimum leaf water potential (Barnard and Ryan 2003) and increased reliance on stored water (Goldstein et al. 1998; Phillips et al. 2003a) and increasing whole-tree capacitance (Meinzer et al. 2004). These compensatory mechanisms are neither universal nor consistent across species (Ryan et al. 2006). For tropical trees it has been shown that total daily use of stored water increases with tree size (Goldstein et al. 1998; Meinzer et al. 2004). It is relatively straightforward to conclude that total daily use from water storage compartments is directly related to tree size: the bigger the tree the greater the volume of stored water used. Nevertheless, it is less intuitive to think about the way in which contributions of stored water as a proportion of total daily water consumption and hydraulic capacitance depend on tree size.

The relative contribution of stored water (mainly in stems) to total daily transpirational losses has been estimated in several species and varies widely, ranging from 10% to 50% among different tree species and ecosystems (e.g. Waring and Running 1978; Schulze et al. 1985; Tyree and Yang 1990; Holbrook and Sinclair 1992; Goldstein et al. 1998; Meinzer et al. 2003; Phillips et al. 2003a; Scholz et al. 2008). Total daily withdrawal of water from internal storage increased with trunk diameter from 0.34 to 0.98 m among four species of a seasonally dry tropical forest, but its relative contribution to the daily water budget remained constant at about 10% (Meinzer et al. 2004). The limitation of stored water use to a nearly constant fraction of total daily water use was related to a high degree of coordination between leaf hydraulic resistance and stomatal resistance. A decline in stomatal conductance from its early morning maximum limited transpiration and xylem tension, thereby preventing further discharge from the water reservoirs into the transpiration stream, and causing time courses of crown transpiration and basal sap flux to converge. In contrast, Scholz et al. (2007) found that the relative contribution of stored water to transpiration in small tropical savanna trees increased from 10% to 31% as stem diameter increased from 6 to 14 cm. Similarly, Phillips et al. (2003a) reported that daily reliance on stored water increased from about 7% to 25% in Douglas-fir trees 15–60 m tall, from 13% to 23% in white-oak trees 10–25 m tall and from 4% to 20% in ponderosa pine trees 12–36 m tall. These findings are mostly consistent (with notable exceptions; e.g. McDowell et al. 2002; Phillips et al. 2003b) with a reduced leaf area to sapwood volume ratio, which may be an adaptation to the hydraulic constraints imposed by increasing tree size that allows greater use of stored water relative to total daily water use. For savanna trees the contribution of stored water estimated by two independent approaches, (Scholz et al. 2007, 2008) was positively associated with apparent soil-to-leaf hydraulic conductance (transpiration/[$\Psi_{soil} - \Psi_{leaf}$]), implying that apparent hydraulic resistance could be transiently reduced by an increasing capacitive effect of stored water on transpirational water flux through the xylem.

Non-universal scaling of the relative contribution of stored water with tree size among trees from different ecosystems is probably a consequence of the strong dependence of stored water use on environmental conditions and the fact that water storage capacity is a non-intrinsic property (see Sect. 1). Even in the same tree, the contribution of stored water can change seasonally, being higher during drier periods (e.g. Phillips et al. 1997, 2003a; Chapotin et al. 2006; Bucci et al. 2008). Nevertheless, absolute daily reliance on stored water apparently increases with tree size in a similar manner across diverse species regardless of whether they are angiosperms or conifers or whether they are tropical or temperate (Fig. 13.1).

Tall trees experience increased risk of xylem embolism as tension in their water column increases with height because of path-length resistance and gravity (Ryan and Yoder 1997; Koch et al. 2004; Domec et al. 2008; Mencuccini et al. 2011). In addition to xylem structural modifications that can decrease the risk of embolism in tall trees (Domec et al. 2008), as discussed above, a higher capacitance could also contribute to avoidance of embolism via reliance on transient release of stored water to constrain transpiration-induced fluctuations in xylem tension (Meinzer et al. 2009). Consistent with this, the data for 35 species in Fig. 13.2 imply that larger or

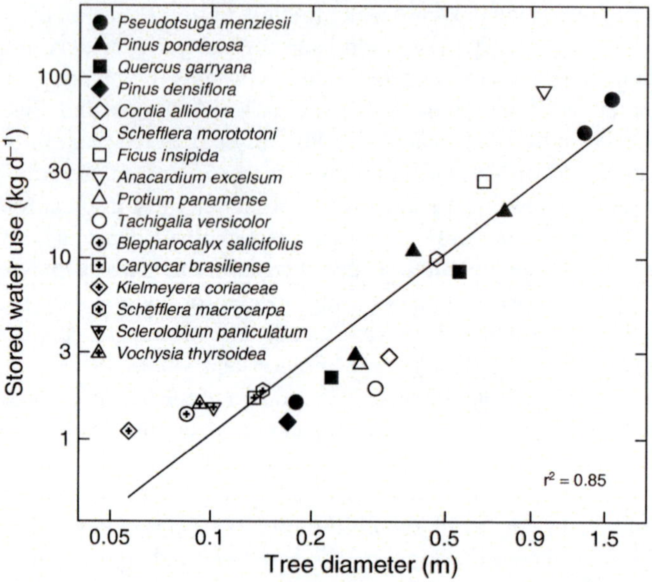

Fig. 13.1 Size-dependence of daily stored water use. The *solid line* represents a linear regression fitted to the log transformed data ($P < 0.0001$) (Data from Phillips et al. 2003a; Čermák et al. 2007; Maherali and DeLucia 2001; Kobayashi and Tanaka 2001; Meinzer et al. 2004; F.G. Scholz et al., 2008 and F.C. Meinzer 2004)

taller trees, independent of their total water storage capacity, can exchange a greater volume of water over the normal operating range of water potential of their storage tissues than smaller or shorter trees or that water in the sapwood of large trees may be released to the transpiration stream under less negative water potential than in smaller trees (Domec and Gartner 2001). Intrinsic *C* in Fig. 13.2 varied between 20 kg m^{-3} MPa^{-1} in apple trees (Landsberg et al. 1976) to more than 500 kg m^{-3} MPa^{-1} in *Pseudotsuga menziesii* (Waring and Running 1978; Meinzer et al. 2006). It seems quite remarkable, however, that specific capacitances in angiosperms and conifers span a similar range of values despite dramatic differences in their wood anatomy, particularly in the fraction of wood volume comprised of conduit lumens (Fig. 13.2).

As expected, trees with greater intrinsic *C* tend to exhibit higher (less negative) minimum branch water potential than trees with lower *C* (Fig. 13.3). The 18 tropical forest and savanna species represented in Fig. 13.3 span a height range of 2–38 m. Individuals with the greatest *C* and least negative values of branch Ψ also tended to be the tallest, implying that within limits, increasing *C* may completely or partially compensate for the impact of gravity and path-length resistance on maximum xylem tension. It is important to note that to study the relationship between *C* and daily minimum Ψ, they both must be assessed at the same or similar axial position in the plant. Such data in large trees are scarce. Nevertheless, axial trends of declining sapwood Ψ and *C* with increasing height in 34- to 45-m-tall ponderosa pine and Douglas-fir trees suggest that the positive correlation between stem Ψ and *C* shown

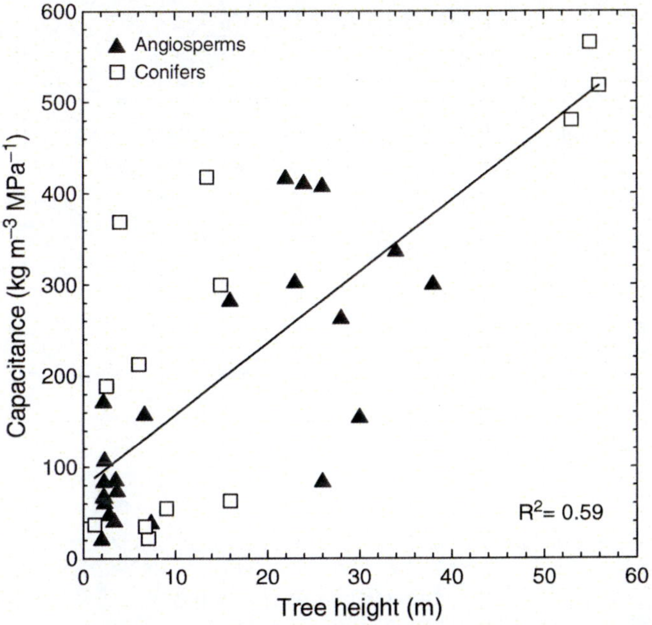

Fig. 13.2 Relationship between tree height and stem intrinsic hydraulic capacitance for conifers (*open symbols*) and angiosperms (*closed symbols*). The *solid line* represents the linear regression fitted to the data ($y = 79 + 8x$; $R^2 = 0.59$; $p < 0.0001$) (Landsberg et al. 1976; Waring and Running 1978; Waring et al. 1979; Tyree and Yang 1990; Domec and Gartner 2001; Meinzer et al. 2003, 2008a; Phillips et al. 2004; Scholz et al. 2007; Schulze et al. 1985)

in Fig. 13.3 may prove to be fairly general when values of Ψ are corrected for height-related trends in its gravitational component (Domec and Gartner 2001; Domec et al. 2005). The behavior depicted in Fig. 13.3 raises the possibility that two distinct mechanisms could be involved in providing adequate hydraulic safety margins under the dynamic conditions that prevail over the course of a typical day: (a) xylem structural features that enhance resistance to embolism and (b) transient capacitive discharge of water that prevents minimum xylem pressure from falling below thresholds that would provoke excessive embolism. A survey of several coniferous and angiosperm species showed that three types of hydraulic safety margins decreased with increasing species-specific daily minimum stem Ψ and stem C. These trends suggest a continuum of relative reliance on release of stored water to buffer fluctuations in xylem pressure versus xylem structural features that reduce vulnerability to embolism measured as P_{50}, the xylem pressure at 50% loss of hydraulic conductivity (Meinzer et al. 2009). In this context, data extracted from several studies reveal an inverse relationship between RWC-based capacitance and P_{50} (Fig. 13.4) and seem to reinforce the notion that above a minimum threshold value, the tree survives by using capacitance to provide hydraulic safety by buffering fluctuations in tension, rather by relying on xylem structural features that directly reduce vulnerability to embolism.

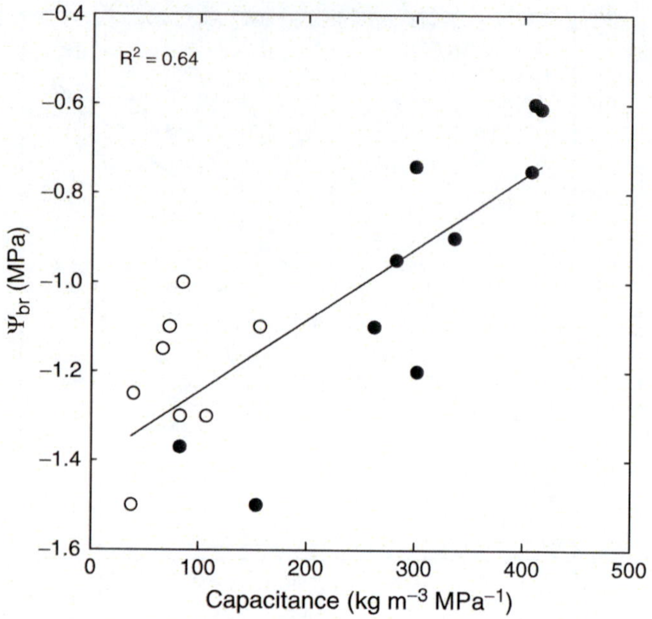

Fig. 13.3 Relationship between sapwood capacitance and minimum branch water potential (Ψ_{br}) for 18 tropical forest (*closed symbols*) and savanna (*open symbols*) species. Ψ_{br} was estimated with a pressure chamber from balance pressures of covered, non-transpiring leaves. The line is the linear regression fitted to the data ($y = -1.4 + 0.002x$; $p < 0.0001$) (Meinzer et al. 2003, 2008b; Scholz et al. 2007)

3 Extending Ohm's Law Analogy: The Time Constant for Water Transport (*RC*)

Several studies have considered the effect of hydraulic capacitance in driving transient responses between rates of change in water flux in tree crowns and trunks (Edwards et al. 1986; Hunt and Nobel 1987; Hunt et al. 1991; Landsberg et al. 1976; Powell and Thorpe 1977; Running 1980a, b; Phillips et al. 1997, 2008; Goldstein et al. 1998; Meinzer et al. 2003; Scholz et al. 2007). Study of the hydrodynamic behavior of trees allows for in vivo diagnosis of *R* and *C* that complements studies on excised tissues, and also allows for investigation of how stomatal dynamics may be coordinated with hydraulic dynamics (Whitehead and Teskey 1995; Phillips et al. 2004). Transient responses can be predicted using a time constant (τ) for water transport, which is the time (hours) required for water flow elsewhere in the plant to reach 67% of its steady state value following a step change in crown transpiration (Phillips et al. 1997). Extending an Ohm's law analogy to include capacitive elements, time constants associated with lags between changes in environmental variables that affect transpiration and changes in water flow within the plant can be estimated as the product of the hydraulic resistance and hydraulic capacitance ($R \times C$) between

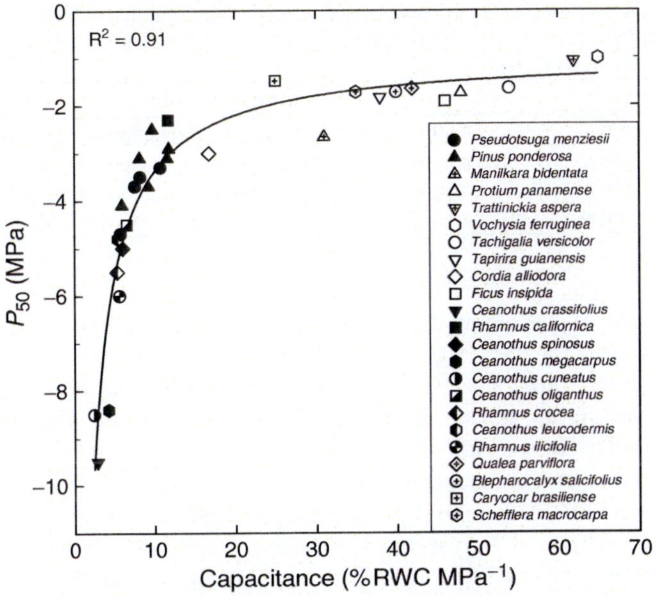

Fig. 13.4 Relationship between capacitance and the xylem pressure causing 50% loss of stem hydraulic conductivity (P_{50}) for several tree species (conifers and angiosperms) from different ecosystems (Domec and Gartner 2001, 2003; Bucci et al. 2006; Pratt et al. 2007; Scholz et al. 2007; Meinzer et al. 2008b; Sperry et al. 2008)

the soil and a reference point within the plant, usually the leaves. As changes in tissue hydration cause *C* and possibly *R* to vary and possibly co-vary, the whole-tree time constant for environmentally-driven changes in water flow may be a dynamic rather than fixed parameter characterizing water transport along the soil-to-atmosphere continuum (Meinzer et al. 2004, 2008b; Scholz et al. 2007). An analysis of τ for trees of different sizes and architectures could provide insights concerning how *C* and *R* may capture branch or whole-tree hydrodynamic behavior as well as how they may mutually adjust and compensate to affect water transport dynamics as plants grow larger.

In a 12-year-old *Pinus taeda* stand (diameter range: 5.3–11.8 cm; height range: 5.9–7.9 m) Phillips et al. (1997) found that while the time constants did not vary significantly within the range of tree sizes studied ($\tau \approx 0.8$h), hydraulic resistance and capacitance terms were individually dependent on stem cross-sectional area: whole-tree capacitance increased and resistance decreased with stem volume. This result may indicate an inverse adjustment of resistance and capacitance to maintain a similar time constant over the range of tree sizes studied. Hunt et al. (1991), analyzing steady-state and non-steady-state water flow models at various hierarchical scales for plants of different biomes suggested that resistances and capacitances may be inversely related as plants become larger. They also reported similar time constants (τ = 1.5–2.3 h) for plants as different as the shrub *Encelia farinosa*, a

Table 13.1 Relative resistances, capacitances, and time constants of different life forms, illustrating how similar hydrodynamics (plants with "medium" τ) may occur in plants of vastly different size and underlying hydraulic properties

R	*C*	τ	Example life form
Small	Small	Small	Phreatophyte (Hultine et al. 2003)
Small	Large	Medium	Seasonally dry tropical tree (Tyree et al. 1991)
Large	Small	Medium	Desert shrub (Hunt and Nobel 1987)
Large	Large	Large	Desert succulent (Hunt and Nobel 1987)

lodgepole pine tree and the hardwood tree *Malus pumila*. Similarly, Meinzer et al. (2004) determined for four large tropical forest canopy tree species comprising a substantial range of tree size (0.34–0.98 m in diameter), that time constants for changes in transpiration and sap flow in response to changing environmental conditions were similar among individuals (~0.53 h), indicating that *R* and *C* co-varied with tree size in an inverse manner. Results of Perämäki et al. (2005) from a modeling approach describing the dynamics of xylem water tensions suggested that the time constant was nonlinear with respect to both stem length and sapwood volume for modeled trees from 15 to 58 m tall and that at a given stem length, an increase in sapwood volume increased the time constant. Their modeling results were in good agreement with field measurements at three different heights on a 27-m-tall Scots pine (*Pinus sylvestris L.*) tree.

We analyzed relationships between τ and plant height using data for different plant forms over a broad range of τ (0.08–3.5 h) and height (0.3–56 m) (Landsberg et al. 1976; Powell and Thorpe 1977; Running 1980a, b; Nobel and Jordan 1983; Edwards et al. 1986; Hunt and Nobel 1987; Loustau et al. 1996; Phillips et al. 1997; Meinzer et al. 2004) and found no significant correlation between these variables. The scarcity of data on τ and the large variation in τ in the available data make it difficult to identify general rules that could help to predict and understand the role of *C* across a gradient of tree size in different species. This is not surprising because *R* and *C* can differ substantially among life forms, leading to variation in τ that is not simply related to plant size (Table 13.1).

4 Direct and Indirect Methods, Experimental and Theoretical Approaches for Estimating Tissue and Whole Plant Water Storage Capacity and Capacitance

Plant WSC and *C* may be estimated directly on excised plant material, or can be inferred indirectly from their effects, in combination with *R*, on water flow dynamics within trees. Direct and indirect approaches are complementary because their strengths and limitations are relatively non-overlapping. Direct methods are good 'differentiators' of tissue-specific *C* but are difficult to apply to all important parts of trees, whereas indirect (time series analysis) methods, while poor 'differentiators', are good 'integrators' of whole tree hydrodynamics and impacts of bulk-scale *R* and *C*, at least between any two points within a tree that are bounded by time series water flow measurements. Moreover, direct and indirect methods provide important checks on each other. Direct measurements of *R* and *C* are done under steady state (*R*) or dynamic equilibrium (*C*) conditions, with their product providing a predicted τ in spite of the fact that no transient measurements were made; conversely, analysis of water transport time series, while indirectly revealing effects of *R* and *C*, directly assess those impacts, under actual non steady state conditions in the field.

Results of studies with sap flow measurements show that plant utilization of water stored in their tissues can contribute significantly to overall water loss and therefore exert strong hydraulic controls on transport dynamics (Phillips et al. 1997, 2004; Wullschleger et al. 1998; Meinzer et al. 2001, 2003; James et al. 2003; Scholz et al. 2007, 2008; Daley et al. 2008). Use of electronic dendrometers to monitor diurnal fluctuations in stem, branch or root radius constitute another important tool in *C*-related studies. Irvine and Grace (1997) showed that the relationship between diameter fluctuations and water tension is linear, direct and immediate. In the same way, all observed patterns of wood shrinkage and expansion in the study of Offenthaler et al. (2001) were consistent with the hypothesis that xylem water tension leads to an elastic contraction of xylem conduits. However, measuring diameter fluctuations at different heights in a single tree Perämäki et al. (2001) inferred that there was no significant water storage due to elastic shrinkage between the evaporating needles and the base of the stem of a 37-year-old Scots pine in southern Finland. Utilizing electronic dendrometers together with stem psychrometers it was possible to determine that stem *C* contributes to homeostasis in the diurnal and seasonal water balance of representative Brazilian savanna trees of different size (Scholz et al. 2007). Combined use of sap flow probes (Čermák and Nadezhdina 2011), electronic dendrometers and model simulations indicate that stored water plays an important role not only during drought periods, but whenever water transport occurs within 6-year-old Norway spruce trees (Zweifel et al. 2001). From an analytical point of view it is important to point out that whereas capacitance is often considered to have a constant value, (Tyree et al 1991; Williams et al. 1996) there is an increasing body of evidence suggesting nonlinear relationships between *C* and Ψ_L or water storage depletion/renewal rates; consequently state-dependent capacitance treatments are required to capture the true stored water dynamics (Meinzer et al. 2003; Scholz et al. 2007; Pettijohn et al. 2009).

5 Sources of Capacitance

5.1 Role of Xylem Embolism

As previously indicated, stored water could play a role in avoiding embolism, or in refilling embolized conduits. Alternatively, embolism could contribute to total storage capacity for the maintenance of plant water status despite the apparent negative impact of embolism on hydraulic conductivity (Sperry 1995). It is known that water storage in plants can occur by three mechanisms: elastic storage, capillary storage and release via embolism (Tyree and Yang 1990; Sperry and Pockman 1993. Measurements to determine the contribution of each mechanism indicate that capacitance is much higher under tensions at which embolism occurs than under smaller tensions associated with elastic deformation. Tyree and Yang (1990) indicated that in very dry environments water released by embolism events can represent a substantial fraction of the total water transpired and therefore be important to the short term survival of trees. Lo Gullo and Salleo (1992) also concluded that the amount of water withdrawn from cavitated xylem conduits in twigs of *Populus deltoides* was sufficient to dampen diurnal fluctuations in leaf water deficit when stress was moderate. Stem water content varies seasonally in trees (e.g. Waring and Running 1978; Scholz et al. 2007) and, at least in conifers, this change is associated with embolized conduits (Holtta et al. 2009). In addition, the results of Zwieniecki and Holbrook (1998) and Bucci et al. (2003) suggest that a substantial fraction of the functional xylem conduits is dynamically emptied and refilled throughout the day. Ambrose et al. (2009) also suggested that the water released from embolism in *Sequoia sempervirens* and *S. giganteum* could be substantial considering the large size of the trees. However, Meinzer et al. (2008b) estimated that in several tropical tree species water released via embolism of vessels accounted for only about 15% of total daily reliance on *C*. The remaining 85% of daily capacitive discharge was apparently derived from parenchyma, fibers and intercellular spaces. Consistent with this, an analysis of the time at which stored water enters the transpiration stream and the time at which the highest loss of xylem specific conductivity is reached in two Brazilian savanna species (Bucci et al. 2003; Scholz et al. 2008) revealed that most of the daily withdrawal from storage occurred before substantial embolism occurred (Fig. 13.5). Although the greatest use of water from internal reservoirs occurred between 830 and 930 h for *Schefflera macrocarpa* and between 900 and 1,200 h for *Caryocar brasiliense*, maximal embolism-induced loss of xylem conductivity occurred at 1,100 and 1,300 h, respectively. More recently, a dynamic model simulation revealed that water released via embolism could play a significant buffering role in xylem water relations (Holtta et al. 2009). This positive buffering effect increased in importance when the ratio of the total amount of water in xylem conduits to transpiration rate increased, when the ratio between the volume of water freed to whole plant hydraulic conductance loss increased, when embolism occurred closer to leaves, or if the associated loss of hydraulic conductance can be repaired over a short time interval. Holtta et al. (2009) also suggested that the importance of embolism as a source of capacitance could be greater in conifers which have a lower proportion of living cells in their xylem than angiosperms. The capacitive release of water via embolism does not necessarily imply a tradeoff involving reduced total water transport efficiency. For example in the conifer Douglas-fir Domec and Gartner (2002) found that within a growth ring, earlywood had a lower water storage capacity than latewood and was much less vulnerable to embolism at water potentials experienced in the field under normal conditions. It was hypothesized that for the natural range of water potential experienced in plants, earlywood

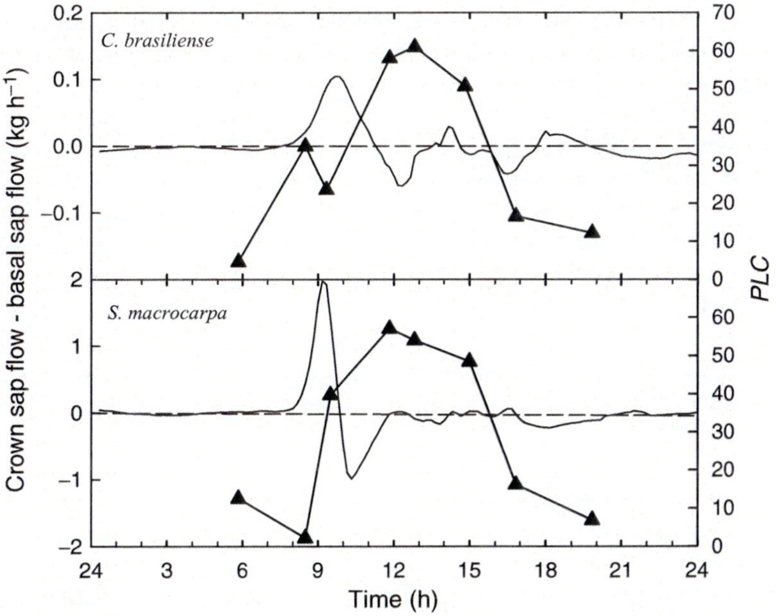

Fig. 13.5 Daily time courses of percentage loss of specific conductivity, *PLC* (*line* and *symbol*) and difference between basal and crown sap flow (*solid line*) in *Caryocar brasiliense* and *Schefflera macrocarpa*. Negative values of crown minus basal sap flow indicate time periods when water transpired was preferentially withdrawn from stem water storage, and positive values indicate time periods when water from the soil was refilling stem storage (Modified from Bucci et al. 2003; Scholz et al. 2008)

would best serve the tree by conducting water efficiently and safely and latewood by providing an important amount of stored water because the specific conductivity of earlywood was about an order of magnitude greater than that of latewood (Domec and Gartner 2002). On the other hand, in diffuse-porous angiosperms the capacitive effect of the high proportion of the living axial and ray parenchyma in relation to total xylem volume could substitute for embolism-induced water release.

5.2 Water Stored in Leaves and Roots

Leaves can also act as a water storage compartment. Water can be stored in the epidermis, mesophyll, specialized cells or in extracellular mucilage (Morse 1990; Sack et al. 2003). Leaf capacitance expressed as the water released per change in leaf water potential per leaf area (C_{leaf}) can vary between 150 and 1,600 mmol m^{-2} MPa^{-1} across temperate and tropical species (e.g. Tyree et al. 1991; Machado and Tyree 1994; Hao et al. 2008; Johnson et al. 2009, S.J. Bucci et al. 2011). Brodribb et al. (2005) compared tropical angiosperms and fern species from Costa Rica with temperate angiosperms and conifers from Chile and found that conifers exhibit substantial leaf capacitance with leaves able to function for long periods without a hydraulic connection to the plant. Hao et al. (2008) compared the leaf capacitance in congeneric species of savannas and forests and found that the former have higher C_{leaf} and were better buffered from rapid changes in leaf water potentials compared to forest species. The contribution of leaf water storage to total daily transpiration is in general lower than that of stem water storage. For example for two tropical species *Ochroma pyramydale* and *Pseudobombax septenatum*, stem water storage contributed 20% to the daily transpiration while leaf water storage contributed less than 1% (Machado and Tyree 1994). Similarly, diurnal use of water stored in stems and leaves in *Thuja occidentalis* was 16% and 5.5%, respectively, of the daily transpiration (Tyree et al. 1991). For succulent species like *Agave deserti*, leaf water storage can contribute as much as 50% to the total flow during maximum transpiration (Smith et al. 1987).

Although the water stored in leaves generally accounts for only a small fraction of daily transpiration, it may play an important role in buffering changes in leaf water potential associated with transpiration-induced water movement along the continuum of resistances from roots to leaves (Sack and Tyree 2005). C_{leaf} is positively correlated with leaf hydraulic conductance (K_{leaf}) in temperate and tropical tree species (Sack et al. 2003; Brodribb et al. 2005; Sack and Tyree 2005; Hao et al. 2008; Domec et al. 2009), thereby affecting the dependence between K_{leaf} and stomatal conductance (Brodribb et al. 2005). In addition to the buffering effect on daily fluctuations in leaf water potential, C_{leaf} may help to replace losses from cuticular transpiration when stomata are closed (Lamont and Lamont 2000). C_{leaf} could play an increasingly important role in buffering fluctuations in leaf water potential as path-length resistances and the magnitude of the gravitational component of water potential increase with tree height. However, no vertical trends in C_{leaf} were detected over a range of 40 m in a study of height-related trends in leaf hydraulic architecture and stomatal control in *Pseudotsuga menziesii* (Woodruff et al. 2007). Similarly, no vertical trend in C_{leaf} was found across four major plant life forms ranging across more than an order of magnitude in size (Hunt et al. 1991). Yet within leaf tissue, as in whole plants as described earlier, variation in leaf hydraulic resistance (R_{leaf}) and C_{leaf} among life forms were inversely correlated to produce relatively uniform τ_{leaf} (Coefficient of Variation 19%, 125% and 68%, respectively, Hunt et al. 1991). For the open grown plants studied in Hunt et al. (1991), this result may indicate that leaves of plants of similar exposure but different size need similar *time* to react to changing environmental conditions and not simply different *quantities* of stored water in leaves.

Roots not only constitute the pathway for water entry into the plant, but also can act as intermediate water sources replacing early morning transpirational water loss from leaves and stems. Root capacitance generally is ignored in studies because it is usually assumed to be small relative to that of the stems and leaves. However, for four dominant woody species of a Neotropical savanna in Brazil, root capacitance varied between 47 and 171 kg m^{-3} MPa^{-1} (Domec et al. 2006) while stem capacitance varied between 67 and 107 kg m^{-3} MPa^{-1} (Scholz et al. 2007). In these savannas, tree roots show daily fluctuations in their cross-sectional area similar to those observed in branches and stems, suggesting daily discharge and refilling of root water storage tissues (Scholz et al. 2008). Root capacitance in savanna trees increases with rooting depth. For example, *Kielmeyera coriaceae*, a species with lateral roots extending to a depth of 30 cm has higher root capacitance (171 kg m^{-3} MPa^{-1}) than *Blepharocalyx salicifolius* (47 kg m^{-3} MPa^{-1}) with lateral roots at 5 cm soil depth. But when soil-to-root water potential gradients are taken into account, the daily water release per volume of root xylem is similar between species (7 kg m^{-3}). High root capacitance in plants with deep roots may play a role in minimizing temporal imbalances between water supply and demand resulting in a lower percentage of root conductivity loss when the soil is dry than when soil is relatively wet (Domec et al. 2006). Although high root capacitance appears to have a cost associated with high vulnerability to embolism, root embolism in Brazilian savanna species is reversible on a daily basis (Domec et al. 2006). Studies are needed to learn whether root capacitance has any functional link to tree size or if the immediate rhizosphere environment is the main determinant of root capacitance.

5.3 Other Potential Water Storage Compartments

In vascular tissues of woody plants water can be stored in different living compartments in addition to the xylem conduits and extracellular spaces. Sapwood contains xylem parenchyma and ray cells among other cell types that can function in water storage and exhibit high capacitance (Holbrook 1995; Meinzer et al. 2003, 2006; Scholz et al. 2007). Additionally, some plants develop conspicuous fleshy tissues external to the sapwood that can also be used as water storage compartments. For example, stems of several Brazilian savanna woody species have outer parenchyma layers that can constitute more than 30% of the stem cross-sectional area (Scholz et al. 2007). Despite the relatively large cross sectional area of these tissues and although their RWC varied up to 12% diurnally, their contribution to total stem capacitance was relatively low. The role of water stored in these tissues may be more related to other functions such as protection from frequent fires than to water transport dynamics. Similarly, Sevanto et al. (2002) found no evidence that the bark of Scots pine functioned as a water storage compartment. On the other hand, water stored in the bark of *Picea abies* trees contributed directly to total transpiration and the relative contribution was independent of the age of the trees studied (Zweifel 1999; Zweifel et al. 2000; Zweifel and Häsler 2001).

Parenchymatous cortical and pith tissues can also be effective as water reservoirs. For example, the well-developed inner and outer parenchyma tissues of *Fouquieria columnaris*, a tree species endemic to arid and semi-arid regions of Mexico, can supply the transpiring leaves and stems with sufficient water for several weeks during long dry seasons (Franco-Viscaino et al. 1990; Nielsen et al. 1990). Caulescent giant rosette species of the genus *Espeletia* growing in tropical alpine environments have voluminous succulent pith consisting of parenchyma cells located in the centre of their woody stem. Release of stored water from the pith into the transpiration stream mitigates leaf water deficits in the early morning when soil water is frozen and root water uptake is impaired by low temperatures (Goldstein et al. 1984).

6 Final Remarks and Open Questions

A number of persistent difficulties need to be overcome in order to gain a more complete understanding of the role of water storage capacity and *C* in maintaining the integrity of the plant water transport system and specifically in buffering daily fluctuations in xylem pressure in intact plants. These include the general scarcity of data on hydraulic capacitance, a lack of relevant data for several types of ecosystems, and fewer studies on temperate trees compared with tropical trees. Additionally, with the exception of studies of Domec and Gartner (2002, 2003), information on *C*, *R*, and wood density for a single tree over time or even for trees of the same species at different heights and/or depths in the sapwood is lacking. Moreover, as emphasized earlier, sap flow techniques constitute a potentially valuable methodology for studies of *C* and utilization of stored water. Despite the fact that sap flow dynamics at different depths in the xylem are often recorded, this information has rarely been used to characterize corresponding radial variation in withdrawal and recharge of water stored in the sapwood (e.g. Domec and Gartner 2002, 2005; James et al. 2003; Phillips et al. 2003a; Ford et al. 2004) As a result, we could be losing information on spatially-differentiated water use within xylem and how it varies during soil drying cycles. Thus there is need for more information not only on total sapwood area and corresponding fluxes in an integrated way but also an understanding of whether all sapwood area is similarly functional in its capacity or contribution of stored water to the dynamics of water movement.

Progress in understanding the physiological role of *C* in trees is also impeded by non-uniformity in the way *C* is measured and expressed, preventing much of the available information from being synthesized. This becomes especially crucial when trying to analyze available information in a tree size context. To this end, we suggest using whenever possible the intrinsic hydraulic capacitance, *C* normalized by tissue volume as defined here, and WSC as defined by Waring et al. (1979). As simple as this recommended standardization is, if adopted widely, it should lead rapidly to amassing a large body of comparable and integrative data which could allow the identification of general rules for water storage and capacitance in a tree size context.

As indicated in the introduction, our main intent was to raise awareness of capacitance as an underappreciated property of the water transport system that can have a large impact on daily water transport dynamics, especially in relation to avoidance of hydraulic failure in intact trees. The traditional emphasis on studies of hydraulic architecture carried out on small plant segments subjected to steady or quasi-steady state conditions may have led to over-interpretation of their implications for intact plants (Meinzer et al. 2010). Determination of hydraulic traits such as P_{50} and specific conductivity of a terminal branch segment is certainly useful but provides only a partial understanding of hydraulic architecture. Non-steady state approaches must be employed to capture or predict the dynamics of water flow through diurnal variations of water uptake and leaf water potential. There is a need to link information from more traditional studies of hydraulic architecture with new information on capacitance and water transport dynamics at different points along the root-to-leaf hydraulic continuum in order to understand how trees solve the problem of maintaining long-distance water transport under dynamic conditions and what sorts of adjustments occur to overcome challenges related to increasing size and complexity of the vascular system in large trees.

References

Ambrose AR, Sillett SC, Dawson E (2009) Effects of tree height on branch hydraulics, leaf structure and gas exchange in California redwoods. Plant Cell Environ 32:743–757

Barnard HR, Ryan MG (2003) A test of the hydraulic limitation hypothesis in fast-growing *Eucalyptus saligna*. Plant Cell Environ 26:1235–1245

Barnard HR, Salmon Y, Kodama N, Sorgel K, Holst J, Rennenberg H, Gessle A, Buchmann N (2007) Evaporative enrichment and time lags between δ18O of leaf water and organic pools in a pine stand. Plant Cell Environ 30:539–550

Brodribb TJ, Holbrook NM, Zwieniecki MA, Palma B (2005) Leaf hydraulic capacity in ferns, conifers and angiosperms: impacts on photosynthetic maxima. New Phytol 165:839–846

Bucci SJ, Scholz FG, Goldstein G, Meinzer FC, Franco AC, Campanello PI, Villalobos-Vega R, Bustamante M, Miralles-Wilhelm F (2006) Nutrient availability constrains the hydraulic architecture and water relations of savanna trees. Plant Cell Environ 29:2153–2167

Bucci SJ, Scholz FG, Goldstein G, Meinzer FC (2003) Dynamic diurnal changes in hydraulic specific conductivity in petiole of two savanna tree species: identifying factors and mechanisms contributing to the refilling of embolized vessels. Plant Cell Environ 26:1633–1645

Bucci SJ, Scholz FG, Goldstein G, Meinzer FC, Franco AC, Zhang Y, Hao GY (2008) Water relations and hydraulic architecture in Cerrado trees: adjustments to seasonal changes in water availability and evaporative demand. Braz J Plant Physiol 20:233–245

Burgess SSO, Pittermann J, Dawson TE (2006) Hydraulic efficiency and safety of branch xylem increases with height in *Sequoia sempervirens* (D. Don) crowns. Plant Cell Environ 29:229–239

Čermák J, Nadezhdina N (2011) Instrumental approaches to studying tree water relations along gradients of tree size and forest age. In: Meinzer FC, Lachenbruch B, Dawson TE (eds) Size- and age-related changes in tree structure and function. Springer, Dordrecht

Čermák J, Kucera J, Bauerle WL, Phillips N, Hinckley TM (2007) Tree water storage and its diurnal dynamics related to sap flow and changes of stem volume in old-growth Douglas-fir trees. Tree Physiol 27:181–198

Chapotin S, Razanameharizaka JH, Holbrook NM (2006) Baobab trees (*Adansonia*) in Madagascar use stored water to flush new leaves but not to support stomatal opening prior to the rainy season. New Phytol 169:549–559

Cowan IR (1972) An electrical analogue of evaporation from, and flow of water in plants. Planta 106:221–226

Daley MJ, Phillips NG, Pettijohn JC, Hadley J (2008) Hydraulic responses to environmental perturbations in *Tsuga canadensis* and *Betula lenta*. Tree Physiol 28:1341–1348

Domec J-C, Gartner BL (2001) Cavitation and water storage capacity in bole xylem segments of mature and young Douglas-fir trees. Trees 15:204–214

Domec J-C, Gartner BL (2002) How do water transport and water storage differ in coniferous earlywood and latewood? J Exp Bot 53:2369–2379

Domec J-C, Gartner BL (2003) Relationship between growth rates and xylem hydraulic characteristics in young, mature and old-growth ponderosa pine trees. Plant Cell Environ 26:471–483

Domec J-C, Pruyn ML, Gartner BL (2005) Axial and radial profiles in conductivities, water storage and native embolism in trunks of young and old-growth ponderosa pine trees. Plant Cell Environ 28:1103–1113

Domec J-C, Scholz FG, Bucci SJ, Meinzer FC, Goldstein G, Villalobos-Vega R (2006) Diurnal and seasonal variation in root xylem embolism in neotropical savanna woody species: impact on stomatal control of plant water status. Plant Cell Environ 29:26–35

Domec J-C, Lachenbruch B, Meinzer FC, Woodruff DR, Warren JM, McCulloh KA (2008) Maximum height in a conifer is associated with conflicting requirements for xylem design. Proc Natl Acad Sci USA 105:12069–12074

Domec JC, Palmroth S, Wardi E, Maier CA, Therezien M, Oren R (2009) Acclimation of leaf hydraulic conductance and stomatal conductance of *Pinus taeda* (loblolly pine) to long-term growth in elevated CO2 (free-air CO2 enrichment) and N-fertilization. Plant Cell Environ 32:1500–1512

Edwards WRN, Jarvis PG, Landsberg JJ, Talbot H (1986) A dynamic model for studying flow of water in single tree. Tree Physiol 1:309–324

Ewers BE, Bon-Lamberty B, Mackay DS (2011) Consequences of stand age and species' functional trait changes on ecosystem water use of forests. In: Meinzer FC, Lachenbruch B, Dawson TE (eds) Size- and age-related changes in tree structure and function. Springer, Dordrecht

Ford CR, Goranson CE, Mitchell RJ, Will RE, Teskey RO (2004) Diurnal and seasonal variability in the radial distribution of sap flow: predicting total stem flow in Pinus taeda trees. Tree Physiol 24:951–960

Franco-Viscaino E, Goldstein G, Ting JP (1990) Comparative gas exchange of leaves and bark in three stem succulents of Baja California. Am J Bot 77:1272–1278

Goldstein G, Meinzer FC, Monasterio M (1984) The role of capacitance in the water balance of Andean giant rosette species. Plant Cell Environ 21:397–406

Goldstein G, Andrade JL, Meinzer FC, Holbrook NM, Cavelier J, Jackson P, Celis A (1998) Stem water storage and diurnal patterns of water use in tropical forest canopy trees. Plant Cell Environ 21:397–406

Hao G, Hoffmann WA, Scholz FG, Bucci SJ, Meinzer FC, Franco AC, Cao K-F, Goldstein G (2008) Stem and leaf hydraulics of congeneric tree species from adjacent tropical savanna and forest ecosystems. Oecologia 155:405–415

Holbrook NM (1995) Stem water storage. In: Gartner BL (ed) Plant stems: physiology and functional morphology. Academic, San Diego, pp 151–174

Holbrook NM, Sinclair TR (1992) Water balance in the arborescent palm, Sabal palmetto. II. Transpiration and stem water storage. Plant Cell Environ 15:401–409

Holtta T, Cochard H, Nikinmaa E, Mencuccini M (2009) Capacitive effect of cavitation in xylem conduits: results from a dynamic model. Plant Cell Environ 32:10–21

Hultine KR, Williams DG, Burgess SSO, Keefer TO (2003) Contrasting patterns of hydraulic redistribution in three desert phreatophytes. Oecologia 135:167–175

Hunt ER, Nobel PS (1987) Non-steady-state flow for three desert perennials of different capacitances. Aust J Plant Physiol 14:363–375

Hunt ER, Running SW, Federer CA (1991) Extrapolating plant water flow resistances and capacitances to regional scales. Agric For Meteor 54:169–195

Irvine J, Grace J (1997) Continuous measurements of water tensions in the xylem of trees based on the elastic properties of wood. Planta 202:455–461

Ishii H (2011) How do changes in leaf/shoot morphology and crown architecture affect growth and physiological function of large, old trees? In: Meinzer FC, Lachenbruch B, Dawson TE (eds) Size- and age-related changes in tree structure and function. Springer, Dordrecht

James SA, Meinzer FC, Goldstein G, Woodruff D, Jones T, Reston T, Mejia M, Clearwater M, Campanello P (2003) Axial and radial water transport and internal water storage in tropical forest canopy trees. Oecologia 134:37–45

Johnson DM, Woodruff DR, McCulloh KA, Meinzer FC (2009) Leaf hydraulic conductance, measured in situ, declines and recovers daily: leaf hydraulics, water potential and stomatal conductance in four temperate and tropical tree species. Tree Physiol 29:879–887

Kobayashi Y, Tanaka T (2001) Water flow and hydraulic characteristics of Japanese red pine and oak trees. Hydrol Process 15:1735–1750

Koch GW, Sillett SC, Jennings GM, Davis SD (2004) The limits to tree height. Nature 428:851–854

Lai CT, Ehleringer JR, Bond BJ, Kyaw Tha Paw U (2006) Contributions of evaporation, isotopic non-steady state transpiration and atmospheric mixing on the δ18O of water vapour in Pacific Northwest coniferous forests. Plant Cell Environ 29:77–94

Lamont BB, Lamont HC (2000) Utilizable water in leaves of 8 arid species as derived from pressure-volume curves and chlorophyll fluorescence. Physiol Plant 110:64–71

Landsberg JJ, Blanchard TW, Warrit B (1976) Studies on the movement of water through apple trees. J Exp Bot 27:579–596

Lo Gullo MA, Salleo S (1992) Water storage in the wood and xylem cavitation in 1-year-old twigs of *Populus deltoides* Bartr. Plant Cell Environ 15:431–438

Loustau D, Berbigier P, Roumagnac P, Arruda-Pacheco C, David JS, Ferreira MI, Pereira J, Tavares R (1996) Transpiration of a 64-year-old maritime pine stand in Portugal. 1. Seasonal course of water flux through maritime pine. Oecologia 107:33–42

Machado JL, Tyree MT (1994) Patterns of hydraulic architecture and water relations of two tropical canopy trees with contrasting leaf phenologies: *Ochroma pyramidale* and *Pseudobombax septenatum*. Tree Physiol 14:219–240

Maherali H, DeLucia EH (2001) Influence of climate-driven shifts in biomass allocation on water transport and storage in ponderosa pine. Oecologia 129:481–491

McDowell NG, Barnard H, Bond BJ, Hinckley T, Hubbard R, Ishii H, Köstner B, Meinzer FC, Marshall JD, Magnani F (2002) The relationship between tree height and leaf area: sapwood area ratio. Oecologia 132:12–20

Meinzer FC, Campanello PI, Domec J-C, Gatti MG, Goldstein G, Villalobos Vega R, Woodruff DR (2008) Constraints on physiological function associated with branch architecture and wood density in tropical forest trees. Tree Physiol 28:1609–1617

Meinzer FC, Clearwater MJ, Goldstein G (2001) Water transport in trees: current perspectives, new insights and some controversies. Environ Exp Bot 45:239–262

Meinzer FC, James SA, Goldstein G, Woodruff D (2003) Whole-tree water transport scales with sapwood capacitance in tropical forest canopy trees. Plant Cell Environ 26:1147–1155

Meinzer FC, James SA, Goldstein G (2004) Dynamics of transpiration, sap flow and use of stored water in tropical forest canopy trees. Tree Physiol 24:901–909

Meinzer FC, Brooks JR, Domec J-C, Gartner BL, Warren JL, Woodruff D, Bible K, Shaw DC (2006) Dynamics of water transport and storage in conifers studied with deuterium and heat tracing techniques. Plant Cell Environ 29:105–114

Meinzer FC, Woodruff DR, Domec J-C, Goldstein G, Campanello PI, Gatti MG, Villalobos-Vega R (2008) Coordination of leaf and stem water transport properties in tropical forest trees. Oecologia 156:31–41

Meinzer FC, Johnson DM, Lachenbruch B, McCulloh KA, Woodruff DR (2009) Xylem hydraulic safety margins in woody plants: coordination of stomatal control of xylem tension with hydraulic capacitance. Funct Ecol 23:922–930

Meinzer FC, McCulloh KA, Lachenbruch B, Woodruff DR, Johnson DM (2010) The blind men and the elephant: the impact of context and scale in evaluating conflicts between plant hydraulic safety and efficiency. Oecologia 164:287–296

Mencuccini M, Hölttä T, Martinez-Vilalta J (2011) Comparative criteria for models of the vascular transport system of tall trees. In: Meinzer FC, Lachenbruch B, Dawson TE (eds) Size- and age-related changes in tree structure and function. Springer, Dordrecht

Morse SR (1990) Water balance in *Hemizonia luzulifolia:* the role of extracellular polysaccharides. Plant Cell Environ 13:39–48

Nielsen ET, Sharifi MR, Rundel PW, Forseth IN, Ehleringer J (1990) Water relations of stem succulent trees in north central Baja California. Oecologia 82:299–303

Nobel PS, Jordan PW (1983) Transpiration stream of desert species: resistances and capacitances for a C3, a C4, and a CAM plant. J Exp Bot 34:1379–1391
Offenthaler I, Hietzand P, Richter H (2001) Wood diameter indicates diurnal and long term patterns of xylem water potential in Norway spruce. Trees 15:215–221
Perämäki M, Nikinmaa E, Sevanto S, Ilvesniemi H, Siivola E, Hari P, Vesala T (2001) Tree stem diameter variations and transpiration in Scots pine: an analysis using a dynamic sap flow model. Tree Physiol 21:889–897
Perämäki M, Vesala M, Nikinmaa E (2005) Modeling the dynamics of pressure propagation and diameter variation in tree sapwood. Tree Physiol 25:1091–1099
Pettijohn G, Salvuccia D, Phillips NG, Daley MJ (2009) Mechanisms of moisture stress in a mid-latitude temperate forest: implications for feedforward and feedback controls from an irrigation experiment. Ecol Model 220:968–978
Phillips N, Nagchaudhuri A, Oren R, Katul G (1997) Time constant for water transport in loblolly pine trees estimated from time series of evaporative demand and stem sapflow. Trees 11:412–419
Phillips NG, Ryan MG, Bond BJ, McDowell NG, Hinckley TM, Èermák J (2003a) Reliance on stored water increases with tree size in three species in the Pacific Northwest. Tree Physiol 23:237–245
Phillips NG, Bond BJ, McDowell NG, Ryan MG, Schauer A (2003b) Leaf area compounds height-related costs of water transport in Oregon white oak trees. Funct Ecol 17:832–840
Phillips NG, Oren R, Licata J, Linder S (2004) Time series diagnosis of tree hydraulic characteristics. Tree Physiol 24:879–890
Phillips NG, Scholz FG, Bucci SJ, Goldstein G, Meinzer FC (2008) Using branch and basal trunk sap flow measurements to estimate whole-plant water capacitance: comment on Burgess and Dawson (2009). Plant Soil 315:315–324
Powell DBB, Thorpe MR (1977) Dynamic aspects of plant water relations. In: Landsberg JJ, Cutting CV (eds) Environmental aspects of crop physiology. Academic, London, pp 57–73
Pratt RB, Jacobsen AL, Ewers FW, Davis SD (2007) Relationships among xylem transport, biomechanics and storage in stems and roots of nine Rhamnaceae species of the California chaparral. New Phytol 174:787–798
Running SW (1980a) Relating plant capacitance to the water relations of *Pinus contorta*. For Ecol Manage 2:237–252
Running SW (1980b) Environmental and physiological control of water flux through *Pinus contorta*. Can J For Res 10:82–91
Ryan MG, Yoder BJ (1997) Hydraulic limits to tree height and tree growth. Bioscience 47:235–242
Ryan MG, Phillips N, Bond BJ (2006) The hydraulic limitation hypothesis revisited. Plant Cell Environ 29:367–381
Sack L, Tyree MT (2005) Leaf hydraulics and its implications in plant structure and function. In: Holbrook MN, Zwieniecki MA (eds) Vascular transport in plants. Elsevier Academic Press, Oxford, pp 93–114
Sack L, Cowan PD, Jaikumar N, Holbrook NM (2003) The 'hydrology' of leaves: coordination of structure and function in temperate woody species. Plant Cell Environ 26:1343–1356
Scholz FG, Bucci SJ, Goldstein G, Meinzer FC, Franco AC, Miralles-Wilhelm F (2007) Biophysical properties and functional significance of stem water storage tissues in neo-tropical savanna trees. Plant Cell Environ 30:236–248
Scholz FG, Bucci SJ, Goldstein G, Meinzer FC, Franco AC, Miralles-Wilhelm F (2008) Temporal dynamics of stem expansion and contraction in savanna trees: withdrawal and recharge of stored water. Tree Physiol 28:469–480
Schulze ED, Cermák J, Matyssek M, Penka M, Zimmermann R, Vasícek F, Gries W, Kucera J (1985) Canopy transpiration and water fluxes in the xylem of the trunk of *Larix* and *Picea* trees – a comparison of xylem flow, porometer and cuvette measurements. Oecologia 66:475–483
Sevanto S, Vesala T, Perämäki M, Nikinmaa E (2002) Time lags for xylem and stem diameter variations in a Scots pine tree. Plant Cell Environ 25:1071–1077

Smith JAC, Schulte PJ, Nobel PS (1987) Water flow and water storage in *Agave deserti*: osmotic implications of Crassulacean acid metabolism. Plant Cell Environ 10:639–648

Sperry JS (1995) Limitations on stem water transport. In: Gartner BL (ed) Plant stems: physiology and functional morphology. Academic, San Diego, pp 105–124

Sperry JS, Pockman WT (1993) Limitation of transpiration by hydraulic conductance and xylem cavitation in *Betula occidentalis*. Plant Cell Environ 16:279–287

Sperry JS, Meinzer FC, McCulloh KA (2008) Safety and efficiency conflicts in hydraulic architecture: scaling from tissues to trees. Plant Cell Environ 31:632–645

Stratton L, Goldstein G, Meinzer FC (2000) Stem water storage capacity and efficiency of water transport: their functional significance in a Hawaiian dry forest. Plant Cell Environ 23:99–106

Tyree MT (1988) A dynamic model for water flow in a single tree: evidence that models must account for hydraulic architecture. Tree Physiol 4:195–217

Tyree MT, Yang S (1990) Water storage capacity of *Thuja*, *Tsuga* and *Acer* stems measured by dehydration isotherms: the contribution of capillary water and cavitation. Planta 182:420–426

Tyree MT, Zimmermann MH (2002) Xylem structure and the ascent of sap. Springer, Berlin

Tyree MT, Snyderman DA, Wilmot TR, Machado JL (1991) Water relations and hydraulic architecture of a tropical tree (*Schefflera morototoni*). Plant Physiol 96:1105–1113

Waring RH, Running SW (1978) Sapwood water storage: its contribution to transpiration and effect upon water conductance through the stems of old-growth Douglas-fir. Plant Cell Environ 1:131–140

Waring RH, Whitehead D, Jarvis PG (1979) The contribution of stored water to transpiration in Scots pine. Plant Cell Environ 2:309–317

Whitehead RD, Teskey RO (1995) Dynamic response of stomata to changing irradiance in loblolly pine (*Pinus taeda* L.). Tree Physiol 15:245–251

Williams M, Rastetter EB, Fernandes DN, Goulden ML, Wofsy SC, Shaver GR, Melillo JM, Munger JW, Fan SM, Nadelhoffer KJ (1996) Modelling the soil–plant-atmosphere continuum in a *Quercus-Acer* stand at Harvard Forest: the regulation of stomatal conductance by light, nitrogen and soil/plant hydraulic properties. Plant Cell Environ 19:911–927

Woodruff DR, Meinzer FC (2011) Size-dependent changes in biophysical control of tree growth: the role of turgor. In: Meinzer FC, Lachenbruch B, Dawson TE (eds) Size- and age-related changes in tree structure and function. Springer, Dordrecht

Woodruff DR, McCulloh KA, Warren JM, Meinzer FC, Lachenbruch B (2007) Impacts of tree height on leaf hydraulic architecture and stomatal control in Douglas-fir. Plant Cell Environ 30:559–569

Wullschleger SD, Meinzer FC, Vertessy RA (1998) A review of whole-plant water use studies in trees. Tree Physiol 18:499–512

Zweifel R (1999) The rhythm of trees – water storage dynamics in subalpine Norway spruce. Ph.D. dissertation, Swiss federal institute of technology-Zurich

Zweifel R, Häsler R (2001) Dynamics of water storage in mature subalpine *Picea abies*: temporal and spatial patterns of change in stem radius. Tree Physiol 21:561–569

Zweifel R, Item H, Häsler R (2000) Stem radius changes and their relation to stored water in stems of young Norway spruce trees. Trees 15:50–57

Zweifel R, Item H, Häsler R (2001) Link between diurnal stem radius changes and tree water relations. Tree Physiol 21:869–877

Zwieniecki MA, Holbrook NM (1998) Diurnal variation in xylem hydraulic conductivity in white ash (*Fraxinus americana* L.), red maple (*Acer rubrum* L.) and red spruce (*Picea rubens* Sarg.). Plant Cell Environ 21:1173–1180

Chapter 14
Size-Dependent Changes in Biophysical Control of Tree Growth: The Role of Turgor

David R. Woodruff and Frederick C. Meinzer

Abstract Xylem pressure at the tops of trees decreases with increasing tree height as a result of gravitational force and friction associated with water movement through the xylem. Unless altered by osmotic adjustment, the turgor of cells in the apical and vascular meristems will decline proportionally with xylem pressure. In the tallest coniferous species on earth osmotic adjustment is insufficient to compensate for the negative impact of vertical gradients of xylem pressure on turgor of developing and mature tissue. This chapter synthesizes the research addressing turgor-related constraints on growth at the cellular, tissue, and whole tree levels. We examine the role of turgor in plant growth and foliar and stem structural and functional characteristics as they are influenced by height-related trends in turgor. We also evaluate height-related trends in turgor as a cause of observed patterns of tree height and tree biomass growth with an emphasis on how reductions in turgor can influence sink/source relationships for photosynthate in tall trees. Height-related reductions in cell turgor generate a cascade of direct and indirect effects on shoot extension and expansion. The multiple consequences of turgor-limited growth may act in a synergistic manner to further constrain shoot extension as trees approach species- and site-specific height limits. Turgor-limited tissue expansion has permanent anatomical and morphological consequences that govern height-dependent changes in an array of growth-related physiological processes. These include the impact of reduced xylem conduit dimensions on efficiency of water supply to growing tissues at the terminal portions of the water transport pathway, reduced growth efficiency associated with increased leaf mass per area, and increased resistance to CO_2 diffusion through the mesophyll to the sites of carboxylation.

D.R. Woodruff (✉) • F.C. Meinzer
USDA Forest Service, Pacific Northwest Research Station,
3200 SW Jefferson Way, Corvallis, OR 97331, USA
e-mail: dwoodruff@fs.fed.us

F.C. Meinzer et al. (eds.), *Size- and Age-Related Changes in Tree Structure and Function*, Tree Physiology 4, DOI 10.1007/978-94-007-1242-3_14,

1 Introduction

A substantial reduction in aboveground wood production following canopy closure in forest stands is a well documented phenomenon (Turner and Long 1975; Ryan and Waring 1992; Weiner and Thomas 2001). Studies involving the use of grafting techniques have provided evidence that the decline in growth is more a function of tree height than age (Mencuccini et al. 2005; Bond et al. 2007; Abdul-Hamid and Mencuccini 2009). Although no definitive cause for this decline in productivity has been identified, evidence has been found to support a number of proposed mechanisms. These include, but are not limited to: an increase in respiration relative to photosynthesis (Yoda et al. 1965; Whittaker and Woodwell 1967), reductions in leaf area resulting from nutrient limitations (Gower et al. 1996), increased stomatal limitation of CO_2 uptake due to height-related increases in hydraulic resistance (Mencuccini and Grace 1996; Ryan and Yoder 1997; McDowell et al. 2002, 2005) and reduced cell expansion resulting from height-related reductions in turgor (Marshall and Monserud 2003; Koch et al. 2004; Woodruff et al. 2004). Other studies involving analyses of non-structural storage carbohydrates and lipids suggest that growth in mature and tall trees is often not limited by constraints on carbon assimilation or carbon supply (Körner 2003; Körner et al. 2005; Würth et al. 2005; Millard et al. 2007; Sala and Hoch 2009; Sala et al. 2011).

In the absence of mitigating factors, an increase in the path length from roots to transpiring foliage is associated with an increase in hydraulic resistance and a consequent decrease in the pressure component of xylem water potential (Richter 1973; Hellkvist et al. 1974). If the increase in path length is in opposition to the forces of gravity, as in an upright tree, it is also associated with a static gravitational reduction in xylem pressure of 0.01 MPa for every 1 m increase in height (Scholander et al. 1965; Zimmermann 1983). The combined dynamic and static components of the height-related trend in xylem pressure typically amount to approximately –0.017 MPa per meter increase in height (Bauerle et al. 1999; Woodruff et al. 2004; Domec et al. 2008). Unless altered by osmotic adjustment, the turgor of cells in the apical and vascular meristems will decline proportionally with xylem water potential (Boyer and Silk 2004). Turgor drives cell expansion (Lockhart 1965) and has also been implicated as a factor in the regulation of cell division (Kirkham et al. 1972). In view of the relationships between tree height, xylem water potential, turgor and processes involved in cell expansion and division, it was hypothesized that height-related trends in stem elongation and foliar morphological characteristics may be largely due to the influence of vertical gradients in water potential on turgor (Marshall and Monserud 2003; Woodruff et al. 2004). According to this hypothesis, growth is limited by reduced sink strength of expanding tissues rather than carbon supply. Reduced turgor can also impose additional constraints on physiological processes such as stomatal opening and phloem transport (Thompson and Holbrook 2003; Buckley 2005; Woodruff et al. 2010). Secondary impacts of reduced turgor on leaf and stem physiology may also arise from anatomical and structural constraints associated with reduced cell and tissue expansion (Hanba et al. 1999; Thompson 2006; Mullin et al. 2009; Woodruff et al. 2008, 2009).

This chapter synthesizes the research addressing turgor-related constraints on growth at the cellular, tissue, and whole tree levels. We examine the role of turgor in plant growth and foliar and stem structural and functional characteristics as they are influenced by height-related trends in turgor. We also evaluate height-related trends in turgor as a cause of observed patterns of tree height and tree biomass growth with an emphasis on how reductions in turgor can influence sink/source relationships for photosynthate in tall trees. We do not address the potential roles that height, height – stress interactions might have on plant growth substances (i.e. hormones), nor the impact of these signaling chemicals on growth or resource allocation.

2 Carbon Limitation vs. Turgor Limitation

Plants accumulate non-structural carbohydrates during periods of high productivity and deplete them when the rate of utilization outpaces the rate of production. Shifts in internal demands for carbohydrates lead to reversals in direction of carbohydrate movement and changes in rates of carbohydrate transport as relative sink strengths shift on daily, seasonal, and ontogenetic time scales (Wolswinkel 1985; Marcelis 1996). The storage of non-structural carbohydrates is an indicator of greater carbon uptake (source) than use of carbon (sink) in trees. Although studies investigating whether height-related constraints on tree growth involve carbon limitation have provided evidence for reduced carbon assimilation with increasing tree size, the concept of carbon limitation as the basis for the primary cause of height-related growth constraints is inconsistent with the emerging evidence suggesting that mature and taller trees are often not limited by carbon supply (Körner 2003; Körner et al. 2005; Würth et al. 2005; Millard et al. 2007; Sala and Hoch 2009; Sala et al. 2011). Körner (2003) examined tree growth in four climatic zones (tropical, Mediterranean, temperate, and boreal/mountain) and found that carbon reserves increased during periods of reduced growth suggesting that constraints on growth were not associated with carbon or assimilation. Würth et al. (2005) found that non-structural carbohydrate reserves remained high in a mature tropical forest throughout the year and increased during the dry season, suggesting that growth was not carbon limited and that water-stress restricted growth more than assimilation in these trees. Sala and Hoch (2009) examined storage of non-structural carbohydrates in ponderosa pine trees across a height gradient and found non-structural carbohydrates increased instead of decreased with tree height, suggesting greater reductions of carbon sink relative to carbon assimilation in taller trees. The implications of these studies are that, at least in many cases, growth constraints in mature and tall trees appear to be more associated with sink activity than with carbon availability or assimilation. The degree to which this phenomenon of increased storage of non-structural carbohydrates with increased tree size is universal across forest ecosystems is yet to be determined. Reduced cell expansion is associated with not only a reduction in cell size, but a reduction in cell division (see Sect. 5) and a feedback inhibition of photosynthesis (see Sect. 7.3). Constraints on cell expansion due to height-related reductions in cell turgor therefore represent a mechanism that could explain observed patterns of reduced growth with increased tree size that is consistent with the mounting evidence for height-related trends in productivity being independent of carbon assimilation or carbon availability. It is noteworthy that elevated carbon dioxide levels will not benefit trees if they are turgor limited as opposed to carbon limited. Furthermore, increased temperatures and/or droughts are likely to exacerbate turgor-related constraints on growth.

3 The Biophysics of Plant Growth

Early studies of biophysical control of plant growth focused on giant algal cells as model systems (e.g. Green 1968). These have been followed by several decades of studies on vascular plants, mostly herbaceous crop species. The scarcity of information on biophysical regulation of tree growth is likely attributable to factors such as their size and structural complexity, technical difficulties associated with isolating and manipulating growing tissues, and in many tree species, the highly seasonal nature of growth, which is often restricted to a few weeks of leaf expansion and shoot extension annually.

Whereas expansion and contraction of cells and organs can occur in response to temporary changes in water content, growth strictly involves irreversible enlargement of cells or organs, which requires changes in the cell wall properties and structure and an influx of water into the cell. Although turgor is necessary for driving cell expansion, cell wall yielding, and therefore a marginal reduction in turgor, is also required for generating the necessary driving force for water to enter the expanding cell. The steps involved in cell growth can be described as follows: water enters the cell as a result of a water potential gradient from the outside to the inside of the cell. This gradient is typically created by a decrease in osmotic potential of the cell contents resulting from the accumulation of solutes within the cell. The water within the cell membrane exerts hydrostatic (turgor) pressure on the cell walls and stretches them. Irreversible cell wall expansion (growth) occurs as load-bearing elements of the cell walls yield in response to the hydrostatic pressure, hemicellulose tethers are cleaved and wall components are synthesized (Harold 1990). Thus, a number of balances must be struck in the cell during growth. In order for growth to occur the cell wall must yield, leading to a drop in turgor pressure that reduces the hydrostatic forces driving cell wall expansion. If turgor pressure is insufficient, the cell wall will not expand, if turgor in the cell is too high, the water potential gradient across the cell wall may decline leading to a reduction in water absorption by the cell. The proper structural integrity of the cell wall must also be maintained during growth. The cell wall must maintain enough rigidity to allow turgor to be maintained, yet it must also yield in order to allow the cell to enlarge. As water enters the cell, the osmotic status of the cell must also be regulated because any increase in cell volume will lead to an increase in water potential (i.e. becoming less negative). For continued cell expansion to occur, osmotic regulation is necessary in order to avoid the dilution of the cell's osmotic content and the relaxation of turgor to the wall yielding threshold. The dynamic interactions between wall yielding, water absorption and osmotic regulation during cell expansion illustrate the complex nature of the processes involved in cell growth.

While there have been many notable advances over the years in understanding the biochemical aspects of cell growth (Cosgrove 2005 and references within), the biophysical controls on cell and tissue expansion are generally still conceptualized using long established growth equations of Lockhart (1965). Because the vast majority of cell volume increase is the result of water gain, for sustained growth to occur water must continue to flow into the cell. Equation 14.1 focuses on the conductance of the hydraulic pathway into the expanding cell,

$$dV / V\mathrm{d}t = L_{\mathrm{m}} A_{\mathrm{c}} (\sigma \Delta \Psi_{\pi} - \Psi_{\mathrm{p}}) / \mathrm{V} = L_{\mathrm{v}} (\sigma \Delta \Psi_{\pi} - \Psi_{\mathrm{p}}) / \mathrm{V} \qquad (14.1)$$

where V = cell volume, t is time, L_{m} = hydraulic conductance of the cell membrane, A_{c} = cell surface area, $\Delta\Psi_{\pi}$ = the osmotic potential step across membrane tissues, Ψ_{p} = turgor pressure, L_{v} = volumetric hydraulic conductance and σ = a membrane reflection coefficient that describes the degree to which the membrane functions as an ideal osmotic membrane ($\sigma = 1$ when the membrane blocks all solute transport and yet permits water movement). Because wall expansion is dependent on physical properties of the cell wall that are related to expansion, and can be reversible if wall stress does not exceed a critical threshold, Lockhart adopted the concepts of the extensibility *(m)* and yield threshold (*Y*). Equation 14.2 relates irreversible cell wall expansion to wall characteristics and Ψ_{p},

$$dV / V\mathrm{d}t = m(\Psi_{\mathrm{p}} - Y) \qquad (14.2)$$

The coefficient *m* is a measure of the plasticity of the cell wall, or the ability of the microfibrils within the cell wall to shift position for irreversible growth, and *Y* is the turgor pressure above which cell wall extension becomes irreversible. When the term ($\Psi_{\mathrm{p}} - Y$) is positive it describes the turgor pressure that results in cell expansion and it is termed "growth effective turgor". Neither *m* nor *Y* is constant, and they are both influenced by a number of factors including environmental stresses, growth regulators and wall metabolism (Green 1968; Cosgrove 1993). For example, water deficit reduces the plasticity of cell walls in leaves, thereby limiting leaf expansion in the absence of rapid osmotic adjustment (Matthews et al. 1984; Hsiao and Xu 2000), but causes rapid osmotic adjustment and increases cell wall plasticity in roots, allowing them to maintain growth (Spollen and Sharp 1991; Hsiao and Xu 2000; Wu and Cosgrove 2000). Because turgor exists as a primary variable in both the water uptake model of cell expansion (Eq. 14.1) and the wall expansion model of cell expansion (Eq. 14.2), turgor pressure acts as the connection that coordinates cell wall yielding with water uptake.

As the cell wall yields it creates disequilibrium between the water potential of the expanding cell and its surroundings, driving water influx into the cell. The effective water potential disequilibrium ($\Psi_{\mathrm{eff}} = \sigma\Delta\Psi_{\pi} - \Psi_{\mathrm{p}}$) is the driving force for water entry into the cell, and it is determined by the relative values of L_v and m (Cosgrove 1985):

$$(\Psi_{eff}) = \left(m / \left(m + L_v\right)\right)\left(\sigma\Delta\Psi_\pi - Y\right) \tag{14.3}$$

The term $(m/(m + L_v))$ within this relationship allows for the estimation of the relative limitations of growth by extensibility (which is a measure of wall plasticity) and water transport. When L_v is substantially less than m, the ratio in this term approaches one, Ψ_p approaches Y and growth is limited by L_v. When L_v is substantially larger than m, $(m/(m + L_v))$ approaches zero, (Ψ_{eff}) approaches zero and growth is limited by m. In the case of individual cells, studies investigating the relative roles of m and L_v in growth restrictions indicate that the wall yielding process is the primary limiting factor for individual cell growth and that water transport does not represent a rate-limitation (Cosgrove 1985, 1993). The hydraulic conductance of the cell (L_v) can vary in response to physical and metabolic changes that may occur during cell growth. Ψ_π can also vary, not only due to osmotic adjustment, but because it is a function of water uptake and will depend on the hydraulic and mechanical resistances of the cell.

4 Turgor in Tissues and Whole Plants

Equations 14.1–14.3 pertain to growth of single cells and additional caveats and complexities apply when modeling growth in tissues and whole organs such as leaves, stems or roots. Although the reflection coefficient σ of cell membranes is close to 1 for most physiological solutes (Steudle 1989), cell walls tend to perform less like ideal osmotic membranes. The overall σ value for multiple cells types within a tissue is therefore likely to be substantially less than 1. In the case of individual cells L_v is essentially determined by L_m. Growth dynamics in leaves, stems and roots will vary as a result of the high variability of different tissue types within these organs, the effects of this variability on L_v, and the substantially greater distances involved in determining L_v.

As opposed to cells in non-vascular tissue, xylem cells are highly adapted to transporting water. The amount of water that passes through the xylem during transpiration is typically many orders of magnitude greater than what is used for tissue expansion (Boyer and Silk 2004). Expanding cells must in effect compete with transpiring stomata for water. When water potentials are plotted in different regions of a non water-stressed plant's growth zone using a pressure probe and nanoliter osmometry, gradients are apparent that indicate highest (least negative) water potentials typically occurring within the xylem conducting tissues, and water potentials declining with distance from xylem conducting tissue (Nonami and Boyer 1993; Nonami et al. 1997). When tensions develop within the vascular system, water potential declines within the xylem such that the gradient between xylem and surrounding tissues may be reversed. Because the water necessary for tissue expansion is derived from the xylem, a positive water potential gradient is needed from expanding tissues to the xylem, and the capacity for tissue expansion is correlated with the magnitude of the gradient in water potential from the xylem to the expanding tissue. This relationship presents the possible scenario of high tensions within the xylem resulting in the cessation of water movement from the xylem to expanding tissues if osmotic potentials within expanding cells are not sufficiently negative. Passioura and Boyer (2003) developed a mathematical model that correlated turgor-driven growth with spatial patterns of water potential and water uptake in order to examine the relationship between rheological properties of the growing cell wall to the transport of water in growing tissue. The model examined the degree to which changes in water potential adjacent to the xylem could rapidly affect growth in expanding tissue. Their results indicated that growth rates in expanding tissue responded immediately to changes in xylem water potential. A reduction in the water potential of xylem water within a stem or leaf can reverse the water potential gradient in a few cells adjacent to the xylem, thus blocking water flow to expanding tissues and restricting cell expansion.

5 The Role of Turgor in Cell Division

Leaf development occurs in stages in which cell proliferation initially dominates, followed by a secondary phase in which the relative rate of cell division decreases and cells advance to the phase of post-mitotic expansion (Granier and Tardieu 1998, 1999). Meristematic cells tend to expand uniformly such that they reach a relatively constant size before entering the second phase where expansion is the dominant growth factor (Jacobs 1997; Beemster et al. 2003). It has been proposed that the impact of water stress on cell division in leaves could be in large part due to the inhibition of meristematic cell expansion and that reductions in leaf cell division associated with water stress are due to the early advancement of the leaf out of the initial stage of leaf development (dominated by cell division) and into the subsequent stage (dominated by cell expansion) (Alves and Setter 2004). Because cell division in-situ typically occurs during a limited period of time as determined by water availability and phenological development, uniformity in cell expansion and size at the meristematic stage, suggests that reductions in turgor would have the effect of limiting cell division in addition to cell expansion.

In a study investigating the influence of turgor on cell expansion and cell division in isolated radish cotyledons, Kirkham et al. (1972) found evidence for a greater sensitivity in cell division to turgor pressure than between cell enlargement and turgor pressure. Although turgor is typically thought of as a biophysical factor in leaf development, a number of studies have provided evidence for turgor also functioning as a triggering mechanism for gene expression. Studies of transport mechanisms in bacteria have determined that expression of Kdp, a transport enzyme with high-affinity for K^+, is regulated by changes in turgor pressure (Epstein 1992; Malli and Epstein 1998). In a study investigating expression patterns of membrane proteins in plants, Jones and

Mullet (1995) described the developmental expression of a turgor-responsive gene that encodes intrinsic membrane proteins in peas and tobacco. The expression pattern of this gene suggests that it facilitates water transport in tissues. Aldehyde dehydrogenase (ALDH) is a family of turgor-responsive proteins derived from vascular plants and first identified in *Pisum sativum* (Guerrero et al. 1990). Stroeher et al. (1995) isolated a gene from *Brassica napus* that is activated in response to water stress and that appears to be triggered by reductions in turgor. The presence of genes in plants that are regulated by changes in turgor pressure, along with evidence for a causative relationship between constraints on turgor and reductions in cell division during leaf growth, indicate the existence of a mechanistic link between turgor and cell division that has yet to be explored in trees.

6 Osmotic Adjustment

Osmotic adjustment is the active accumulation of solutes in the symplast resulting in a reduction in water potential and an increase in the driving force for water movement into plant cells. A number of studies have documented the occurrence of osmotic adjustment as an adaptation to drought and salinity stress in plants (Hsiao et al. 1976; Osonubi and Davies 1978; Turner and Jones 1980; Morgan 1984; Ranney et al. 1991; Rieger 1995; Silveira et al. 2009), and some osmotic adjustment to partially compensate for vertical gradients in Ψ_l has been observed in conifer trees (Koch et al. 2004; Woodruff et al. 2004; Meinzer et al. 2008). Osmotic adjustment can lead to complete restoration of turgor following drought; however, turgor maintenance in leaves may be inhibited if yield turgor and tissue extensibility have been altered as a result of water stress (Roden et al. 1990; Lu and Neumann 1998). The solutes required for osmotic adjustment to maintain turgor represent an expenditure of resources, both in terms of the photosynthate diverted to produce organic solutes such as low molecular weight sugars and sugar alcohols, and in terms of the uptake, accumulation and storage of inorganic solutes used for osmotic adjustment such as potassium and other inorganic solutes. The rapid decrease in tissue plastic extensibility of Douglas-fir needles following bud-break (Meinzer et al. 2008) implies that the opportunity to facilitate turgor-driven leaf expansion through osmotic adjustment may be confined to a short time period. Seasonal patterns related to phenological events may lead to scenarios in which photosynthates are abundant during some periods of the year and limited during others, leading to insufficient availability of organic solutes for osmotic adjustment to sustain turgor-driven cell expansion. Seasonal trends in both foliar sugar content and leaf osmotic potential in some woody species appear to be related to phenological cycles (Hinckley et al. 1983; Teskey et al. 1984; Kubiske and Abrams 1991). In some conifers the solute content of mature foliage reaches a low point in late spring during early leaf expansion and then increases during late spring and summer (Tyree et al. 1978; Ritchie and Shula 1984; Woodruff et al. 2004). During early spring bud swell and flushing in Douglas-fir, the production of sugars does not keep pace with the rate at which they are metabolized: sugar concentration in foliage increases nearly twofold from spring (when they are most-needed for osmotic adjustment) to fall (Billow et al. 1994). Sap velocities of 2.4–3.3 m per day in 50- to 60-m-tall conifers during periods of peak transpiration (Meinzer et al. 2006) suggest that potassium and other inorganic solutes that are taken up from the soil and transported to developing tissues near the tops of trees may need to be stored in areas close to expanding shoots in order to be available for osmotic adjustment when evaporative demand and xylem transport rates are low.

Osmotic potential of Douglas-fir needles varied seasonally with no evidence of osmotic adjustment along a height gradient during the spring to compensate for the gravitational gradient of Ψ_l (Woodruff et al. 2004). After leaf expansion was complete, vertical osmotic potential gradients between June and August were about −0.009 MPa m^{-1} at dawn and −0.011 MPa m^{-1} at midday, approximating the gravitational gradient, but insufficient to compensate for the combined gravitational and frictional gradient of about −0.017 MPa m^{-1}. The resulting vertical gradient in turgor of mature needles was about −0.0056 MPa m^{-1}. Similarly, turgor of foliated shoots of *Sequoia sempervirens* (D. Don) Endl. declined linearly with height at a rate of about 0.006 MPa m^{-1} (Koch et al. 2004) indicating that in this tall species osmotic adjustment was also insufficient to fully compensate for the vertical gradient in Ψ_l. The rapid loss of a height-related gradient in osmotic potential in conifer foliage during tissue expansion suggests that there are substantial constraints on solute availability during this key phenological stage of foliar growth and development. Understanding solute generation (either via conversion or formation) and transport from short- and distant sources is still an important, but largely unknown topic.

7 Growth, Anatomy and Physiological Processes Influenced by Turgor

As indicated above, Ψ declines with increasing tree height due to the hydrostatic constraints imposed by gravity and from frictional resistance during transpiration. In the absence of compensatory mechanisms such as osmotic adjustment, or changes in yield turgor or cell wall extensibility, the turgor of meristematic and expanding cells will decrease in proportion to the vertical gradient in Ψ (xylem tension gradient). These relationships between height, Ψ, and turgor; and the dependence of growth on turgor, imply a causal relationship between tree height and turgor-driven growth (Fig. 14.1). Turgor can also have direct impacts on physiological processes such as phloem transport as influenced by source-to-sink pressure differentials and translocation of photosynthate as affected by sink strengths (Fig. 14.1). The impacts of turgor on cell and tissue anatomical and morphological characteristics that influence physiological processes and behavior suggest a cascade of secondary effects of height-related trends in turgor on leaf and stem physiology (Fig. 14.1).

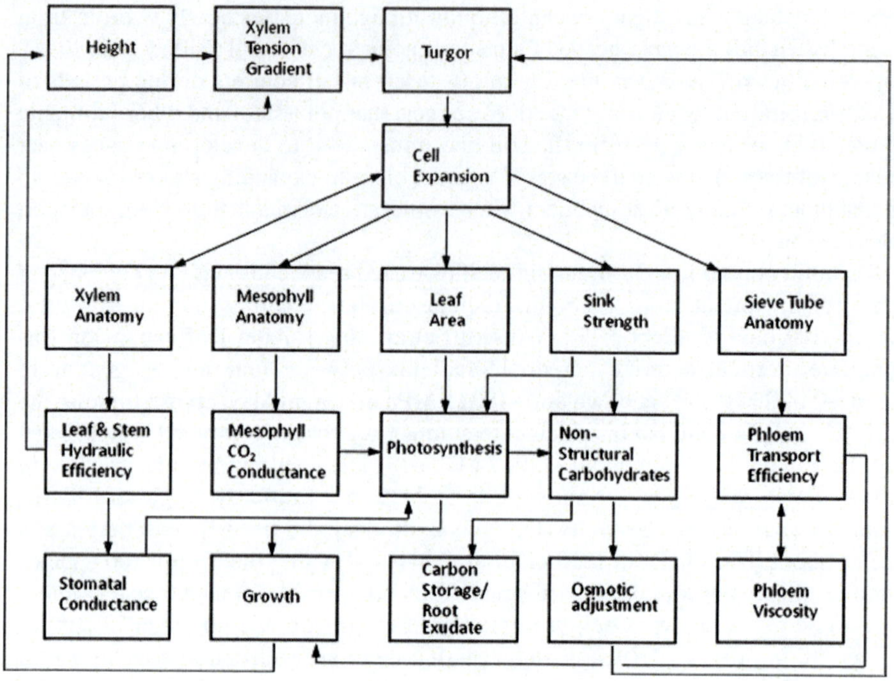

Fig. 14.1 Schematic illustrating relationships between tree height, turgor, growth and anatomical and functional characteristics (see Sect. 7 for explanation)

7.1 Foliar and Shoot Structure

Trends in leaf morphological characteristics with canopy position have been well documented, but were traditionally attributed to vertical gradients in irradiance (Ellsworth and Reich 1993; Sprugel et al. 1996; Niinemets and Tenhunen 1997). Height-related trends in shoot and foliar morphology that are consistent with reduced turgor and cellular expansion have been observed across a broad range of tree species, genera and families (Marshall and Monserud 2003; Koch et al. 2004; Woodruff et al. 2004; Anfodillo et al. 2006; England and Attiwill 2006; Ishii et al. 2008; Oldham et al. 2010; Ishii 2011). Using overlapping saw cuts in *Taxus baccata* to reduce foliar water potential, Richter (1974) was able to create sun foliage morphology in highly shaded environments. Branch elongation decreased linearly with increasing height in western red cedar (*Thuja plicata* J. Donn ex D. Don), despite inconsistent trends in light reduction ratios along a height gradient (Edelstein and Ford 2000). Marshall and Monserud (2003) used branch length as a correlate of shade in a study examining the relative influences of height and irradiance on foliar anatomy of multiple conifers. They found that specific leaf area (SLA) declined with increasing branch height in Douglas-fir, western white pine (*Pinus monticola* Dougl. ex D. Don), and ponderosa pine (*Pinus ponderosa*

Dougl. ex P. and C. Laws.) and that there was no significant trend in SLA with branch length. In conjunction with a linear reduction in turgor pressure in foliage during bud-break and foliar expansion in Douglas-fir along a height gradient from ~14 to ~56 m, Woodruff et al. (2004) found height-related increases in SLA, and height-related reductions in needle width, needle length and shoot extension. Trends in SLA for this study contrasted with vertical profiles of irradiance within the same stand, providing evidence that the observed trends in foliar structure were driven more by hydrostatic gradients in Ψ_l than by light level. In a study examining growth constraints associated with increased tree height Koch et al. (2004) measured turgor and SLA of foliage sampled along a height gradient in redwoods (*Sequoia sempervirens*). Both midday and nighttime turgor decreased with increasing height. They used direct site factor (an index of solar radiation based on hemispherical photos) to compare the relative influence of irradiance versus Ψ_l on foliar anatomical characteristics. They reported that there was a several-fold increase in leaf mass per area (LMA) over a ~50m vertical span within the redwood forest canopy and that, relative to Ψ_l, the influence of light on LMA was only minimal. An additional observation made in this study was that when branches were cut from the tops of tall trees and replanted in moist soil and high light, the newly produced foliage was substantially more expanded than the existing foliage on the same branch. They also observed that leaves of a 2-m-tall epiphytic redwood rooted in soil near the top of a 95-m-tall redwood were substantially more expanded than those of the host tree. In other studies, mass per area ratios of sun exposed foliage and shoots sampled along a height gradient from about 30 to 110 m in redwood and 38 to 88 m in giant sequoia (*Sequoiadendron giganteum*) increased with height; and shoot mass per area (SMA) correlated more closely with height than with canopy openness (Ishii et al. 2008; Ambrose et al. 2009; Mullin et al. 2009). Ishii et al. (2008) also reported a linear reduction in shoot length with increasing height in redwood. England and Attiwill (2006) reported that in eucalyptus (*Eucalyptus regnans* F. Muell) SLA decreased and thickness increased in leaves sampled from the upper canopy of trees along a height gradient from 10 to 80 m. All leaves in this study were sampled from outer branches from the upper crown and were therefore considered to be sun leaves. In a study examining anatomical properties of palm tree (*Washingtonia robusta*) fronds sampled along a height gradient from open-grown trees 2–34 m tall (Renninger et al. 2009), epidermal cell size decreased with increasing height. Minimum leaf water potentials in this study were correlated with tree height with the taller palms generally having more negative minimum leaf water potentials than shorter palms. In a study investigating height-related variation in leaf anatomy of *Sequoia sempervirens*, Oldham et al. (2010) examined paired samples of foliage from inner (shaded) and outer (sunlit) portions of tree crowns in order to separate the effects of gradients in Ψ and light on foliar morphology. They found significant height-related trends in a number of foliar anatomical characteristics, and that the hydrostatic gradient explained over 75% of the variation found in key anatomical traits, while light availability failed to correlate more closely than height to any foliar anatomical characteristic.

7.2 Hydraulic Efficiency and Water Transport

Xylem conduit dimensions typically decrease with increasing height in trees, either in stem tissue along an axial gradient (Carlquist 1975; James et al. 2003; Anfodillo et al. 2006) or in foliage sampled from the tops of trees along a height gradient (Woodruff et al. 2008). Turgor-related constraints on cell expansion can limit xylem hydraulic conductance in both leaves and stems due to the effect of conduit lumen diameter on hydraulic conductance. Even small variations in conduit lumen diameter can lead to major changes in transport efficiency as illustrated by the fourth power relationship between conduit diameter (*D*) and hydraulic conductance (K_t, m^4 MPa^{-1} s^{-1}) in the Hagen–Poiseuille equation:

$$K_t = \pi\rho D^4 / 128\eta \tag{14.4}$$

where ρ is the density of water (5.55×10^7 mmol m^{-3}) and η is the viscosity of water (1.002×10^{-9} MPa s at 20°C). Conduit length can also have a substantial influence on xylem hydraulic conductance in both vessel and tracheid-bearing species (Comstock and Sperry 2000; Pittermann et al. 2006). In a study examining conduit anatomy along a height gradient from 0.5 to 44 m in 50 different species (Anfodillo et al. 2006), mean hydraulically weighted xylem conduit diameter was found to decrease with increasing height in all trees. There was a strong correlation between degree of xylem conduit tapering (the narrowing of xylem conduits from the bottom to the tops of trees) and tree height, independent of any other variables such as age, site and altitude. In a study investigating height-related trends in xylem anatomy and hydraulic characteristics of Douglas-fir leaves, Woodruff et al. (2008) found that maximum hydraulic conductance of foliated shoots declined with increasing tree height. The height-related decline in shoot hydraulic efficiency was strongly correlated with a number of foliar anatomical characteristics that also decreased with height; including total tracheid lumen area per needle cross-section, hydraulic mean diameter of leaf tracheid lumens and leaf tracheid length. Xylem conductivity at the tops of trees can be reduced by changes in other conduit structural features, such as pit aperture diameter and margo pore diameter; both of which are linked to turgor-related constraints on overall cell expansion. Pit aperture conductance of Douglas-fir sapwood decreased sharply with increasing height and was projected to approach zero over a height range consistent with maximum heights reported for the species (Domec et al. 2008). In Douglas-fir, *Pinus palustris* and *Pinus elliottii*, tracheid bordered pit conductance decreased along an axial gradient from trunks to upper branches (Domec et al. 2006; Gonzalez-Benecke et al. 2010).

Diminished hydraulic conductance with increasing tree height implies a further constraint on the capacity for cell expansion in developing tissues. This is because reduced hydraulic conductance represents a constraint upon the delivery of water to growing tissue unless sufficient osmotica are available to generate water potential gradients large enough to sustain rates of water uptake while simultaneously maintaining turgor. Substantial hydraulic resistance within plants occurs at the terminal portion of the pathway from the soil to the expanding cells, i.e. within the non-vascular tissue (Boyer 1974; Molz and Boyer 1978; Kramer and Boyer 1995; Boyer and Silk 2004). Water transport through non-vascular tissue can occur through two main pathways: within the cell wall (apoplastic pathway), or from cell to cell (symplastic pathway) via plasmodesmata. Diffusivity for water movement within non-vascular plant tissue (D_c, $m^2\ s^{-1}$) is related directly to the permeability of the cell membranes and is approximated using measurements of half-time water potential equilibrations ($T_{½}$) between individual cells and their surroundings (Steudle 1992; Nonami et al. 1997). Less expanded, smaller non-vascular cells imply that water must be transported across a greater number of membranes per unit distance. The following equation from Nonami (1998) illustrates the relationship between D_c and cell size in transport of water in non-vascular tissue:

$$D_c = \frac{a_{cc} \cdot \Delta x^2 \cdot \ln 2}{2A} \cdot \frac{1}{T_{½}} \tag{14.5}$$

where a_{cc} = is the cross sectional area of the cell normal to the flow of water (m^2), Δx is the width of the cell in the direction of flow (m) and A is the surface area of the cell (m^2). The passage of water through two cell walls per cell is represented by the factor of two in the denominator. Nonami (1998) found high correlations between cell size and diffusivity for water in different regions of the zone of elongation in soybean stems. Although the impact of cell size is normally considered in the context of vascular tissue, its impact on the rate of water transport in non-vascular tissue is also substantial; thus highlighting the role of turgor-driven cell expansion in hydraulic conductance of non-vascular tissue.

7.3 *Photosynthesis*

Reduced expansion and division in leaf cells leads to a reduction in photosynthetically active leaf area and thus a reduced potential for production of photosynthate. Phenologically driven variability in sink-source relationships (Wiemken and Ineichen 2000) along with compartmentalization of mobile carbohydrates in tissues where they cannot be retrieved (Chapin et al. 1990; Millard et al. 2007) suggest that even plants that store excess carbon in the form of non-structural carbohydrates during periods of excess production of photosynthate may experience reduced growth due to constraints on photosynthesis during peak periods of growth. Increased LMA resulting from reduced cell expansion implies an additional reduction in growth due to the increased investment of photosynthate necessary to produce a given amount of leaf area. Increased LMA is associated with a reduction in intercellular airspace relative to mesophyll surface area (Hanba et al. 1999) and a consequent reduction in conductance of CO_2 within the mesophyll (g_m) to the sites of assimilation within the leaf (Parkhurst 1994; Niinemets 1999; Flexas et al. 2008). Studies examining trends in g_m along a height gradient in tall trees have shown correlations between reduced leaf expansion and reduced g_m with increasing height (Woodruff et al. 2009; Mullin et al. 2009;

Steppe et al. 2011). In a study examining height-related trends in leaf structure and gas-exchange of detached shoots of Douglas-fir, both needle thickness and mesophyll thickness increased with tree height, and g_m and CO_2 assimilation in ambient [CO_2] decreased with tree height (Woodruff et al. 2009). Mullin et al. (2009) examined correlations between height-related trends in physiological and morphological characteristics of *Sequoia sempervirens* foliage. They found that shoot mass per area increased with height, and that maximum mass-based photosynthesis (A_{mass}), standardized photosynthesis (A_{std}, c_i = 247 ppm) and g_m decreased with height. In a study examining the effects of tree height on leaf structure and gas exchange properties of *S. sempervirens*, Ambrose et al. (2009) found that A_{mass} decreased with height. Within these studies height-related trends in gas exchange characteristics consistently parallel vertical trends in foliar structural characteristics that are likely to influence gas exchange and that are influenced by variation in turgor-driven cell expansion. Turgor-related reductions in phloem transport can also lead to feedback constraints on photosynthesis (see Sect. 7.4). Evidence for regulation of photosynthesis by end-product accumulation has been reported for a wide range of higher plants (Foyer 1987; Goldschmidt and Huber 1992; Wiemken and Ineichen 2000; Paul and Foyer 2001) and in-vitro experiments indicate that mesophyll carbohydrate concentration can alter the expression of photosynthetic gene promoters (Jang and Sheen 1994; Koch et al. 1996), leading to a reduction in photosynthesis. Myers et al. (1999) found a dramatic increase in sugar concentrations and reductions in a number of photosynthetic parameters including maximum photosynthesis, carboxylation rate and electron transport rate in branches distal to where phloem had been girdled on branches.

7.4 Phloem Transport

Limitations on turgor can influence phloem transport in several ways, including the influence of turgor on the pressure gradient that drives phloem transport. The pressure flow model for phloem transport first proposed by Münch (1927) states that phloem transport through sieve elements is driven by an osmotically generated pressure gradient created by the active loading of phloem at the sources and the active unloading of phloem at the sinks. The Hagen–Poiseuille equation illustrates the relationship between rate of fluid transport through conduits and it has been used to describe phloem transport (J_v) in sieve elements (van Bel and Hafke 2005),

$$J_v = \Delta\Psi_p \pi r^4 / 8\eta L \tag{14.6}$$

where Ψ_p = turgor, r = sieve tube radius, η = phloem sap viscosity and L = the distance between source and sink. Any reduction in Ψ_p will have a direct impact on J_s by reducing the pressure gradient from source to sink locations. The development of a pressure gradient in response to phloem uploading is dependent upon the influx of water into the sieve tubes. As tissue Ψ and Ψ_p are reduced with increasing tree height, a given $\Delta\Psi_p$ will reflect a progressively more negative value of Ψ_π within the phloem tissue and a progressively greater concentration of photosynthate. Phloem viscosity increases with increasing sucrose concentration, resulting in a further reduction in J_s as illustrated in Eq. 14.6. However, the influences of pressure gradient and viscosity on phloem transport are possibly overshadowed by the effect of phloem anatomy on J_s. Equation 14.7 from Thompson and Holbrook (2003) more precisely estimates the influence of sieve tube anatomical characteristics on phloem specific conductivity (k) using a model that incorporates sieve element radius (r) and length (l), sieve pore radius (r_p) and length (l_p), and the number of pores per sieve plate (N_p),

$$k = \left[\frac{8N_p r_p^4 l}{8N_p r_p^4 (l - l_p) + (8l_p + 3\Pi r_p) r^4} \right] \frac{r^2}{8} \tag{14.7}$$

Just as xylem cell anatomical characteristics are largely influenced by turgor during cell development, primary growth of phloem tissue is heavily influenced by turgor-driven cell expansion that will have a direct impact on sieve tube radius and sieve tube length. A height-related reduction in turgor is thus likely to have manifold impacts upon phloem tissue anatomy and phloem transport. Such reductions in phloem transport cannot only limit growth by constraining the transport of photosynthate to developing tissues, they can also lead to feedback constraints on photosynthesis (see Sect. 7.3).

7.5 Sink Reduction

Growing tissues are a sink for photosynthate. Sink strength is largely determined by a tissue's ability to import assimilates and is thus strongly influenced by growth (e.g. cell division and turgor-driven cell expansion). Reductions in photosynthetic capacity and reduced oxygen sensitivity of photosynthesis can result from sink limitations due to constraints on storage and translocation of photosynthate (Neales and Incoll 1968). In addition to reduced photosynthetic capacity, sink strength reduction in growing tissues may also lead to increased loss of photosynthate to the soil. Plant roots release a range of different compounds into the rhizosphere, and the release of carbon compounds from living plant roots into the soil is a widespread phenomenon (Curl and Trueglove 1986; Högberg et al. 2008; Jones et al. 2009). Root exudate alone represents a substantial carbon cost to plants with estimates ranging from 5% to 21% of all photosynthetically fixed carbon being transferred to the rhizosphere through root exudates (Marschner 1995). Root exudates are primarily free sugars (e.g. glucose and sucrose), amino acids (e.g. glycine and glutamine) and organic acids (citrate, malate, and oxalate) that are passively lost from the root to the soil. Although plasma membrane permeability and spatial location of solutes in the root tissue influence rates of loss of root exudate, the primary factor influencing exudation is the root to soil concentration gradient of the individual compounds

(Jones et al. 2009). Any sink reductions in growing tissues due to constraints on turgor-driven cell expansion could thus to lead to increased accumulations of "unused" photosynthate and thus increased losses of carbon to the rhizosphere in the form of root exudate. Studies investigating stand productivity along chronosequences show evidence for an age-related shift in allocation towards a greater proportion of gross primary productivity being diverted belowground (Ryan and Waring 1992; Barnard and Ryan 2003). Despite this evidence for increased allocation to belowground with increasing tree size, study results do not support the argument that declining productivity is offset by increasing belowground root production (Vogt et al. 1987; Smith and Resh 1999). One explanation for the observed patterns is a tree height-related increase in the deposition of photosynthate into the soil. Despite the substantial amount of carbon estimated to be lost via root exudation, and the considerable amount of interest in quantifying carbon budgets and productivity in forest ecosystems, very little is known of the ontogenetic dynamics of root exudation of trees.

8 Conclusions and Directions for Future Research

Results obtained largely from studies conducted on tall conifers imply that height-related reductions in cell turgor generate a cascade of direct and indirect effects on shoot extension and expansion. The multiple consequences of turgor-limited growth may act in a synergistic manner to further constrain shoot extension as trees approach species- and site-specific height limits. Turgor-limited tissue expansion has permanent anatomical and morphological consequences that govern height-dependent changes in an array of growth-related physiological processes. These include the impact of reduced xylem conduit dimensions on efficiency of water supply to growing tissues at the terminal portions of the water transport pathway, reduced growth efficiency associated with increased LMA (e.g. Zhang et al. 2009), and increased resistance to CO_2 diffusion through the mesophyll to the sites of carboxylation. Additionally, height-related changes in phloem structure and function can lead to inhibition of photosynthesis, reduced transport of photosynthate to growing tissues and shunting of assimilate belowground.

In the tallest coniferous species on earth osmotic adjustment is insufficient to compensate for the impact of vertical gradients of xylem tension on turgor of developing and mature tissue. However, the osmotic potentials in leaves of the tallest conifers are not extreme compared with those observed in leaves of arid zone and Mediterranean climate woody species so the apparent inability of tall conifers to undergo sufficient osmotic adjustment during critical periods of shoot growth is surprising. The nature of the constraints on osmotic adjustment in tall conifer species thus merits further study. It would be useful to know the identities and sources of the principal osmotica involved in generating the turgor necessary for growth. It is likely that the seasonal decline in sugars and other solutes in mature foliage of conifers during the late winter and early spring results from a mobilization and translocation of solutes necessary to generate sufficient turgor to drive expansion and extension growth at shoot tips. A fraction of exported sugars is undoubtedly used for synthesis of cell walls and other cell constituents. Regardless of its direct impact on turgor and expansion of new tissue, loss of solutes from mature leaf tissue will lower its turgor, potentially impeding export of current photosynthate to growing tissues.

Although we have emphasized the direct and indirect impacts of reduced turgor on shoot expansion in tall trees, the contribution of other factors such as nutrient availability and hormone synthesis cannot be excluded and should be evaluated. Established growth equations may prove to be valuable conceptual tools in this context. Scant published data for trees suggest that properties such as cell wall plastic extensibility and the minimum turgor necessary for irreversible wall yielding (yield threshold) may not be important constraints in rapidly expanding leaves (e.g. Metcalfe et al. 1991; Meinzer et al. 2008). However, the relative role of hydraulic bottlenecks upstream from expanding tissue has not been examined. In species showing a height-related decline in xylem conductivity, increasingly steep axial xylem tension gradients would be required to sustain water transport to shoot tips. This implies an even greater demand for osmotica to drive growth-induced water uptake.

Significant progress has been made in understanding the nature of limitations on shoot extension in tall conifers, but comprehensive understanding of the mechanisms responsible for height-related constraints on tree growth is impaired by a scarcity of published data for angiosperms and a broad range of woody growth forms exhibiting differing degrees of apical dominance and seasonal growth periodicity. In shorter-statured growth forms lacking strong apical dominance, the gravitational component of xylem tension may not play a dominant role in limiting turgor and extension growth, but even in short trees, axial gradients in xylem conductivity and xylem tension may constrain growth-induced water uptake at branch tips. Size-related constraints on shoot extension are also likely to differ among species with differing degrees of seasonal growth periodicity such as species showing more or less continuous leaf production and shoot extension throughout the entire growing season versus species whose entire seasonal shoot growth is determined by the contents of overwintering buds. Advances in understanding of size-related biophysical constraints on tree growth will improve site- and species-specific predictions of tree performance and will provide an essential mechanistic tool for predicting forest responses to changing climate.

References

Abdul-Hamid H, Mencuccini M (2009) Age- and size-related changes in physiological characteristics and chemical composition of *Acer pseudoplatanus* and *Fraxinus excelsior* trees. Tree Physiol 29:27–38

Alves AAC, Setter TL (2004) Response of Cassava leaf area expansion to water deficit: cell proliferation, cell expansion and delayed development. Ann Bot 94:605–613

Ambrose AR, Sillett SC, Dawson TE (2009) Effects of tree height on branch hydraulics, leaf structure and gas exchange in California redwoods. Plant Cell Environ 32:743–757

Anfodillo T, Carraro V, Carrer M, Fior C, Rossi S (2006) Convergent tapering of xylem conduits in different woody species. New Phytol 169:279–90

Barnard HR, Ryan MG (2003) A test of the hydraulic limitation hypothesis in fast-growing *Eucalyptus saligna*. Plant Cell Environ 26:1235–1245

Bauerle WL, Hinckley TM, Cermak J, Kucera J, Bible K (1999) The canopy water relations of old-growth Douglas-fir trees. Trees 13:211–217

Beemster GTS, Fiorani F, Inzé D (2003) Cell cycle: the key to plant growth control? Trends Plant Sci 8:154–158

Billow C, Matson P, Yoder B (1994) Seasonal biochemical changes in coniferous forest canopies and their response to fertilization. Tree Physiol 14:563–574

Bond BJ, Czarnomsky N, Cooper C, Day ME, Greenwood MS (2007) Developmental decline in height growth in Douglas-fir. Tree Physiol 27:441–453

Boyer JS (1974) Water transport in plants: mechanisms of apparent changes in resistance during absorption. Planta 117:187–207

Boyer JS, Silk WK (2004) Hydraulics of plant growth. Funct Plant Biol 31:761–773

Buckley TN (2005) The control of stomata by water balance. New Phytol 168:276–292

Carlquist S (1975) Ecological strategies of xylem evolution. University of California Press, Berkeley

Chapin FS, Shulze ED, Mooney HA (1990) The ecology and economics of storage in plants. Annu Rev Ecol Syst 21:423–447

Comstock JP, Sperry JS (2000) Theoretical considerations of optimal conduit length for water transport in vascular plants. New Phytol 148:195–218

Cosgrove DJ (1985) Cell wall yield properties of growing tissues. Evaluation by in vitro stress relaxation. Plant Physiol 78:347–356

Cosgrove DJ (1993) Wall extensibility – its nature, measurement and relationship to plant cell growth. New Phytol 124:1–23

Cosgrove DJ (2005) Growth of the plant cell wall. Nat Rev Mol Cell Biol 6:850–861

Curl EA, Trueglove B (1986) The rhizosphere. Springer, Berlin

Domec J-C, Lachenbruch B, Meinzer FC (2006) Bordered pit structure and function determine spatial patterns of air-seeding thresholds in xylem of Douglas-fir (*Pseudotsuga menziesii*; *Pinaceae*) trees. Am J Bot 93:1588–1600

Domec J-C, Lachenbruch B, Meinzer FC, Woodruff DR, Warren JM, McCulloh KA (2008) Maximum height in a conifer is associated with conflicting requirements for xylem design. Proc Natl Acad Sci USA 105:12069–12074

Edelstein ZR, Ford ED (2000) Branch and foliage morphological plasticity in old-growth *Thuja plicata*. Tree Physiol 23:649–662

Ellsworth DS, Reich PB (1993) Canopy structure and vertical patterns of photosynthesis and related traits in a deciduous forest. Oecologia 96:169–178

England JR, Attiwill PM (2006) Changes in leaf morphology and anatomy with tree age and height in the broadleaved evergreen species. *Eucalyptus regnans* F. Muell. Trees 20:79–90

Epstein W (1992) Kdp, a bacterial P-type ATPase whose expression and activity are regulated by turgor pressure. Acta Physiol Scand 146:193–199

Flexas J, Ribas-Carbó M, Diaz-Espejo A, Galmés J, Medrano H (2008) Mesophyll conductance to CO_2: current knowledge and future prospects. Plant Cell Environ 31:602–621

Foyer CH (1987) The basis for source-sink interaction in leaves. Plant Physiol Biochem 25:649–657

Goldschmidt EE, Huber SC (1992) Regulation of photosynthesis by end-product accumulation in leaves of plants storing starch, sucrose, and hexose sugars. Plant Physiol 99:1443–1448

Gonzalez-Benecke CA, Martin TA, Peter GF (2010) Hydraulic architecture and tracheid allometry in mature *Pinus palustris* and *Pinus elliottii* trees. Tree Physiol 30:361–375

Gower ST, McMurtrie RE, Murty D (1996) Aboveground net primary production decline with stand age: potential causes. Trends Ecol Evol 11:378–382

Granier C, Tardieu F (1998) Spatial and temporal analyses of expansion and cell cycle in sunflower leaves – a common pattern of development for all zones of a leaf and different leaves of a plant. Plant Physiol 116:991–1001

Granier C, Tardieu F (1999) Water deficit and spatial pattern of leaf development. Variability in responses can be simulated using a simple model of leaf development. Plant Physiol 119:609–619

Green PB (1968) Growth physics in *Nitella*: a method for continuous in vivo analysis of extensibility based on a micro-manometer technique for turgor pressure. Plant Physiol 43:1169–1184

Guerrero FD, Jones JT, Mullet JE (1990) Turgor-responsive gene transcription and RNA levels increase rapidly when pea shoots are wilted. Sequence and expression of three inducible genes. Plant Mol Biol 15:11–26

Hanba YT, Miyazawa SI, Terashima I (1999) The influence of leaf thickness on the CO_2 transfer conductance and leaf stable carbon isotope ratio for some evergreen tree species in Japanese warm-temperate forests. Funct Ecol 13:632–639

Harold FM (1990) To shape a cell: an inquiry into the causes of morphogenesis of microorganisms. Microbiol Rev 54:381–431

Hellkvist J, Richards GP, Jarvis PG (1974) Vertical gradients of water potential and tissue water relations in Sitka spruce trees measured with the pressure chamber. J Appl Ecol 11:637–667

Hinckley TM, Duhme F, Hinckley AR, Richter H (1983) Drought relations of shrub species: assessment of the mechanisms of drought resistance. Oecologia 59:344–350

Högberg P, Högberg MN, Göttlicher SG, Betson NR, Keel SG, Metcalfe DB, Campbell C, Schindlbacher A, Hurry V, Lundmark T, Linder S, Näsholm T (2008) High temporal resolution tracing of photosynthate carbon from the tree canopy to forest soil microorganisms. New Phytol 177:220–228

Hsiao TC, Xu L-K (2000) Sensitivity of growth of roots versus leaves to water stress: biophysical analysis and relation to water transport. J Exp Bot 51:1595–1616

Hsiao TC, Acevedo E, Fereres E, Henderson DW (1976) Stress metabolism; water stress, growth and osmotic adjustment. Philos Trans R Soc Lond B Biol Sci 273:479–500

Ishii H (2011) How do changes in leaf/shoot morphology and crown architecture affect growth and physiological function of large, old trees? In: Meinzer FC, Lachenbruch B, Dawson TE (eds) Size- and age-related changes in tree structure and function. Springer, Dordrecht

Ishii HT, Jennings GM, Sillett SC, Koch GW (2008) Hydrostatic constraints on morphological exploitation of light in tall *Sequoia sempervirens* trees. Oecologia 156:751–763

Jacobs T (1997) Why do plant cells divide? Plant Cell 9:1021–1029

James SA, Meinzer FC, Goldstein G, Woodruff DR, Jones T, Restom T, Mejia M, Clearwater M, Campanello P (2003) Axial and radial water transport and internal water storage in tropical forest canopy trees. Oecologia 134:37–45

Jang JC, Sheen J (1994) Sugar Sensing in Higher Plants. Plant Cell 6:1665–1679

Jones JT, Mullet JE (1995) Developmental expression of a turgor-responsive gene that encodes an intrinsic membrane protein. Plant Mol Biol 28:983–996

Jones DL, Nguyen C, Finlay RD (2009) Carbon flow in the rhizosphere: carbon trading at the soil–root interface. Plant Soil 321:5–33

Kirkham MB, Gardner WR, Gerloff GC (1972) Regulation of cell division and cell enlargement by turgor pressure. Plant Physiol 49:961–962

Koch K, Wu Y, Xu J (1996) Sugar and metabolic regulation of genes for sucrose metabolism: potential influence of maize sucrose synthase and soluble invertase responses on carbon partitioning and sugar sensing. J Exp Bot 47:1179–1185

Koch GW, Sillett SC, Jennings GM, Davis SD (2004) The limits to tree height. Nature 428:851–854

Körner C (2003) Carbon limitation in trees. J Ecol 91:4–17

Körner C, Asshoff R, Bignucolo O, Hättenschwiler S, Keel SG, Pelaez-Riedl S, Pepin S, Siegwolf RTW, Zotz G (2005) Carbon flux and growth in mature deciduous forest trees exposed to elevated CO_2. Science 309:1360–1362

Kramer PJ, Boyer JS (1995) Water relations of plants and soil. Academic, San Diego

Kubiske ME, Abrams MD (1991) Seasonal, diurnal and rehydration-induced variation of pressure–volume relationships in *Pseudotsuga menziesii*. Physiol Plant 83:107–116

Lockhart JA (1965) An analysis of irreversible plant cell elongation. J Theor Biol 8:264–275

Lu Z, Neumann PM (1998) Water-stressed maize, barley and rice seedlings show species diversity in mechanisms of leaf growth inhibition. J Exp Bot 49:1945–1952

Malli R, Epstein W (1998) Expression of the Kdp ATPase is consistent with regulation by turgor pressure. J Bacteriol 180:5102–5108
Marcelis L (1996) Sink strength as a determinant of dry matter partitioning in the whole plant. J Exp Bot 47:1281–1291
Marschner H (1995) Mineral nutrition of higher plants, 2nd edn. Academic, London
Marshall JD, Monserud RA (2003) Foliage height influences specific leaf area of three conifer species. Can J For Res 33:164–170
Matthews MA, Van Volkenburgh E, Boyer JS (1984) Acclimation of leaf growth to low water potentials in sunflower. Plant Cell Environ 7:199–206
McDowell NG, Phillips N, Lunch CK, Bond BJ, Ryan MG (2002) An investigation of hydraulic limitation and compensation in large, old Douglas-fir trees. Tree Physiol 22:763–774
McDowell NG, Licata J, Bond BJ (2005) Environmental sensitivity of gas exchange in different-sized trees. Oecologia 145:9–20
Meinzer FC, Brooks JR, Domec JC, Gartner BL, Warren JM, Woodruff DR, Bible K, Shaw DC (2006) Dynamics of water transport and storage in conifers studied with deuterium and heat tracing techniques. Plant Cell Environ 29:105–114
Meinzer FC, Bond BJ, Karanian JA (2008) Biophysical constraints on leaf expansion in a tall conifer. Tree Physiol 28:197–206
Mencuccini M, Grace J (1996) Developmental patterns of above-ground hydraulic conductance in a Scots pine (*Pinus sylvestris* L.) age sequence. Plant Cell Environ 19:939–948
Mencuccini M, Martínez-Vilalta J, Vanderklein D, Hamid HA, Korakaki E, Lee S, Michiels B (2005) Size-mediated ageing reduces vigour in trees. Ecol Lett 8:1183–1190
Metcalfe JC, Davies WJ, Pereira JS (1991) Control of growth of juvenile leaves of *Eucalyptus globulus*: effects of leaf age. Tree Physiol 9:491–500
Millard P, Sommerkorn M, Grelet GA (2007) Environmental change and carbon limitation in trees: a biochemical, ecophysiological and ecosystem appraisal. New Phytol 175:11–28
Molz FJ, Boyer JS (1978) Growth-induced water potentials in plant cells and tissues. Plant Physiol 62:699–704
Morgan JM (1984) Osmoregulation and water stress in higher plants. Annu Rev Plant Physiol 35:299–319
Mullin LP, Sillett SC, Koch GW, Tu KP, Antoine ME (2009) Physiological consequences of height-related morphological variation in *Sequoia sempervirens* foliage. Tree Physiol 29:999–1010
Münch E (1927) Versuche über den Saftkreislauf. Ber Dtsch Bot Ges 45:340–356
Myers DA, Thomas RB, DeLucia EH (1999) Photosynthetic responses of loblolly pine (Pinus taeda) needles to experimental reduction in sink demand. Tree Physiol 19:235–242
Neales TF, Incoll LD (1968) The control of leaf photosynthesis rate by the level of assimilate concentration in the leaf: a review of the hypothesis. Bot Rev 34:107–125
Niinemets Ü (1999) Components of leaf dry mass per area – thickness and density – alter leaf photosynthetic capacity in reverse directions in woody plants. New Phytol 144:35–47
Niinemets Ü, Tenhunen JD (1997) A model separating leaf structural and physiological effects on carbon gain along light gradients for the shade-tolerant species *Acer saccharum*. Plant Cell Environ 20:845–866
Nonami H (1998) Plant water relations and control of cell elongation at low water potentials. J Plant Res 111:373–382
Nonami H, Boyer JS (1993) Direct demonstration of a growth-induced water potential gradient. Plant Physiol 102:13–19
Nonami H, Wu YJ, Boyer JS (1997) Decreased growth-induced water potential – primary cause of growth inhibition at low water potentials. Plant Physiol 114:501–509
Oldham AR, Sillet SC, Tomescu AMF, Koch GW (2010) The hydrostatic gradient, not light availability, drives height-related variation in *Sequoia sempervirens* (*Cupressaceae*) leaf anatomy. Am J Bot 97:1087–1097
Osonubi O, Davies WJ (1978) Solute accumulation in leaves and roots of woody plants subjected to water stress. Oecologia 32:323–332
Parkhurst DF (1994) Diffusion of CO_2 and other gases inside leaves. New Phytol 126:449–479

Passioura JB, Boyer JS (2003) Tissue stresses and resistance to water flow conspire to uncouple the water potential of the epidermis from that of the xylem in elongating plant stems. Funct Plant Biol 30:325–334

Paul MJ, Foyer CH (2001) Sink regulation of photosynthesis. J Exp Bot 52:1383–1400

Pittermann J, Sperry JS, Hacke UG, Wheeler JK, Sikkema EH (2006) Inter-tracheid pitting and the hydraulic efficiency of conifer wood: the role of tracheid allometry and cavitation protection. Am J Bot 93:1265–1273

Ranney TG, Bassuk NL, Whitlow TH (1991) Osmotic adjustment and solute constituents in leaves and roots of water stressed cherry (*Prunus*) trees. J Am Soc Hortic Sci 4:684–688

Renninger HJ, Phillips N, Hodel DR (2009) Comparative hydraulic and anatomic properties in palm trees (*Washingtonia robusta*) of varying heights: implications for hydraulic limitation to increased height growth. Trees 23:911–921

Richter H (1973) Frictional potential losses and total water potential in plants – re-evaluation. J Exp Bot 24:983–994

Richter H (1974) Elevated suction tensions and morphological alterations caused by transverse cuts in a *Taxus* stem. Flora 163:291–309

Rieger M (1995) Offsetting effects of reduced root hydraulic conductivity and osmotic adjustment following drought. Tree Physiol 15:379–385

Ritchie GA, Shula RG (1984) Seasonal changes of tissue water relations in shoots and root systems of Douglas-fir seedlings. For Sci 30:538–548

Roden J, Van Volkenburgh E, Hinckley TM (1990) Cellular basis for limitation of poplar leaf growth by water deficit. Tree Physiol 6:211–219

Ryan MG, Waring RH (1992) Stem maintenance and stand development in a subalpine lodgepole pine forest. Ecology 73:2100–2108

Ryan MG, Yoder BJ (1997) Hydraulic limits to tree height and tree growth. Bioscience 47:235–242

Sala A, Hoch G (2009) Height-related growth declines in ponderosa pine are not due to carbon limitation. Plant Cell Environ 32:22–30

Sala A, Fouts W, Hoch G (2011) Carbon storage in trees: does relative carbon supply decrease with tree size? In: Meinzer FC, Lachenbruch B, Dawson TE (eds) Size- and age-related changes in tree structure and function. Springer, Dordrecht

Scholander PF, Hammel HT, Bradstreet ED, Hemmington EA (1965) Sap pressure in vascular plants. Science 148:339–346

Silveira JAG, Araujo SAM, Lima JPMS, Viegas RA (2009) Roots and leaves display contrasting osmotic adjustment mechanisms in response to NaCl-salinity in *Atriplex nummularia*. Environ Exp Bot 66:1–8

Smith FW, Resh SC (1999) Age-related changes in production and below-ground carbon allocation in *Pinus contorta* forests. For Sci 45:333–341

Spollen WG, Sharp RE (1991) Spatial-distribution of turgor and root-growth at low water potentials. Plant Physiol 96:438–443

Sprugel DG, Brooks JR, Hinckley TM (1996) Effects of light on shoot geometry and needle morphology in *Abies amabilis*. Tree Physiol 16:91–98

Steppe K, Niinemets Ü, Teskey RO (2011) Tree size- and age-related changes in leaf physiology and their influence on carbon gain. In: Meinzer FC, Lachenbruch B, Dawson TE (eds.) Size- and age-related changes in tree structure and function. Springer, Dordrecht

Steudle E (1989) Water-flow in plants and its coupling to other processes – an overview. Method Enzymol 174:183–225

Steudle E (1992) The biophysics of plant water: compartmentation, coupling with metabolic processes, and flow of water in plant roots. In: Somero GN, Osmond CB, Bolis CL (eds) Water and life: comparative analysis of water relationships at the organismic, cellular, and molecular levels. Springer, Heidelberg

Stroeher VL, Boothe JG, Good AG (1995) Molecular cloning and expression of a turgor-responsive gene in *Brassica napus*. Plant Mol Biol 27:541–551

Teskey RO, Grier CC, Hinckley TM (1984) Change in photosynthesis and water relations with age and season in *Abies amabilis*. Can J For Res 14:77–84

Thompson MW (2006) Phloem: the long and the short of it. Trend Plant Sci 11:26–32
Thompson MW, Holbrook NM (2003) Application of a single-solute non-steady-state phloem model to the study of long-distance assimilate transport. J Theor Biol 220:419–455
Turner NC, Jones M (1980) Turgor maintenance by osmotic adjustment: a review and evaluation. In: Turner NC, Kramer PJ (eds) Adaptation of plants to water and high temperature stress. Wiley-Interscience, New York, pp 87–103
Turner J, Long JN (1975) Accumulation of organic matter in a series of Douglas-fir stands. Can J For Res 5:681–690
Tyree MT, Cheung YNS, MacGregor ME, Talbot AJB (1978) The characteristics of seasonal and ontogenetic changes in the tissue-water relations of *Acer, Populus, Tsuga and Picea*. Can J Bot 56:635–647
van Bel AJE, Hafke JB (2005) Physicochemical determinants of phloem transport. In: Holbrook NM, Zwieniecki M (eds) Long-distance transport systems in plants: integration and coordination. Elsevier Science, New York
Vogt KA, Vogt DJ, Moore EE, Babatunde AF, Redlin MA, Edmonds RL (1987) Conifer and angiosperm biomass in relation to stand age and site productivity in Douglas-fir forests. J Ecol 75:857–870
Weiner J, Thomas SC (2001) The nature of tree growth and the "age-related decline in forest productivity". Oikos 94:374–376
Whittaker RH, Woodwell GM (1967) Surface area relations of woody plants and forest communities. Am J Bot 54:931–939
Wiemken V, Ineichen K (2000) Seasonal fluctuations of the levels of soluble carbohydrates in spruce needles exposed to elevated CO_2 and nitrogen fertilization and glucose as a potential mediator of acclimation to elevated CO_2. J Plant Physiol 156:746–750
Wolswinkel P (1985) Phloem unloading and turgor-sensitive transport: factors involved in sink control of assimilate partitioning. Physiol Plant 65:331–339
Woodruff DR, Bond BJ, Meinzer FC (2004) Does turgor limit growth in tall trees? Plant Cell Environ 27:229–236
Woodruff DR, Meinzer FC, Lachenbruch B (2008) Height related trends in leaf xylem anatomy and hydraulic characteristics in a tall conifer: safety versus efficiency in foliar water transport. New Phytol 180:90–99
Woodruff DR, Meinzer FC, Lachenbruch B, Johnson DM (2009) Coordination of leaf structure and gas exchange along a height gradient in a tall conifer. Tree Physiol 29:261–272
Woodruff DR, Meinzer FC, McCulloh KA (2010) Height-related trends in stomatal sensitivity to leaf-to-air vapour pressure deficit in a tall conifer. J Exp Bot 61:203–210
Wu Y, Cosgrove DJ (2000) Adaptation of roots to low water potentials by changes in cell wall extensibility and cell wall proteins. J Exp Bot 51:1543–1553
Würth MKR, Pelaez-Riedl S, Wright SJ, Körner C (2005) Non-structural carbohydrate pools in a tropical forest. Oecologia 143:11–24
Yoda K, Shinozaki K, Ogawa H, Hozumi K, Kira T (1965) Estimation of the total amount of respiration in woody organs of trees and forest communities. J Biol Osaka City Univ 16:15–26
Zhang Y-J, Meinzer FC, Hao G-Y, Scholz FG, Bucci SJ, Takahashi FSC, Villalobos-Vega R, Giraldo JP, Cao K-F, Hoffmann WA, Goldstein G (2009) Size-dependent mortality in a Neotropical savanna tree: the role of height-related adjustments in hydraulic architecture and carbon allocation. Plant Cell Environ 32:1456–1466
Zimmermann MH (1983) Xylem structure and the ascent of sap. Springer, Berlin

Chapter 15
Instrumental Approaches for Studying Tree-Water Relations Along Gradients of Tree Size and Forest Age

Jan Čermák and Nadezhda Nadezhdina

Abstract There is an increasing demand for data on tree structural and functional attributes that can be collected simultaneously at multiple sites and integrated across landscapes. Here, we present several examples of approaches applicable for measuring tree and stand structures and for characterizing spatial and temporal variation in important functions of the above- and belowground parts of both large and small trees. Special attention is given to explanations of the theoretical basis for several sap flow techniques and to the types of information that can be gleaned from carefully planned measurements of sap flow in stems and roots. A variety of approaches for characterizing the structure and function of root systems: the hidden half of trees, are also described.

1 Introduction

Excellent studies of tree structure and function have been published, but most of these have been confined to a few selected and well-equipped sites. However, there is increasing demand for data on tree structural and functional attributes that can be collected simultaneously at multiple sites and integrated across landscapes. Here, we present several examples of approaches applicable for measuring tree and stand structures and for characterizing spatial and temporal variation in important functions of the above- and belowground parts of both large and small trees. Data obtained from these approaches include a number of biometric characteristics of the leaf and root distribution, sap flow and effective functional parameters of crowns and roots.

J. Čermák (✉) • N. Nadezhdina
Faculty of Forestry and Wood Technology, Mendel University of Agriculture and Forestry Brno, Brno 61300, Czech Republic
e-mail: Cermak@mendelu.cz

F.C. Meinzer et al. (eds.), *Size- and Age-Related Changes in Tree Structure and Function*, Tree Physiology 4, DOI 10.1007/978-94-007-1242-3_15,

2 Biometric Measurements at the Whole-Tree Level

2.1 Leaf Area Distribution Based on Allometric Relationships and Optical Methods

Several methods are commonly used to estimate the leaf area index (LAI) in forest stands; these methods include optical methods such as fish-eye hemispherical photography (Lee et al. 1983; Brown et al. 2000; Jonckheere et al. 2004). When more detailed leaf distribution patterns are required, photographic images combined with image analysis usually provide reasonable results (Čermák et al. 2007b). Images of entire crowns and of individual branches are calibrated in terms of leaf density (counted manually) and their measured area on sub-samples. Both the vertical and radial leaf distributions are calculated based on general geometrical forms of tree crowns. The leaf distribution data derived by this model for large trees correspond well with the measured leaf distribution data, as shown in Fig. 15.1. The vertical leaf distribution can serve as a background for evaluation of different physiological processes (such as transpiration and photosynthesis (De Lorenzi et al. 2009), whereas the radial leaf distribution (indicating differences in LAI within the crown over a range from 1 to 7) provides a background for more detailed evaluation of remote sensing images.

Comparable data of leaf distribution can also be obtained by applying the "cloud" method, based on measurement of the spatial distribution of leaves along the branches (Čermák 1989, 1998; Čermák et al. 1998, 2008a; Morales et al. 1996a, b, c; Tognetti et al. 2000; Urban et al. 2009). The position of each leafed volume on branches ("cloud") is measured three dimensionally. Additionally, the total leaf dry mass and leaf mass per area are measured on sub-samples for each leafed volume. The data are subsequently entered into a spreadsheet and evaluated in terms of vertical and radial (or spatial) distributions. Knowledge of spatial patterns of leaf dry mass per area, which is more or less linearly dependent on the long-term integral of illumination during leaf growth, allows the derivation of the relative illumination of foliage (expressed as a percentage of that above the canopy) and also permits calculation of the fractions of sunlit and shaded leaves (Čermák 1989, 1998). These parameters serve as a background for physiological and other studies. Methods involving destructive sampling for analyses of leaf distribution analysis and leaf mass per area require manual work and therefore are a bit cumbersome, but provide data on any type of stand and also trees with different social positions within stands.

Methods similar to those applied to large trees can be even more easily used for leaf distribution studies in small individuals of woody species, such as seedlings or saplings as shown on example of *Rhododendron* shrubs (Nadezhdina et al. 2004a). An example of a European beech (*Fagus sylvatica*) sapling photographed in a dense shrub canopy from two perpendicular sides against a white blanket is shown in Fig. 15.1. Each photo was doubled from perpendicular directions and taking the right and left sides of the crown separately if the sapling was large and its crown did not fit within the width of the blanket). The data were evaluated by image

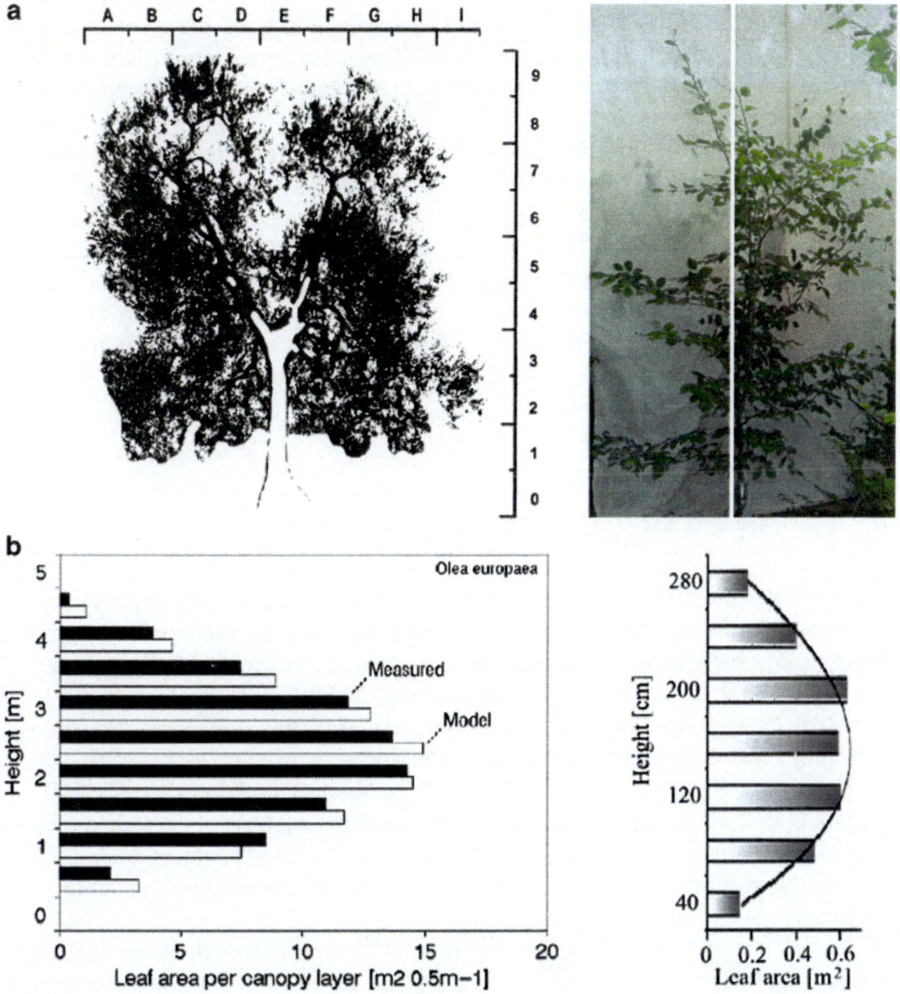

Fig. 15.1 *Left panels*: Leaf distribution in old olive trees (*Olea europaea*) in the orchard (southern Italy) estimated by different methods: (**a**) Photographic image of a tree (*upper panel*), elaborated using an image analyzer system and original leaf area measured manually in tree crowns (**b**) modeling approach (*white columns*) based on measurements (*black columns*) of main biometric data (stem diameter, height of tree top and crown base and projected crown area (*lower panel*, modified according to Čermák et al. 2007b). *Right panels*: Example of the photographic image of beech (*Fagus sylvatica*) sapling (the image of a larger sapling is composed of two pictures – *left and right crown side here*), analyzed on the basis of (**a**) similar photos taken from two perpendicular sides (*upper panel*) and simultaneously (**b**) through evaluation of spatial patterns of leaf dry mass per area. Pictures were analyzed by the image analyzer system. Columns show results of images and line results of mass per area (*lower panel*)

analysis and on the basis of leaf count over the whole sapling and the corresponding leaf sub-samples of leaf dry mass per area. This parameter, which is easier to determine on a small scale than in large trees, yields data comparable to that of the image analysis.

Modern and more sophisticated approaches include lidar technology, which consists of a measuring device containing a rotating laser system, are able to acquire a high-resolution (5 cm) crown image. A complex 3D image is obtained when the tree crown images are taken from at least three different sides (Van Der Zande et al. 2009). This system works well with solitary trees, trees growing on stand edges or those growing in sparse canopies, but there are problems with its application in dense canopies. A laser scanning system has also been applied within the framework of remote sensing technologies and is suitable for characterizing crowns on a large-scale basis.

2.2 *The Hidden Half of Trees: Root Systems*

Studies of the root systems of adult trees are especially challenging because their complex architecture includes both fine absorbing roots as well as coarse conducting roots (e.g., Zygurovskaya 1958; Jeník and Sen 1964; Schuurman and Goedwaagen 1965; Kolesnikov 1972; Bohm 1979; Sutton and Tinus 1983; Steele et al. 1997; Coutts et al. 1999; Johnson et al. 2001). Coarse (skeleton) tree roots were classically analyzed by manually excavating them from the surrounding soils (Jeník 1957, 1978; Kutschera 1960; Kolesnikov 1972; Vyskot 1976; Mauer and Palátová 1996, 2003; Tatarinov et al. 2008). Applying additional microscopic analysis of excavated roots yields the most comprehensive type of data set (Kutschera and Lichtenegger 2002). Although this information is very detailed and valuable, the required methods are laborious, time-consuming and destructive.

Because plants take up soil water and mineral elements through absorbing root surfaces, their size and permeability are crucial parameters for plant survival. Several methods have been applied for estimating fine and thin root densities. Indirect methods include surface adsorption of acid or basic organic dyes (Wilde and Voigt 1949; Dunham 1958; Carman 1982; Costa et al. 2001), gravimetrically estimated loss of concentrated salt solutions (Carley and Watson 1966), absorption and desorption of nitrite (Ansari et al. 1995) and photoelectric measurements. In contrast, direct methods of root surface measurement (Lopez et al. 2001) are based on sequential soil core sampling techniques or the use of root-free in-growth cores and minirhizotrons (McQueen 1968; Persson 1983; Persson et al. 1995; Persson and Ahlstrom 2002; Steele et al. 1997; Helmisaari and Hallbacken 1999; Janssens et al. 1999). Special software and image analyses are often used, including fractal modeling and sophisticated allometric analysis (Coutts et al. 1999; Drexhage et al. 1999; West et al. 1999; Coomes and Grubb 2000; Bond-Lamberty et al. 2002; Salas et al. 2004), but the use of such techniques usually precludes the study of entire tree root systems. Several types of non-destructive methods have been used to estimate the functional extent and depth of root systems of forest trees including radioactive tracers (Woods 1969) or combined measurements of soil water content and sap flow (Čermák et al. 1980; Čermák and Kučera 1990a, b; Nadezhdina and Čermák 2003a, b). Because these methods provide little or no resolution of root structure, an alternative method suitable for root quantification in situ is of great importance. In particular, studies of absorbing root surfaces are necessary to better understand plant physiological processes such as water transport and nutrient uptake (Smerthust and Comerford 1993; Steudle 1995; Comerford et al. 1995; Eckhard and Horst 1996).

2.3 Root Excavation with Supersonic Air Stream (the "Air-Spade")

Entire root systems can be excavated using a supersonic air stream (Air-Spade 150/90, made by Concept Engineering Group, Inc., Verona, PA, USA). This system is based on air flow from a strong compressor (Ingersoll-Rand VHP 400, giving an air flow rate of up to 12 m^3 min^{-1} at an operating pressure of up to 1.2 MPa) that produces a "laser-like" thin air stream with a velocity of Mach two when connected to the special nozzle (Rizzo and Gross 2000; Nathenson and Jarabak 2001; Nadezhdina and Čermák 2003a, b; Čermák et al. 2008c). The method relies on the velocity of the stream and not on its pressure. The air stream is not harmful when it touches a smooth object (such as stones, roots or feet), but if it enters any small pores the air becomes compressed within, causing them to explode. Soil contains such pores and is therefore dispersed during micro-explosions into small particles and blown away (Fig. 15.2). Whereas fine roots (around 1 mm in diameter) can be exposed without major damage, micron-sized hairs of mycorrhiza cannot. Sufficient soil moisture is a pre-condition for proper application of this system. Dry soil is dispersed into relatively large pieces (up to the size of a fist) that hold fine roots too tightly, potentially damaging them when whole soil pieces are moved away. Roots can also be damaged by fast-moving pieces of sand or gravel. If the root network does not allow these hard objects to be moved away, it is better to remove them manually.

For optimal data analysis, the excavated root systems must be further evaluated and quantified based on photo images, using an image analyzer. Photographic images of excavated root systems using the central projection (of the photographic lenses) must be transformed to the orthogonal projection to find the correct geometrical positions of the analyzed objects. Generation of a square network when applying a grid (e.g., 1 × 1 m) to the field before taking images allows maintenance of a proper scale when composing particular pictures into a mosaic of the area studied. Interactive graphical software (e.g., KOKEŠ v. 6.57, Gepro Praha) has been used in the transformation procedure. The resulting transformed images are connected into a seamless photo-mosaic, where the mean position error is ±2 cm and the mean raster accuracy is 0.5 cm (Čermák et al. 2008c).

Fig. 15.2 *Upper panel*: excavation of tree root systems using a supersonic air stream (the "air-spade") in a Norway spruce (*Picea abies*) forest. Sufficiently moist soil is dispersed and blown away by the stream, while smooth objects (roots, stones, etc.) remain undamaged. If the whole root system is to be exposed, the trees must be felled beforehand for operator safety reasons. *Lower panel*: the air-spade system can also be used for excavating rather deep roots (here 2.5 m). Such roots of very different sizes are visible along transect over the root system of a large *Platanus orientalis* tree growing close to the house wall in the city

2.4 Three-Dimensional Visualization of Root Systems Based on the Ground-Penetrating Radar (Georadar)

Coarse roots can be visualized using ground-penetrating radar (GPR, Hruška et al. 1999; Čermák et al. 2000; Wielopolski et al. 2000; Stokes et al. 2002). Root systems were visualized using a PulseEKKO 1000 geo-radar (Sensors and Software, Inc., Canada) by applying shielded antennas and a frequency of 900 MHz. This assures an accuracy of

2 cm when specifying the location, a sampling depth of 1.5 m, and direct detection of roots with diameters over 1.1 cm (location, depth and root diameter are estimated using direct detection). Only larger roots can be visualized in deeper soil layers when application of modified antennas and frequency is required. In principle, radar cannot see fine roots, which can only be detected indirectly on the basis of diffraction markers indicating location and depth, but not root diameter. Our measurements were done along a network of perpendicular profiles using a 0.25-m grid with a 0.05-m step. The effective velocity (v_{ef}) was applied to convert the temporal radar data into the corresponding depth, and $v_{ef} = 0.1$ m ns^{-1} was suitable for weathered granodiorite. A large tree root system can be covered by ~400 m of profiles using ~10,000 geo-radar point measurements (Hruška et al. 1999; Hruška 2008).

Individual geo-radar records were elaborated using the software package EKKO TOOLS (4.23 from Sensors and Software), which improved the signal-to-noise ratio, filtered interfering signals and frequencies and eliminated external disturbing sources. The individual records were used to produce a series of "slides," from which we constructed horizontal and vertical maps of root systems below the areas measured. The 3D pictures are more precise when a dense grid is applied (i.e., 10 × 10 cm), while larger errors appear in sparser grids (i.e., 50 × 50 cm), because positions of individual roots must be reconstructed between the "slides", and larger distances do not allow correct execution of this process. Geo-radar also cannot "see" vertically growing roots that run parallel to the electromagnetic beam. Although this system works with acceptable errors on homogeneous soils, it does not work on stony soils, as individual stones cause many false reflections that are difficult to distinguish from roots.

3 Measurements of Whole Tree Function – Water Relations

3.1 Features of Integration, Excessive Energy Load, Leaves and Landscape Cooling

There are a limited number of variables that can be routinely measured at the whole-tree level to integrate the behavior of large numbers of leaves and roots amounting tens to hundreds of thousands per broadleaf tree and hundreds of thousands to several millions of needles in coniferous species. In fact, the integrating effect of the variable may be more important than the measurable variable itself. Practically, the most comprehensive variable applicable for these purposes is sap flow, which links roots and leaves and provides important information about the relationships between trees, water and the physiological processes responsible for the transformation of the largest amounts of solar energy. The energy needed for photosynthesis and other functions only accounts for a fraction of a percent of that incident on the plant. A much larger part of this energy is absorbed by exposed surfaces, thereby increasing their temperature. Therefore, a huge amount of water is needed for cooling plant surfaces to prevent overheating and to allow other processes to work normally. The entire landscape is cooled this way if covered sufficiently by vegetation (Makarieva et al. 2006, 2008; Makarieva and Gorshkov 2008). No other phenomenon is more effective for cooling than the evaporation of water. The importance of water is greatest when it becomes scarce. Outside of the mostly frozen arctic regions, lack of water (drought) is the most frequent limiting factor for plant growth. Large plants, especially trees, have to transport massive amounts of water through their bodies to leaves distant from roots, especially through stem sapwood consisting of specialized conducting tissue (xylem).

3.2 *Main Principles of Sap Flow Measurements*

Following pioneering work in the last century (Huber 1932), many types of sap flow measurement methods based on a variety of principles (e.g., thermodynamic, electric, magneto-hydrodynamic, nuclear magnetic resonance, etc.) have been described. However, only a few of these, particularly those based on thermodynamics, are widely used in field-grown trees. A variety of thermodynamic methods are available for calculating the mass flow of sap in trees using heat as a tracer. Measurements can be made in any conductive organ of the tree, including branches and roots. Depending on the method, measurements are made in a localized part of the conductive organ immediately surrounding the sensors, or around the entire perimeter of the conductive organ. Some methods integrate the sap flow in the whole sapwood, while others provide information for calculating sap flow at different depths behind the cambium. Some methods are invasive because the sensors are located within the sapwood; others are non-invasive, with the sensors located outside of the conductive organ, although in close contact. Some methods are suitable for stems of very small diameters where only non-invasive ones are applicable, while others can be used in large trees. A subset of these methods measures continuously, while others measure discontinuously. Some methods cannot measure below a certain threshold rate of sap flow, whereas other methods are useful for measuring very low sap flow rates and even those occurring during reverse flow. Therefore, the user must be aware of the potential and limitations of the different available methods in order to select the most suitable one. In the following sections, we mention the currently most widespread methods for estimating sap flow from heat measurements with references to key papers where details about each method are given. However, the measurements of sap flow itself represents only one of many aspects of sap flow studies. Additional important items to consider include evaluation of errors and spatial integration of data measured by individual sensors to the whole-tree and stand levels. These aspects are discussed below.

3.2.1 Classical Heat-Pulse and Cohen's T-Max Methods

The first reference to a heat pulse velocity method (HPV) was by Huber (1932), whose system was based on two kinds of probes: a heater and two temperature sensors above it. The time of the first appearance of heat at the upper downstream sensor after passing the lower one was assumed to be the same as the time taken for the sap to move this distance. Huber recognized the importance of distinguishing between convective heat transfer by the moving sap and the transport of heat by thermal conduction. To separate these two effects, Huber and Schmidt (1936) developed an early version of the 'compensation' heat-pulse (CHP) method in which one sensor was downstream and the other sensor upstream of the heater. The time of peak warming of the upstream sensor as compared to the downstream one was used to 'compensate' for the effects of thermal conduction. Later, Marshall (1958) proposed a new probe arrangement and developed a theoretical framework for the compensation heat-pulse method. This analytical theory, with further improvements by Swanson (1965, 1970, 1971), was used by Cohen et al. (1981) to develop the heat-pulse method known as the T-max method. This method uses a heater and two probes. One of the probes, with a single temperature sensor, is inserted downstream of a line heater to record the time delay for the maximum temperature increase at the sensor location. A second probe, also with a temperature sensor, is located upstream and used as a reference.

3.2.2 Green's HPV System

The heat-pulse velocity (HPV) system developed by Green and Clothier (1988) and Green (1998) is based on the CHP method (Swanson and Whitfield 1981) The HPV system can be used for either the compensation or T-max methods (Green et al. 2003). For the compensation method, two temperature probes are placed asymmetrically on either side of a line heater that is inserted radially into the sapwood. Each temperature probe has several thermocouples at different depths behind the cambium. Following the application of a brief 1- to 2-s heat pulse, the time delay for an equal temperature increase at the two sensors placed at the same depth below the cambium, upstream and downstream from the heater, is used to calculate the heat pulse velocity. A theoretically derived correction factor is then used to correct the heat pulse measurements for any probe-induced effects of wounding and to calculate volumetric rates of sap flow. For the T-max method, the upstream probe is simply shifted to a position 40 mm from the heater so that it does not sense the heat pulse. This second reference probe is used to compensate for any background changes in stem temperature that may occur during the T-max measurement. Correction factors are needed to derive actual sap flow values from the recorded T-max values.

3.2.3 The Heat-Dissipation (HD) Method

The heat-dissipation method, developed by Ittner (1968), Radchenko et al. (1982) and especially Granier (1985), is based on the detection of convective heat transport (heat carried with the sap stream). A thermocouple is mounted on each of two thin needles inserted radially in the sapwood with a vertical distance of about 10 cm between them. The upper needle contains a heating element operated at constant power. The sap flow density is calculated from the temperature difference between the two needles. The maximum temperature difference occurs when flow is zero, and decreases with increasing sap flow velocity. This is the most widespread method because of its simplicity and relatively low cost of sensors, although recording of very low flows (e.g., resaturating at night) and high flows (e.g., around midday in rapidly transpiring trees) are difficult.

3.2.4 Heat Balance with External Heating or Stem Heat Balance (SHB) Method

Vieweg and Ziegler (1960) developed a method where a heater band is wrapped around the conductive organ. A small quantity of heat is applied continuously to increase the temperature of the enclosed stem to a certain level above that of the non-heated parts below. The heat balance method was first described by Daum (1967), who measured the natural flow of heat and did not apply any artificial heating. A similar method was developed by Sakuratani (1981) and advanced by Valancogne and Nasr (1989) as well as by Weibel and Vos (1994). Further versions were applied by Baker and Van Bavel (1987), Steinberg et al. (1990), Ishida et al. (1991), Senock and Ham (1993) and Chandra et al. (1994), among others. All of these methods are based on the calculation of an energy balance between the energy put into the stem and energy losses from temperature measurements made on top of or just under the bark. The stem heat balance gauge actually comprises a flexible heater several centimeters in width that is wrapped around the stem and enclosed in layers of insulating and weatherproof materials. Heat is applied to the entire circumference of the stem encircled by the heater and the mass flow of sap is obtained from the balance of the fluxes of heat into and out of the heated section of stem. The SHB method is suitable for both herbaceous and woody plants. However, with large stem sizes, both the heater band and energy requirements become too large, and the calculation of the stored heat term is difficult. Therefore, these methods are rarely used for stem diameters larger than 16–17 cm.

3.2.5 The Heat-Ratio Method (HRM)

The methods of Cohen and Green are reported to be of limited value for measuring low flow rates (Becker 1998). The heat-ratio method (HRM) is a heat-pulse-based method developed by Burgess et al. (1998, 2000a) (using some theory from Marshall (1958) and Barrett et al. (1995)) that is able to accurately measure low rates of sap flow. The HRM measures the heat pulse velocity V_h (cm s^{-1}) from the ratio of the temperature increase following the release of a pulse of heat at points equidistant downstream and upstream from a line heater:

$$V_{\mathrm{h}} = \left[4kt\ln(v_1/v_2) - \left(x_2^{\,2}\right) + \left(x_1^{\,2}\right)\right] / \left[2t\left(x_1 - x_2\right)\right] \qquad (15.1)$$

where k is the thermal diffusivity of fresh wood (cm^2 s^{-1}), t is time of measurements, v_1 and v_2 are increases of temperature relative to ambient, and x_1 and x_2 are upstream and downstream distances of the probes from the line heater. As in other similar methods units of V_h can be written as sap flow density ($cm^3\ cm^{-2}\ s^{-1}$), which allows integrating sap flow values for whole trees, when using sapwood area. The HRM configuration corrected for wounding as well as other operational details are given by Burgess et al. (2001a). The HRM is very suitable for fieldwork, because it is sensitive to the direction of sap flow and therefore is also able to measure reverse flow in stems, roots and other conductive organs.

3.2.6 Tree Trunk Heat Balance (THB) Method

The original trunk heat balance (THB) method, characterized by direct electric heating and internal sensing of temperature, was designed for large trees (Čermák et al. 1973, 1976b, 1982, 2004; Kučera et al. 1977; Tatarinov et al. 2005). The variant used for tree trunks with diameters over 12–15 cm consists of three to five metal plates inserted in parallel into the sapwood, at a distance of 2 cm from each other. Rather wide sensors (altogether 4–8 cm) are used to represent significant parts of stems, especially when more are installed together around the stem. The plates are used as the electrical poles of the heater. An AC voltage source isolated from ground is connected to the plates, and the current running through the wood between the electrodes generates rather even heating between them (Fig. 15.3). The current is automatically distributed in the xylem along the electrodes, reaching the heartwood edge even when the electrodes in the sapwood do not reach it. Furthermore, the heartwood is not heated even if the electrodes are inadvertently inserted deeper than the sapwood/heartwood boundary, because the heartwood is usually relatively dry and therefore nearly non-conductive. Therefore sap flow is integrated over the entire sapwood. Either the power (P), which is directly proportional to sap flow or the temperature difference (dT), which is indirectly proportional, can be held constant by electronic circuits while the other variable is being recorded. This method calculates the heat balance of a defined heated space. As in all heat-balance systems, the temperature difference between the heated and non-heated parts of stem segments is measured and used to calculate the mass flow. The constant dT variant of the system has the advantage of better dynamic properties and of avoiding eventual overheating of the segment in which it is located.

Basically, the input energy has to be split between the conductive heat losses and the warming of passing water according to the following simple equation:

$$P = QdTc_w + dT\lambda \tag{15.2}$$

where P is the heat input power (W), Q is the sap flow rate (kg s^{-1}), dT is the temperature difference in the measuring point (K), c_w is the specific heat of water (J kg^{-1} K^{-1}) and λ is the coefficient of heat losses from the measuring point (W K^{-1}). The magnitude of λ is clearly evident in the sap flow data as the so-called "fictitious flow" ($Q_{w\ fic}$), which is recorded even when the actual flow is zero (under constant P, dT reaches its maximum). When calculating the actual sap flow (Q_w), it is necessary

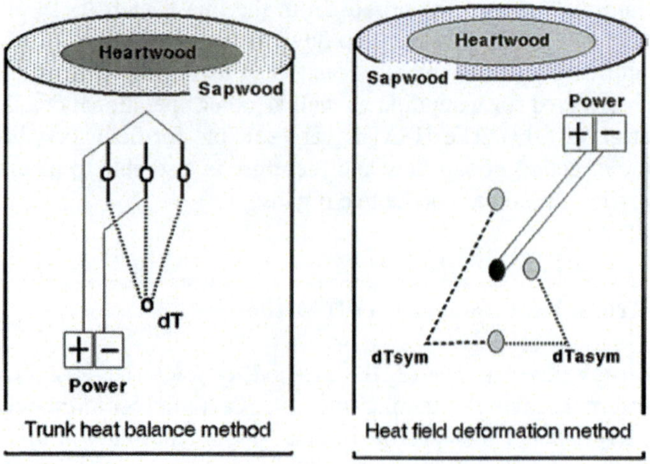

Fig. 15.3 Schemes of sensor geometry for the trunk heat balance, THB, (*left*) and heat field deformation, HFD, (*right*) methods. *THB* electrodes supplying the alternating electric current isolated from the ground are shown as thick vertical lines and thermocouples (placed in 1-mm-diameter stainless steel hypodermic needles) are marked as circles. *dT* represents the measured temperature difference. The distance between edge electrodes delineates the width of the measured stem section (here 40 mm). The total sap flow per section or per tree is measured. An additional compensating pair of thermocouples installed on both sides of the electrodes (not shown) helps to correct for the impact of natural temperature gradients. *HFD* the linear heater of the HFD sensor is shown as a black circle and thermocouples measuring axial, symmetrical temperature differences (dT_{sym}) as well as tangential, asymmetrical, temperature differences (dT_{asym}) are shown as gray circles. The width of the measured stem section determines the distance between the heater and tangential thermocouples. This width can be doubled by installing the tangential thermocouples on both sides of the heater (not shown). A series of thermocouples installed in each needle allows measurement of both the total sap flow and its radial pattern across the entire sapwood depth

to subtract $Q_{w\ fic}$ (estimated periodically when the actual flow approaches zero, such as in the early morning after a prolonged rain) from the recorded flow data ($Q_{w\ rec}$).

$$Q_w = Q_{w\ rec} - Q_{w\ fic} \tag{15.3}$$

The impact of temperature gradients between dT probes can be compensated for by installation of additional pairs of thermocouples outside the heat field of the heater (Čermák and Kučera 1981). This modification has also been adapted for use with heat dissipation sensors (Silvestre and Ferreira 1998).

In addition, "baby" sensors (applicable for conducting organ diameters 0.6–2 cm) are also available. These are significantly improved versions of sensors based on flexible external heating and sensing as described by Čermák et al. (1984) and Lindroth et al. (1995). Heat is supplied by a resistance wire fastened to a soft resilient insulating tissue, assuring good contact with non-cylindrical stem surfaces, even if the plant grows in diameter by up to about 50% during the period of measurement.

The THB method is robust and provides reliable data for long-term measurements in trees with diameters over 15 cm in a range of species, size and environmental conditions. The exact estimation of very low flow (values approaching the fictitious flow) is difficult because such flow interferes with the changes in heat loss. The THB measuring system has good dynamic properties (Čermák and Kučera 1991), although some inertia can occur. Due to the plane form of the electrodes, installation in stems with highly curved, spiral water conductive pathways (Kozlowski and Wignet 1963; Waisel et al. 1972) may be difficult because a larger sapwood area may be cut than that corresponding to the electrode thickness. Partially rotten stems with wet heartwood, stems with very deep sapwood, or stems of some tropical species with sapwood located deep inside may cause problems. The THB sensors integrate the radial sap flow profile by technical averaging within the wide stem sections when considering power distribution as mentioned above. This method was even applied as a standard when testing other methods (Offenthaler and Hietz 1998; Schubert 1999).

3.2.7 Heat Field Deformation (HFD) Method

The heat field deformation (HFD) method is based on measurement of the deformation of the heat field around a needle-like linear heater inserted into the stem in a radial direction (Nadezhdina et al. 1998, 2002, 2006; Čermák et al. 2004, Fig. 15.3). The frontal view of the heat field around the linear heater looks like a symmetrical ellipse due to different heat conductivities of the stem in the axial and tangential directions under zero flow conditions (the stem must be considered as a complex material consisting of a solid xylem as well as water and air) and takes a form of a gradually prolonging deformed ellipsoid under increasing flows (Nadezhdina et al. 2004b). Using the HFD method, sap flow is calculated from the ratio of temperature gradients around the linear heater in the axial direction and symmetrical position (dT_{sym}) as well as in the tangential direction and asymmetrical position (dT_{asym}). The gradients are measured by two pairs of thermocouples inserted (similarly to the heater) in stainless steel hypodermic needles of 1.5 mm diameter. Sap flow is then calculated from the temperature differences. dT_{sym} is also known as the sap flow index (SFI), which can be used as a qualitative stress indicator (Nadezhdina 1999). Experimentally, the above ratio is proportional to the sap flow rate, Q_w (Nadezhdina et al. 1998). Constants in the corresponding equation also include the geometry of the measuring point and physical properties of the conducting system (Nadezhdina et al. 2006):

$$q_i = 3600 \times D \times \frac{dT_0 + dT_{sym} - dT_{asym}}{dT_{asym}} \times \frac{Z_{ax}}{Z_{tg} \times L_{sw}}, \qquad (15.4)$$

where q_i is sap flux density (cm^3 cm^{-2} h^{-1}) estimated at a thermocouple junction *i*, *D* is the thermal diffusivity of fresh wood (cm^2 s^{-1}), dT_{sym} (°C) is the temperature difference between the axial thermocouple junctions, dT_{asym} is the temperature difference between the tangential thermocouple junctions, Z_{ax} is the distance between the upper junction of the axial thermocouple and the heater (usually = 1.5 cm), Z_{tg} is the distance between the upper junction of the tangential thermocouple and the heater (usually = 0.5 cm between needle surfaces) and L_{sw} is sapwood depth (cm). A nominal value for D of 0.0025 cm^2 s^{-1} was chosen, according to Marshall (1958). The temperature difference under conditions of zero flow (dT_0) was determined from the relationship: $dT_0 = (dT_{sym} - dT_{asym})_0$ and the ratio (dT_{sym}/dT_{asym}) when taking the point, where the extended line intersects the ordinate (Nadezhdina et al. 2006). The HFD method is the most sensitive measuring system presently available, responds quickly (with minimum inertia). Reverse sap flow associated with hydraulic redistribution (Daum 1967; Štěpánek et al. 1972; Burgess et al. 1998) can also be directly measured with this method (Nadezhdina et al. 2009, 2010) as well as basipetal and nighttime resaturating flows (Kunia 1955; Daum 1967; Burgess et al. 2000a; Brooks et al. 2002), which provide valuable information about plant water status.

The HFD method can be fully exploited in large trees only when applying multi-point sensors, which allow measurements of radial patterns of sap flow (Čermák and Nadezhdina 1998; Nadezhdina and Čermák J 1998; Nadezhdina et al. 2009, 2010) Some uncertainties were found during measurements of very high sap flow densities, such as those measured just below the crown of tall trees (N. Nadezhdina, unpublished data). This may be a common problem for most methods using needle-like heaters. Multi-point HFD sensors integrate the cross-sectional area of the whole stem sapwood by measuring the flow at different depths and allow three-dimensional representation of flow (Čermák et al. 2004). Single-point sensors with needle heating can be used in small stems, roots or branches (over 10 mm in diameter).

3.2.8 Integration of Flow from Individual Sensors to the Whole-Tree Level

Measurement of sap flow itself represents only one aspect of sap flow studies. Additional important items to consider include evaluation of errors, integration of data measured by a given sensor for the whole stems (e.g. Marshall 1958; Shackel et al. 1992; Hatton et al. 1990, 1995; Allsheimer et al. 1998; Clearwater et al. 1999; Nadezhdina et al. 2002) and spatial variation of flow within trees (Čermák et al. 1984, 1992, 2008b; Čermák and Kučera 1990a, b; Granier et al. 1994; Čermák and Nadezhdina 1998; Jimenez et al. 2000; Nadezhdina and Čermák 2003a, b; Nadezhdina et al. 2007). The interior of a tree is far from homogenous. As trees increase in girth, old knots and limbs are grown over and often become invisible at the surface, causing sap flow to be divided around the knot, leaving non-conducting voids immediately above and below. In general, it is not possible to avoid such areas. Careful attention to probe placement, use of the correct sampling design and careful analysis of the collected data are the only ways to insure representative and unbiased estimates of sap flow (Swanson 1970, 1994). The variability in flow should be considered, especially over the sapwood radial profile and around the trunk circumference at a given measurement height (Čermák et al. 1992, 2003; Čermák et al. 1992, 2004; Anfodillo et al. 1993; Phillips et al. 1996; Nadezhdina et al. 2002). A series of measuring points at different xylem depths is needed to characterize

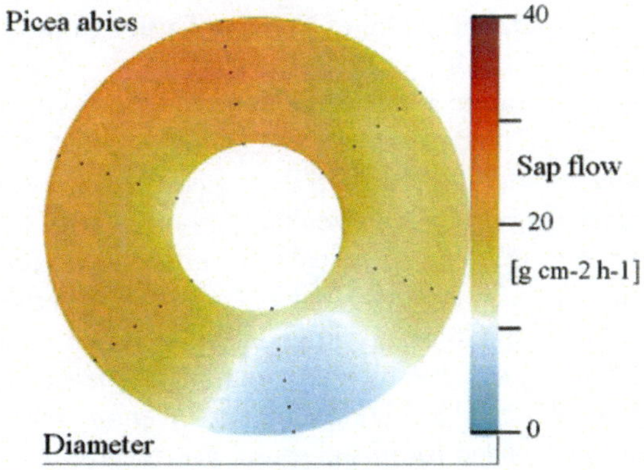

Fig. 15.4 Sap flow density across the stem at breast height (1.3 m) in a large Norway spruce tree measured by a series of six multi-point sensors based on the heat field deformation method. Each of the multi-point sensors is marked by a radial series of dots showing positions of 30 individual sensors. Density is represented by colors according to the scale on the right. The image was made using software of D. Nečas available at http://trific.ath.cx/software/dendrofit/ (2003)

each radial and circumferential pattern. In the example shown in Fig. 15.4 six multi-point sensors were installed, comprising a total of 30 measuring points for characterizing spatial variation in flow in a *Picea abies* stem cross section. It may also be necessary to consider that the specific flow and velocity may increase or decrease axially from the stem base to thin branches depending on species (Huber 1932). If water storage is not considered (e.g., when taking daily totals), data for the whole stem at the base of the live crown should be practically the same as that at the tree base (Schulze et al. 1985; Čermák and Kučera 1990a).

As shown in Fig. 15.4, there are two approaches to obtain true integrated radial patterns of sap flow: application of integrating sensors (such as THB) or multi-point sensors (such as HFD) situated over stem radii. Several of these sensors must be installed around the stem circumference to record the flow variation at different azimuths. Values for flow density must be multiplied by the corresponding sapwood cross-sectional areas (e.g., over individual growth rings or distinctly defined annuli) to obtain data for the whole stem. Misleading results can be obtained if a series comprising only single-point sensors is installed at different depths and simultaneously at various parts of the stem circumference. Changing the position of such sensors can yield different data. Positioning of non-integrating sap flow sensors only at a single point provides data that cannot be reliably applied for integration over the whole sapwood depth (i.e., the cross-sectional area), because the operator does not know the point at which the measured data are representative of the whole sapwood depth. If such a sensor is positioned at any other place, it will substantially overestimate or underestimate the representative value (Nadezhdina et al. 2002; Cermák et al. 2004). This problem is more important for highly variable radial flow densities, when errors of tens of percent can easily occur. Much smaller errors can also occur in rather homogeneous flow patterns, when the most representative point is misjudged by even a very short distance.

3.2.9 Upscaling of Data from Series of Trees to Entire Stands

Interpretation of results of studies at the whole-tree level is often clear, but sometimes (e.g., in landscape-oriented studies) the data need to be scaled up from a series of sample trees to entire stands. Relating sap flow measured in a series of carefully selected trees encompassing a range of size that is representative of the stand can do this. The "quantiles of total" method helps to identify such trees simultaneously based on size and stand density (Čermák and Kučera 1990b; Čermák and Michálek 1991). Each selected tree should represent approximately the same fraction of the selected stand parameter. A suitable biometric parameter measurable at the sample tree as well as at the whole-stand level (e.g., forestry inventory based on DBH, basal area, leaf or root area) is needed for this purpose, irrespective of whether the stand is heterogeneous or homogeneous.

The equation generalizing values of all sample trees ("the scaling curve"; Fig. 15.5, upper panel) is applied to derive values for individual DBH classes. These values are then multiplied by the numbers of trees in the classes and summarized to obtain stand values (e.g. Morikawa 1974, 1986; Loustau et al. 1998; Čermák and Kučera 1987, 1990a; Meiresonne et al. 1999, 2003; Čermák et al. 2004). The resulting error is usually acceptable if a sufficient number of sample trees (roughly a dozen per species) are measured (Fig. 15.5, lower panel). If the data are to be scaled to even higher levels of biological organization (such as a watershed or forest plantation), a series of well-selected experimental plots representing that area must be incorporated. The selection is based on the forest site classification ("forest types") or other site classifications available in particular areas at the whole-stand and landscape levels. Detailed hydrologic modeling or detailed remote sensing are suitable for such purposes. Data obtained from selected sites have been successfully utilized for verification of results from models calculated from the same sites (Balek et al. 1985a, b; Čermák and Kučera 1991; Oltchev et al. 2002a, b; Chiesi et al. 2002).

4 Sap Flow Measurements in Studies of Hydraulic Architecture

4.1 Whole-Tree Water Storage

Tree water storage confers a very important capacity to buffer extremes in xylem tension (Waring and Running 1978; Waring et al. 1979) that allows the tree to minimize hydraulic dysfunction associated with excessive embolism when soil water supply is limiting. The total amount of water available for withdrawal from storage in tree tissues can be estimated on the basis of tissue sampling (Čermák and Nadezhdina 1998; Kravka et al. 1999). As shown in earlier studies, fractions of stored water used on individual days can be estimated based on sap flow measurements

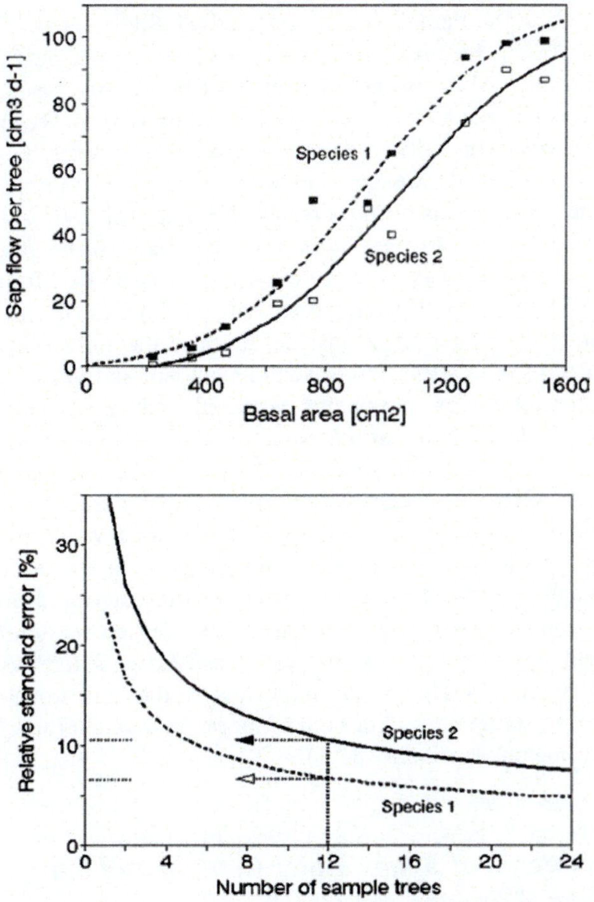

Fig. 15.5 The upper panel shows relationships between values of daily or diurnal sap flow for individual trees and the biometric parameter basal area, which was measurable at the sample tree and stand levels. Points represent the individual sample trees in a relatively homogeneous stand (Species 1) and a more heterogeneous stand (Species 2). The calculated "scaling curve" was applied for derivation of flow rates for mean trees of different DBH classes. Stand level data were obtained as mean values of trees in classes multiplied by the number of trees in the classes and summed. The lower panel shows the corresponding standard errors of stand sap flow (transpiration) measurements based on 12 sample trees for each species

(Landsberg et al. 1976; Čermák et al. 1976a, 1982). The frequently applied and most straightforward approach for such estimation is to measure changes in stem volume using sensitive electric dendrometers or by diurnal courses of transpiration of foliage (or sap flow in small branches close to foliage) and shifted sap flow time series in the stem at its base or at different heights above ground (Čermák et al. 1982, 2007a;

Phillips et al. 1997, 2003; Peramaki et al. 2001, 2005; Steppe et al. 2005; Meinzer et al. 2003, 2006, 2008). Ideally, both dimensional changes and sap flow should be measured concurrently. Shifts in the time series of branch and basal sap flow are typically brief in small plants, (e.g., only about 1.5 min in reed, Rychnovská et al. 1980), but are often substantially more pronounced in large trees (Goldstein et al. 1998; Phillips et al. 2003). However heat storage at the measuring points must be taken into account to prevent distortion of results (Sevanto et al. 2009). Measurements on giant old-growth Douglas fir trees on summer days have shown good correspondence between estimates of daily reliance on stored water derived from branch and basal sap flow differences and dendrometer-based estimates of changes in stem volume (Čermák et al. 2007a). These analyses revealed that only a small amount of water was extracted from elastic tissues such as phloem and young wood, with a significantly larger volume of water being extracted from rigid conducting tissues such as sapwood. This differs in various parts of the tree crown, as little transpired water (3%) was extracted from relatively dry tissues of the upper crown, whereas much more (up to 35%) was obtained from the rest of the tree. Whole-tree transpiration can be maintained with stored water for about a week, but the treetop itself can supply water only for several hours. Relationships of the recorded changes in stem volume and cumulated values of sap flow differences at different heights (or transpiration and sap flow) show their linear dependence over almost the whole day (although with only a 0.5-h shift at the tree top and 3 h at the tree base), but not at night (Čermák et al. 2007a). This pattern indicates that the stem grows significantly only at night during sunny days with high evaporative demand (and probably for longer period of time on warm rainy days).

4.2 *Radial Patterns of Xylem Water Content and Sap Flow Related to Sapwood Depth*

The sapwood cross-sectional area is a simple biometric parameter widely used for scaling up transpiration data between trees and forest stands and for estimation of free water storage in stems. However, it is not always clear how the sapwood area is best determined. One of the most common methods for distinguishing sapwood and heartwood is based on xylem water content (also called wetwood), which is typically about $50\%_{vol}$ in sapwood (of this, about $40\%_{vol}$ is free water) and about $10\%_{vol}$ in heartwood where water is physically bound and unavailable for withdrawal from storage. The higher water content of sapwood is also observable in its more translucent appearance when looking at it against a light source. Sapwood has also been distinguished according its color or the presence of living cells. All of these characteristics can be used to estimate the potential size (depth, area or volume) of sapwood, which might not correspond to the xylem layers where sap flow actually takes place. The radial pattern of sap flow and estimated heartwood boundary (where flow density reaches zero) may correspond to the boundary estimated by the above-mentioned methods, but the functional sapwood is sometimes smaller due to drought or other factors. Furthermore, the flow density can differ significantly along the sapwood depth, so the potential size of sapwood is not always reliable for purposes of scaling up the sap flow when it is measured over a narrow range of sapwood depth (Čermák and Nadezhdina 1998). The potential size of sapwood differs according to species and the size and age of trees (Kravka et al. 1999). The actual radial patterns of sap flow can also be quite different even among individuals having similar sapwood depths. The entire stem cross-sectional area is usually water conducting in small (young) trees with the relative sapwood depth decreasing as the stem grows, and a larger fraction of the xylem radius is gradually converted from sapwood to heartwood (Fig. 15.6).

4.3 Water-Conducting Systems in Stems

It has long been suggested that species differ in the degree of sectoriality of their xylem and that they can be divided in two large groups: sectored and integrated. Knowledge of this aspect of tree hydraulic architecture becomes more important for larger trees. Ring-porous species and some conifers (e.g., *Thuja*) tend to be more sectored. Very steep spiral pathways occurring in some species (e.g., often in *Syringa vulgaris*) are rare. Xylem of diffuse-porous species is regarded as usually more integrated. Some authors have found sectored flow to consist of direct connections between certain parts of the root system with only a certain part of the foliage situated on the same side of the stem (Lanner 1984; Larson et al. 1994). However, other studies (Roach 1939; Vite and Rudinsky 1959; Zimmerman 1978; Teskey et al. 1985) have observed the fraction conducting tissues in the stem xylem cross-sectional area to increase toward the top of the tree. Sap flow measurements may provide the easiest way to study the degree of integration or sectoriality of water pathways in trees. This method is especially effective when applying experimental manipulations such as branch or root severing (Fig. 15.7). For example, sap flow in all xylem layers of an oak stem decreased to zero after the large branch above it was severed during our experiments (Figs. 15.7 and 15.8). The sensor on the opposite side of the stem recorded no flow decrease after branch severing. The sap flow also decreased in all xylem layers of a maple tree on the side where the branch was severed (Figs. 15.7 and 15.8). The largest flow decrease was recorded in the outer xylem. However, the flow did not decrease to zero, and some low homogeneous flow occurred in all xylem layers below the cut branch. At the same time, the flow did not decrease in the outer xylem layers on the opposite side of the stem, but the flow decreased in the inner xylem, especially in the deepest part of it (42 mm below the cambium). This example indicates that water pathways in oak just below the crown were directed to a certain part of the crown only (sectored), whereas in the more integrated maple, water pathways were directed to the whole crown from each side of the stem, presumably to the nearest branch in the outer xylem and also to branches on the opposite side through the inner stem xylem.

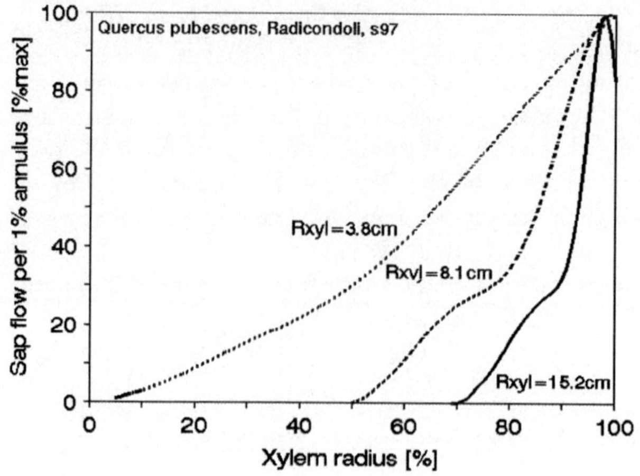
Quercus pubescens, Radicondoli, s97
Rxyl=3.8cm
Rxyl=8.1cm
Rxyl=15.2cm
Sap flow per 1% annulus [%max]
Xylem radius [%]
0
20
40
60
80
100

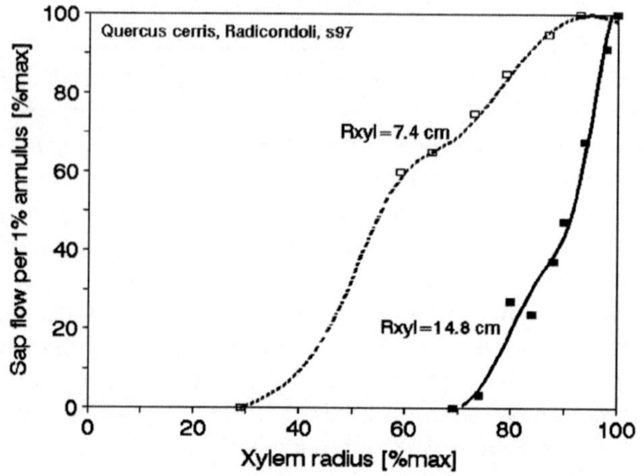
Quercus cerris, Radicondoli, s97
Rxyl=7.4 cm
Rxyl=14.8 cm
Sap flow per 1% annulus [%max]
Xylem radius [%max]
0
20
40
60
80
100

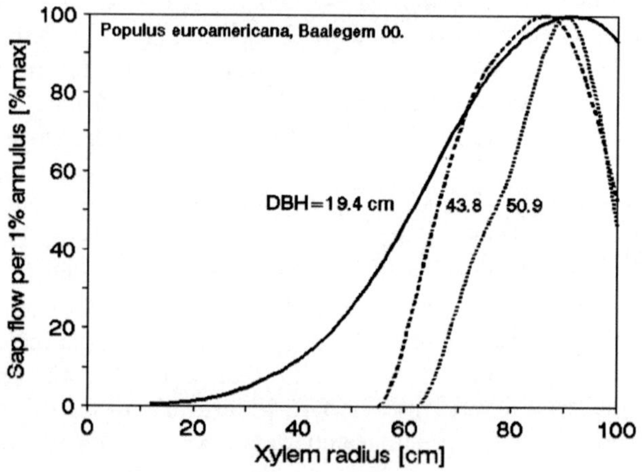
Populus euroamericana, Baalegem 00.
DBH=19.4 cm
43.8
50.9
Sap flow per 1% annulus [%max]
Xylem radius [cm]
0
20
40
60
80
100

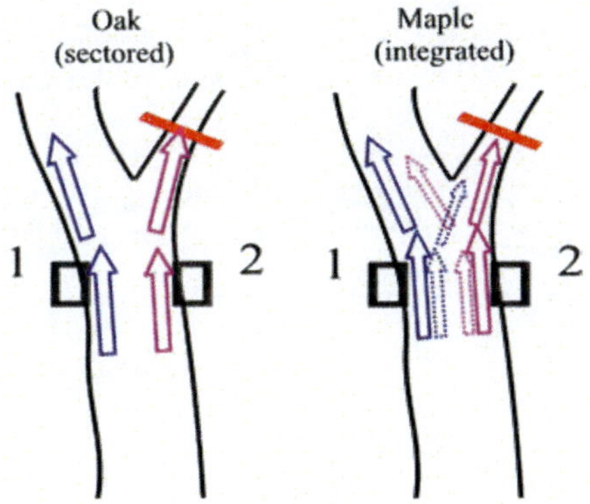

Fig. 15.7 Redistribution of water in trees: scheme of sap flow sensor installation in the upper trunk of sample trees where the branch was cut during the field experiment. *Arrows* indicate model directions of sap flow when branches were still present. Transport in oak (*Quercus robur*, *left panel*) occurs within restricted pathways (sectorial), whereas more integrated transport occurs in maple (*Acer platanoides*, *right panel*). *Red lines* indicate where branches were cut; *solid arrows* show main direction of sap flow in the outer stem xylem and *dashed arrows* model additional flow direction in the inner stem xylem. *Black boxes* represent sensors marked by their numbers. These numbers correspond to numbers on graphs in Fig. 15.8

4.4 Studies of Water Redistribution in Trees: Soil–Root Compartment

Horizontal distribution of water in soil caused by high tree water demand for transpiration plays an important role in tree survival, especially in low hydraulic conductive heavy soils (Čermák et al. 1993; Čermák and Prax 2001). Redistribution of water in trees is also very important. Sap flow methods with symmetrically placed thermometers (e.g., some versions of heat balance, heat ratio and heat field deformation methods) opened the possibility of directly detecting the hydraulic redistribution in soil mediated by tree roots (Caldwell and Richards 1989; Burgess et al. 1998, 2000a, b, 2001b; Scholz et al. 2002; Moreira et al. 2003; Hultine et al. 2003; Warren et al. 2005; Nadezhdina et al. 2006; Brooks et al. 2006). Passive hydraulic redistribution in soil determined by gradients of water potential should be common for trees of any size (and other plants including grasses) if the moisture is heterogeneously

---

Fig. 15.6 Changes in radial sap flow patterns representing the actual conducting sapwood (in contrast to the potential conducting sapwood according to xylem water content or its color) in relation to tree size as indicated by xylem radius (R_{xyl}). Examples of ring-porous species *Quercus pubescens and Q. cerris* of are shown in the upper and middle panels; respectively, and the diffuse porous species *Populus euroamericana* is shown in the lower panel. Almost all of the stem cross-section area is conductive in young trees (*upper panel*); and this fraction gradually decreases with increasing tree size and age (*middle and lower panels*) irrespectively of the species

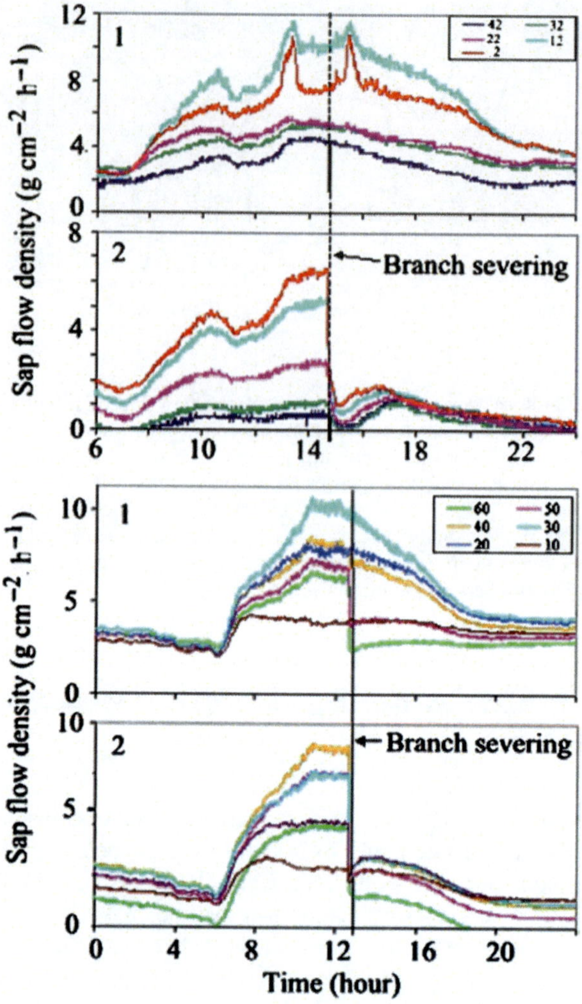

Fig. 15.8 Sap flow dynamics at different xylem depths on opposite sides of tree stems recorded before and after branch severing. The vertical line indicates the moment of severing. The numbers 1 and 2 on figure panels correspond to positions of sensors shown in Fig. 15.7. The numbers inside legend boxes show depths below the cambium (mm) where sap flow was measured. *Upper pair of panels*: In *Quercus robur*, a large drop in flow (nearly to zero) occurred below the severed branch (2), but not on the opposite side of the stem (1). Lower pair of panels: In *Acer platanoides*, the drop in flow that occurred below the severed branch was large (2), but the flow never completely stopped. Flow also dropped in the inner xylem of the opposite stem side (1)

distributed in bulk soil occupied by roots. Hydraulic redistribution within the aboveground part of a tree (Daum 1967; Štěpánek et al. 1972; Burgess and Bleby 2006; Nadezhdina et al. 2006, 2009) may occur, especially in large trees where it can play an important role in the tree water balance.

5 Effective Whole Tree Parameters: Crowns

5.1 Evaluation of the Effective Size of Tree Crowns

Crown conductance is often calculated from the ratio of transpiration estimated from sap flow to vapor pressure deficit (VPD). Similarly, the effective projected crown area (A_{c_eff}) can be calculated based on sap flow measurements. For such calculations, it is assumed that the tree crown behaves like a given fraction of the area of a standard stand whose total area is equivalent to that used for calculation of potential evapotranspiration (PET), which is a more complex parameter than VPD that includes solar radiation. The magnitude of A_{c_eff} is derived from the sap flow rate per tree (Q_{wt} in l per day) divided by the value of evapotranspiration (PET in l per m^2 per day) for the same fine day: ($A_{c\_eff} = Q_{wt}/PET$), yielding results in m^2 per tree (Čermák et al. 1982). This area is smaller than the projected crown area, but differences are sufficiently small to permit the calculated data to be roughly checked by comparison with simple measurements of projected crown area (A_{c_pro}). By applying similar calculations using the same variables already scaled to the stand level, the relative transpiration can be obtained. Theoretically, it is also possible to calculate effective projected root area (A_{r_eff}) this way, if the daily decreases in soil moisture measured over the entire rooting zone of a particular tree are available (Čermák and Kučera 1990a).

5.2 Evaluation of Conducting Systems by Combined Macro- and Micro-level Studies

Sap flow technology has also been successfully applied for evaluation of the efficiency of the conducting system in trees: theoretical flow in stems and leaves was calculated using the Hagen–Poiseuille equation on the basis of anatomical data, measured spatial distribution of water potential in crowns and compared to actually measured sap flow. Examples are provided by the floodplain forest composed mostly of pedunculate oak (*Quercus robur* L.) and the laurel forest composed mostly of *Laurus azorica* L. Leaf petioles were sampled over the whole crowns and sapwood was sampled from tree stem bases. Floodplain and laurel forests differ significantly in their leaf area indices, leaf density per unit stand area and basal area, however the ratio of theoretical maximum flows in stems to that in petioles and maximal measured via sap flow are very similar in both species. Maximum measured sap flow reached an efficiency of about 95% in both species (approaching theoretical values of 100% in petioles), but the stems have large reserves (theoretical flow in stems was about 40 times higher than the maximum measured sap flow). This mechanism evidently allows better survival of long-lifespan trees, if their stems can work over much longer periods of time as compared to short-lifespan leaves and their petioles (Table 15.1).

Table 15.1 Evaluation of the efficacy of conducting systems based on laboratory anatomical measurements and field sap flow studies

Species	LAI (m^2m^{-2})	Leaf/petiole density ($pc\ ha^{-1}$)	Vessels per: Stems ($pc\ ha^{-1}$)	Vessels per: Petioles ($pc\ ha^{-1}$)	Theoretical flow in: Stems ($mm\ h^{-1}$)	Theoretical flow in: Petioles ($mm\ h^{-1}$)	Max. measured sap flow ($mm\ h^{-1}$)
Quercus	5.0	20×10^6	–	–	10.1	0.25	0.24
Laurus	7.8	40×10^6	676×10^6	8.2×10^9	30.1	0.72	0.68

The main stems of trees at breast height as well as leaf petioles were analyzed for *Quercus robur* in floodplain forests in southern Moravia (Krejzar and Kravka 1998) and for *Laurus azorica* in foggy laurel forests of Tenerife in the Canary Islands (Morales et al. 2002; Čermák et al. 2002). Leaf or petiole density is expressed in pieces per stand area unit (1 ha). Vessel density is given separately per stems and per petioles, in both cases in pieces per stand area unit. Calculated theoretical flow is also given separately per stems and petioles in mm (=$dm^3\ m^{-2}$) per hour. The highest (max.) measured sap flow over the growing season is expressed in the same units

5.3 *Form of Crowns (Branch Distribution) and Variation of Sap Flow Around Stems*

When sap flow sensors are installed near the crown, they are close to branches and the sap flow better characterizes the behavior of the branches, as illustrated for a rather large solitary cork oak tree (*Quercus suber*) with a relatively short stem (Fig. 15.9; upper panel). The measuring system recorded seasonal and diurnal changes of sap flow. The sap flow was higher mostly below both major branches, under which the sapwood was also deeper. However, the magnitude differed according to the intensity of illumination on the currently illuminated side and shaded on the opposite side, so a certain diurnal alternation appeared (Fig. 15.9; lower panel). The dynamics of shading and light suppression effects at the whole tree level can be quantified this way (N. Nadezhdina, unpublished).

5.4 *Effective Shape of Tree Crowns Based on Crown Shade*

Comparisons of the daily total sap flow and PET are valuable, but still include some uncertainties. The solar radiation term, which is one of the inputs for the PET calculation, is usually measured on a flat horizontal surface area, while tree crowns and their leaf distribution usually have much more complex spatially distributed surfaces. When working with individual leaves, this makes exact geometric descriptions very difficult. It is questionable whether obtaining the viewpoint of the whole tree this way is really necessary, when simplification is possible by eliminating very large numbers of leaves (tens of millions per hectare) using a suitable integrating method. A similar method evaluates the radiation geometry of crowns based on shade areas of different forms projected onto a horizontal surface (Monteith 1975). This technique works rather well with crowns of solitary trees growing in the open,

Fig. 15.9 Circumferential variation in sapwood depth and sap flow recorded in the afternoon in a *Quercus suber* tree with sap flow sensors positioned close to the base of the crown. Sap flow was scaled to the whole tree based on data recorded from sensors installed in azimuths represented by the *black squares*. The relationship between circumferential variation in sap flow and sapwood depth changed throughout the day according to the positioning of the sun and mutual shading of branches

but it is not directly applicable in dense forest stands. However, the basic equations described for direct, diffuse and total radiation for different types of leaf distributions (such as vertical, spherical, conical, etc.) are useful.

5.5 Evaluation of the Effective Shape of Tree Crowns Based on Diurnal Dynamics of Sap Flow

One approach is based on comparison of the measured course of sap flow in individual trees with the diurnal course of radiation measured on a flat horizontal surface or derived courses of radiation for spherical or other geometrical shapes (Fig. 15.10, upper panel). Diurnal courses of radiation and their differences can be shown on a normalized scale (Fig. 15.10, middle panel). Daily total circumglobal radiation is higher by 70% than that of global radiation. When diurnal courses of sap flow in small and large Douglas fir trees (*Pseudotsuga menziessii*) during an almost cloudless day were also normalized (Fig. 15.10, lower panel) the shape of the time course for the large tree was a bit closer to that of circumglobal radiation, whereas the sap flow time course of the small tree was more similar to that of global radiation (Fig. 15.10, lower panel). The sap flow time course lags behind that of radiation in the morning hours due to utilization of internal water storage and continues to lag in the afternoon and night hours partially due to persistently high vapor pressure deficit and especially because of flow associated with tissue rehydration. Acceptable results can be obtained when working during cloudless or almost cloudless days and by avoiding the decreasing portion of sap flow curve distorted by resaturating flows and the impact of vapor pressure deficit. Further analysis of this promising approach is ongoing.

5.6 Estimates of Fractions of Tree Crowns Based on Radial Patterns of Sap Flow

The illumination of trees of different social positions (and therefore usually of different sizes in closed stands) and of variable crown density substantially differs, especially under high flow conditions around mid-day. This is made most visible by the sap flow radial patterns. If the crowns are sparse, light penetrates more uniformly into the canopy, and the sap flow at different sapwood depths is more similar. If the

Fig. 15.10 Effective form of tree crowns assessed from measurements of sap flow dynamics. *Upper*: scheme of global radiation measured on a horizontal disc (R_g) and circumglobal radiation measured on a sphere (R_{cg}) of the same diameter. *Middle panel*: Diurnal courses of R_g and R_{cg} (using a normalized scale) and the difference between them. This difference can reach up to 70% on a non-normalized scale (Modified according to Čermák and Kučera 1990b). *Lower panel*: normalized diurnal courses of sap flow in small suppressed and large dominant Douglas fir (*Pseudotsuga menziesii*) trees on an almost cloudless day. The form of the curve of the large tree is closer to that of R_{cg}, while the form of small tree curve is closer to that of R_g (when time shifts associated with storage not taken into account). This means, that the form of the small trees is effectively flat and that of large one is more spatially pronounced (effectively closer to a sphere)

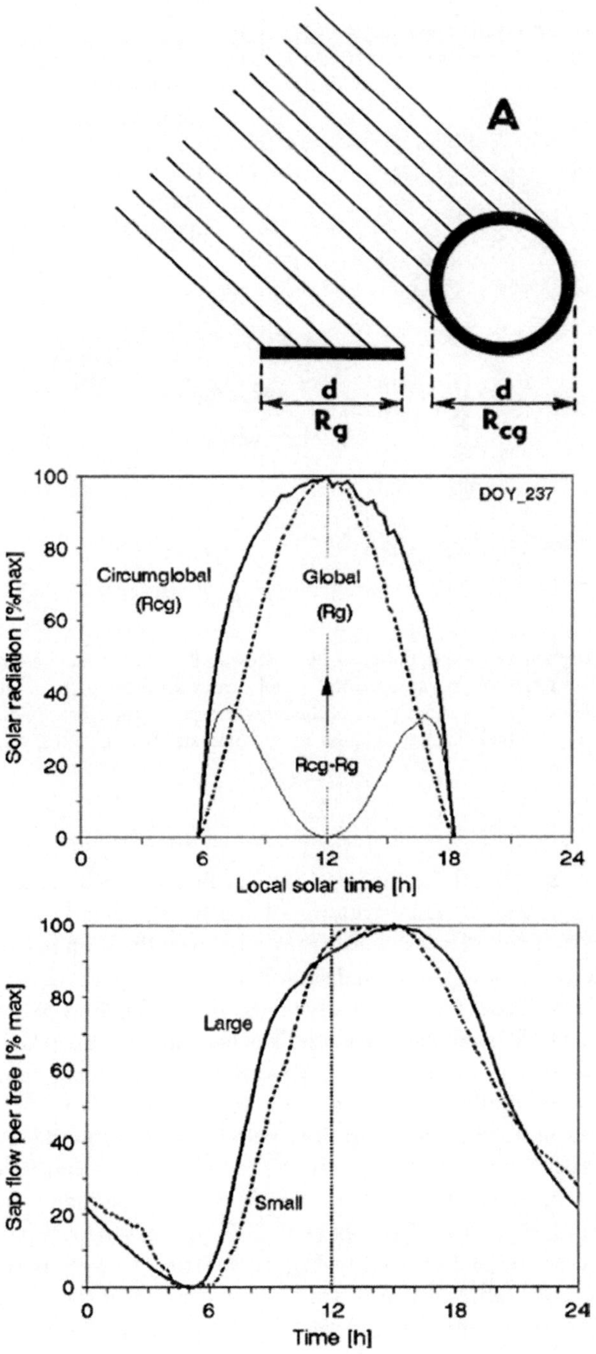
A
d
Rg
d
Rcg
100
80
60
40
20
0
Solar radiation [%max]
DOY_237
Circumglobal
(Rcg)
Global
(Rg)
Rcg-Rg
0
6
12
18
24
Local solar time [h]
100
80
60
40
20
0
Sap flow per tree [% max]
Large
Small
0
6
12
18
24
Time [h]

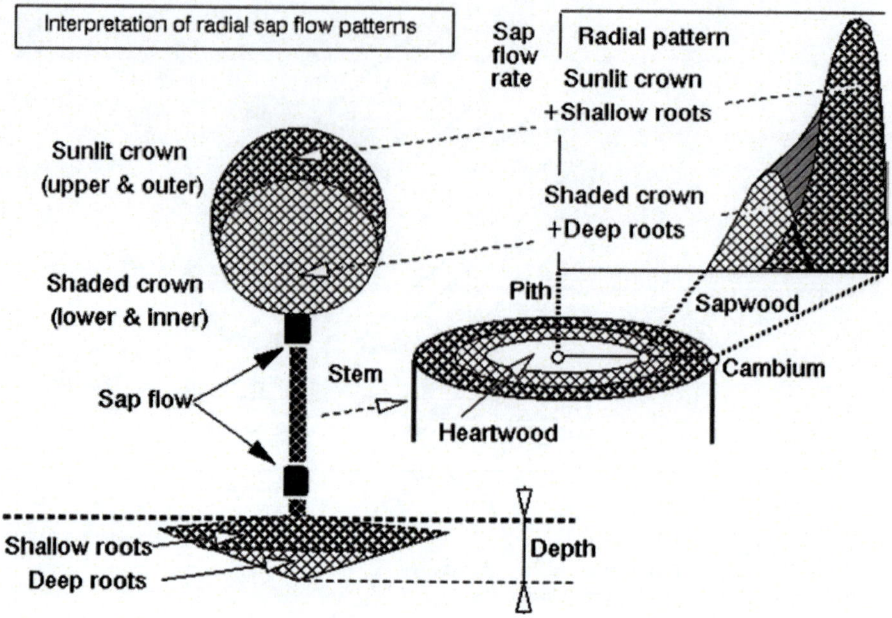

Fig. 15.11 General scheme of the radial patterns of the sap flow in the main stems of trees and their interpretation in terms of prevailing water supply from different parts of the root system (superficial and sinker roots) to various parts of the crown: exposed (*mostly upper and outer*) and shaded (*mostly lower and inner*). Interpretation is approximate, because all layers are partially overlapping

crowns are dense (usually at the top), greater sap flow in outer sapwood layers is observed. The amplitude of radial patterns of sap flow is clearly much higher in dominant then in suppressed trees. Sensors installed near the stem base provide data that better characterize root systems and sensors installed higher on stems (e.g., just below the green crown) provide data mostly on the crown. Sap flow in outer sapwood layers measured near the stem base characterizes the water supply presumably from superficial roots and shallow soil layers, while flow in the deeper sapwood layers represents the water supply from sinker and tap roots and therefore deeper soil layers. When measuring the same sap flow patterns just below the green crown, outer sapwood layers supply water mostly to the treetop and exposed outer layers of crowns, while inner sapwood layers supply water presumably to lower and inner canopy layers (Fig. 15.11, Čermák et al. 2008b). These results were based on experimental studies focused on root or branch severing as well as soil irrigation experiments in several coniferous and diffuse porous broadleaf species (Nadezhdina et al. 2006, 2009). It is more difficult to interpret radial patterns of sap flow in narrow and principally asymmetric sapwood only in ring-porous species. Flow dynamics in different sapwood layers can be very different, based on changes during the daytime and sensitive responses to changes in the tree water supply. The ability to evaluate the effective, functional size and shape of tree crowns as well as water supply in different parts of crowns under field conditions is useful for assessing stand optical roughness, amounts of absorbed radiant energy, eventual defoliation and tree health.

6 Effective Whole Tree Parameters – Roots

6.1 *Effective Root Absorbing Surface Area Determined by the Modified Earth Impedance Method*

Several methods based on their electric properties were applied for root system studies. In herbaceous plants they were based e.g. on root capacitance (Chloupek 1972; Dalton 1995). Good results were also obtained with bare roots of small woody species submerged in a water solution with electric impedance (Cao et al. 2010a, b). We have applied a method based on a modification of the earth electrical conductance method in small as well as large trees (Staněk 1997; Aubrecht et al. 2006). The method is based on the principle, that if a plant root system is submerged in a heterogeneous aqueous solution (such as soil water) and connected to a simple electric circuit, the electrical current from the external source (represented by ion flow) enters the plant through the electrically conducting/ion-absorbing root surfaces as well as lenticels or damaged parts of roots. This is measured by a series of electrodes inserted in the stems and soil. Suberized non-absorbing root surfaces, irrespective of their total area, barely conduct electric current. Considering the different conductivity of tree tissues and soil, the effective conducting or absorbing root surface area (A_{root_ef}) in m^2 per sampled tree can be estimated as:

$$A_{root\_ef} = (\rho_{wood} L_{mean} \xi \eta \zeta) / R_{rm} \tag{15.5}$$

where ρ_{wood} is resistivity of woody tissue (Ω), L_{mean} is the distance (m) between interfaces of measured rooted segments (not the length of roots), ξ is a dimensionless coefficient characterizing mutual shielding of roots (causing a small negative error), η is the coefficient characterizing possible mechanical injury (causing a small positive error), ζ is a negative error in estimation of root area caused by current passing outside the measured segment (both of these coefficients can approach 1), and R_{rm} (Ω) is the value taken from the applied ratio-meter. More details and an analytical description of the principle are provided in the above-mentioned papers. A study of average root distribution measured around Sitka spruce trees growing in wet soil along ridges (Fig. 15.12) indicated that about 2/3 of effective absorbing roots were developed along ridges and only about 1/3 across ridges. Windthrown trees were always lying across ridges, revealing shorter coarse roots previously oriented across ridges (Butler et al. 2010). The method has been tested using allometric relationships in hundreds of trees of several species spanning a greater than two order of magnitude range in diameter (Čermák et al. 2006). Further testing continues to improve the technology and also to better specify eventual differences in ion and water pathways, the role of mycorrhizae and other variables.

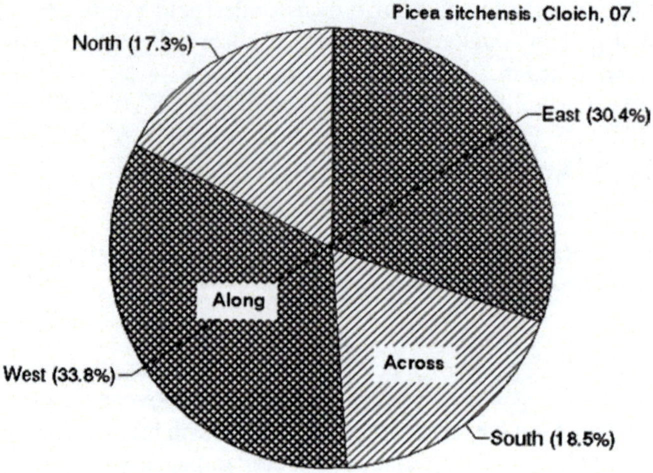

Fig. 15.12 Mean spatial distribution of effective absorbing roots in Sitka spruce (*Picea sitchensis*) trees measured by the modified earth impedance method for trees growing along ridges in rather wet soil. Differences in root distribution along and across the ridges are shown (Results associated to the study of Butler et al. 2010)

6.2 *Effective Absorbing Root Distribution Derived from Radial Sap Flow Patterns*

Radial patterns of sap flow measured from different sides of stems and expressed as the corresponding curves of flow density are usually asymmetrical and composed of two peaks, corresponding to the outer and inner sapwood layers. Based on previous experiments (Nadezhdina and Čermák 1999; Nadezhdina et al. 2003, 2006) that measured the sap flow in stems rather close to the soil surface, the curves can be interpreted in terms of effective root distribution, which can be better specified using mathematical methods (Čermák et al. 2008b). This can be observed for an adult European beech tree growing on stony soil under drought stress conditions (Fig. 15.13, upper panel). Patterns of water supply from different soil layers around the tree indicated, that sinker roots were about twice as active in the northwest direc-

Fig. 15.13 *Upper*: Radial patterns of sap flow in the stems of an adult European beech tree (*Fagus sylvatica*) expressed for different annuli along the stem radius in terms of percentage of total flow values (*thin dashed line*). This curve is asymmetrical and composed of two peaks corresponding to the outer and inner sapwood layers, as found experimentally (Nadezhdina et al. 2007). These peaks can be differentiated using mathematical methods (Čermák et al. 2008b). *Middle*: Sinker roots were about twice as active in the NW direction when compared to the South direction, while the water supply from superficial roots was about 2.5-fold higher in the West than in the East. *Lower*: Examples of branch size distribution around the crown in a tree growing on stony soil under drought stress conditions. The crown radius reached maximal values in the South

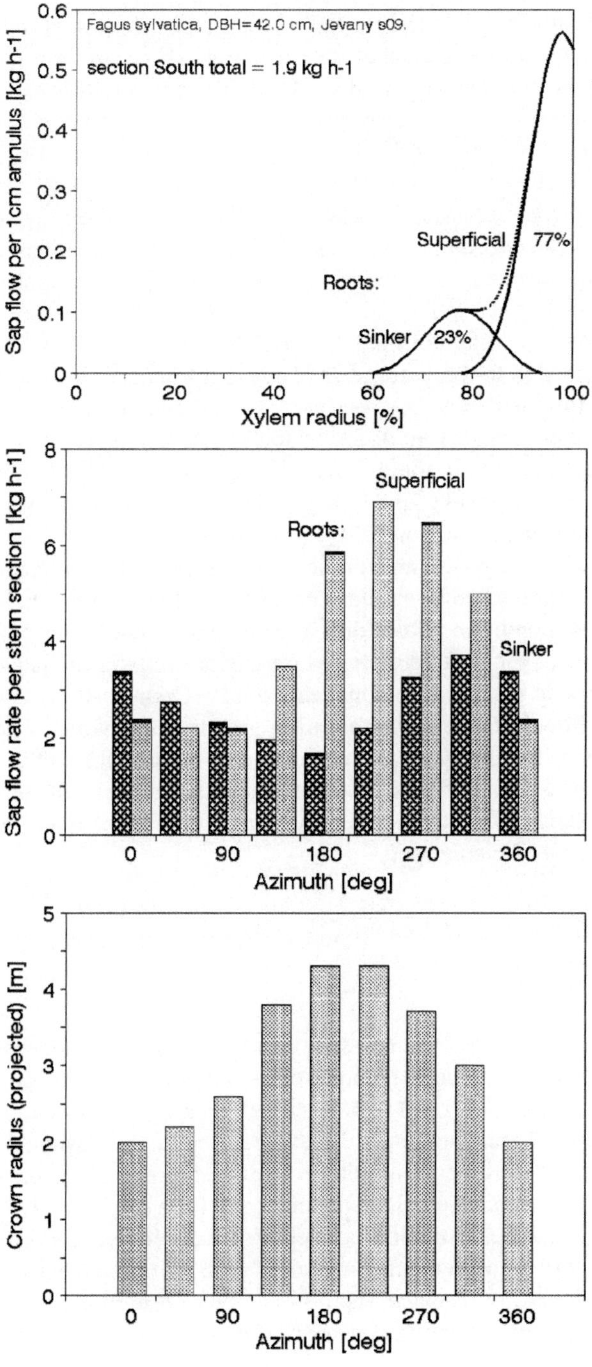
Fagus sylvatica, DBH=42.0 cm, Jevany s09.
section South total = 1.9 kg h-1
Sap flow per 1cm annulus [kg h-1]
Superficial
77%
Roots:
Sinker
23%
Xylem radius [%]
Sap flow rate per stem section [kg h-1]
Superficial
Roots:
Sinker
Azimuth [deg]
Crown radius (projected) [m]
Azimuth [deg]

tion as in the south direction, while the water supply from superficial roots was 2.5-fold greater in the west than in the east (Fig. 15.13, middle panel). Branch distribution around the tree stem indicated that the crown radius reached maximal values in the south (Fig. 15.13, lower panel).

The responses of sap flow to localized watering and root-severing experiments provided experimental verification that the sap flow in the outer layers of stem sapwood comes from superficial roots, while the flow in the inner layers comes from sinker roots (Nadezhdina and Čermák 2003a, b). The circumferential variation in the rate of sap flow near the base of the trunk also reflects the positions of large coarse roots. The lowest sap flow as well as fewer lateral roots were found on the Northeastern side of the stem of mountain ash (Stokes et al. 2002). The sap flow in the stem corresponded to the presence of large roots beneath the sap flow sensors. The sap flow in the stem above the root zone was lower when no or few active roots were present. Water supply from different soil layers can be also interpreted as the effective amounts of roots in these layers. For example, a larger fraction of functional sinker roots (32%) was found in Scots pine trees growing on a deeper layer of sandy soil, and a smaller fraction (22%) was observed in the shallower soil layer. The fraction of effective roots can significantly differ for different azimuths around the stems. This variation can be similar to the branch distribution (Nadezhdina et al. 2007) under some conditions. Root distribution also differed in leaning Scots pine trees, which have lower water supply by superficial roots in the leaning direction than in the opposite direction (against the wind) (Čermák et al. 2008b). Radial patterns of sap flow in *Pinus rotundata* trees growing on a wooded peat bog were characterized by a clear single peak indicating root distribution in the very shallow soil (Kučerová et al. 2010). These results suggest that patterns of root distribution cannot be generalized due to the usually high variation of data and should be verified for each specific condition.

7 Conclusions

Routine studies of large trees in forest stands outside of well-equipped experimental plots are possible with appropriately modified laboratory equipment, but such studies are often logistically difficult and require numerous assumptions and extrapolation of data. Therefore instrumentation and methods specifically designed for field use and able to make measurements simultaneously in multiple trees spanning a broad range of sizes are recommended. Such methods provide a substantially more realistic view of how physiological function is integrated from individual organs to whole trees than laboratory measurements on excised parts of trees or extrapolating from juvenile to adult trees. For instance, sap flow can be measured in any part of trees of any size. In addition to allowing estimation of processes such as transpiration and water storage, it can provide new and important insights concerning tree behavior, effective structure-related tree parameters, redistribution of water in the root-soil compartment, and simultaneously occurring bilateral flows in roots and stems.

Several methods for biometric (size and structure) measurements of crowns (leaf area and spatial distribution) and root systems (coarse and fine roots) are also available, although more experience is needed for different species and various conditions to improve their application.

Acknowledgements This work has been partially done within the framework of the Research Invention No. MSM 6215648902 and project NAZV QC 60063. The authors are grateful to American Journal Experts for language corrections.

References

Allsheimer M, Kostner B, Falge F, Tenhunen JD (1998) Temporal and spatial variation of transpiration of Norway spruce stands within a forested catchment of the Fichtelgebirge, Germany. Ann Sci For 55:103–123

Anfodillo T, Sigalotti GB, Tomasi M, Semenzato P, Valentini R (1993) Application of thermal imaging in the study of sap flow in woody species. Plant Cell Environ 16:997–1001

Ansari SA, Kumar P, Gupta BN (1995) Root surface measurements based on adsorption and desorption of nitrite. Plant Soil 175:133–137. Kluwer, Netherlands

Aubrecht L, Stanek Z, Koller J (2006) Electric measurement of the absorbing surfaces in whole tree roots by the earth impedance method – I theory. Tree Physiol 26:1105–1112

Baker JM, van Bavel CHM (1987) Measurement of mass flow of water in the stems of herbaceous plants. Plant Cell Environ 10:777–782

Balek J, Čermák J, Kučera J, Palouš M, Prax A (1985a) Remote sensing measurements of regional transpiration distribution as an input for the deterministic water balance simulation. In: Proceedings of the 4th international hydrology symposium on multivariate analysis of hydrological processes, Colorado State University, Colorado, pp 1–4

Balek J, Čermák J, Kučera J, Palouš M, Prax A (1985b) The possibilities to estimate transpiration by remote sensing (in Czech.). Vodohospodarský casopis 33:497–505

Barrett DJ, Hatton TJ, Ash JE, Ball MC (1995) Evaluation of the heat pulse velocity technique for measurement of sap flow in rainforest and eucalypt forest species of south-eastern Australia. Plant Cell Environ 18:463–469

Becker P (1998) Limitations of a compensation heat pulse velocity system at low sap flow: implications for measurements at night and in shaded trees. Tree Physiol 18:177–184

Bohm W (1979) Methods of studying root systems. Springer-Verlag, Berlin, 188 p

Bond-Lamberty BC, Wang C, Gower ST (2002) Aboveground and belowground biomass and sapwood area allometric equations for six boreal tree species of northern Manitoba. Can J For Res 32:1441–1450

Brooks JR, Meinzer FC, Coulombe R, Gregg J (2002) Hydraulic redistribution of soil water during summer drought in two contrasting Pacific Northwest coniferous forests. Tree Physiol 22:1107–1117

Brooks JR, Meinzer FC, Warren JM, Domec JC, Coulombe R (2006) Hydraulic redistribution in a Douglas-fir forest: lessons from system manipulations. Plant Cell Environ 29:138–150

Brown N, Jennings S, Wheeler P, Nabe-Nielsen J (2000) An improved method for the rapid assessment of forest understorey light environments. J Appl Ecol 37:1044–1053

Burgess SSO, Bleby TM (2006) Redistribution of soil water by lateral roots mediated by stem tissues. J Exp Bot 57:3283–3291

Burgess SSO, Adams MA, Turner NC, Ong CK (1998) The redistribution of soil water by tree root systems. Oecologia 115:306–311

Burgess SSO, Adams MA, Bleby TM (2000a) Measurement of sap flow in roots of woody plants: a commentary. Tree Physiol 20:909–913

Burgess SSO, Pate JS, Adams MA, Dawson TE (2000b) Seasonal water acquisition and redistribution in the Australian woody phteratophyte *Banksia prionotes*. Ann Bot 85:215–224

Burgess SSO, Adams MA, Turner NC, Beverly CR, Ong CK, Khan AAH, Bleby TM (2001a) An improved heat pulse method to measure low and reverse rates of sap flow in woody plants. Tree Physiol 21:589–598

Burgess SSO, Adams MA, Turner NC, White DA, Ong CK (2001b) Tree roots: conduits for deep recharge of soil water. Oecologia 126:158–165

Butler AJ, Barbier N, Cermak J, Koller J, Thornily C, Micevoy C, Nicoll B, Grace J, Meir P (2010) Estimates and relations between aboveground and belowground resource exchange surface areas in a Sitka spruce managed forest. Tree Physiol 30:705–714

Caldwell MM, Richards JH (1989) Hydraulic lift: water efflux from upper roots improves effectiveness of water uptake by deep roots. Oecologia 79:1–5

Cao Y, Repo T, Silvennoinen R, Lehto T, Pelkonen P (2010a) An appraisal of the electric resistance method for assessing root surface area. J Exp Bot 61:2491–2497

Cao Y, Repo T, Silvennoinen R, Lehto T, Pelkonen P (2010b) Analysis of the willow root system by electric impedance spectroscopy. J Exp Bot 62:351–358

Carley HE, Watson RD (1966) A new gravimetric method for estimating root surface areas. Soil Sci 102:289–291

Carman JG (1982) A non-destructive stain technique for investigating root growth dynamics. J Appl Ecol 19:873–879

Čermák J (1989) Solar equivalent leaf area as the efficient biometric parameter of individual leaves, trees and stands. Tree Physiol 5:269–289

Čermák J (1998) Leaf distribution in large trees and stands of the floodplain forests in southern Moravia. Tree Physiol 18:727–737

Čermák J, Kučera J (1987) Transpiration of fully grown trees and stands of spruce (*Picea abies* (L.) Karst.) estimated by the tree-trunk heat balance method. In: Swanson RH, Bernier PY, Woodward PD (eds) Proceedings of forest hydrology and watershed measurements, Vancouver, Canada, Aug 1987. Publ. No.167, IAHS-AISH, Wallingford, UK, pp 311–317

Čermák J, Kučera J (1990a) Water uptake in healthy and ill trees under drought and hypoxia and non-invasive assessment of the effective size of root systems. In: Persson H (ed) Proceedings of the COST 612 workshop "Above and belowground interactions in forest trees in acidified soils", Simlangsdalen, 21–23 May 1990, Sweden, pp 185–195

Čermák J, Kučera J (1990b) Scaling up transpiration data between trees, stands and watersheds. Silva Carelica 15:101–120

Čermák J, Kučera J (1991) Extremely fast changes of xylem water flow rate in mature trees, caused by atmospheric, soil and mechanical factors. In: Raschi A, Borghetti M (eds) Proceedings of the CEC international workshop "Methodologies to assess the impacts of climatic changes on vegetation: analysis of water transport in plants and cavitation of xylem transport in plants and cavitation of xylem conduits", Firenze, Italy, 29–31 May 1991, pp 181–190

Čermák J, Nadezhdina N, Martinek J (2003) 3D visualization of complex sap flow patterns across tree stems – a methodical note. In: Proceedings of the 5th international workshop on field techniques for environmental physiology, Tenerife, Canary Islands, Spain, 16–22 Mar 2003, pp 45–51

Čermák J, Kučera J (1981) The compensation of natural temperature gradient in the measuring point during the sap flow rate determination in trees. Biol Plant 23:469–471

Čermák J, Michálek J (1991) Selection of sample trees in forest stands using the "quantils of total" (in Czech). Lesnictvi (Forestry) 37:49–60

Čermák J, Nadezhdina N (1998) Sapwood as the scaling parameter – defining according to xylem water content or radial pattern of sap flow? Ann Sci For 55:509–521

Čermák J, Prax A (2001) Water balance of the floodplain forests in southern Moravia considering rooted and root-free compartments under contrasting water supply and its ecological consequences. Ann Sci For 58:1–12

Čermák J, Deml M, Penka M (1973) A new method of sap flow rate determination in trees. Biol Plant 15:171–178

Čermák J, Palát M, Penka M (1976a) Transpiration flow rate in fully-grown tree *Prunus avium* L. by heat balance method estimated, in connection with some meteorological factors. Biol Plant 18:111–118

Čermák J, Kučera J, Penka M (1976b) Improvement of the method of sap flow rate determination in adult trees based on heat balance with direct electric heating of xylem. Biol Plant 18:105–110

Čermák J, Huzulák J, Penka M (1980) Water potential and sap flow rate in adult trees with moist and dry soil as used for the assessment of the root system depth. Biol Plant 22:34–41

Čermák J, Úlehla J, Kučera J, Penka M (1982) Sap flow rate and transpiration dynamics in the full-grown oak (*Quercus robur* L.) in floodplain forest exposed to seasonal floods as related to potential evapotranspiration and tree dimensions. Biol Plant 24:446–460

Čermák J, Jeník J, Kučera J, Židek V (1984) Xylem water flow in a crack willow tree (*Salix fragilis* L.) in relation to diurnal changes of environment. Oecologia 64:145–151

Čermák J, Cienciala E, Kučera J, Lindroth A, Hallgren JE (1992) Radial velocity profiles of water flow in stems of spruce and oak and response of spruce tree to severing. Tree Physiol 10:367–380

Čermák J, Matyssek R, Kučera J (1993) Rapid response of large, drought stressed beech trees to irrigation. Tree Physiol 12:281–290

Čermák J, Riguzzi F, Ceulemans R (1998) Scaling up from the individual trees to the stand level in Scots pine: 1. Needle distribution, overall crown and root geometry. Ann Sci For 55:63–88

Čermák J, Hruška J, Martinková M, Prax A (2000) Urban tree root systems and their survival near houses analyzed using ground penetrating radar and sap flow techniques. Plant Soil 219(1–2):103–115

Čermák J, Jimenez MS, Gonzales-Rodriguez AM, Morales D (2002) Laurel forests in Tenerife, Canary Islands: efficiency of water conducting system in *Laurus azorica* trees. Trees 16:538–546

Čermák J, Kučera J, Nadezhdina N (2004) Sap flow measurements with two thermodynamic methods, flow integration within trees and scaling up from sample trees to entire forest stands. Trees 18:529–546

Čermák J, Ulrich R, Staněk Z, Koller J, Aubrecht L (2006) Electric measurement of the absorbing surfaces in whole tree roots by the earth impedance method – II verification based on allometric relationships and root severing experiments. Tree Physiol 26:1113–1121

Čermák J, Kučera J, Bauerle WL, Phillips J, Hinckley TM (2007a) Tree water storage and its diurnal dynamics related to sap flow and changes of trunk volume in old-growth Douglas-fir trees. Tree Physiol 27:181–198

Čermák J, Gašpárek J, De Lorenzi F, Jones HG (2007b) Stand biometry and leaf area distribution in an old olive grove at Andria, southern Italy. Ann For Sci 64:491–501

Čermák J, Tognetti R, Nadezhdina N, Raschi A (2008a) Stand structure and foliage distribution in *Quercus pubescens* and *Quercus cerris* forests in Tuscany (central Italy). For Ecol Manage 255:1810–1819

Čermák J, Nadezhdina N, Meiresonne L, Ceulemans R (2008b) Scots pine root distribution derived from radial sap flow patterns in stems of large leaning trees. Plant Soil 305:61–75

Čermák J, Ulrich R, Culek I, Čermák M (2008c) Visualization of root systems by the supersonic air stream, pp 89–95. In: Neruda J (ed) Determination of damage to soil and root system of forest trees by the operation of logging machines. Mendel University of Agriculture and Forestry Publishing House, Brno, p 138 p

Chandra S, Lindsey PA, Bassuk NL (1994) A gauge to measure the mass flow rate of water in trees. Plant Cell Environ 17:867–874

Chiesi M, Maselli F, Bindi M, Fibbi L, Bonora L, Raschi A, Čermák J, Nadezhdina N (2002) Calibration and application of forest-BCG in a Mediterranean area by the use of conventional and remote sensing data. Ecol Model 154:251–262

Chloupek O (1972) The relationship between electric capacitance and some other parameters of plant roots. Biol Plant 14:227–230

Clearwater MJ, Meinzer FC, Andrale JL, Goldstein G, Holbrook NM (1999) Potential errors in measurement of non-uniform sap flow using heat dissipation probes. Tree Physiol 19:681–687

Cohen Y, Fuchs M, Green GC (1981) Improvement of the heat-pulse method for determining sap flow in trees. Plant Cell Environ 4:391–397

Comerford NB, Smerthurst PJ, Escamilla JS (1995) Nutrient uptake by woody root systems. New Zeal J For Sci 24:195–212

Coomes DA, Grubb PJ (2000) Impact of root competition in forests and woodlands: a theoretical framework and review of experiments. Ecol Monogr 70:171–207

Costa C, Dwyer LM, Hammel C, Muamba DF, Wang XL, Nantais L, Smith DL (2001) Root contrast enhancement for measurement with optical scanner-based image analysis. Can J Bot 79:23–29

Coutts MP, Nielsen CC, Nicoll BC (1999) The development of symmetry, rigidity and anchorage in the structural root systems of conifers. Plant Soil 217:1–15

Dalton FN (1995) In-situ root extent measurements by electrical capacitance methods. Plant Soil 173:157–165

Daum CR (1967) A method for determining water transport in trees. Ecology 48:425–431

De Lorenzi F, Nadezhdina N, Čermák J, Nadezhdin V, Pitacco A (2009) Sap flow in a mature olive tree: dynamics and quantification in trunk and branches. In: 7th international workshop on sap flow, Seville (Spain), 21–24 Oct 2008. Acta Horticulturae (ISHS) 846:315–322

Drexhage M, Chauviere M, Colin F, Nielsen CNN (1999) Development of structural root architecture and allometry of *Quercus petraea*. Can J For Res 29:600–608

Dunham CW (1958) Use of methylene blue to evaluated rooting of cuttings. J Am Soc Hort Sci 63:450–453

Eckhard G, Horst M (1996) Nutrient and water uptake by roots of forest trees. Pflanzenernährung Bodenkultur 159:11–21

Goldstein G, Andrade JL, Meinzer FC, Holbrook NM, Cavelier J, Jackson P, Celis A (1998) Stem water storage and diurnal patterns of water use in tropical forest canopy trees. Plant Cells Environ 21:397–406

Granier A (1985) Une nouvelle méthode pour la mesure du flux de sève brute dans le tronc des arbres. Ann Sci For 42:193–200

Granier A, Anfodillo T, Sabatti M, Cochard H, Dreyer E, Tomasi M, Valentini R, Breda N (1994) Axial and radial water flow in the trunks of oak trees: a quantitative and qualitative analysis. Tree Physiol 14:1383–1396

Green SR (1998) Measurements of sap flow by the heat-pulse method. An Instruction Manual for the HPV system. HortResearch internal Report IR98, Palmerston North, New Zealand

Green SR, Clothier BE (1988) Water use of kiwifruit vines and apple trees by the heat-pulse technique. J Exp Bot 39:115–123

Green SR, Clothier BE, Jardine B (2003) Theory and practical application of heat-pulse to measure sap flow. Agron J 95:1371–1379

Hatton TJ, Catchpole EA, Vertessy RA (1990) Integration of sap flow velocity to estimate plant water use. Tree Physiol 6:201–209

Hatton TJ, Moore SJ, Reece PH (1995) Estimating stand transpiration in Eucalyptus populea woodland with the heat pulse method: measurement errors and sampling strategies. Tree Physiol 12:219–227

Helmisaari HS, Hallbacken L (1999) Fine-root biomass and necromass in limited and fertilized Norway spruce (*Picea abies* (L.) Karst.) stands. For Ecol Manage 119:99–110

Hruška J (2008) Visualization of root systems by georadar, pp 85–88. In: Neruda J (ed) Determination of damage to soil and root system of forest trees by the operation of logging machines. Monograph. Mendel University of Agriculture and Forestry Publishing House, Brno, 138 p

Hruška J, Čermák J, Šustek S (1999) Mapping of tree root systems by means of the ground penetrating radar. Tree Physiol 19:125–130

Huber B (1932) Beobachtung und Messung pflanzlicher Saftstrome. Ber Deut Bot Ges 50:89–109

Huber B, Schmidt E (1936) Weitere thermoelektrische Untersuchungen uber den Transpirationsstrom der Baume. Tharandter Forstl Jahrsblad 87:369–412

Hultine KR, Williams DG, Burgess SSO, Keefer TO (2003) Contrasting patterns of hydraulic redistribution in three desert phteratophytes. Oecologia 135:167–175

Ishida T, Campbell GC, Calissendorff C (1991) Improved heat balance method for determining sap flow rate star open. Agr Forest Meteorol 56:35–48

Ittner E (1968) Der Tagesgang der Geschwindikkeit des Transpirationsstromes im Stamm einer 75-yahringer Fichte. Oecol.Plant. III:177–183

Janssens IA, Sampson DA, Čermák J, Meiresonne L, Riguzzi F, Overloop S, Ceulemans R (1999) Above- and belowground phytomass and carbon storage in a Belgian Scots pine stand. Ann Forest Sci 56:81–90

Jeník J (1957) Root systems of oak trees. Quercus robur L. and Q.petraea Liebl. (in Czech). Rozpravy České Akademie Věd 67:1–88

Jeník J (1978) Roots and root systems in tropical trees: morphogenic and ecologic aspects. In: Tomlinson PB, Zimmerman MH (eds) Tropical trees as living systems. Cambridge University Press, Cambridge, pp 323–349

Jeník J, Sen DN (1964) Morphology of root systems in trees: a proposed terminology. In: Proceedings of the 10th international botanical congress. Edinburgh, pp 393–394

Jimenez MS, Nadezhdina N, Čermák J, Morales D (2000) Radial variation in sap flow rate in five laurel forest tree species in Tenerife, Canary Islands. Tree Physiol 20:1149–1156

Johnson MG, Tingey DT, Phillips DL, Storm MJ (2001) Advancing fine root research with minirhizotrons. Environ Exp Bot 45:263–289

Jonckheere I, Fleck S, Nackaerts K, Muys B, Coppin P, Weiss M, Baret F (2004) Review of methods for *in situ* leaf area index determination. Part I. Theories, sensors, and hemispherical photography. Agr Forest Meteorol 121:19–35

Kolesnikov VA (1972) Methods of studying root systems of woody plants (in Russian). Monograph. Lesnaya Promyshlennost, Moscow, 152 p

Kozlowski TT, Wignet CH (1963) Patterns of water movement in forest trees. Bot Gaz 124:301–311

Kravka M, Krejzar T, Čermák J (1999) Water content in stem wood of large pine and spruce trees in natural forests in central Sweden. Agr Forest Meteorol 98–99:555–562

Krejzar T, Kravka M (1998) Sap flow and vessel distribution in annual rings and petioles of large oaks. Lesnictvi-Forestry 44:193–201

Kučera J, Čermák J, Penka M (1977) Improved thermal method of continual recording the transpiration flow rate dynamics. Biol Plant 19:413–420

Kučerová A, Čermák J, Nadezhdina N, Pokorný J (2010) Transpiration of *Pinus rotundata* on a wooded pet bog in Central Europe. Trees 24:919–930

Kunia Y (1955) Studies on the sap streaming in plants by the thermoelectric method. Sci Rep Tokyo Univ (Biol) 21:153–178

Kutschera L (1960) Wurzelatlas mitteleuropaischer Ackerunkrauter und Kulturpflanzen. DLG-Verlag, Frankfurt am Main, 574 p

Kutschera L, Lichtenegger E (2002) Wurzelatlas, mitteleuropaischer Waldbaume und Straucher. Leopold Stocker Verlag, Graz-Stuttgart, 604 p

Landsberg JJ, Blanchard TW, Warrit B (1976) Studies on movement of water through apple trees. J Exp Bot 27:579–596

Lanner RM (1984) Trees of the Great Basin. University of Nevada Press, Reno

Larson DW, Doubt J, Matthes-Sears U (1994) Radially sectored hydraulic pathways in the xylem of *Thuja occidentalis* as revealed by use of dyes. Int J Plant Sci 155:569–582

Lee YJ, Alfaro RI, Van Sickle GA (1983) Tree-crown defoliation measurements from digitized photographs. Can J For Res 13:956–961

Lindroth A, Čermák J, Kučera J, Cienciala E, Eckersten H (1995) Sap flow by heat balance method applied to small size *Salix*-trees in a short-rotation forest. Biomass Bioenergy 8(1):7–15

Lopez B, Sabate S, Gracia CA (2001) Vertical distribution of fine root density, length density, area index and mean diameter in a *Quercus ilex* forest. Tree Physiol 21:555–560

Loustau D, Domec JC, Bosc A (1998) Interpreting the variations in xylem sap flux density within the trunk of maritime pine (*Pinus pinaster* Ait.): application of a model for calculating water flows at tree and stand levels. Ann Sci For 55:29–46

Makarieva AM, Gorshkov VG (2008) The forest biotic pump of river basins. Russ J Ecol 39:537–540

Makarieva AM, Gorshkov VG, Li B-L (2006) Conservation of water cycle on land via restoration of natural closed-canopy forests: implications for regional landscape planning. Ecol Res 21:897–906

Makarieva AM, Gorshkov VG, Li B-L (2008) Precipitation on land versus distance from the ocean: evidence of a forest pump of atmospheric moisture. Ecol Complex 179:1–6

Marshall DC (1958) Measurement of sap flow in conifers by heat transport. Plant Physiol 33:385–396

Mauer O, Palátová E (1996) Morphogenesis of the Norway spruce (*Picea abies* (L.) Karst.) root system from natural regeneration up to 30 years of stand age. Lesnictvi-Forestry 42:116–127

Mauer O, Palátová E (2003) The role of root system in silver birch (*Betula pendula* Roth) dieback in the air-polluted area of Krušné hory Mts. J For Sci 49:191–199

McQueen DR (1968) The quantitative distribution of absorbing roots in *Pinus sylvestris* and *Fagus sylvatica* in a forest succession. Oecol Plant 3:83–99

Meinzer FC, James SA, Goldstein G, Woodruff D (2003) Whole-tree water transport scales with sapwood capacitance in tropical forest canopy trees. Plant Cell Environ 26:1147–1155

Meinzer FC, Brooks JR, Domec JC, Gartner BL, Warren JM, Woodruff DR, Bible K, Shaw DC (2006) Dynamics of water transport and storage in conifers studied with deuterium and heat tracing techniques. Plant Cell Environ 29:105–114

Meinzer FC, Woodruff DR, Domec JC, Goldstein G, Campanello PI, Gatti MG, Villalobos-Vega R (2008) Coordination of leaf and stem water transport properties in tropical forest trees. Oecologia 156:31–41

Meiresonne L, Nadezhdina N, Čermák J, VanSlycken J, Ceulemans R (1999) Transpiration of a poplar stand: model calibration and validation by soil water and sap flow measurements. In: Feyen J, Wiyo K (eds) Modeling of transport processes in soils at various scales in time and space, International workshop of EurAgEng's field of interest on soil and water, Leuven, Belgium, 24–26 Nov 1999, pp 1–10

Meiresonne L, Sampson DA, Kowalski AS, Janssens IA, Nadezhdina N, Čermák J, Van Slycken J, Ceulemans R (2003) Water flux estimates from a Belgian Scots pine stand: a comparison of different approaches. J Hydrol 270:230–252

Monteith JL (1975) Principles of environmental physics. Edward Arnold, Kent, 242 p

Morales D, Gonzalez-Rodriguez AM, Čermák J, Jimenez MS (1996a) Laurel forests in Tenerife, Canary Islands: the vertical profiles of leaf characteristics. Phyton (Austria) 36:1–13

Morales D, Jimenez MS, Gonzalez-Rodriguez AM, Čermák J (1996b) Laurel forests in Tenerife, Canary Islands: I. The site, stand structure and leaf distribution. Trees 11:34–40

Morales D, Jimenez MS, Gonzalez-Rodriguez AM, Čermák J (1996c) Laurel forests in Tenerife, Canary Islands: II. Leaf distribution patterns in individual trees. Trees 11:41–46

Morales D, Jimenez MS, Gonzalez-Rodriguez AM, Čermák J (2002) Laurel forests in Tenerife, Canary Islands: vessel distribution in stems and in petioles of *Laurus azorica* trees. Trees 16:529–537

Moreira MZ, Scholz FG, Bucci SJ, Sternberg LS, Goldstein G, Meinzer FC, Franco AC (2003) Hydraulic lift in a neotropical savanna. Funct Ecol 17:573–581

Morikawa Y (1974) Sap flow in *Chamaecyparis obtusa* in relation to water economy of woody plants. Bull Tokyo Univ For 66:251–297

Morikawa Y, Hattori S, Kiyono Y (1986) Transpiration of a 31-year-old *Chamaecyparis obtusa* Endl. stand before and after thinning. Tree Physiol 2:105–114

Nadezhdina N (1999) Sap flow index as an indicator of plant water status. Tree Physiol 19:885–891

Nadezhdina N, Čermák J (1998) Response of sap flow rate along tree stem and coarse root radii to changes of water supply. In: Stokes A (ed) Proceedings of the international symposium. The supporting roots – structure and function, Bordeaux, France, 20–24 July 1998, p 81

Nadezhdina N, Čermák J (1999) Responses of sap flow rate along tree stem and coarse root radii to changes of water supply. Plant and Soil 12:1–12. In: Stokes A (ed) Proceedings of the supporting roots of trees and woody plants: form, function and physiology, pp 227–238. Kluwer, Dordrecht, Boston, London, 430 p

Nadezhdina N, Čermák J (2003a) Instrumental methods for studies of structure and function of root systems in large trees. J Exp Bot 54:1511–1521

Nadezhdina N, Čermák J (2003b) Instrumental methods for studies of structure and function of root systems in large trees. In: Proceedings of the 5th international workshop on field techniques for environmental physiology, Tenerife, Canary Islands, Spain, 16–22 Mar 2003, pp 23–33

Nadezhdina N, Čermák J, Nadyezhdin V (1998) Heat field deformation method for sap flow measurements. In: Cermák J, Nadezhdina N (eds) Measuring sap flow in intact plants. IUFRO, Publishing House of Mendel University, Brno, Czech Republic, pp 72–92

Nadezhdina N, Čermák J, Ceulemans R (2002) Radial pattern of sap flow in woody stems related to positioning of sensors and scaling errors in dominant and understorey species. Tree Physiol 22:907–918

Nadezhdina N, Gašpárek J, Nadyezhdin V, Čermák J (2003) Sap flow dynamics in terms of competition between overstorey spruce and understorey beech in southern Moravia. International conference, the question of conversion of coniferous forests – ConForest, Freiburg im Breisgau, Germany, 27 Sept–02 Oct 2003

Nadezhdina N, Tatarinov F, Ceulemans R (2004a) Leaf area and biomass of *Rhododendron* understorey in a stand of Scots pine. For Ecol Manage 187:235–246

Nadezhdina N, Čermák J, Tributsch H (2004b) Infrared images of sap flow in stems of lime trees under natural and experimental conditions. Ann Sci For 61:203–213

Nadezhdina N, Čermák J, Gašpárek J, Nadyezhdin V, Prax A (2006) Vertical and horizontal water redistribution inside Norway spruce (*Picea abies*) roots in the Moravian upland. Tree Physiol 26:1277–1288

Nadezhdina N, Čermák J, Meiresonne L, Ceulemans R (2007) Transpiration of Scots pine in Flanders growing on soil with irregular substratum. For Ecol Manage 243:1–9

Nadezhdina N, Steppe K, De Pauw DJW, Bequet R, Čermák J, Ceulemans R (2009) Stem-mediated hydraulic redistribution in large roots on opposing sides of a Douglas-fir tree following localized irrigation. New Phytol 184:932–943

Nadezhdina N, David TS, David JS, Ferreira MI, Dohnal M, Tesar M, Gartner K, Leitgeb E, Nadezhdin V, Čermák J, Jimenez MS, Morales D (2010) Trees never rest: the multiple facets of hydraulic redistribution. Ecohydrology 3:431–444

Nathenson R, Jarabak A (2001) The evolution of air-tools for use in arboriculture. Tree Care Industry 2001:29–41

Offenthaler I, Hietz P (1998) A comparison of different methods to measure sap flow in spruce. In: Cermák J, Nadezhdina N (eds) Measuring sap flow in intact plants. Proceedings of 4th international workshop, Židlochovice, Czech Republic. IUFRO, Publishing House of Mendel University, Brno, Czech Republic, pp 55–64

Oltchev A, Čermák J, Nadezhdina N, Tatarinov F, Tischenko A, Ibrom A, Gravenhorst G (2002a) Transpiration of a mixed forest stand: field measurements and simulation using SVAT models. Boreal Environ Res 7:389–397

Oltchev A, Čermák J, Gurtz J, Tischenko A, Kiely G, Nadezhdina N, Zappa M, Lebedeva N, Vitvar T, Albertson JD, Tatarinov F, Tischenko D, Nadyezhdin V, Kozlov B, Ibrom A, Vygodskaya N, Gravenhorst G (2002b) The response of the water fluxes of the boreal forest region at the Volga's source area to climatic and land-use changes. Phys Chem Earth 27:675–690

Peramaki M, Nikinmaa E, Sevanto S, Ilvesniemi H, Siivola E, Hari P, Vesala T (2001) Tree stem diameter variations and transpiration in Scots pine: an analysis using a dynamic sap flow model. Tree Physiol 21:889–897

Peramaki M, Vesala T, Nikinmaa E (2005) Modeling the dynamics of pressure propagation and diameter variation in tree sapwood. Tree Physiol 25:1091–1099

Persson H (1983) The distribution and productivity of fine roots in boreal forests. Plant Soil 71:87–101

Persson H, Ahlstrom K (2002) Fine root response to nutrient supply in nitrogen manipulated Norway spruce catchments areas. For Ecol Manage 168:29–41

Persson H, Vonfircks Y, Majdi H, Nilsson LO (1995) Root distribution in a Norway spruce (*Picea abies* (L.) Karst.) stand subjected to drought and rammonium-sulfate application. Plant Soil 169:161–165

Phillips N, Oren R, Zimmermann R (1996) Radial patterns of xylem sap flow in non-, diffuse- and ring-porous tree species. Plant Cell Environ 19:983–990

Phillips N, Nagchaudhuri A, Oren R, Katul G (1997) Time constant for water transport in loblolly pine trees estimated from time series of evaporative demand and stem sap flow. Trees 11:412–419

Phillips NG, Ryan MG, Bond BJ, McDowell NG, Hinckley TM, Čermák J (2003) Reliance on stored water with tree size in three species in the Pacific Northwest. Tree Physiol 23:237–245

Radchenko SS, Semin VS, Berdin VS, Ilnitsky OA (1983) Determination of the rate of water transpirt along the xylem of plants (in Russian). Fiziologiya i Biokhimiya kulturnykh rasteniy 15(3):291–296

Rizzo DM, Gross R (2000) Distribution of *Armillaria melea* on pear root systems and comparison of excavation techniques. In: Stokes A (ed) The supporting roots of trees and woody plants: form, function and physiology. Developments in plant and soil sciences, vol 87. Kluwer, Dordrecht, pp 305–311

Roach WA (1939) Plant injection as a physiological method. Ann Bot 3:155–226

Rychnovská M, Čermák J, Šmíd P (1980) Water output in a stand of *Phragmites communis* Trin. A comparison of three methods. Acta Sci Nat (Brno) 14:1–27

Sakuratani T (1981) A heat balance method for measuring water flux in the stem of intact plants. J Agric Meteorol 37:9–17

Salas E, Ozier-Lafontaine H, Nygren P (2004) A fractal model applied for estimating root biomass and architecture in two tropical legume tree species. Ann For Sci 61:337–345

Scholz FG, Bucci SJ, Goldstein G, Meinzer FC, Franco AC (2002) Hydraulic redistribution of soil water by neotropical savanna trees. Tree Physiol 22:603–612

Schubert H (1999) Qualitative und quantitative Untersuchungen verschiedene Methoden der Xylemflussmessung an Baumen. Diplomarbeit der Forstwissensschaftlichen Faultat der Ludwig-Maxmilians-Universitat Munchen, Lehrstuhl fuer Forstbotanik. Leiter: Prof. Dr. R. Matyssek, wissenschaftliche Betreuung: Dr. K. H.Haberle und Dr. B.Gotz

Schulze ED, Čermák J, Matyssek R, Penka M, Zimermann R, Vašícek F, Gries W, Kučera J (1985) Canopy transpiration and flow rate fluxes in the xylem of the trunk of *Larix* and *Picea* trees – a comparison of xylem flow, porometer and cuvette measurements. Oecologia 66:475–483

Schuurman JJ, Goedwaagen MAJ (1965) Methods for the examination of root systems and roots, 2nd edn. Center for Agricultural Publishing and Documentation, Wageningen, 86 p

Senock RS, Ham JM (1993) Heat balance sap flow gauge for small diameter stems. Plant Cell Environ 16:593–601

Sevanto S, Holtta T, Nikinmaa E (2009) The effect of heat storage during low flow rates on the output of Granier-type sap-flow sensors. Acta Hortic 846:45–52

Shackel KA, Johnson RS, Medawar CK, Phene CJ (1992) Substantial errors in estimates of sap flow using the heat-balance technique on woody stems under field conditions. J Am Soc Hort Sci 117:351–356

Silvestre J, Ferreira MI (1998) Sap flow measurements with the Granier method on a vintage in central Portugal during a drying period. In: Cermák J, Nadezhdina N (eds) Proceedings of the 4th international workshop on measuring sap flow in intact plants, Zidlochovice, Czech Republic, 3–5 Oct 1998, pp 127–131

Smerthust PJ, Comerford RB (1993) Simulating nutrient uptake by single or competing and contrasting root system. Soil Sci Soc Am J 57:1361–1367

Staněk Z (1997) Physical aspects of resistivity measurements in plants from viewpoint of their ecological applications (in Czech). Habilitation thesis, Department of Physics, Technical University in Prague, Czech Republic, 166 p

Steele SJ, Gower ST, Vogel JG, Norman JM (1997) Root mass, net primary production and turnover in aspen, jack pine and black spruce forests in Saskatchewan and Manitoba, Canada. Tree Physiol 17:577–587

Steinberg SL, van Bavel CHM, McFarland MJ (1990) Improved sap flow gauge for woody and herbaceous plants. Agron J 82:851–854

Štěpánek V, Penka M, Čermák J (1972) Physiology of nourishment of woody species – transpiration and shoot water content in oak (in Czech). Res Rep, Forest Fac, Agric Univ Brno, VU: VI-5-1. 45 p

Steppe K, De Pauw DJW, Lemeur R, Vanrolleghem A (2005) A mathematical model linking tree sap flow dynamics to daily stem diameter fluctuations and radial stem growth. Tree Physiol 26:257–273

Steudle E (1995) Water transport across roots. Plant Soil 167:79–90

Stokes A, Fourcaud T, Hruška J, Čermák J, Nadezhdina N, Nadyezhdin V, Praus L (2002) An evaluation of different methods to investigate root system architecture of urban trees *in situ*. I. Ground penetrating radar. J Arboricul 28–1:1–9

Sutton RF, Tinus RW (1983) Root and root system terminology. For Sci 24:138 p

Swanson RH (1965) Seasonal course of transpiration of lodgepole pine and Engelmann spruce. In: Proceedings of the international symposium on forest hydrology, Pennsylvania State University, 29 Aug–10 Sept 1965, pp 417–432

Swanson RH (1970) Sampling for direct transpiration estimates. New Zeal J Hydrol 9:72–77

Swanson RH (1971) Velocity distribution patterns in ascending xylem sap during transpiration. In: Symposium on flow – its measurement and control in science and industry. Canadian Forestry Service Paper No.4/2/171, 11 p

Swanson RH (1994) Significant historical development in thermal methods for measuring sap flow in trees. Agric For Meteorol 72:113–132

Swanson RH, Whitfield DWA (1981) A numerical analysis of heat-pulse velocity theory and practice. J Exp Bot 32:221–239

Tatarinov FA, Kučera J, Cienciala E (2005) The analysis of physical background of tree sap flow measurements based on thermal methods. Meas Sci Technol 16:1157–1169

Tatarinov FA, Urban J, Čermák J (2008) The application of "clump technique" for root system studies of *Quercus robur* and *Fraxinus excelsior*. For Ecol Manage 255:495–505

Teskey RO, Grier CC, Hinckley TM (1985) Relation between root system size and water inflow capacity of *Abies amabilis* growing in a sub alpine forest. Can J For Res 15:669–672

Tognetti R, Raschi A, Nadezhdina N, Čermák J (2000) Stand structure and foliage distribution patterns in individual trees of *Quercus pubescens* and *Quercus cerris* forests in Tuscany (central Italy). In: Tognetti R, Raschi A (eds) 5th international workshop on measuring sap flow in intact plants, Firenze, Italy, 9–10 Nov 2000, pp 1–27. Fondazione per la Meteorologie Applicata, 2003

Urban J, Tatarinov F, Nadezhdina N, Cermák J, Ceulemans R (2009) Crown structure and leaf area of the understorey species *Prunus serotina*. Trees 23:391–399

Valancogne C, Nasr N (1989) Measuring sap flow in the stem of small trees by a heat balance method. HortScience 24:383–385

Van der Zande D, Mereu S, Nadezhdina N, Cermák J, Muys B, Coppin P, Manes F (2009) 3D up scaling of transpiration from leaf to tree using ground based LIDAR: application on a Mediterranean Holm oak (*Quercus ilex* L.) tree. Agric For Meteorol 149(10):1573–1583. doi:10.1016/j.agrformet.2009.04.010

Vieweg GH, Ziegler H (1960) Thermoelektrische Registrierung der Geschwindigkeit des Transpirationsstromes I. Deutsche Botanical Geselschaft Berlin 73:221–226

Vite JP, Rudinsky JA (1959) The water conducting systems in conifers and their importance to the distribution of trunk injected chemicals. Contrib Boyce Thompson Inst 20:27–38

Vyskot M (1976) Tree story biomass in lowland forests in South Moravia. Rozpravy CSAV 86(10), Academia Praha, 186 p

Waisel Y, Lipschnitz N, Kuller Z (1972) Patterns of water movement in trees and shrubs. Ecology 53:520–523

Waring RH, Running SW (1978) Sapwood water storage: its contribution to transpiration and effect upon water conductance through the stems of old-growth Douglas fir. Plant Cell Environ 1:131–140

Waring RH, Whitehead D, Jarvis PG (1979) The contribution of stored water to transpiration in Scots pine. Plant Cell Environ 2:309–317

Warren JM, Meinzer FC, Brooks JR, Domec JC (2005) Vertical stratification of soil water storage and release dynamics in Pacific Northwest coniferous forests. Agric For Meteorol 130:39–58

Weibel FP, Vos JA (1994) Transpiration measurements on apple trees with an improved stem heat balance method. Plant Soil 166:203–219

West GB, Brown JH, Enquist BJ (1999) A general model for the structure and allometry of plant vascular systems. Nature 400:664–667

Wielopolski L, Hendrey G, McGuigan M, Daniels J (2000) Imaging tree root systems *in situ*. In: Proceedings of the 8th international conference on ground penetrating radar, Gold Coast Australia, pp 642–646

Wilde SA, Voigt GK (1949) Absorption – transpiration quotient of nursery stock. J Forest 47:643–645

Woods FW (1969) Root extension of forest trees: a method of investigation using radioactive traces. In: Whittington WJ (ed) Root growth. Butterworths, London, pp 413–417

Zimmerman MH (1978) Hydraulic architecture of some diffuse-porous trees. Can J Bot 56:2286–2295

Zygurovskaya LN (1958) Anatomical and physiological research of absorbing, growing and conducting roots of woody species (in Russian). Trudy Instituta Lesa 41:5–31

Part VI
Scaling from Individual Traits to Stands

Chapter 16
Ecophysiology of Secondary Succession in Tropical Moist Forest: Scaling from Individual Traits to Whole-Plant Performance

Niels P.R. Anten and N. Galia Selaya

Abstract Secondary forests have probably become the dominant forest type in the tropics and provide many of the same ecosystem function as primary forests. There is thus an urgent need for a further understanding of their ecology and functioning. Yet the mechanisms underlying species replacement during secondary succession in tropical forest are not clearly understood. What traits enable the early successional species to initially achieve dominance and by what means are the later successional species able to initially persist in the vegetation and become dominant later on? We review ecophysiological studies that have addressed this question for tropical moist forests. We discuss that while most research has focused on the trade-off between shade tolerance and other life history traits, a similar tradeoff with respect to water and nutrient availability also plays a role. We also strongly believe that the emphasis of the research should shift from comparing individual traits between different successional groups to the development of quantitative links between these traits and ecological performance measures at the whole-plant level (i.e., growth and survival), and point out that this inevitably entails a need to understand and quantify size related effects on tree functioning.

N.P.R. Anten (✉)
Ecology and Biodiversity Group, Institute of Environmental Biology, Utrecht University, P.O. Box 800.84, 3508 TB Utrecht, The Netherlands
e-mail: N.P.R.Anten@uu.nl

N.G. Selaya
Ecology and Biodiversity Group, Institute of Environmental Biology, Utrecht University, P.O. Box 800.84, 3508 TB Utrecht, The Netherlands

Programa Manejo de Bosques de la Amazonía Boliviana (PROMAB), Casilla 107, Riberalta, Bolivia

F.C. Meinzer et al. (eds.), *Size- and Age-Related Changes in Tree Structure and Function*, Tree Physiology 4, DOI 10.1007/978-94-007-1242-3_16,

1 Introduction

In the tropics human induced forest disturbance, i.e. timber extraction and conversion of forest land for agriculture and other uses, is leading to an increase in secondary forest area (Brown and Lugo 1990; Houghton 1994). In addition, in many tropical countries shifts from agrarian to industrial economies have led to large-scale abandonment of agricultural and pasture lands (see Marin-Spiotta et al. 2007). As a result it has been estimated that secondary forests are now the dominant forest cover in the tropics (Houghton 1994). Secondary forests can provide many of the same ecosystem goods and services as primary forest, including C sequestration, erosion control and watershed management (Brown and Lugo 1990; Guariguata and Ostertag 2001). There is thus an urgent need for further understanding of the ecology and functioning of these forest types.

Secondary succession is the long-term directional change in community composition following a disturbance. In tropical forests this is a complex process that can be driven by propagule availability, random demographic factors, as well as the interplay between differences in functional traits and changes in the physical environment as they occur during succession (see Chazdon 2009). The pattern of succession can also be strongly influenced by the type of disturbance and land-use prior to field abandonment (Lawrence 2003).

Nevertheless, some generalizations can be made. Early successional stages, shortly after field abandonment, are dominated by fast growing light demanding pioneer species whereas later stages are dominated by shade tolerant slower growing species, but which can ultimately attain a larger maximum height than the pioneers (Bazzaz and Pickett 1980; Whitmore 1996; Turner 2001). Under conditions where the soil is not overly degraded and where propagule availability is not limiting, secondary succession in moist tropical forests has been described as proceeding in the following general pattern (Finegan 1996). Shortly after field abandonment various herbs, shrubs and climbers rapidly colonize the site. Short-lived pioneer trees (SLP) establish early in succession and dominate the canopy throughout the first 10–20 years. Long-lived pioneer trees (LLP) and some late successional species are also present from the early stage onwards, but they replace the SLP in the canopy only later in succession. The LLP will typically dominate the forest for the next 75–150 years. During this phase there is a gradual transition with later successional species becoming increasingly dominant. But how is species replacement during secondary succession regulated? What traits enable the early successional species to initially achieve dominance and by what means are the later successional species able to initially persist in the vegetation and become dominant later on?

Smith and Huston (1989) developed a conceptual model based on variation in shade tolerance of different species. They proposed that where competition for light is strong, a trade-off between shade tolerance and other life history traits may be the primary mechanism determining species replacement in forest succession. Though this model was initially developed for succession in temperate forest, much of the work in tropical rain forest has (usually implicitly) been based on this model. Most studies on secondary succession in tropical rain forest thus implicitly assume that a trade-off between shade tolerance and the ability for rapid growth at high light is a primary driving force determining species replacement. But other factors may limit plant growth in tropical rain forest as well: soils are generally limited in nutrient availability (Lawrence 2003) while periodic droughts are common even in hyper wet forests (Allen et al. 1996). The degree of water and nutrient limitation probably changes as secondary forests mature. In addition, adaptations to low nutrient or water availability might come at the expense of adaptations to shade (Smith and Huston 1989).

Much of the work on secondary forest species has focused on comparing their individual traits (e.g. Paz 2003; Santiago and Wright 2007) without quantitatively linking these traits to whole-plant performance, e.g. in terms of carbon gain, growth or survival, in natural conditions in the forest. Without such analyses, the relative importance of interspecific differences in trait values in determining species replacement during succession cannot be assessed. Scaling from individual to whole-plant performance introduces the problem of tree size (e.g. height). LLPs and later successional canopy species have greater maximum height than early successional ones (i.e., short-lived pioneers Turner 2001) and forest height increases with forest age. Taller trees may have more respiration relative to their photosynthesis (Givnish 1988), and may be subjected to larger hydrological stress than short trees (Tyree and Sperry 1988). Thus, potential size-effects on tree functioning should be considered.

In this chapter we first discuss how plant growth conditions, light availability, temperature and soil water and nutrient contents change during the course of secondary succession. The subsequent section deals with differences in plant traits between early and late successional species and discusses how these are associated with adaptations to different light, nutrient and water availabilities. In the final section we scale up from individual traits to whole plant performance to address the question to which extent individual traits and trait correlation can be quantitatively associated with the success of different species at different stages of succession.

We limit this chapter to moist evergreen lowland tropical forests (tropical rain forests, hereafter). They constitute 30% of the terrestrial net primary productivity (Whittaker 1975) and hold about half of the world's plant species (Terborgh 1992). They generally differ from dry deciduous forests (occurring in drier areas) by being taller, and by having a greater species richness, basal area and leaf area index (Chazdon 2009). But where information about tropical forests is lacking we sometimes extrapolate from other forest types. We also limit the nomenclature for functional groups by indicating a gradual scale from earlier successionals (for simplicity denoted as pioneers) to later successionals.

2 Physical Growth Conditions

During succession vegetation structure changes dramatically. Vegetation height, standing biomass and leaf area index all increase while the vertical structure becomes more complex. Concomitantly litter production and root activity in the soil increase. These changes may have profound consequences for the environmental conditions such as light availability, temperature, humidity, and soil water and nutrient availabilities in the forest (Bazzaz and Pickett 1980). Yet in spite of the importance of these factors for plant growth, relatively few studies provide quantitative measurements of these conditions for secondary tropical forest.

2.1 Light Availability, Temperature and Air Humidity

Changes in light availability are the most conspicuous environmental change that occurs during secondary succession. During the initial phase of succession both the LAI and vegetation height increase rapidly creating an increasingly steep light gradient in the canopy. Depending on prior use and level of clearing almost full sunlight can reach the soil shortly after field abandonment. But this readily declines as plants develop their canopies and can drop to less than 1% of day light in mature forests (Chazdon 1986). This decline in understory light availability tends to taper off in older forest (Fig. 16.1a). For example, Peña-Claros (2003) found that the relative amounts of light reaching the forest floor did not differ significantly between 3, 10 and 20 year old successional forest stands. Similar results were obtained in other studies (e.g. Nicotra et al. 1999; Montgomery and Chazdon 2001).

Variation in light availability on a horizontal plane, due to tree-fall gaps or leaf clustering in the canopy, occurs in every forest type (Niinemets and Anten 2009). Understory light conditions in a tropical forest can typically vary between 1–2% and 30% in large treefall gaps (Chazdon 1986). The death rate of mature trees and the concomitant occurrence of gaps increase during the course of succession. Consequently, the light availability at the forest floor tends to become more heterogeneous during the later stages of succession (Nicotra et al. 1999).

Due to the reduction in radiation, day time temperatures in the lower strata of the canopy should also be expected to decrease during succession (Fetcher et al. 1985; Lebrija-Trejos 2008). However, real data for tropical moist forest are lacking. As a result of this cooling and due to the transpiration of plants, water vapor pressure deficits strongly decrease (Fetcher et al. 1985, Fig. 16.1b). These effects obviously only apply for the understory and sub canopy layers and not for plants with leaves at the top of vegetation. Temperatures and VPDs at the top of canopy would probably not change very much with forest age.

2.2 Soil Characteristics: Nutrient and Water Availability

Most tropical soils are limited in nutrients especially phosphorus, and thus nutrient dynamics probably play an important role in determining secondary succession (Lawrence 2003). The classical view is that large scale disturbances due increases in soil temperature, aeration and associated microbial activity cause a flush in nutrient availability (Bazzaz and Pickett 1980; Guariguata and Ostertag 2001; Lawrence

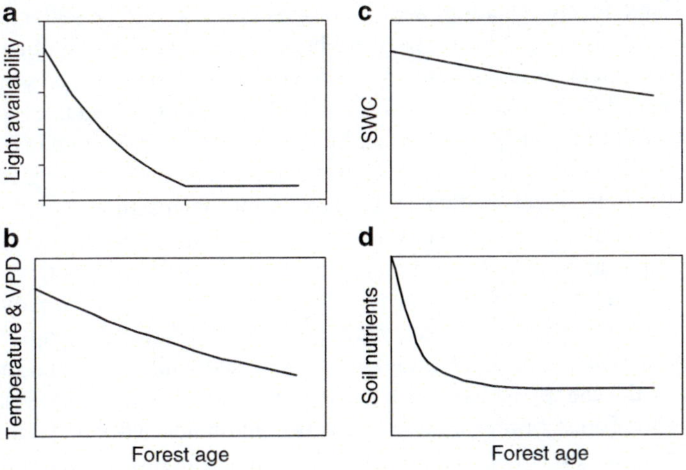

Fig. 16.1 Hypothesized qualitative patterns of changes in environmental conditions in forest understory with forest age during secondary succession in tropical rain forest: (**a**) light availability, (**b**) air temperature and vapor pressure deficit, (**c**) soil water content and (**d**) soil nutrient availability. Note that these are very general and putative patterns that will not generally apply to every site

et al. 2005, Fig. 16.1d). But depending on the intensity and duration of subsequent use nutrient losses may then occur through harvesting, burning, erosion or leaching (Dos Santos 1987; Lu et al. 2002). It has been documented for example that soil nutrient stocks rapidly decline in the first years of pasture use making that sites with light pasture-use have larger nutrients stocks at the time of field abandonment than those with heavy use (Buschbacker et al. 1988).

In theory forest growth after field abandonment should, at least partly, reverse these losses. Vegetation plays an important role in providing and holding nutrients in the soil and maintaining soil structure (Lavelle 1987), for example by protecting the soil from drying, compaction and leaching (Lu et al. 2002). In association with mycorrhizae plants can also acquire nutrients, especially phosphorus, from large areas and soil volumes (Lambers et al. 1998). On the other hand, growing trees take up nutrients and thus tend to deplete soil nutrient stocks.

Nutrient measurements showed that aboveground nutrient pools (in plant tissue) increased with forest age concomitant with the increase in aboveground biomass, as expected (e.g. Hughes et al. 1999). Results for soil pools vary between sites and between the nutrients in question, but the general trend appears to be constant or declining nutrient stocks with forest age (e.g. Buschbacker et al. 1988; Johnson et al. 2001; Feldpausch et al. 2004, Fig. 16.1d). In some cases soil nitrogen availability was an exception to this trend in that it increased with forest age possibly due to the N fixation by certain vegetation components (Feldpausch et al. 2004).

Soil properties may also vary depending on prior land use (pastures, crop cultivation, agroforestry, slash and burn agriculture) and this is believed to strongly affect local patterns of succession. Pastures for example often entail the use of heavy equipment, and in combination with cattle this may result in soil compaction (Guariguata and Ostertag 2001). Slash and burn agriculture is at the other extreme as it involves small areas and less manipulation of the soil and thus has a less severe impact on the soil (Ferguson and Griffith 2004). The impact of land use type on soil nutrient availability however is not clear. Lawrence et al. (2005) found N, P, and K levels in the soil to be higher at sites that had experienced more shifting cultivation cycles. On the other hand, neither maize cultivation nor pasture-use affected soil nutrient stocks in a moist secondary forest in Mexico (Hughes et al. 1999).

It so far unclear how water availability in the soil (soil water potential, SWP) changes during succession. Several factors determine SWP, i.e., precipitation, water uptake by plants (transpiration) and evaporation. As the forest develops transpiration initially increases with increasing LAI of the stand while more precipitation is intercepted by the canopy which thus tends to reduce SWP (e.g. Becker et al. 1988). However, as the forest further matures trees become taller and the stand becomes dominated by later successional species. As discussed below, both trends can be associated with lower transpiration rates per unit leaf area, while LAI tends to level off, resulting in an overall reduction of the canopy-level transpiration (Ewers et al. 2005). The importance of these latter effects has to our knowledge not been quantified for tropical forests. In addition, lower VPD of the air and lower wind velocity in the understory at later successional stages, slow down evaporation positively affecting SWP. The relationship between successional age and SWP is thus hard to predict. Declines in SWP following disturbances have been observed in both temperate (Gray et al. 2002) and tropical forest (see Ostertag 1988) suggesting a tendency for SWC to decline with forest age (Fig. 16.1c).

Overall, an increasingly tall and steep light gradient with declining light levels at the forest floor is probably the most important environmental change that occurs during secondary succession in tropical rainforests. But other factors change as well: the air in the understory becomes cooler and more humid, while soil nutrient and water availability tend to decrease. In general however much more research is needed especially on factors other than light.

3 Difference in Plant Traits Between Pioneers and Late Successional Species

As discussed in the previous section environmental conditions change drastically during the course of succession in tropical rain forest. The changes in the occurrence and relative dominance of species with forest age could thus reflect their specific adaptations to these conditions. Though, as noted in the introduction, other factors such as dispersal limitations and random demographic factors, also affect this pattern. In this section we will discuss plant traits associated with adaptations to different levels of light, nutrient and water availability. An overview of the expected differences in trait values between early and later successional tree species is given in Table 16.1.

Table 16.1 Overview of the individual plant traits discussed in this chapter and the degree to which they differ between tropical rain forest trees of different successional status

Trait	Symbol	Cue (Plasticity)	Changes with successional status
Biomass allocation			
Leaf mass fraction	LMF		Decreases
Stem mass fraction	SMF		Neutral
Root mass fraction	RMF		Increases
Leaf traits			
Photosynthetic capacity per unit leaf area	A_{max}		Decreases
Photosynthetic capacity per unit leaf mass	A_{mass}		Decreases
Leaf nitrogen content per unit area	N_{area}		Neutral
Leaf nitrogen content per unit mass	N_{mass}		Decreases
Leaf dark respiration	r_d		Decreases
Specific leaf area	SLA		Decreases
Leaf longevity			Increases
Leaf toughness			Increases
Stem and crown characteristics			
Crown depth			Increases
Wood dry mass density			Increases
Wood modulus of rupture			Increases
Hydraulic conductivity			Decreases
Cavitation resistance			Increases
Root traits			
Specific root length	SRL		Decreases
Specific absorption rate	SAR		Decreases
Root toughness			Increases
Plasticity traits			
SLA		Light	Decreases
Stem elongation		Light quality	Decreases
Leaf turnover		Light	Decreases

'Symbol' refers to the symbol or abbreviation used in the text (not all traits were represented as such). The 'cue' refers to the main environmental factor that induces a given plastic response. In the column 'changes with successional status', 'increases' implies that later successional species have high trait values than pioneers, 'decreases' denotes the opposite and 'neutral' denotes that there is no clear trend

3.1 Different Light Availabilities

Huston and Smith (1987) proposed that the tradeoff between shade tolerance and other life-history traits might be the primary factor that underlies species replacement during secondary succession. Their contention was that growing fast at high light conditions somehow goes at the expense of performance in the shade and vice versa.

3.1.1 Plant Growth and Light Availability

The traditional view regarding variation in shade tolerance is that light-demanding species grow faster than shade tolerant ones at high light availability but are incapable of growing in the shade (Bongers and Popma 1990). The underlying argument is that traits associated with rapid growth at high light (e.g. a high rate of light-saturated photosynthesis, A_{max}) are correlated with high rates of respiration, which at low light would lead to an unfavorable carbon balance. To test this idea a large number of experiments have been done comparing growth of tropical forest trees of different successional status (reviewed by Veneklaas and Poorter (1998) and Poorter (2005)). The general trend was that pioneers indeed grew considerably faster – they had a greater relative growth rate (RGR) – than later successional species at high light, but surprisingly this was also the case in the shade (Fig. 16.2a and see Chazdon et al. 1996).

RGR can be analyzed in terms of its components mean growth per unit leaf area (unit leaf rate, ULR) and leaf area per unit mass (LAR) (RGR = LAR * ULR). ULR in turn arises from the balance between photosynthesis and respiration and is thus related to leaf physiology. The ULR of pioneers was greater than that of shade tolerant species at high light but not in the shade (Fig. 16.2b). This difference is probably partly due to the fact that pioneer species tend to have higher A_{max} values than later successional species (e.g. Ellsworth and Reich 1996). Differences in A_{max} may arise from differences in leaf nitrogen content (N_{area}), nitrogen being the main component of the photosynthetic system, or from differences in leaf photosynthetic nitrogen-use efficiency (PNUE, $A_{max} = N_{area} * PNUE$). The greater A_{max} of pioneers appears to be associated with a larger PNUE rather than with a larger N_{area} (e.g. Poorter and Bongers 2006; Selaya and Anten 2010).

The LAR is the product of the specific leaf area (SLA) and the leaf mass fraction (LMF) and thus reflects both biomass allocation and leaf structure. At both light conditions pioneers were found to have a larger LMF and especially SLA and thus a larger LAR (Fig. 16.2). The pioneers also exhibited a greater plastic increase in SLA at low light intensity. A large LAR could be beneficial at low light conditions as it contributes to a greater light acquisition per unit plant mass (Hirose and Werger 1995).

The fact that in growth experiments pioneers were found to initially grow faster in the shade than later successionals evidently seems to be counterintuitive. If pioneers are shade intolerant, why then do they grow faster than shade tolerant species at low light? Several explanations have been forwarded for this. The first is that characteristics that facilitate rapid growth may correlate negatively with those that make plants resistant to either physical or biotic damage. The second explanation is related to plant size: rapid growth results rapid size increment and thus greater respiratory costs.

3.1.2 Leaf Turnover

It is generally found that pioneers have considerably lower leaf longevities than later successional species (e.g. Reich et al. 1997; Poorter and Bongers 2006;

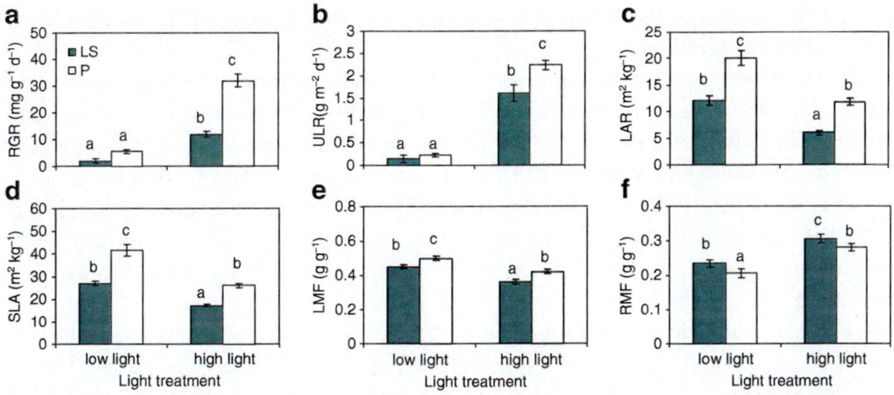

Fig. 16.2 (**a**) Relative growth rate (RGR), (**b**) unit leaf rate (ULR), (**c**) leaf area ratio (LAR), (**d**) specific leaf area (SLA), (**e**) leaf mass fraction (LMF) and (**f**) root mass fraction of pioneers (P) and later successional species (LS) at low (<3% of full sun) and high irradiance (20–50% of full sun). Different letters indicate significant differences between groups (P < 0.05, LSD test) in a two way anova with light and functional group as fixed factors, and bars indicate standard errors of mean (n = 38–164) (Redrawn from Poorter 2005)

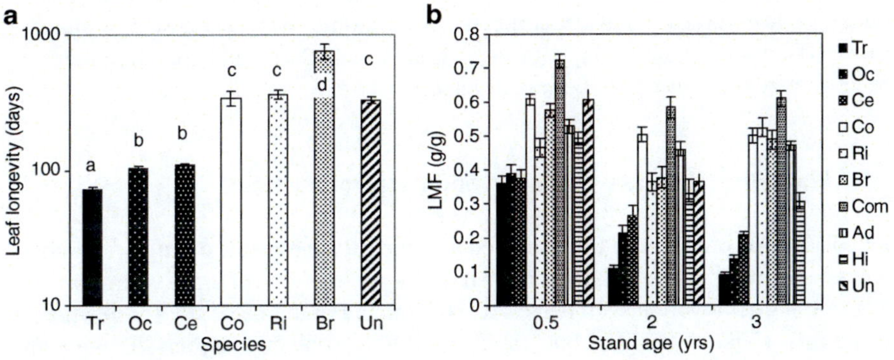

Fig. 16.3 (**a**) Mean leaf longevity and (**b**) mean leaf mass fraction (LMF) of different tree species growing in a young secondary forest stand in the Bolivian Amazon. LMF was measured on trees in three differently aged forest stands. *Different letters* in panel (**a**) denote significant differences between species at $P < 0.05$ based on a one-way ANOVA with species as fixed factors and *bars* denote standard error. Results of a similar ANOVA for LMF are given in Selaya et al. (2008). *Tr* denotes *Trema micrantha*, *Oc Ochroma pyramidale*, *Ce Cecropia ficifolia*, all pioneers. *Co* denotes *Couratari guianensis*, *Ri Rinereocarpus uleii*, *Br Brosimum lactescens*, *Ps Pseudolmedia laevis Ad Adenocalymma impressum.*, *Com Combretum gracilis.* and *Un Uncaria guianensis*, all later successional species (LS). For details of methodology see Selaya et al. (2008)

Selaya et al. 2008, Fig. 16.3). Three reasons can be forwarded to explain this difference. First, thin, highly productive leaves with high protein contents might be more vulnerable to damage than thicker leaves that are less productive. Leaf traits that facilitate high rates of net carbon gain (SLA, leaf N contents, A_{max} and PNUE) tend to be negatively correlated with leaf longevity, these correlations having been demonstrated across thousands of species globally (Reich et al. 1997, 1998; Wright et al. 2004).

Leaf mass per area (LMA; the inverse of SLA) is the product leaf thickness L_T of a large leaf mass density L_D (LMA = L_T*L_D). L_D, in turn, largely reflects the amount cell wall dry mass per unit volume which is a strong determinant of leaf toughness (Read and Stokes 2006). Among 19 Bolivian tropical tree species L_D and toughness but not L_T were positively correlated with leaf life span (Alvarez-Clare and Kitajima 2007; Kitajima and Poorter 2010), suggesting that traits associated with physical strength, probably partly through their effect on herbivory avoidance, are key in facilitating long leaf life span.

Second, as noted above the high photosynthetic capacity of pioneer species tends to be associated with high leaf respiration (Givnish 1988; Lambers et al. 1998). Such leaves therefore have higher light requirements to achieve positive net carbon gain, and are therefore dropped sooner as light levels diminish (Givnish 1988). Third rapid leaf turnover might be associated with the necessity for pioneers to compete for light in the top of the canopy. Leaf area growth tends to be limited by nutrient and particularly nitrogen availability (Lambers et al. 1998). When leaves drop part of the nutrients is resorbed while the rest is lost. The resorbed nutrients may contribute to additional stem elongation and leaf area production in the top of the canopy, and experimental work with transgenic tobacco suggests that plants with a fast leaf turnover have a competitive advantage over those with a slower turnover (Boonman et al. 2006). But the degree, to which this phenomenon plays a role in tropical trees, needs to be investigated.

3.1.3 Plant Size, Biomass Allocation and Respiratory Costs

Poor survival of pioneers in the shade could be partly the result of an unfavorable biomass allocation. Larger plants tend to have larger fractions of biomass in support structures and relatively less in leaves. This is related to the necessity for mechanical stability (Niklas 1992). Thus rapid growth is probably associated with an increased respiratory load.

Variation in plastic growth responses to shade may play an additional role. Plants can respond to the presence of neighbors through increased stem elongation. This response, typically known as the shade avoidance syndrome, appears to be induced by changes in the spectral composition of light that occurs when radiation is transmitted or reflected off plant tissue (Smith 1982). The increase in stem elongation may come at the expense of the allocation to leaves (Anten and Hirose 1998). It has been documented in temperate forests that early successional trees tend to show a stronger shade avoidance response than late successional ones (Gilbert et al 2001). This could explain the fact that while pioneers exhibited a greater mass allocation to leaves in experiments where plants were grown in isolation (Poorter 2005), studies on plants growing in dense natural forest seem to show the opposite result; pioneers having considerably lower leaf mass fractions than later successional species (Selaya et al. 2007, 2008, Fig. 16.3).

3.1.4 Crown Structure

We define crown structure as the overall size and shape of the crown including the geometrical arrangement of branches and leaves. It was proposed that shade tolerant species should produce broad shallow crowns with a minimal leaf overlap thus minimizing self shading and improving light acquisition. Conversely light-demanding species should produce narrow crowns with more layers. This increases the efficiency of height increment, improving their ability to compete for light at the top of the canopy (Horn 1971; King 1990).

Field studies on tropical trees however, do not confirm this prediction. Crowns of shade tolerant tree species tended to be deeper with more layers than those of pioneers (e.g. Kitajima et al. 2005). The production of a broad crown with minimal leaf overlap while maintaining an efficient angular leaf display with respect to the prevailing light conditions may require substantial investment in support (additional branches), which might be prohibitively expensive for trees growing in the shade (Valladares et al. 2002). For light demanding species investing biomass for height growth improves access to light but may entail continuous remobilization of resources from older to younger leaves (see Sect. 3.1.2, Valladares et al. 2002). For these reasons such species should be expected to have shallower crowns.

3.1.5 Wood Mass Density and Tree Allometry

Wood specific gravity (amount of dry matter per unit volume of fresh wood) is believed to play a key role in determining light acquisition strategies of trees (King et al. 2006; Chave et al. 2009; Anten and Schieving 2010; King 2011). Light-demanding tree species tend to have lower wood specific gravities than shade tolerant ones (Putz et al. 1983; Welden et al. 1991; King et al. 2006).

It is generally believed that low wood specific gravity is energetically more efficient – less energy being expended per unit stem volume – but that it also entails a greater mechanical risk: trees with low wood specific gravity being more prone to mechanical failure (e.g. Putz et al. 1983; Muller-Landau 2004). This notion is based on the strong positive correlation that exists between tissue strength and wood specific gravity (e.g. King et al. 2006; Sterck et al. 2006). However, differences in the mechanical safety of trees arise from the combined effects of the length–diameter allometry of stems and branches, wood strength and rigidity, and wood dry- and fresh-mass density, as well as the direction of the forces that plants are exposed to. Tree species with low wood specific gravity can compensate for their lower tissue strength by having a relatively thicker stem, but this entails additional mass investment. Anten and Schieving (2010) showed that trees with low wood specific gravity needed less mass to produce a vertical stem of given length and given mechanical stability (expressed in terms of its ability to carry the weight of the tree or to resist wind forces), contradicting the general notion that these trees should be more prone to mechanical failure (Fig. 16.4). Low wood specific gravity is thus both mechanically and energetically

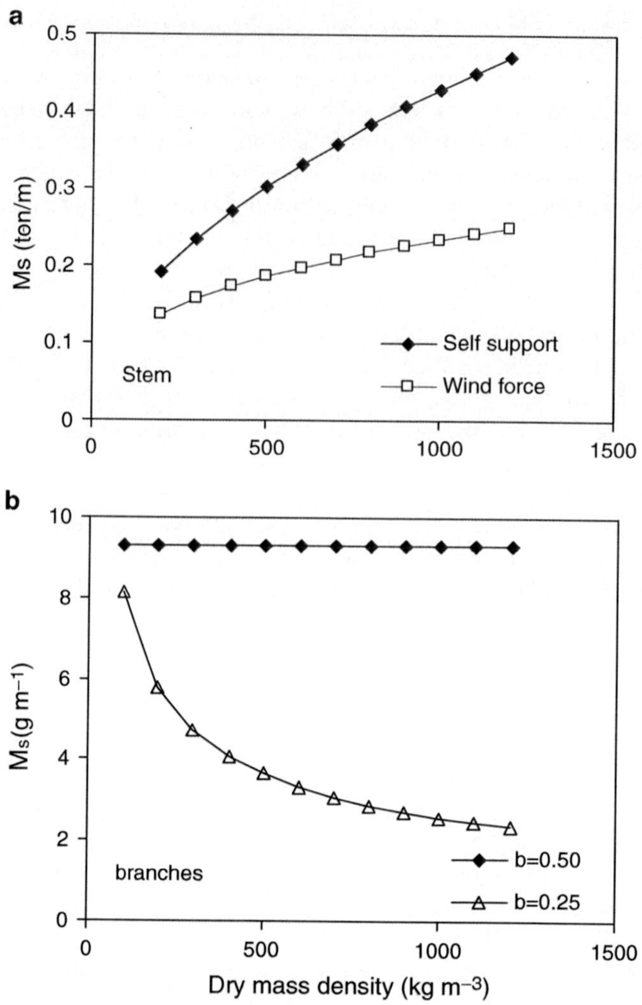

Fig. 16.4 Results of an analytical engineering model simulating the amount of mass needed per unit length of **(a)** stem and **(b)** branch to maintain a given degree mechanical stability (resistance to mechanical failure) plotted as function of the wood dry mass density. The modes of mechanical failure considered are **(a)** the ability of the stem to support the weight of the plant 'self support' and its ability to resist wind forces 'wind force' without breaking and **(b)** the ability of the branch to carry its own weight. In the case of branches this depends critically on whether fresh mass density increases more ($b = 0.5$) or less ($b = 0.25$) steeply with dry mass density. For details see Anten and Schieving (2010)

more efficient in terms of height growth and is probably an important trait enabling light demanding trees to compete for light at the top of the canopy.

What then is the advantage of high density wood in shade tolerant species? First, high wood specific gravity tends to be more efficient in terms biomass expenditure to grow a horizontal branch of given length and resistance against rupture (Sterck et al 2006; Anten and Schieving 2010). To persist in the shade, efficient horizontal crown expansion, reducing self-shading and thus improving light acquisition might be at least as important as efficient height growth (see Sect. 3.1.4). In addition high density wood is probably more resistant to pathogen attacks (Augspurger 1984) as well the occurrence of xylem cavitation (see Sect. 3.3). Finally, wood with high specific gravity is more resistant to crushing under static loads (King et al. 2006). Later successional canopy trees may achieve greater maximum height and weight than early successional ones, and are thus more prone to failure through crushing.

3.2 Differences in Nutrient Availability

As noted above, tropical forests are considered to be deficient in nutrients. Yet, compared to the extensive research on differences in light requirements between pioneers and late successionals, their nutrient requirements have received much less attention.

3.2.1 Growth Experiments

Lawrence (2003) conducted a meta-analysis of experiments wherein seedlings of pioneers and later successional species, grown at high light, were subjected to different nutrient availabilities. Interestingly, the magnitude of the growth stimulation by nutrient addition was considerably greater for the pioneer species than for the shade tolerant ones. This suggests that, pioneers are less tolerant to low nutrient availability than later successional species.

The distinctive responses of pioneers and later successionals to nutrient addition may partly be related to their respective leaf traits. For herbaceous temperate species it was documented that those from nutrient rich habitats exhibit traits that facilitate high productivity (e.g. high SLA, leaf N content, photosynthetic capacity) but low biomass longevity while those from nutrient poor habitats exhibited traits with the opposite effect (Aerts 1996), a distinction that is similar to that between pioneers and later successional species in tropical rain forest (see above).

3.2.2 Root Characteristics

Adaptations to differences in nutrient availability are probably largely mediated by the size and characteristics of the root system. Plants tend to allocate biomass to resources to those organs that capture the most limiting resources; fractional allocation to roots increases with water and nutrient limitations, and decreases in the shade. One could thus argue that adaptations to shade may conflict with adaptations to low nutrient availability (Smith and Huston 1989). Contrary to this assumption, however, seedlings of late successional species have been found to allocate larger fractions of their mass to roots (Paz 2003; Coll et al. 2008).

Early successional species tend to produce thinner roots with a greater length per unit mass (Specific Root Length, SRL) than later successional ones (Paz 2003; Coll et al. 2008). It could be argued that these results may reflect a tradeoff between the shade tolerance of plants and their ability to compete for nutrients. Thin roots tend to translate into a relatively large surface and associated ability to absorb nutrients efficiently (Lambers et al. 1998), which could be important for nutrient competition and rapid growth at high light. Under shaded conditions survival of plants may depend critically on their ability to store carbohydrates belowground as well as on the resistance of the root system to physical damage (Kitajima et al 1996; Kobe 1997). Some tradeoffs between these two functions may exist if an increased storage capacity or mechanical strength of roots entails a smaller area to mass volume.

Differences in root characteristics between tropical trees or seedlings of different successional status might also be associated with differences in their leaf characteristics. As noted, pioneers tend to have leaves with a higher mass-based photosynthetic capacity (A_{mass}), a greater photosynthetic nitrogen-use efficiency (PNUE) and a higher mass-based nitrogen content (N_{mass}). Hikosaka and Osone (2009) developed a model demonstrating that this relationship can only exist if pioneer species also have larger root-specific N absorption rates (SAR). Why is this? A greater A_{mass} and PNUE will result in more C uptake relative to N uptake and all else being equal should result in a lower N_{mass}. However, a positive A_{mass}–N_{mass} relationship may still be possible if high A_{mass} plants also allocate more N to leaves, less biomass to leaves or have roots with a greater SAR. Out of these three traits only SAR seems to be positively correlated with A_{mass} at least among herbaceous species and temperate tree species (Garnier 1991; Hikosaka and Osone 2009). While SAR values have not been compared among tropical tree species, the greater SRL of pioneer species compared to later successionals are likely associated with similarly larger SAR (Reich et al. 1998).

3.3 *Water Availability*

As noted in Sect. 1, tropical moist forests can experience an annual dry season of up to 3 months in addition to periodic droughts, such as those associated with the El Nino Southern Oscillation, and such dry spells can have a strong effect on the population dynamics of tropical rain forest species (Slik 2004).

Given the reductions in light availability and increase in air humidity during succession, drought stress, if it occurs, is probably more severe in earlier than later successional forest. In addition because of the tradeoff in shoot versus root allocation, shade tolerance should come at the expense of drought tolerance (Smith and Huston 1989). This could lead one to argue that pioneer species should be more drought tolerant than later successionals. But empirical studies that have measured mortality rates during drought events have found results that oppose this notion; increases in mortality rates during droughts tend to be larger among pioneers than among later successionals (Kochummen and Ng 1977; Swaine and Hall 1983; but see Condit et al. 2005).

Plant adaptations to dry conditions can roughly be categorized into three categories: (1) maintaining water uptake; (2) maintaining water transport from roots to leaves and (3) reducing transpiration losses. Below we discuss for each category possible differences between tropical trees of different successional status.

3.3.1 Maintaining Water Uptake

During dry spells the upper soil layers dry out faster than deeper layers (Lambers et al. 1998). A large size and depth of the rooting system may therefore enable plants to acquire water under dry conditions. As noted above, later-successional species have been found to maintain larger root mass fractions but their roots are also thicker with a smaller SRL. Owing to their limited size and thus smaller rooting systems, seedlings are particularly vulnerable to losing access to water during dry conditions. Seedlings of later successional species tend to have larger cotyledons, thus more carbohydrate reserves, and allocate more to root growth both of which would favor their survival (Bazzaz and Pickett 1980; Kitajima 2003). On the other hand, rapid growth in the wet season may enable seedlings of pioneers to produce a sufficiently large rooting system to survive the dry season. The relative importance of these factors probably depends on the timing of seed germination relative to occurrence of a dry spell.

3.3.2 Maintaining Water Transport from Roots to Leaves

Water transport from roots to leaves is possible as long as xylem vessels maintain their function and as long as leaf water potential is more negative than that of the soil (Lambers et al. 1998). Regarding the former, xylem embolism and the associated loss of hydraulic conductivity is probably one of the most important problems that trees deal with under dry conditions.

Among dry forest species, pioneers were found to be more prone to embolism than later successional species, losing their hydraulic conductivity at considerably less negative xylem water potentials (Markesteijn 2010). This difference was probably associated with the larger wood dry mass density of later successional species. A high wood mass density can result from narrow vessels (Hacke et al. 2001), though this correlation is not always evident (Chave et al. 2009). Narrow, thick-walled vessels tend to have smaller pit membrane pores (Hacke et al. 2001), which is associated with a greater resistance to embolism (Lambers et al. 1998). The drawback of narrow vessels is their larger hydraulic resistance. Indeed later successionals in tropical dry forest were reported to have stems with a lower hydraulic conductivity than pioneers (Markesteijn 2010). It remains to be seen however whether a similar pattern of hydraulic wood properties with successional status of trees exists in tropical moist forest.

Water transport through the plant is strongly related to tree size and age (Tyree and Sperry 1988). The negative effect of gravity on water potential (Ψ) increases with tree height (Zimmermann 1983). This gravitational effect interacts with hydraulic path-length resistance (hydrodynamic tension) to further lower Ψ during transpiration. The extent to which hydraulics thus limit tree height is currently debated (Becker et al. 2000; Ryan et al. 2006). Trees can mitigate hydrodynamic effects (though not the gravitational effects) of height on Ψ in various ways including: an increase in the sapwood area to leaf area ratio, reduction in the leaf osmotic potential, increased root surface area, greater capacity for water storage and/or stomatal closure (see Ewers et al. 2005), all of which however have carbon costs.

As later successional canopy species tend to have greater maximum height than earlier ones, negative height effects on Ψ and their potential mitigation, may play a role in determining patterns of species succession and ecosystem functions during secondary forest succession. Yet most research on this topic has been in the form of single species studies.

3.3.3 Reducing Transpiration

The most direct way of reducing transpiration is deciduousness, but is not considered in this chapter as we limit ourselves to evergreen forests. Other ways of reducing water losses are a tight stomatal control and, possibly in association with this, a high photosynthetic water-use efficiency (WUE). The latter entails that a given rate of photosynthetic carbon gain can be achieved with lower rate of transpiration (Lambers et al. 1998). Poorter and Bongers (2006) found that WUEs did not differ significantly between trees of different successional status. Pioneers exhibited higher photosynthetic rates (see earlier sections) but also higher rates of stomatal conductance and transpiration. This larger demand on water transport could partly explain why they have wood with a larger hydraulic conductivity (see previous section).

Overall it appears that pioneers species are less drought tolerant than later successionals. This suggests the periodic dry spells may play a role in shaping succession by causing relatively large increases in mortality among pioneers and by changing light availability on the forest floor (Slik 2004).

4 Scaling up from Individual Traits to the Whole-Plant Level

The studies discussed so far have focused on individual traits and correlation between traits. This may provide valuable hypotheses about the role of differences in plant traits in determining successional dynamics. However, it does not provide any quantitative indication of how different trait correlations determine whole-plant performance under natural conditions at different stages of succession. Two approaches have been taken to address this issue: (1) empirically comparing species trait values to their estimated whole-plant performance measures (growth and survival) under field conditions, and (2) by using physiological models that quantitatively relate individual traits to whole-plant performance. In this section we discuss how such studies provide insights into relationship between the interspecific trait differences (discussed in the previous section) between trees of different successional status, and their competitive ability, photosynthesis, growth and survival at different stage of succession.

4.1 Empirical Studies Relating Traits to Whole-Plant Performance

Several studies have evaluated whole-plant growth of tropical tree species in relation to their physiological and morphological traits but most of them were conducted under controlled conditions (see reviews by Veneklaas and Poorter 1998 and Poorter 2005). Their results, though valuable, cannot be directly extrapolated to plants growing under natural forest conditions where plants are exposed to different and more fluctuating conditions as well as the presence of a larger number of pathogens and herbivores (Kitajima et al. 1996). More recent work is starting to focus on plant performance measures determined in the field. Comparing the height growth rate of 35 co-existing rain forest species, Coley (1988) found that species with fast leaf turnover rates have rapid growth height growth in gaps. Reich et al (1992) reported that height growth in gaps was particularly correlated with photosynthetic capacity but their study entailed plants from different biomes.

More recently, Poorter and Bongers (2006) compared leaf traits, height growth in gaps and low light survival rates across 53 Bolivian tropical tree species. Overall leaf traits were good predictors of plant performance. Fast growth in high light was associated with a high SLA, photosynthetic capacity and stomatal conductance and negatively with leaf longevity. Crown light exposure of juvenile plants (CE_{juv}), a proxy for the light requirement for regeneration, was similarly correlated with these traits and also positively correlated with leaf respiration. Conversely survival in the shade was positively correlated with leaf lifespan and negatively with SLA, A_{max} and respiration. Kitajima and Poorter (2010) analyzed traits underlying SLA – leaf thickness, leaf mass density and toughness – in relation to herbivory, growth in gaps and sapling survival in the shade. They found that leaf mass density, and toughness but not leaf thickness were positively correlated with survival and negatively correlated with herbivory and growth at high light, suggesting that physical strength of leaves plays an important role in determining shade tolerance (Table 16.2).

Baltzer and Thomas (2007) took a slightly different approach and determined the whole-plant light compensation point for growth and survival for 20 species in a

Table 16.2 Pair-wise correlations of lamina size and mechanical traits with ecological performance measures of saplings of 19 tree species from a Bolivian tropical forest (Redrawn from Kitajima and Poorter 2010)

	Palatability index		Leaf life span		Survival rate		Growth rate		CE_{juv}	
Trait	Coefficient	Sig	Coefficient	Sig	Coefficient	Sig	Coefficient	Sig	Coefficient	Sig
LMA	**–0.83**	***	**0.81**	***	**0.79**	***	**–0.72**	***	**–0.72**	***
Lamina thickness	–0.36	NS	0.30	NS	0.19	NS	–0.04	NS	–0.12	NS
Lamina density	**–0.61**	**	**0.62**	**	**0.57**	*	**–0.68**	***	**–0.60**	***
Lamina fracture toughness	**–0.65**	**	**0.82**	***	**0.69**	***	**–0.73**	***	**–0.76**	***
Lamina work-to-shear	**–0.83**	***	**0.87**	***	**0.77**	***	**–0.53**	*	**–0.67**	***

LMA leaf mass per unit area (LMA = 1/SLA), palatability index indicates the vulnerability to herbivory, CE_{juv} the mean light exposure indicating the light requirement for regeneration. For all variables Pearson correlation coefficients are shown except survival where Spearman rank correlation was shown. Significant correlation are shown in bold and Sig indicates significance levels

NS not significant ($P > 0.05$)

*$P < 0.05$; **$P < 0.01$; ***$P < 0.001$

Malaysian rain forest and related these whole-plant values to leaf traits. Contrary to the pot experiments discussed above, they found a cross-over in the relationship between growth and light availability, with shade tolerant species exhibiting more rapid growth at low light. Interestingly, they also found that out of four leaf traits (A_{max}, leaf N content, leaf longevity and respiration) only respiration was significantly correlated with the whole-plant light compensation point for growth. This work seems to suggest that shade tolerance of later-successional species is associated more with low respiration and a low light compensation point for growth than with minimizing damage and maximizing survival.

4.2 *Physiological Growth Models*

Physiological growth models entail a second means of scaling from individual traits to whole-plant performance. In such models, characteristics of plants measured in the field are mechanistically linked to plant photosynthesis, carbon balance and growth and by comparing model prediction to field observation specific hypotheses about tree functioning can be tested. Relatively few studies however have applied physiological models to tropical forest succession.

During early secondary succession in tropical forests, vegetation density, height and LAI rapidly increase (see above), and light competition therefore most likely plays an important role in determining the pattern of species replacement that occurs during this phase of succession. Whole-plant photosynthesis (P_{mass}) or growth (RGR) per unit mass in relation to light acquisition can be factorized as follows;

$$P_{plant} \text{ or } RGR = \Phi_{mass} * LUE \qquad (16.1)$$

where Φ_{mass} is the whole-plant light capture per unit mass and LUE (light-use efficiency) either photosynthesis or growth per unit captured light (Hikosaka et al. 1999). Φ_{mass} in turn is the product of light capture per unit leaf area (Φ_{area}) and leaf area per unit mass (LAR; $\Phi_{mass} = \Phi_{area}$ *LAR) (Hirose and Werger 1995). LAR reflects allocation traits (SLA and LMF), Φ_{area} traits associated with position of leaves in the gradient in the vegetation (crown structure and tree stem allometry) and LUE physiological traits (photosynthesis and respiration) (Hikosaka et al. 1999).

Selaya et al. (2007, 2008) applied this approach to young secondary forest stands in the Bolivian Amazon, ranging between 0.5 and 3 years in age. They found that during the first 4 years of succession pioneers grew disproportionally faster than later successionals, rapidly dominating the top layers of the canopy. Calculations with a canopy photosynthesis model showed that they also had higher whole-plant photosynthetic rates; associated with both a higher Φ_{area} and LUE but with a lower LMF and associated LAR than that of the later successionals. This suggested that the early competitive advantage of pioneers was associated with physiological traits (particularly a higher A_{max} values) as well as with a more favorable conversion of

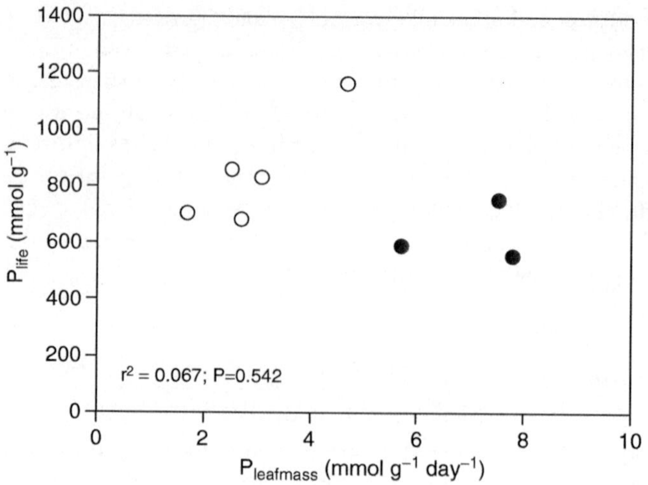

Fig. 16.5 The life time carbon gain per unit leaf mass (P_{life}) as a function of the daily whole-plant photosynthesis divided by the leaf mass of that plant ($P_{leafmass}$). Individual points indicate mean values per species in a 0.5 year old secondary forest stand. *Closed circles* are pioneers and *open circles* later successional species. For details see Selaya and Anten (2010)

biomass to height, probably due to their considerably lower wood specific gravity. As noted in the section on individual traits the lower LMF of pioneers was either a direct consequence of plant size – taller plants must invest disproportionately more mass in support structures to maintain mechanical stability – and/or a consequence of their shorter leaf longevity.

Using the same data set and model approach, Selaya and Anten (2010) calculated the life time net photosynthesis per unit leaf mass, by integrating daily photosynthesis per unit leaf mass ($P_{leafmass} = P_{plant}$/whole-plant leaf mass) over leaf life span taking account of the change in the position in the light gradient that plants may experience as succession proceeds. This life-time photosynthesis was not very different between pioneers and late successional species (Fig. 16.5). This indicated that despite the large differences in traits that exist between pioneers and later successional species both groups produced leaves with about the same returns on investment. This probably partly explains how later successional species are able to coexist with the pioneers during the early stages of succession. It also implicates leaf longevity as a key trait involved in the ability of later successional species to persist early in succession.

But photosynthetic carbon gain is not equivalent to growth. Sterck et al. (2005) developed a complex growth model that integrates the spatial production of metamers (usually an internode with one or more leaves, and several meristems) with leaf physiological traits and leaf turnover to calculate the growth and three-dimensional structures of trees. In a set of simulations they compared trees that differed in correlated leaf traits; comparable to the pioneer vs. late successional dichotomy discussed earlier. Trees had leaves with high photosynthetic capacity (A_{mass}), respiration

(r_d) and SLA but low leaf longevity ('short leaf-life trees' similar to pioneer species), intermediate trait values ('intermediate leaf-life trees') or low A_{mass}, r_d and SLA but long leaf life span ('long leaf-life trees' comparable to late-successional species). Simulations were run for trees growing in open conditions, large and small gaps, and under a closed forest canopy.

In both the small gap and understory the 'short leaf-life' tree readily stopped growing and died. The 'intermediate leaf-life' tree performed somewhat better; it was able to grow in small gaps but under the forest canopy it persisted for only 2 years. The long leaf-life tree on the other hand was able to grow in all sites. This result can be attributed to the fact that large carbon costs due to either high respiratory rates or fast leaf turnover are disadvantageous under shaded conditions.

In the larger gap and the open site the 'short leaf-life' tree initially grew faster than the other two types but it stopped increasing in height after about 6 years. At this stage it also produced fewer leaves than it lost, resulting in a reduction in standing leaf area. Both the 'intermediate' and 'long leaf-life' trees on the other hand were able to continue growing with the former achieving a greater maximum tree height than the short leaf-life tree. The mechanism underlying these results can be explained as follows. A high photosynthetic capacity and SLA facilitates rapid growth. But as plants grow bigger, their photosynthetic carbon gain levels off due to self shading while respiration continues to increase. Thus the net assimilation declines and a point can be reached where this not enough to compensate for biomass losses. This point is reached much earlier in tree species with highly productive but short-lived leaves than in trees that exhibit the opposite trend. This in turn may partly explain why pioneers are initially able to gain dominance but can only maintain this for a limited number of years. As they start losing leaf area and ultimately start dying, more light penetrates to lower vegetation layers providing opportunities for later successional species. These in turn, owing to their less productive but more persistent leaf traits are able to reach a greater size, have larger leaf areas and maintain a dominant position for much longer.

5 Conclusions

Changes in light availability are probably the most important change in environmental conditions that occurs during secondary succession in tropical rain forest. But other factors change as well: the air becomes cooler and more humid, while soil nutrient and water availabilities tend to decrease, but more data are needed to quantify these trends.

Pioneers exhibit traits that facilitate rapid growth and efficient acquisition of resources enabling them to compete for light but also for belowground resources early in succession. This ability to compete partly results from traits (e.g. a strong shade avoidance response and rapid leaf turnover) that maximize the ability to compete with neighbor plants. Through a number of direct and indirect tradeoffs the characteristics of pioneers negatively correlate with durability and long-term efficiency. For example' thin, highly productive leaves are too flimsy to live long. Thin roots with a high absorbing area to mass ratio are not very resistant to mechanical stress nor do they provide sufficient capacity for storage of carbohydrates. Wood anatomy that facilitates mechanically efficient height increment and high rates of water transport may also be vulnerable to disease, impact damage or embolism.

Physiological models provide a strong tool to quantitatively relate these traits and their correlations to whole-plant performance. They have shown that later successional species produce leaves with about the same long-term returns on investment as pioneers, even though they initially grow in the shade. The inherent trade-off between leaf productivity and longevity and its interaction with plant size may also be an important mechanism explaining why pioneers are unable to survive in spite of their faster growth and why they can initially win the competition but are unable to maintain this advantage indefinitely.

Acknowledgements We thank Robin Chazdon and an anonymous reviewer for constructive comments on the manuscript. Part of this work was funded by Programa Manejo de Bosques de la Amazonía Boliviana (PROMAB).

References

Aerts R (1996) Nutrient resorption from senescing leaves of perennials: are there general patterns. J Ecol 84:597–608

Allen R, Lindesay J, Parker DE (1996) El Niño southern oscillation and climate variability. Aubrey Books, Silver Springs

Alvarez-Clare S, Kitajima K (2007) Physical defence traits enhance seedling survival of neotropical tree species. Funct Ecol 21:1044–1054

Anten NPR, Hirose T (1998) Biomass allocation and light partitioning among dominant and subordinate individuals in *Xanthium canadense* stands. Ann Bot 82:665–673

Anten NPR, Schieving F (2010) The role of wood mass density and mechanical constraints on the economy of tree architecture. Am Nat 175:250–260

Augspurger CK (1984) Seedling survival of tropical tree species: interaction of dispersal distance, light gaps and pathogens. Ecology 65:1705–1712

Baltzer JL, Thomas SC (2007) Determinants of whole-plant light requirements in Bornean rain forest tree saplings. J Ecol 95:1208–1221

Bazzaz FA, Pickett STA (1980) Physiological ecology of tropical succession: a comparative review. Annu Rev Ecol Syst 11:287–310

Becker P, Rabenold PE, Idol JR, Smith AP (1988) Water potential gradients for gaps and slops in a Panamanian tropical moist forest's dry season. Biotropica 31:93–101

Becker P, Meinzer FC, Wullschleger SD (2000) Hydraulic limitation of tree height: a critique. Funct Ecol 14:4–11

Bongers F, Popma J (1990) Leaf dynamics of seedlings of rain forest species in relation to canopy gaps. Oecologia 82:122–127

Boonman A, Anten NPR, Dueck TA, Jordi WJRM, van der Werf A, Voesenek JACJ, Pons TL (2006) Functional significance of shade-induced leaf senescence in dense canopies: an experimental test using transgenic tobacco. Am Nat 168:597–607

Brown S, Lugo AE (1990) Tropical secondary forests. J Trop Ecol 6:1–32

Buschbacker R, Uhl C, Serrao EAS (1988) Abandoned pastures in Eastern Amazonia. II. Nutrient stocks in the soil and vegetation. J Ecol 76:682–699

Chave J, Coomes D, Jansen S, Lewis SL, Swenson NG, Zanne AE (2009) Towards a worldwide wood economics spectrum. Ecol Lett 12:351–366

Chazdon RL (1986) Leaf display, canopy structure and light interception of two understory palm species. Am J Bot 72:1493–1502

Chazdon RL (2009) Chance and determinism in tropical forest succession. In: Carson WP, Schnitzer SA (eds) Tropical forest community ecology. Wiley Blackwell, Oxford, pp 384–408

Chazdon RL, Pearcy RW, Lee DW, Fetcher N (1996) Photosynthetic responses of tropical forest plants to contrasting light environments. In: Mulkey SS, Chazdon RL, Smith AP (eds) Tropical forest plant ecophysiology. Chapman & Hall, New York, pp 5–55

Coley PD (1988) Effects of plant growth rate and leaf life-time on the amount and type of anti-herbivore defense. Oecologia 74:531–536

Coll L, Potvin C, Messier C, Delagrange S (2008) Root architecture and allocation patterns of eight native tropical species with different successional status used in open-grown mixed plantations in Panama. Trees 22:585–596

Condit R, Hubbell SP, Foster RB (2005) Neotropical tree and shrub species and the impact of a severe drought. Ecol Monogr 65:419–439

Dos Santos JM (1987) Climate, natural vegetation, and soils in Amazonia: an overview. In: Robert ED (ed) The geophysiology of Amazonia: vegetation and climate interactions. Wiley, New York, pp 25–36

Ellsworth DS, Reich PB (1996) Photosynthesis and leaf nitrogen in five Amazonian tree species during early secondary succession. Ecology 77:581–594

Ewers BE, Gower ST, Bond-Lamberty B, Wang CK (2005) Effects of stand age and tree species on canopy transpiration and average stomatal conductance of boreal forests. Plant Cell Environ 28:660–678

Feldpausch TR, Rondon MA, Fernandez ECM, Riha SJ, Wandelli E (2004) Carbon and nutrient accumulation in secondary forests generating on pastures in central Amazonia. Ecol Appl 14:S164–S176

Ferguson BG, Griffith DM (2004) Tecnologica agricola y conservacion biologica en El Peten, Guatemala. Manejo Integrado Plagas Agroecol 72:72–85

Fetcher N, Oberhauer SF, Strain BR (1985) Vegetation effects on microclimate in lowland tropical forest in Costa Rica. Int J Biometeor 29:145–155

Finegan B (1996) Pattern and process in neotropical secondary rainforest: the first 100 years of succession. Trends Ecol Evol 11:119–124

Garnier E (1991) Resource capture, biomass allocation and growth in herbaceous plants. Trends Ecol Evol 6:126–131

Gilbert IR, Jarvis PG, Smith H (2001) Proximity signal and shade avoidance differences between early and late successional trees. Nature 411:792–795

Givnish TJ (1988) Adaptations to sun and shade: a whole plant perspective. Aust J Plant Physiol 15:63–92

Gray AN, Spies TA, Easter MJ (2002) Microclimatic and soil moisture responses to gap formation in coniferous forests. Can J For Res 32:332–343

Guariguata MR, Ostertag R (2001) Neotropical secondary forest succession: changes in structural and functional characteristics. For Ecol Manage 148:185–206

Hacke UG, Sperry JS, Pockman WT, Davis SD, McCulloh KA (2001) Trends in wood density and structure are linked to prevention of xylem implosion by negative pressure. Oecologia 126: 457–461

Hikosaka K, Osone Y (2009) A paradox of leaf trait convergence: why is leaf nitrogen content higher in species with higher photosynthetic capacity? J Plant Res 122:245–251

Hikosaka K, Sudoh S, Hirose T (1999) Light acquisition and use of individuals competing in a dense stand of an annual herb *Xanthium canadense*. Oecologia 118:388–396

Hirose T, Werger MJA (1995) Canopy structure and photon flux partitioning among species in a herbaceous plant community. Ecology 76:466–474

Horn HS (1971) The adaptive geometry of trees. Princeton University Press, Princeton

Houghton RA (1994) The worldwide extent of land-use change. Bioscience 44:305–313

Hughes RF, Kauffman JB, Jaramillo VJ (1999) Biomass, carbon and nutrient dynamics of secondary forests in a humid tropical region of Mexico. Ecology 80:1892–1907
Huston M, Smith T (1987) Plant succession: life history and competition. Am Nat 130:168–198
Johnson CM, Vieira ICG, Zarin DJ, Frizano J, Johnson AH (2001) Carbon and nutrient storage in primary and secondary forests in eastern Amazonia. For Ecol Manage 147:245–252
King DA (1990) Allometry of saplings in understorey trees of a Panamaniam forest. Funct Ecol 4:27–32
King DA (2011) Size-related changes in tree proportions and their potential influence on the course of height growth. In: Meinzer FC, Lachenbruch B, Dawson T (eds) Size and age related changes in tree structure and function. Springer, Dordrecht
King DA, Davies SJ, Tan S, Noor NS (2006) The role of wood density and stem support costs in the growth and mortality of tropical trees. J Ecol 94:670–680
Kitajima K (2003) Impact of cotyledon and leaf removal on seedling survival in three treespecies with contrasting cotyledon functions. Biotropica 35:429–434
Kitajima K, Poorter L (2010) Tissue-level leaf toughness, but not lamina thickness, predicts sapling leaf lifespan and shade tolerance of tropical tree species. New Phytol 186:708–721
Kitajima K, Mulkey SS, Wright J (1996) Ecophysiology off tropical tree seedlings. In: Mulkey SS, Chazdon RL, Smith AP (eds) Tropical forest plant ecophysiology. Chapman & Hall, New York
Kitajima K, Mulkey SS, Wright SJ (2005) Variation in crown light utilization characteristics among tropical canopy trees. Ann Bot 95:535–547
Kobe RK (1997) Carbohydrate allocation to storage as a basis of interspecific variation in sapling survivorship and growth. Oikos 80:226–233
Kochummen KM, Ng FSP (1977) Natural plant succession after farming in Kepong. Mal For 40:61–78
Lambers H, Chapin FS III, Pons TL (1998) Plant physiological ecology. Blackwell, Malden
Lavelle P (1987) Biological processes and productivity of soils in the humid tropics. In: Robert ED (ed) The geophysiology of Amazonia: vegetation and climate interactions. Wiley, New York, pp 175–222
Lawrence D (2003) The response of tropical tree seedlings to nutrient supply: meta-analysis for understanding a changing tropical landscape. J Trop Ecol 19:1–12
Lawrence D, Suma V, Mochea JP (2005) Changes in species composition with repeated shifting cultivation: limited role of soil nutrients. Ecol Appl 15:1952–1967
Lebrija-Trejos E (2008) Tropical dry forest recovery processes and causes of change. Ph.D. thesis, Wageningen University
Lu D, Moran E, Mausel P (2002) Linking Amazonian secondary forest growth to soil properties. Land Degrad Dev 13:331–343
Marin-Spiotta E, Ostertag R, Silver WL (2007) Long-term patterns in tropical reforestation: plant community composition and aboveground biomass accumulation. Ecol Appl 173:828–839
Markesteijn L (2010) Drought tolerance of tropical tree species. Ph.D. thesis, Wageningen University
Montgomery RA, Chazdon RL (2001) Forest structure, canopy architecture, and light transmittance in tropical wet forests. Ecology 82:2207–2718
Muller-Landau HC (2004) Interspecific and inter-site variation in wood specific gravity of tropical trees. Biotropica 36:20–32
Nicotra AB, Chazdon RL, Iriarte S (1999) Spatial heterogeneity of light and woody seedling regeneration in tropical wet forests. Ecology 80:1908–1926
Niinemets Ü, Anten NPR (2009) Packing the photosynthetic machinery: from leaf to canopy. In: Laisk A, Nedbal L, Govindjee (eds) Photosynthesis in silico: understanding complexity from molecules to ecosystems. Springer, Dordrecht, The Netherlands, pp 363–399
Niklas KJ (1992) Plant biomechanics: an engineering approach to plant form and function. University of Chicago Press, Chicago
Ostertag R (1988) Belowground effects of canopy gaps in a tropical wet forest. Ecology 79:1294–1304
Paz H (2003) Root/shoot allocation and root architecture in seedlings: variation among forest sites, microhabitats and ecological groups. Biotropica 35:318–332

Peña-Claros M (2003) Changes in forest structure and species composition during secondary succession in the Bolivian Amazon. Biotropica 35:450–461

Poorter L (2005) Resource capture and use by tropical forest tree seedlings and their consequences for competition. In: Burslem DF, Pinard MA, Hartley SE (eds) Biotic implication in the tropics: their role in maintenance of species diversity. Cambridge University Press, Cambridge, pp 35–64

Poorter L, Bongers F (2006) Leaf traits are good predictors of plant performance across 53 rain forest species. Ecology 87:1733–1743

Putz FE, Coley PD, Lu K, Montalvo A, Aiello A (1983) Uprooting and snapping of trees: structural determinants and ecological consequences. Can J For Res 13:1011–1020

Read J, Stokes A (2006) Plant biomechanics in an ecological context. Am J Bot 93:1546–1565

Reich PB, Walters MB, Ellsworth DS (1992) Leaf lifespan in relation to leaf, plant, and stand characteristics among diverse ecosystems. Ecol Monogr 62:365–392

Reich PB, Walters MB, Ellsworth DS (1997) From tropics to tundra: global convergence in plant functioning. Proc Natl Acad Sci U S A 94:13730–13734

Reich PB, Walters MB, Tjoelker MG, Vanderklein D, Buschena C (1998) Photosynthesis and respiration rates depend on leaf and root morphology and nitrogen concentration in nine boreal tree species differing in relative growth rate. Funct Ecol 12:395–405

Ryan MG, Phillips N, Bond BJ (2006) The hydraulic limitation hypothesis revisited. Plant Cell Environ 29:367–381

Santiago LS, Wright SJ (2007) Leaf functional traits of tropical forest plants in relation to growth form. Funct Ecol 21:19–27

Selaya NG, Anten NPR (2010) Leaves of pioneer and late successional species have similar life time carbon gain in tropical secondary forest. Ecology 91:1102–1113

Selaya NG, Anten NPR, Oomen RJ, Matthies M, Werger MJA (2007) Aboveground mass allocation and light interception of tropical forest trees and lianas early in succession. Ann Bot 99:121–130

Selaya NG, Oomen RJ, Netten J, Werger MJA, Anten NPR (2008) Biomass allocation and leaf life span in relation to light interception by tropical forest plants during the first years of secondary succession. J Ecol 96:1211–1221

Slik JWF (2004) El Nino droughts and their effects on tree species composition and diversity in tropical rain forests. Oecologia 141:114–120

Smith H (1982) Light quality photoreception and plant strategy. Annu Rev Plant Physiol 33:481–518

Smith T, Huston M (1989) A theory of spatial and temporal dynamics of plant communities. Plant Ecol 83:49–69

Sterck FJ, Schieving F, Lemmens A, Pons TL (2005) Performance of trees in forest canopies: exploration with a bottom-up functional-structural plant growth model. New Phytol 166:827–843

Sterck FJ, van Gelder HA, Poorter L (2006) Mechanical branch constraints contribute to life-history variation across tree species in a Bolivian forest. J Ecol 94:1192–1200

Swaine MD, Hall JB (1983) Early succession on cleared forest land in Ghana. J Ecol 71:601–627

Terborgh J (1992) Diversity and the tropical rainforest. Scientific American Library, New York

Turner IM (2001) The ecology of trees in the tropical rain forest. Cambridge University Press, Cambridge

Tyree MT, Sperry JS (1988) Do woody-plants operate near the point of catastrophic xylem dysfunction caused by dynamic water-stress – answers from a model. Plant Physiol 88:574–580

Valladares F, Skillman JB, Pearcy RW (2002) Convergence in light capture efficiencies among tropical forest understory plants with contrasting crown architectures: A case of morphological compensation. Am J Bot 89:1275–1284

Veneklaas EJ, Poorter L (1998) Growth and carbon partitioning of tropical tree seedlings in contrasting light environments. In: Lambers H, Poorter H, van Vuuren MMI (eds) Inherent variation in plant growth, physiological mechanisms and ecological consequences. Backhuys, Leiden, pp 337–361

Welden CW, Hewett SW, Hubbell SP, Foster RB (1991) Sapling survival, growth, and recruitment: relationships to canopy height in Neotropical forest. Ecology 72:35–50
Whitmore TC (1996) A review of some aspects of tropical rain forest seedling ecology with suggestions for further enquiry. In: Swaine MD (ed) The ecology of tropical forest tree seedlings. Parthenon Publishing Group, Paris, pp 3–39
Whittaker RH (1975) Communities and ecosystems. Macmillan, New York
Wright IJ et al (2004) The worldwide leaf economics spectrum. Nature 428:821–827
Zimmermann MH (1983) Xylem structure and the ascent of sap. Springer, Berlin

Chapter 17
Age-Dependent Changes in Environmental Influences on Tree Growth and Their Implications for Forest Responses to Climate Change

Steven L. Voelker

Abstract Dendrochronology has long used tree-rings to better understand climate-tree-growth relationships within a site or across a region, with tree age often considered to contribute unwanted noise to the signal in question. Here I demonstrate that strong climate-tree-growth relationships exist across sites, on a global scale, and that these changes are correlated with variation in average net primary productivity. The age-specific tree-ring data analyses used here show that the sensitivity of tree growth to environmental variability changes predictably with tree age. Young trees were found to be particularly sensitive to each of the environmental factors investigated. These results are discussed in the context of climate change and established changes in tree morphology and physiological function with tree age or size. I argue that explicitly treating tree or forest age can yield tangible improvements in the projection of terrestrial carbon sink responses to climate change by increasing the accuracy with which forest to non-forest ecosystem boundaries can be projected.

1 Introduction

Understanding the processes that govern terrestrial ecosystem productivity is critical to predicting future concentrations of carbon dioxide in the atmosphere (Canadell et al. 2007). Forests make up a large portion of terrestrial productivity and their contribution to carbon stored in live trees, dead wood and soils keeps increasing long after the age in which maximum tree height is approached (Luyssaert et al. 2008). Since 1850, the total amount of carbon released by cutting and burning of

S.L. Voelker (✉)
Department of Wood Science and Engineering, Oregon State University,
Corvallis, OR 97331, USA
e-mail: steve.voelker@oregonstate.edu

F.C. Meinzer et al. (eds.), *Size- and Age-Related Changes in Tree Structure and Function*, Tree Physiology 4, DOI 10.1007/978-94-007-1242-3_17,

forests is approximately one third of all carbon emissions (Houghton 1999). Indeed, many forests of the northern hemisphere are still recovering from this period of exploitative logging and land use practices, resulting in the northern hemisphere being identified as a major carbon sink over the past two decades (Keeling et al. 1996; Schimel et al. 2001; Nemani et al. 2003). Predictions by each of ten coupled carbon and climate change models suggest that northern forests will continue to be a sink (Friedlingstein et al. 2006; Qian et al. 2009). However, data from the most recent decade suggest that droughts, largely in the southern hemisphere, have reduced global net primary productivity (NPP) despite the continued sink in the northern hemisphere (Zhao and Running 2010).

There are three main processes that determine the portion of global NPP associated with forests: (1) tree mortality rates, (2) tree establishment/colonization rates, and (3) the growth rates of the trees present. These three processes rarely act independently because mortality rates of individual trees are strongly dependent on their radial growth rates (Kobe and Coates 1997; Wyckoff and Clark 2002) while the stand level balance between mortality and establishment greatly influences forest age/size distributions. Ultimately the interaction of these processes can affect the boundaries between forested and non-forested ecosystems. Substantial exceptions to the interdependence of these variables can occur when an exogenous factor such as climate change induces shifts in the likelihood or severity of disturbances or in the favorability of an environment for or against herbivores and plant pathogens to a degree that is not balanced by concomitant changes in tree growth, mortality and establishment rates.

One major challenge is to accurately model NPP responses to changes in climatic conditions for forests that differ in tree age/size distributions. Efforts are underway to accurately scale physiological or growth responses of forests from the tree or stand-level to landscapes (Bradford et al. 2008; Hudiberg et al. 2009; Duane et al. 2010). Low-frequency variation in tree growth or forest productivity is difficult to discern from most empirical data owing to the presence of large and often inaccessible carbon pools as well as the inter-annual variability in carbon fluxes of old forests. A review of forest inventory data suggests that the growth rates of extant forests have increased through the late twentieth century (Boisvenue and Running 2006). Likewise, other studies from tropical and temperate forests have shown increases in aboveground tree growth rates or total aboveground biomass production, although the amount varies by region and species and often parallels an increase in mortality rates (Caspersen et al. 2000; Laurance et al. 2004, 2009; Lewis et al. 2004; Phillips et al. 2008; Thomas et al. 2010). Some evidence suggests that the rate of change in radial growth in response to changing resources levels diminishes with tree age or size (Körner 2006; Voelker et al. 2006).

Another major challenge is to understand how boundaries of ecosystems will shift due to climate change and whether the persistence of forests along these boundaries is affected by tree age/size. Non-forested vegetated lands such as shrub, scrub, alpine, tundra or grassland contribute only about 75% of the terrestrial carbon uptake of forests growing under the same climatic conditions (Del Grosso et al. 2008). Consequently, any climate change-induced shifts in the area of forests versus non-forest on the peripheries of these ecosystems will impact carbon uptake to at least this extent. Differences in tree age or size distributions can affect the three primary processes governing carbon storage in forests mentioned above, but are not explicitly represented in large scale models of forest extent or productivity. Moreover, the physiological mechanisms that determine growth and survival of tree seedlings are largely uncharacterized (Johnson et al. 2011). If tree age or size affects the ability of a tree to survive environmental stresses and thus for forests to sustain adequate regeneration and establishment, there exists some substantial risk that current vegetation models will be unable to project shifting boundaries between forested and non-forested regions. The thresholds determining a transition between these ecosystem types are particularly hard to predict because the amounts and distribution of forest versus non-forest vegetation can influence feedbacks on local water and nutrient availability and in some cases even impact regional weather conditions (Scheffer et al. 2005; Hirota et al. 2010). Therefore a determination of age-related responses in tree radial growth to relevant biophysical variables may shed light on the nature of threshold responses of ecosystem boundaries.

As forests age and increase in stature, they extend root systems deeper into the soil, attain sexual maturity and incur changes in the environmental conditions to which leaves and meristems are exposed. However, it is not well known whether these developmental patterns are accompanied by changes in the sensitivity of growth to biophysical drivers. A lack of insight on the physiological mechanisms determining forest growth and persistence severely restricts our ability to predict future terrestrial carbon uptake at landscape to global scales. Rather than undertake the daunting task of tracking each physiological mechanism over long periods in time, it may be possible to infer which processes are most affected by tree age or size by investigating the effect of different biophysical variables on tree growth at different ages. These broad scale patterns in growth can serve to guide further in-depth studies of tree size/age dependent physiological mechanisms. Towards this end, vast dendrochronological (tree-ring) datasets already exist. In this chapter I use a compilation of transformed tree-ring data to establish overall, age-averaged tree productivity responses to temperature, precipitation, relative humidity and an index of treeline proximity. I then quantify how tree age affects responses to these biophysical controls over productivity and discuss the implications of these results for forest responses to global climate change.

2 Methods

2.1 Tree-Ring Data

Data from 46 temperate and boreal tree species were obtained from the International Tree Ring Data Bank (ITRDB; http://www.ncdc.noaa.gov/paleo/treering.html). These species were largely selected on the basis of data availability from the ITRDB. Some species known to be very shade tolerant (i.e. *Tsuga* spp.) were not used because variation in productivity from site to site was more likely to be confounded by the frequency of canopy disturbances. Within some species used here, data from certain sites were rejected because there were too few trees sampled or when data had obviously been artificially shortened (i.e. where each series began upon a decadal-year). Altogether the dataset used here includes measurements of more than ten million ring-widths from 60,457 ring-width series collected at 1,838 collection sites (Table 17.1).

Conventional dendrochronological methods use a sample population of predominately old trees where all ring-widths are aligned by calendar year after cross-dating verification. Each file from the ITRDB was inspected for data quality and then each ring-width series was manually re-aligned by cambial age at each site (as ring number from the pith) such that wood produced in the same developmental age could be compared within each species. This first step in data standardization is conceptually similar to the RCS (regional chronology standardization) method that has been used to infer other regional to hemisphere-scale climate investigations using tree-rings (Briffa et al. 1996; Esper et al. 2002; Cook et al. 2003). Because RCS-type methods are known to have end-fitting problems at the beginnings and ends of chronologies (Briffa 2011) the methods briefly described herein employ a series of standardizations aimed at eliminating the potential for this sort of bias within the raw ring-width data before they are combined into species-level growth curves plotted by cambial age (Fig. 17.1).

2.2 *Standardization for Off-Center Tree Cores*

Increment cores often do not reach the center of the tree due to imperfect increment borer alignment, heart-rot or eccentric cambial growth patterns. These unreported tree rings affect the perceived cambial age and if all cores are aligned by the earliest tree-ring reported, those with more "missing" rings near the pith would tend to underestimate tree growth. To minimize the potential for a systematic offset from older trees having more missing rings, it was assumed that the number of unreported rings and the resulting effect on productivity can be estimated from the number of rings reported. To determine an age to missing rings relationship, I visually inspected 2,606 increment cores from seven different principal investigators. Those increment cores outside of my own collections (about half) were from archived collections at the University of Arizona Tree-Ring Laboratory (TRL). Data from all TRL cores used here had been measured and the data reported in the ITRDB. TRL cores were collected from locations across the Western United States. The data used for the missing rings relationship were collected for the purposes of climatic reconstruction and were from the following species: *Juniperus occidentalis*, *Pinus flexilis*, *Pinus ponderosa*, *Pinus monophylla*, *Pseudotsuga menziesii* and *Quercus macrocarpa*. In most cases, my own work and the notes of others at the TRL indicated the cores were collected by more than one person, often two to four people at each site. Thereby the cores likely reflect the work of 15–20 or more collectors' individual

Table 17.1 A list of the species used, specified by number of tree-ring series, number of sites sampled and the mean number of tree-rings reported across all ring-width series

Species	Tree-ring series	Sites	Tree-rings reported
Abies alba	1,266	52	127
Abies lasiocarpa	834	22	138
Abies pindrow	115	10	217
Abies spectabilis	600	23	236
Austrocedrus chilensis	759	32	257
Cedrus excelsa	235	7	376
Cedrus lebani	459	16	203
Fitzroya cupressoides	344	10	733
Juniperus occidentalis	1,066	22	379
Juniperus virginiana	531	15	206
Larix decidua	349	14	282
Larix lyallii	306	6	273
Larix sibirica	1,358	48	247
Larix gmelinii	1,178	47	258
Libocedrus bidwillii	1,151	27	298
Nothofagus pumilio	1,077	38	188
Nothofagus solanderi	569	25	163
Picea abies	3,344	130	131
Picea glauca	4,267	112	180
Picea mariana	1,075	33	122
Picea obovata	725	28	181
Picea rubens	506	16	217
Pinus albicaulis	474	13	257
Pinus aristata	193	8	528
Pinus cembra	284	10	181
Pinus contorta	367	12	201
Pinus echinata	1,041	26	122
Pinus edulis	2,434	75	296
Pinus flexilis	1,162	24	319
Pinus jeffreyi	669	17	295
Pinus longaeva	767	15	647
Pinus monophylla	481	13	256
Pinus nigra	851	40	219
Pinus palustris	381	7	88
Pinus ponderosa	4,568	166	246
Pinus resinosa	678	22	215
Pinus rigida	210	3	97
Pinus strobus	610	19	164
Pinus sylvestris	8,459	190	159
Pseudotsuga menziesii	5,765	193	293
Quercus alba	2,574	65	188
Quercus douglassii	611	10	188
Quercus macrocarpa	1,033	38	124
Quercus petraea	240	11	135
Quercus robur	1,812	48	132
Quercus Spp. (European)	473	32	125
Quercus stellata	2,206	48	178

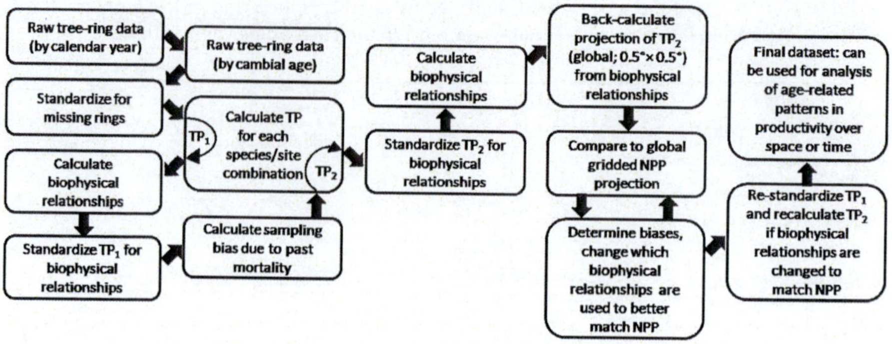

Fig. 17.1 Schematic flowchart showing the main steps used to produce a global tree-ring database that can be used to track tree productivity in space or time. TP is tree productivity; subscripts indicate iterations of tree productivity as affected by standardization scheme

habits. Because data in the ITRDB are also collected for dendroecological studies, included here were data that I and three assistants collected from *Pinus echinata*, *Quercus coccinea*, *Quercus macrocarpa* and *Quercus velutina*. These collections were made to determine stand establishment ages and compare tree growth rates during early development. Altogether these species and purposes for tree-ring collection should be reasonably representative of the larger ITRDB database.

The missing rings to pith were primarily estimated using the geometric calculation of Duncan (1989). In some cases where very few rings were missing or when growth ring arc height could not be accurately measured, the number of missing rings was visually estimated by comparing ring-width patterns and ray angles to "pith locator diagrams" (i.e. sets of concentric rings printed on clear plastic sheets). When available, notes on tree diameter measurements were also used in combination with core length and growth trends to help estimate the missing rings to pith.

From the age estimates for each tree-ring series the number of missing tree-rings to pith was estimated and combined with average ring-widths for each cambial age to calculate separate standardizations for each species (Fig. 17.2). This standardization was applied to each ring-width in the dataset and accounts for the smaller standardizations needed for species with cambial age patterns that varied little such as *Pinus longaeva* versus the larger standardizations needed for species with strong negative exponential patterns such as *Pinus ponderosa*.

2.3 *Climate and Biophysical Standardizations*

Each ring-width in the dataset was also standardized by productivity trends as predicted from three climate variables, and an index of treeline proximity. These standardizations used globally-gridded $0.5° \times 0.5°$ resolution monthly climate data for temperature, precipitation and vapor pressure from the IPCC data distribution center (http://www.ipcc-data.org/), which were originally made available through

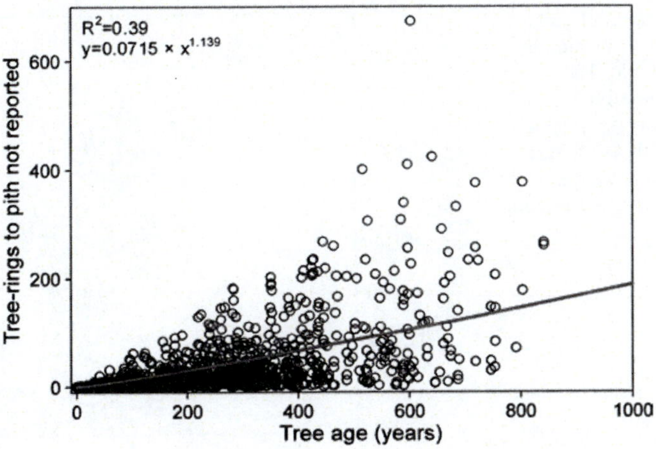

Fig. 17.2 The relationship of estimated number of tree-rings to pith not reported to tree age. Data are from 2,606 increment cores from nine species and numerous collectors (see Sect. 2 for details)

the Climate Research Unit at the University of East Anglia (http://www.cru.uea.ac.uk/cru/). Average growing season temperatures were calculated by weighting monthly mean temperatures with the growing degree days for the same month. Monthly relative humidity was calculated from the algorithm for saturated vapor pressure reported by Paw U and Gao (1988) that used ambient air and dew point temperatures. Monthly relative humidity values were weighted according to growing degree days for that month in each cell to calculate average growing season relative humidity. Mean annual precipitation was used to standardize tree-rings rather than growing season precipitation because of the ability of soil to store a certain amount of water that falls previous to the growing season.

The unique climate conditions at treeline can limit productivity not only through the effects on leaf-level physiology associated with the above climate variables, but also due to changes in plant form. A biophysical index of treeline proximity was calculated to help capture local elevation-related variation in site level productivity. Treeline position was estimated by using high resolution digital imagery and elevation models provided by Google earth software, version 4.2, Google Inc., Mountain View, CA. (Fig. 17.3). To these data was fit the following sigmoidal relationship:

$$\text{Treeline elevation} = \left(\text{LowTLE} \times \text{Latitude}^{-6.032}\right) / \left(\text{HighTLE}^{-6.032} + \text{Latitude}^{-6.032}\right),$$

where LowTLE is the low latitude treeline elevation (the best-fit relationship was 4,153 m), HighTLE is the high latitude treeline elevation (the best-fit relationship was 49 m), and −6.032 is a constant determining the shape of relationship (Fig. 17.3). To calculate treeline proximity the elevation at each tree-ring site was divided by predicted treeline elevation, and then the ratio was multiplied by 100 to yield a percentage.

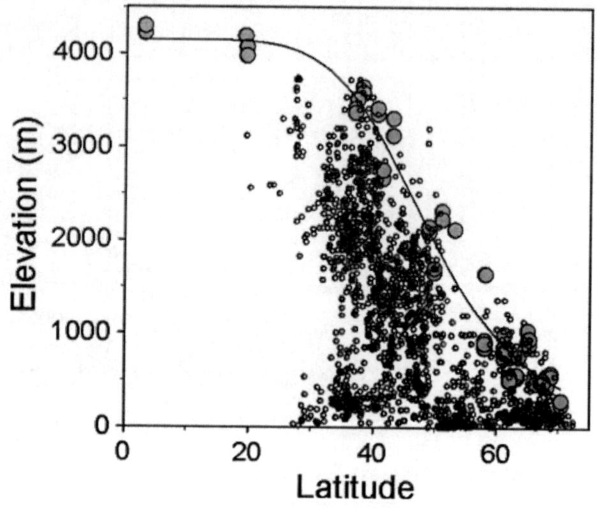

Fig. 17.3 Position of global average treeline elevation in relation to latitude. The sigmoidal regression fit to the *gray circles* was significant ($R^2=0.95$, $P<0.0001$). Small *open circles* indicate the distribution tree-ring sampling locations

2.4 *Standardization for Biases Due to Past Mortality and Tree-Ages Sampled*

The extent to which ecological effects deriving from past stand dynamics or epigenetic differences may allow some slow-growing trees to disproportionately survive to very old ages was uncertain (*sensu* Black et al. 2008). This growth pattern could result in tree growth being overestimated for recent times if sites dominated by young trees were more likely to contain fast-growers that had undergone less selection against that growth strategy. To account for this potential bias in differential-survival, I quantified this pattern only after an initial round of climate and biophysical standardizations were completed. For each species with enough replication of tree-rings formed before 1850 (to avoid anthropogenic effects occurring later on) ring-width data were divided into 10 percentile bins according to relative age. Relative age was determined for each tree by dividing the maximum age of the species in the database by the age of the tree as indicated by the number of rings reported for a ring-width series. Each ring-width value for a progressively older relative age class was divided by the mean ring-width of the same cambial age from the youngest relative age class. For example, growth across the first 80 cambial ages were compared among 10 relative age classes for bur oaks (*Quercus macrocarpa*; max. age=315) while for bristlecone pines (*Pinus longaeva*; max. age=5,591) growth across the first 700 cambial ages were compared among 10 relative age classes. To plot the data as shown, the difference between one and the best-fit linear regression prediction of the Y-intercept for each species was used to reset data so that the slope of the line was the same but the Y-intercept was equal to one (Fig. 17.4). For clarity

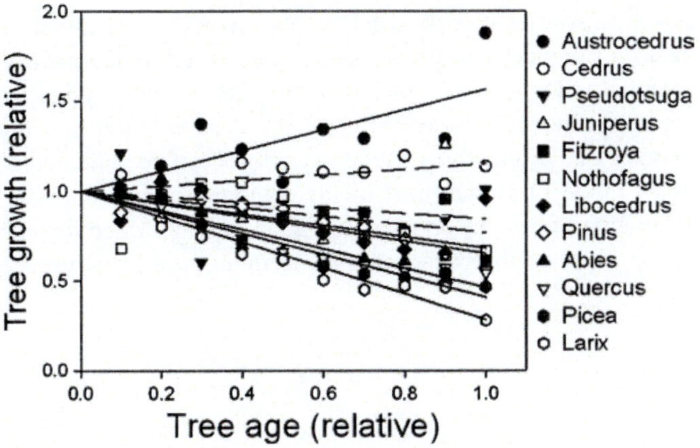

Fig. 17.4 Genera-level estimates for the effect of tree age on growth rates. Relative tree age was calculated from the maximum age in the database set equal to one. Relationships with solid lines were significant ($P<0.05$). Although genera-level estimates are shown for clarity, species-level estimates were used to adjust for age biases within the database where enough data existed

only genera-level relationships were plotted. Therefore each species was standardized according to

$$Y = y \times (1 + G_{max} \times A_r)$$

where Y is the post-standardization ring-width index value, y is the pre-standardization ring-width index value, G_{max} is the predicted growth of a species sampled at its maximum age (i.e. relative age = 100%) and A_r is relative age. Some species (generally those with less than 500 ring-width series reported) were standardized with the genera-level relationships because this paucity of data did not allow for adequate replication across relative age classes.

2.5 *Model Validation with NPP*

Validation is needed to judge the effectiveness of a model. In this case tree productivity can be compared against biophysical variables which may differ in the amount of control they exert over tree productivity across the range of a biophysical variable. It is uncertain, however, whether co-variation of some biophysical variables may cause unwanted biases to be introduced to estimates of tree productivity. To validate a model of tree productivity that derives from empirical relationships between tree growth and biophysical data, a back-prediction of tree productivity across the globe was attempted using age-specific growth in combination with 0.5° × 0.5° climate and elevation data. Gridded elevation data were obtained from the University Corporation for Atmospheric Research (http://dss.ucar.edu/datasets/).

By comparison to similarly gridded NPP data projections (Cramer et al. 1999) spatial patterns in differences can be assessed. These spatial patterns between projections can then help diagnose whether weighting certain biophysical variables or adding others can significantly improve the model so as to better match patterns in NPP. In this manner tree productivity modeling efforts can be validated with spatial data and changes made to reduce bias in the standardizations applied. This will improve the accuracy of future efforts at constructing tree productivity chronologies aimed at detecting whether differences in long-term trends exist within certain subsets of the data.

3 Results

When averaged across ages 1–300 years, tree productivity as affected by temperature, precipitation and treeline proximity all had predicted minimum values (i.e. where trees still grow) that were about 60% of maximum (Fig. 17.5). Minimum predicted tree productivity under low relative humidity conditions was approximately 80% of maximum indicating this variable was less important. Constraints on tree productivity at sites with low annual precipitation formed the steepest relationship shown for these biophysical variables indicating that productivity over this range is most sensitive. This sensitivity to precipitation may have been equaled by treeline proximity, had species-specific trends in tree height been used to more accurately estimate productivity.

To investigate age × environment interactions, sigmoidal curves were fit to each of the 31 ages investigated within each of the four biophysical relationships. When plotted across tree ages, treeline and relative humidity described half as much of the variation for old trees as for young trees (Fig. 17.6). There were no distinct long-term trends in variability of productivity with age for temperature or precipitation. Besides variation within the relationships, age-related trends in the minimum values predicted for productivity for each biophysical variable should also be an indicator of the strength of age × environment interactions. Indeed, as shown by the age-related trends in variation (Fig. 17.6), the minimum values for treeline and relative humidity showed a weakening of their effect on productivity throughout the ages investigated (Fig. 17.7). Temperature and precipitation also showed some weakening in their effect on minimum productivity with age, but this trend was associated with tree ages less than 50 years (Fig. 17.7).

The first round of spatial validation for tree productivity found the back-prediction from the average relationship to the four biophysical variables to show a coarse-scale agreement with NPP (Fig. 17.8a, b). As expected for this initial round of validation, in some regions tree productivity did not correlate well with NPP. Underestimates of NPP by tree productivity generally occurred where high elevations coincided with colder or drier regional climates (Fig. 17.8c). Overestimates of NPP tended to be greatest in regions where annual NPP is dominated by grass, shrub, savanna or woodland ecosystems (Fig. 17.8c).

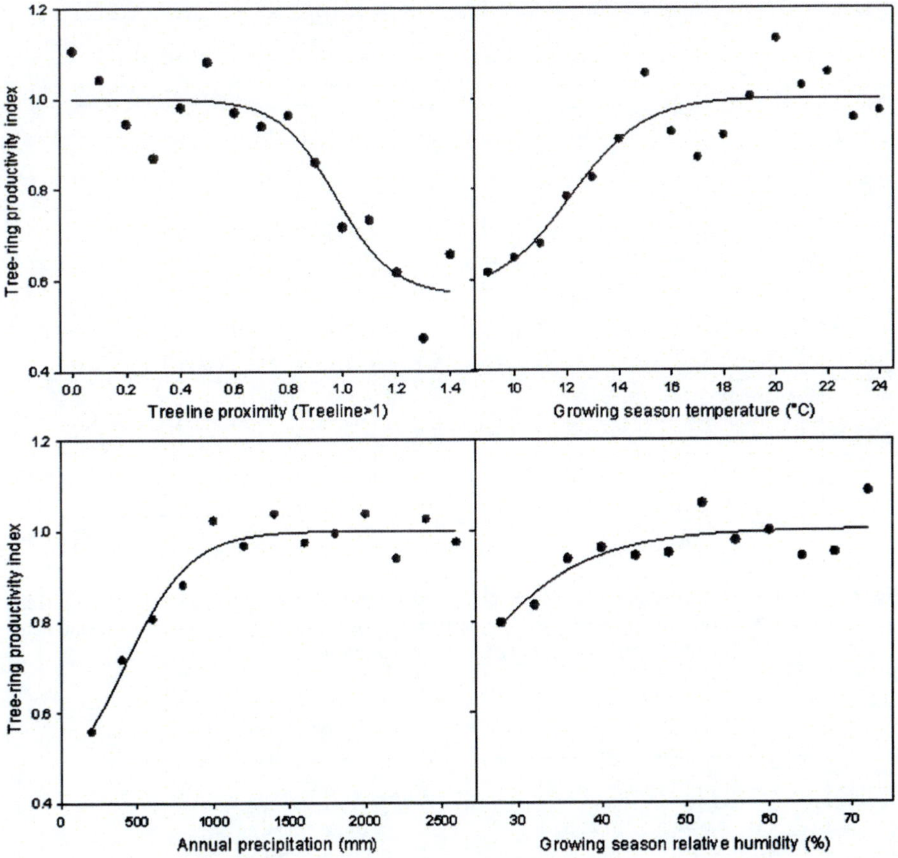

Fig. 17.5 Overall trends in tree productivity plotted against four biophysical variables. Data were binned at intervals of 0.1 for treeline proximity, 1°C for temperature, 200 mm for annual precipitation and 4% for relative humidity

4 Discussion

4.1 Age-Dependence of Tree Productivity to Biophysical Variables

Tree productivity is equivalent to changes in radial growth rate, and biomass increment scales, albeit roughly, with radial growth rate to the third power. Minimum levels of age-averaged tree productivity ranged from about 0.55–0.8 across the range of the biophysical variables investigated (Fig. 17.5). In terms of biomass increment these minimum values are equivalent to a range of 0.55^3–0.80^3, or approximately 17–51% of the maximum biomass increment than at maximum tree productivity equal to one. Tree productivity was found to be strongly related to temperature,

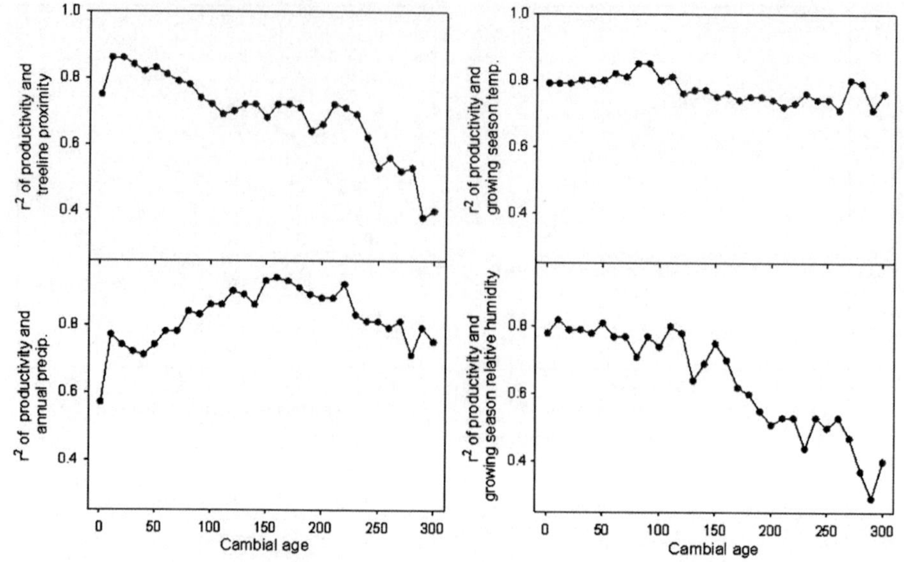

Fig. 17.6 Age-specific changes in the coefficient of determination (r^2) describing the relationship between tree productivity and four biophysical variables. Age specific regressions used to calculate r^2 values were of the same sigmoidal form as those shown in Fig. 17.5

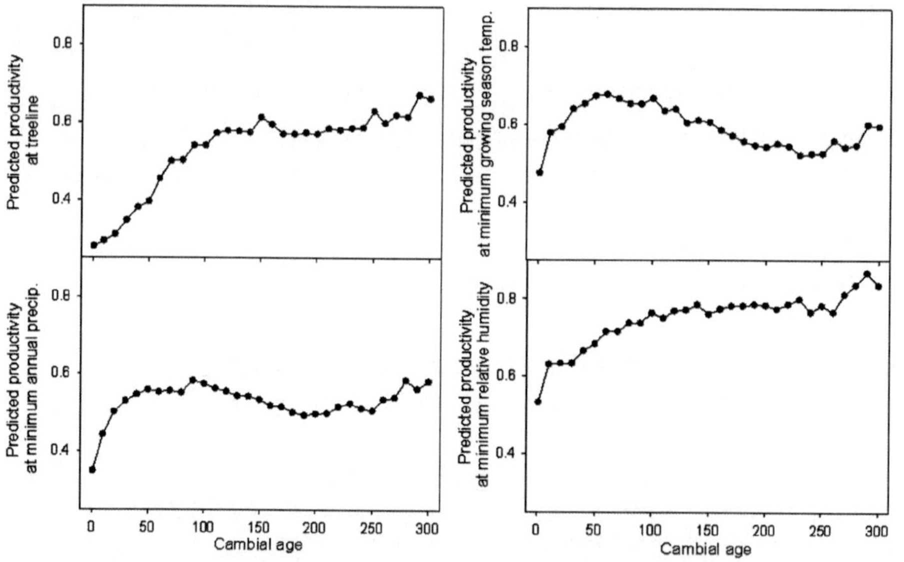

Fig. 17.7 Age-specific changes in the productivity predicted by regressions fit to minimum or maximum values for four biophysical variables. Minimum or maximum values are for the same bins specified in Fig. 17.5. Age specific regressions were of the same sigmoidal form as those shown in Fig. 17.5

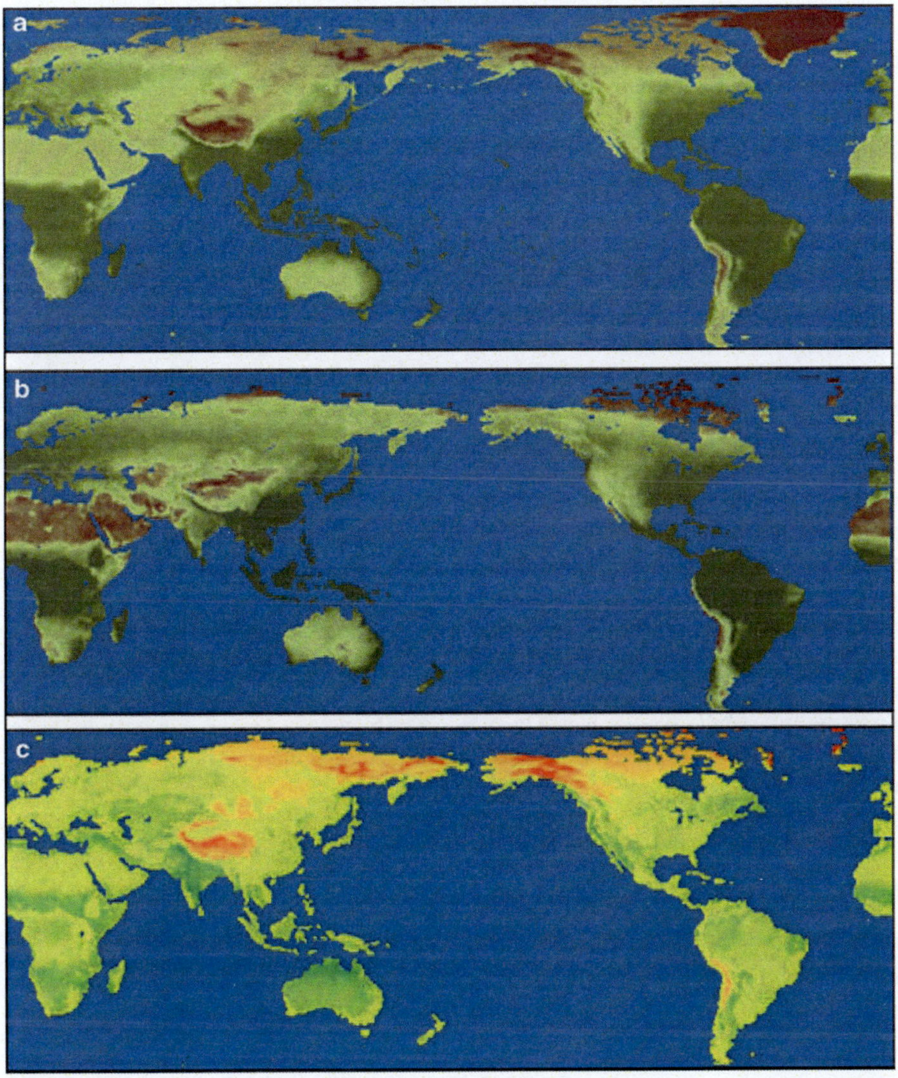

Fig. 17.8 Comparison of 0.5°×0.5° projections of (**a**) tree productivity, (**b**) model averaged annual net primary productivity and (**c**) the differences between the projections. For *panels* a and b, green indicates high productivity and brown indicates low productivity. Gridded NPP in *panel* b, ranging from 0 to 1,270 $C m^{-2}$, and are plotted after Cramer et al. (1999). For *panel* c *green* indicates where tree productivity overestimates modeled NPP and *red* indicates where tree productivity underestimates modeled NPP

precipitation, and treeline proximity while relative humidity had a smaller influence (Fig. 17.5). These age-averaged results are generally similar to the findings of other empirical studies of how temperature and precipitation influences vegetation productivity (Lieth 1972; Schuur 2003; Anderson et al. 2006).

For modeling efforts at landscape to global scales, the effects of tree age have often been assumed to average out. On the other hand, patterns in tree growth or vegetation productivity integrate many complex physiological processes that can be predicted to undergo directional changes as forests grow older and larger. As a general rule, the growth of older trees appears to be more buffered from large variations in climate. However the pattern of this age-dependence as well as the overall strength of the effect differed among biophysical variables (Fig. 17.7).

Accurately predicting the effects of climate change on terrestrial carbon storage will depend on an ability to accurately delimit forest cover and forest productivity at those sites. Compared to other carbon sinks within a tree the allocation of carbon to stem growth is generally a low priority (Oliver and Larson 1996). Hence, trees that chronically grow very slowly for their species and size are much more likely to succumb to competition, pests and diseases (Kobe and Coates 1997; Bigler et al. 2007; Wyckoff and Clark 2002; Voelker et al. 2008). As such, at some minimal level of forest productivity or average tree growth, the likelihood that a forest will persist at that site will be severely diminished. After a transition to non-forest, the re-colonization of that same site may require a more amenable climate for trees than the threshold value that caused the initial ecosystem transition. For example, if a mix of woodlands and forests persisted across a landscape where annual precipitation was 600 mm but was replaced by a grassland/savanna when decades of drought caused average precipitation to drop to 400 mm, the re-colonization of that site may require annual precipitation to significantly exceed the 600 mm that had previously sustained the forests and woodlands. This can occur because of a loss of seed sources, nurse trees or other safe sites for seedling/sapling establishment. Moreover, increased precipitation must persist above a certain threshold for a long enough period that the seedlings can establish as trees and bypass the early growth stages that could otherwise succumb to short-term droughts if their rooting depth and soil water supply are limited. Most models of tree/grass coexistence do not incorporate these tree age/size-related effects (Sankaran et al. 2004) and in most cases it could only be guessed at how long this establishment phase may take or the shape of the curve. It may be nearly impossible to measure the actual shape of this age/size-related hysteresis in sensitivity to climatic variables, but the age-dependent patterns described here should provide a reasonable first-order approximation.

4.2 The Influence of Precipitation

Tree productivity was estimated at sites with annual precipitation that ranged from 200 to 2,600 mm. This range covers the vast majority of forests across the globe. Like tree productivity (Fig. 17.5) NPP saturates above 1,600 mm annual precipitation (Schuur 2003). On the few rare sites where precipitation exceeds 4,000 mm/year NPP is somewhat reduced (Schuur 2003; Del Grosso et al. 2008), likely because the forests are radiation limited. There are also regions that receive less than 100 mm annual precipitation in which both hardwood or conifer trees can grow, albeit very slowly and only as scattered trees (Abdoun et al. 2005; Andersen and Krzywinski 2007). These species are likely relicts of a formerly wetter climate in Northern Africa whose widely scattered trees are sensitive to anthropogenic uses and presumably to continuing aridity. For both of these species the adults persist while seedlings of *Cupressus* are very rare and regeneration of *Acacia* is largely from re-sprouting rather than seed origin (Abdoun et al. 2005; Andersen and Krzywinski 2007).

At larger scales of inference, precipitation is undoubtedly one of the strongest determinants of the rates of successional development, or the amount of forest cover, vegetated cover, leaf area and NPP (Lieth 1972; Gholz 1982; Churkina et al. 1999; Schuur 2003; Sankaran et al. 2005; Scheffer et al. 2005; Anderson-Teixeira et al. 2008). As expected, tree productivity follows a broadly similar pattern (Fig. 17.5). However, these variables are not entirely similar in their sensitivity to precipitation. Average NPP decreases dramatically between 1,500 and 2,000 mm of annual precipitation (Lieth 1972; Schuur 2003) whereas the upper bounds for woody cover decrease sharply at about 650 mm (Sankaran et al. 2005) and vegetation cover is reduced near a threshold of 350 mm annual precipitation (Scheffer et al. 2005). In comparison, maximum tree productivity occurs midway between these thresholds, with an inflection point at approximately 1,000 mm annual precipitation.

For most growing conditions the greatest change affecting the water status of a tree as it grows larger is the development of a more extensive root system that can provide access to water in deep soil layers when upper soil layers are dry (Fritts 1976; Kozlowski and Pallardy 1997; Canadell et al. 1996; Cairns et al. 1997). However the occurrence of branch dieback and mortality after regional droughts has been found to be greater in older and/or taller trees (Voelker et al. 2008; Zhang et al. 2009). Perhaps it is this contrast in sensitivity to drought of younger versus older trees that has prevented any systematic effort to determine a global average age-dependence of responses to climate variables including precipitation. For the data summarized here, trees older than 50 years accrued biomass four times more rapidly than the youngest trees at the minimum annual precipitation sampled, 200 mm (Fig. 17.7). This radical change in sensitivity to precipitation is best explained by changes in maximum rooting depth that would be expected to occur during the first few decades of development for most tree species.

Although a critical rooting depth to access deep soil water is likely reached in the first few decades of a trees life, some trees may increase in height for another 100 years or more. Associated with increasing tree size are greater tensions on the xylem stream which may require a tradeoff between xylem efficiency and safety (Ryan et al. 2006; Domec et al. 2008) as well as changes in leaf morphology and stomatal functioning (Koch et al. 2004; Woodruff et al. 2009, 2010; McDowell et al. 2011; Woodruff and Meinzer 2011). All else equal this height-related change in tree function would impinge on the water status of leaves on the uppermost branches, potentially making the growth of a 200-year-old tree more sensitive to drought stress than a shorter 80-year-old tree. To help compensate for the greater hydrostatic tensions at height, large old trees can alter their wood properties, sapwood area and crown morphology. These processes can benefit the tree by providing a greater whole crown light use-efficiency via the production of epicormic sprouts (Ishii 2011) or promoting increased water storage that can buffer extremes in evaporative demand (Phillips et al. 2003; Domec et al. 2005; Meinzer et al. 2006). The greater reliance on stored water suggests that the oldest trees may need to more conservatively regulate water status as the asymptotic height is approached for that species and site. Taken together, the difference in timing and relative importance of these influences may explain the complex shape of the response to minimum rainfall (Fig. 17.7).

4.3 The Influence of Relative Humidity

Greater vapor pressure deficits can directly cause reductions in stomatal conductance (Leuning 1995). At the global scale of this investigation, stomatal sensitivity of the species investigated should be reasonably coupled to local vapor pressure deficit regimes that are in turn largely influenced by growing season temperatures. Indeed, stomatal sensitivity to vapor pressure deficit scales with average canopy conductance for a species (Oren et al. 1999). However, low canopy conductance can occur in cold, low productivity regions or in hot and arid low productivity regions. Because the temperature dependence of tree growth was already of interest, I investigated the sensitivity of tree productivity to relative humidity to better isolate the effects of atmospheric drought from low temperatures on canopy conductance and thus productivity. A recent study showed that branches excised from increasing tree heights and thus uncoupled with *in situ* xylem tensions still had maximum stomatal conductance values which occurred at greater vapor pressure deficits (Woodruff et al. 2010). This response of stomatal sensitivity is apparently a response to those branches being formed at different osmotic pressures necessary to balance the increasing xylem tensions at greater tree heights. Hence, *in situ* responses of stomatal conductance are expected to be similar among tree heights and ages while age-related decreases in sensitivity to both precipitation and relative humidity (Figs. 17.6 and 17.7) are likely a function of changes to whole plant water status as affected by age-related changes in rooting depth. At sites where trees grew at the minimum average growing season relative humidity, older trees were characterized by biomass increments that were about threefold greater than the youngest trees. The change in relative humidity response was greatest for young trees, less than 50 years old, which corresponds better to expected differences in rooting depth rather than a limitation imposed by tree height which would continue to increase for many decades after this age.

Relative humidity also plays a large part in determining the rates of evapotranspiration and site water balance. Because low summer relative humidity values can occur even in regions with high annual precipitation, variation in relative humidity connotes a likelihood of drought stress that would not be represented by precipitation alone. Ideally, the precipitation and relative humidity variables would be combined with other meteorological data for each site to produce a water balance function, but these calculations and analyses are still underway.

4.4 The Influence of Treeline Proximity

The influence of treeline proximity was investigated because gridded meteorological data do not capture local orographic effects on meteorological variables. Treeline elevation changes on average with latitude (Fig. 17.3) yet is not absolute in a given locality considering the itinerant nature of regional climates at decadal to millennial timescales. Another source of variation in treeline elevation is that at mid-latitude treelines can be substantially lower on average when the mountain range has a maritime climate in comparison to the higher treeline position of drier continental slopes. Despite these considerations, the equation given here for average treeline position agrees well with the upper boundary elevations for tree-ring collection sites (Fig. 17.3).

Reductions in growth with elevation and the related treeline phenomena are often correlated with an array of meteorological, soil and site factors that are known to be temperature dependent (Körner 1998; Sveinbjörnsson and Hofgaard 2002; Holtmeier and Broll 2005; Malanson et al. 2007; Anderson-Teixeira et al. 2008). Some research suggests that variation in photosynthetic assimilation is not important because low temperatures, especially soil temperatures, provide a direct and primary limitation on growth metabolism (Körner and Hoch 2006; Hoch and Körner 2009). However other work suggests that differences in resource availability that affect photosynthetic carbon capture cannot be discounted and that there is likely a gradient between limitation of growth by carbon demand and limitation of growth by carbon supply (Bansal and Germino 2008; Susiloto et al. 2010; Sveinbjörnsson et al. 2010). Moreover, carbon gain of seedlings at some treelines can depend critically on cloudy, high humidity conditions (Johnson and Smith 2008), an aspect of local climate that is poorly understood at present.

Dwarf tree forms or twisted Krummholz forests, are a common sight at many treelines, yet rarely have radial growth rates of trees been accurately determined as they vary across elevation gradients or proximity to treeline. Regional studies of radial growth rates have been undertaken by two research groups, and in each case they established a non-linear effect of altitude (Paulsen et al. 2000; Coomes and Allen 2007). These studies are in agreement with the more general curve provided here for treeline proximity (Fig. 17.5). The age-dependence of growth is stronger at treeline than for any other biophysical variable investigated (Fig. 17.7). Near treeline, the biomass increment of old trees should be approximately 12-fold greater than that of young trees. This suggests that growth is so slow for young trees at treeline that even infrequent disturbances, drought or herbivory have a good chance to kill most trees before they ever reach an effective reproductive size.

As discussed for precipitation, the conditions that limit growth at treeline likely require an extended period of more favorable conditions for substantial treeline expansion to occur. The mechanisms behind this age-dependence are not clear. Observational data have shown the timing of cambial activity can be initiated much later for old- growth versus young montane forests, suggesting age/size-related interactions with temperature thresholds may be important for predicting forest growth responses (Rossi et al. 2008a, b). It also seems likely that the trees themselves ameliorate the microsite conditions that may have previously limited the growth of more exposed trees (i.e. shifts in snowpack depth, air temperature, windspeed or the trapping of plant litter and sediments that can provide water and nutrients). Finally, age-related changes in growth form may play a role as there is some evidence from high-latitude treelines that upright growth forms tend to produce greater tree growth (Devi et al. 2008).

4.5 The Influence of Temperature

Growing season temperatures had a strong, non-linear effect on tree productivity when averaged across all ages (Fig. 17.5). Similar to the gross primary productivity and mean annual temperature data summarized by Lieth (1972) I found that a sigmoidal function had the best fit to tree productivity. This contrasts with other extensive studies of gross primary production rates that show a linear trend with winter minimum temperatures (Gholz 1982) and no saturation at high growing season temperatures (Anderson et al. 2006). The data compiled by Anderson et al. (2006) were all from relatively moist sites which contrasts with the tree-ring database employed here where data from most sites with higher temperatures were from semi-arid regions. The theory provided by Anderson et al. (2006) suggests biomass accumulation should increase linearly with temperature owing to the temperature response of Rubisco carboxylation. However, for plants growing within any particular environment, net photosynthesis declines above a certain threshold temperature. This occurs at approximately 24°C for most C3 species (Long et al. 2006). Therefore some reduction in photosynthesis and growth would be expected for high temperatures alone. In comparison to relative humidity or treeline proximity, the age-dependence of growth responses to temperature were more complex (Figs. 17.6 and 17.7). As observed for the response to precipitation, the minimum predicted growth at low temperatures increased across ages 1–50 (Fig. 17.7), but this response declines again thereafter. It is unclear whether this trend has a biological or ecological basis. However, one could speculate that this complex shape could result from younger trees growing in an open stand (at boreal or sub-alpine sites) being able to trap more heat near the ground level in spring which would locally thaw the soil while the shade of canopy closure could result in an insulating blanket of snow persisting longer into the spring and causing the soil to stay frozen for a longer period.

Leaf temperatures would also be expected to change with tree age or size as leaves are exposed to changes in irradiance, wind speeds, boundary layer conditions and xylem tensions. Evidence of leaf temperatures across a wide latitudinal range suggest that trees modify their form and hydraulic characteristics to promote optimal photosynthesis near leaf temperatures of 21.4°C (Helliker and Richter 2008). Further investigations are needed to determine whether similar morphological changes occur in trees as they age to help keep a consistent leaf temperature as this effect could help explain the complex shape of age-dependence of the growth response to temperature (Fig. 17.7).

4.6 Potential Biases of Tree Productivity Estimates

Estimates of tree productivity do not account for changes in the number of trees per unit area, which can affect NPP independent of leaf area index. For example, as forests transition to savannas and then grasslands the grass component can sustain equivalent leaf area, but NPP is generally lower even if the ecosystem change is due to fire and/or grazing and the climate is very similar (Del Grosso et al. 2008). Therefore the threshold precipitation value causing a rapid decline in NPP (Schuur 2003; Del Grosso et al. 2008) is greater than the threshold causing a similar decline in vegetated cover. The steep decline in tree productivity (Fig. 17.5) occurs between that for the vegetated cover by grasslands and the global average NPP including all vegetation types. This indicates that the decline in NPP is influenced by disturbance (fire, grazing or anthropogenic) that reduces tree frequency before tree growth is affected. Because of this offset, tree productivity would be expected to overestimate NPP for disturbance-dependent ecosystems.

Part of the age-dependence of growth on the four biophysical variables documented here could occur if young trees at low productivity sites consistently allocated a greater fraction of growth belowground compared to older trees (Fig. 17.7). However, this situation seems unlikely because a large datasets on biomass components suggest the ratio of above to belowground allocation is more or less independent of tree size (Enquist and Niklas 2002).

4.7 Comparison of Spatial Patterns of Tree Productivity and NPP

Spatial patterns of tree productivity and NPP show substantial agreement (Fig. 17.8a, b). Regional differences did occur (Fig. 17.8c), but the importance of these differences needs to be assessed with respect to the goals of this research. Rather than being a direct diagnostic of shortcomings of tree productivity estimates, the projected deviations from the NPP projection should be thought of as a guide to potential biases. In this way it can be determined whether using biophysical variables to standardize tree-ring data can back-predict realistic spatial results and thus validate the opposite approach for calculating unbiased estimates of tree productivity back in time.

Tree productivity projections across deserts, tundra and wet tropical forests are bound to be less accurate because no tree-ring data are available for these regions. Nonetheless, it is instructive to compare productivity and NPP projections across all terrestrial ecosystems. Overall, the methods employed here may need little further improvement where differences arise due to differences in disturbance regimes that affect the vegetation types and reduce NPP below what would be expected for the potential vegetation type at the site. These differences (Fig. 17.8c, see green-colored regions) are most evident in portions of sub-Saharan Africa, India and Australia. These regions correspond strongly to savanna or grassland ecosystems (see Del Grosso et al. 2008) where frequent fires and/or grazing reduce NPP from the potential maximum expected for those climatic conditions. For these types of overestimates of NPP no further changes should be required for the purposes of this model.

The largest underestimates of NPP (Fig. 17.8c, see red-colored regions) strongly correspond to ecosystems classified as tundra by Del Grosso et al. (2008). These tend to occur in relatively higher elevations at cooler northerly latitudes or, for example, in high elevation regions that are relatively dry and cold such as the Tibetan Plateau (Fig. 17.8c). The underestimate of NPP for these regions likely results from low precipitation values (<400 mm/year) predicting extremely low tree productivity (Fig. 17.5) even though evapotranspiration and resulting drought stress may rarely limit the vegetation in these cold regions. These differences can likely be corrected to better match NPP by employing a water balance function that uses annual evapotranspiration and precipitation estimates. The rate of reductions in tree productivity due to low temperatures and treeline proximity might also be attenuated by correcting for the number of hours of sunlight received at each site. Finally, the structure of the model right now employs a mean response to the four biophysical variables. Further tree productivity projections should attempt to optimize performance by employing decisions as to which factor is most limiting at each site. For example, treeline proximity was included to differentiate local to regional effects of elevation that might not be captured by temperature alone but this effect may not need to be incorporated at high latitudes where sites are already clearly limited by growing season temperature irrespective of elevation. The incorporation of decision algorithms that differentiate between temperature and precipitation limitations should also implicitly determine forest/grassland transitions while transitions to treeline proximity from another variable should implicitly determine forest/tundra boundaries. By incorporating age-specific responses of tree productivity and comparing current versus predicted future climate projections, the locations and direction of change in NPP as affected by tree age might be coarsely estimated for the first time.

5 Conclusion

The database described here is an improvement over previous empirical modeling because it permits the establishment of global trends in the age dependence of tree growth responses to biophysical variables. These data are important for justifying the need for further characterization of changes to physiological function with tree age to help incorporate mechanisms into predictions of forest cover and rates of future terrestrial carbon sequestration.

The findings presented here argue that the growth of older forests may be more resilient to climate change effects, making old-growth forests all the more valuable considering that they can increase their carbon storage for centuries after peak production is reached (Luyssaert et al. 2008). Although older forests may be initially buffered from the effects of smaller changes in mean temperature and precipitation, the data here cannot speak to whether they are indeed more buffered from an increased incidence of extreme events such as severe droughts. Moreover, it may prove that the ever more dynamic and unpredictable disturbance regimes and cycles of pest and pathogen abundance may have stronger effects on NPP than the direct effects on tree function from shifts in average climate.

Old growth forests receive much attention, but the data here suggest that the growth and establishment of younger trees are more sensitive to shifts in climate. Because establishment can take a number of years, or even decades for slow-growing trees, an increase in the frequency of severe drought events without a change in the mean levels of precipitation could also cause major changes to the populations of sensitive tree species. Consequently, if a changing climate causes species migrations and shifts in ecosystem boundaries, the early growth stages should be thought to represent the greatest potential for bottlenecks in the persistence of forests or colonization by new forests. In turn, these changes can have far-reaching effects on the location of ecosystem boundaries and the resulting differences in carbon storage and other services they provide.

Acknowledgements This study was made possible because of the existence of the International Tree-ring Data Bank. SLV thanks the contributors, volunteers and NOAA staff invested in maintaining the excellent quality and availability of data within the ITRDB. SLV is also indebted to Steve Leavitt and Pearce Paul Creasman for access to the LTRR archives, to Alice Hyde for her help processing tree-ring data and to JR brooks and FC Meinzer for critical reading of earlier drafts of the manuscript. SLV was supported by a grant from the NSF award 0743882 DEB – Ecosystems.

References

Abdoun F, Jull AJT, Guibal F, Thinon M (2005) Radial growth of the Sahara's oldest trees: *Cupressus dupreziana* A. Camus. Trees 19:661–670

Andersen GL, Krzywinski K (2007) Longevity and growth of *Acacia tortilis*; insights from ^{14}C content and anatomy of wood. BMC Ecol 7:4. doi:10.1186/1472-6785-7-4

Anderson KJ, Allen AP, Gillooly JF, Brown JH (2006) Temperature-dependence of biomass accumulation rates during secondary succession. Ecol Lett 9:673–682

Anderson-Teixeira KJ, Vitousek PM, Brown JH (2008) Amplified temperature dependence in ecosystems developing on the lava flows of Mauna Loa, Hawai'i. Proc Natl Acad Sci USA 105:228–233

Bansal S, Germino MJ (2008) Carbon balance of conifer seedlings at timberline: relative changes in uptake, storage, and utilization. Oecologia 158:217–227

Bigler C, Gavin DG, Gunning C, Veblen TT (2007) Drought induces lagged tree mortality in a subalpine forest in the Rocky Mountains. Oikos 116:1983–1994

Black BA, Colbert JJ, Pederson N (2008) Relationships between lifespan and radial growth rate within North American tree species. Ecoscience 15:349–357

Boisvenue C, Running SW (2006) Impacts of climate change on natural forest productivity – evidence since the middle of the 20th century. Glob Change Biol 12:862–882

Bradford JB, Birdsey RA, Joyce LA, Ryan MG (2008) Tree age, disturbance history, and carbon stocks and fluxes in subalpine Rocky Mountain forests. Glob Change Biol 14:2882–2897

Briffa KR (2011) A closer look at regional curve standardization of tree-ring records: justification of the need, a warning of some pitfalls, and suggested improvements in its application. In:

Hughes MK, Swetnam TW, Diaz HF (eds) Dendroclimatology: progress and prospects. Springer, New York, pp 113–145 (pp 340)
Briffa KR, Jones PD, Schweingruber FH, Karlén W, Shiyatov SG (1996) Tree-ring variables as proxy-climate indicators: problems with low-frequency signals. In: Jones PD, Bradley RS, Jouzel J (eds) Climate variations and forcing mechanisms of the last 2,000 years. Springer, New York, pp 9–41 (pp 649)
Cairns MA, Brown S, Helmer EH, Baumgardner GA (1997) Root biomass allocation in the world's upland forests. Oecologia 111:1–11
Canadell J, Jackson RB, Ehleringer JR, Mooney HA, Sala OE, Schulze ED (1996) Maximum rooting depth of vegetation types at the global scale. Oecologia 108:583–595
Canadell JG, Le Quéré C, Raupach MR, Field CB, Buitenheis ET, Ciais P, Conway TJ, Gillet NP, Houghton RA, Marland G (2007) Contributions to accelerating atmospheric CO_2 growth from economic activity, carbon intensity, and efficiency of natural sinks. Proc Natl Acad Sci USA 104:18866–18870
Caspersen JP, Pacala SW, Jenkins J, Hurtt GC, Moorcroft PR, Birdsey RA (2000) Contributions of land-use history to carbon accumulation in U.S. forests. Science 290:1148–1151
Churkina G, Running SW, Schloss AL (1999) Comparing global models of terrestrial net primary productivity (NPP): the importance of water availability. Glob Change Biol 5(S1):46–55
Cook ER, Krusic PJ, Jones PD (2003) Dendroclimatic signals in long tree-ring chronologies from the Himalayas of Nepal. Int J Climatol 23:707–732
Coomes DA, Allen RB (2007) Effects of size, competition and altitude on tree growth. J Ecol 95:1084–1097
Cramer W, Kicklighter DW, Bondeau A, Moore B III, Churkina G, Nemry B, Ruimy A, Schloss AL (1999) Comparing global models of terrestrial net primary productivity (NPP): overview and key results. Glob Change Biol 5(S1):1–15
Del Grosso S, Parton W, Stohlgren T, Zheng D, Bachelet D, Prince S, Hibbard K, Olson R (2008) Global potential net primary production predicted from vegetation class, precipitation, and temperature. Ecology 89:2117–2126
Devi N, Hagedorn F, Moiseev P, Bugmann H, Shiyatov S, Mazepa V, Rigling A (2008) Expanding forests and changing growth forms of Siberian larch at the Polar Urals treeline during the 20th century. Glob Change Biol 14:1581–1591
Domec J-C, Pruyn ML, Gartner BL (2005) Axial and radial profiles in conductivities, water storage and native embolism in trunks of young and old growth ponderosa pine trees. Plant Cell Environ 128:1103–1113
Domec J-C, Lachenbruch B, Meinzer FC, Woodruff DR, Warren JA, McCulloh KA (2008) Ultimate height in a conifer is limited by conflicting requirements for xylem design. Proc Natl Acad Sci USA 109:12069–12074
Duane MV, Cohen WB, Campbell JL, Hudiburg T, Turner DP, Weyermann D (2010) Implications of alternative field-sampling designs on Landsat-based mapping of stand age and carbon stocks in Oregon forests. For Sci 56:405–416
Duncan RP (1989) An evaluation of errors in tree age estimates based on increment cores in Kahikatea (*Dacrycarpus dacrydoides*). NZ Nat Sci 16:31–37
Enquist BJ, Niklas KJ (2002) Global allocation rules for patterns of biomass partitioning in seed plants. Science 295:1517–1520
Esper J, Cook ER, Schweingruber FH (2002) Low-frequency signals in long tree-line chronologies for reconstructing past temperature variability. Science 295:2250–2253
Friedlingstein P, Cox P, Betts R et al (2006) Climate-carbon cycle feedback analysis: results from the (CMIP)-M-4 model intercomparison. J Climate 19:3337–3353
Fritts HC (1976) Tree rings and climate. Academic, London, pp 567
Gholz H (1982) Environmental limits on aboveground net primary production, leaf area and biomass in vegetation zones of the Pacific Northwest. Ecology 63:469–481
Helliker BR, Richter SL (2008) Subtropical to boreal convergence of tree-leaf temperatures. Nature 454:511–513

Hirota M, Nobre C, Oyama MD, Bustamante MMC (2010) The climate sensitivity of the forest, savanna and forest-savanna transition in tropical South America. New Phytol 187:707–719

Hoch G, Körner C (2009) Growth and carbon relations of tree line forming conifers at constant vs. variable low temperatures. J Ecol 97:57–66

Holtmeier F-K, Broll G (2005) Sensitivity and response of northern hemisphere altitudinal and polar treelines to environmental changes at landscape and local scales. Glob Ecol Biogeogr 14:395–410

Houghton RA (1999) The annual net flux of carbon to the atmosphere from changes in land use 1850-1990. Tellus 51B:298–313

Hudiberg T, Law B, Turner DP, Campbell J, Donato D, Duane M (2009) Carbon dynamics of Oregon and northern California forests and potential land-based carbon storage. Ecol Appl 19:163–180

Ishii H (2011) How do changes in leaf/shoot morphology and crown architecture affect growth and physiological function of large, old trees? In: Meinzer FC, Lachenbruch B, Dawson TE (eds) Size- and age-related changes in tree structure and function. Springer, Dordrecht

Johnson DM, Smith WK (2008) Cloud immersion alters microclimate, photosynthesis and water relations in *Rhododendron catawbiense* and *Abies fraseri* seedlings in the Southern Appalachian Mountains, USA. Tree Physiol 28:385–392

Johnson DM, McCulloh KA, Reinhardt K (2011) The earliest stages of tree growth: development, physiology and impacts of climate change. In: Meinzer FC, Lachenbruch B, Dawson TE (eds) Size- and age-related changes in tree structure and function. Springer, Dordrecht

Keeling CD, Chin JFS, Whorf TP (1996) Increased activity of northern vegetation inferred from atmospheric CO_2 measurements. Nature 382:146–149

Kobe RK, Coates KD (1997) Models of sapling mortality as a function of growth to characterize interspecific variation in shade tolerance of eight tree species of northwestern British Columbia. Can J For Res 27:227–236

Koch GW, Sillett SC, Jennings GM, Davis SD (2004) The limits to tree height. Nature 428:851–854

Körner C (1998) A re-assessment of high elevation treeline positions and their explanation. Oecologia 115:445–459

Körner C (2006) Plant CO_2 responses: an issue of definition, time and resource supply. New Phytol 172:393–411

Körner C, Hoch G (2006) A test of treeline theory on a montane permafrost island. Arct Antarct Alp Res 38:113–119

Kozlowski TT, Pallardy SG (1997) Physiology of woody plants, 2nd edn. Academic, San Diego

Laurance WF, Oliveira AA, Laurance SG, Condit R, Nascimento HEM, Sanchez-Thorin AC, Lovejoy TE, Andrade A, D'Angelo S, Ribeiro JE, Dick CW (2004) Pervasive alteration of tree communities in undisturbed Amazonian forests. Nature 428:171–175

Laurance SGW, Laurance WF, Nascimento HEM, Andrade A, Fearnside PM, Rebello ERG, Condit R (2009) Long-term variation in Amazon forest dynamics. J Veg Sci 20:323–333

Leuning R (1995) A critical appraisal of a combined stomatal-photosynthesis model for C_3 plants. Plant Cell Environ 18:339–355

Lewis SL, Phillips OL, Baker TR, Lloyd J, Malhi Y, Almeida S, Higuchi N, Laurance WF, Neill DA, Silva JNM, Terborgh J, Torres Lezama A, Vásquez Martinez R, Brown S, Chave J, Kuebler C, Núñez Vargas P, Vinceti B (2004) Concerted changes in tropical forest structure and dynamics: evidence from 50 South American long-term plots. Philos T Roy Soc B 359:421–436

Lewis SL, Lopez-Gonzalez G, Sonké B, Affum-Baffoe K, Baker TR, Ojo LO, Phillips OL, Reitsma JM, White L, Comiskey JA, Djuikouo KM-N, Ewango CEN, Feldspausch TR, Hladik A, Lloyd J, Lovett JC, Makana J-R, Mahli Y, Mbago FM, Ndangalasi HJ, Peacock J, Peh KS-H, Sheil D, Sunderland T, Swaine MD, Taplin J, Taylor D, Thomas SC, Votere R, Wöll H (2009) Increasing carbon storage in intact African tropical forests. Nature 457:1003–1006

Lieth H (1972) Modeling the primary productivity of the world. Nat Resour 8:5–10

Long SP, Zhu X-G, Naidu SL, Ort DR (2006) Can improvement in photosynthesis increase crop yields? Plant Cell Environ 29:315–330

Luyssaert S, Schulze ED, Börner A, Knohl A, Hessenmöller D, Law BE, Ciais P, Grace J (2008) Old-growth forests as global carbon sinks. Nature 455:213–215

Malanson GP, Butler DR, Fagre DB, Walsh SJ, Tomback DF, Daniels LD, Resler LM, Smith WK, Weiss DJ, Peterson DL, Bunn AG, Heimstra CA, Liptzin D, Bourgeron PS, Shen Z, Millar CI (2007) Alpine treeline of western North Amreica: linking organism-to-landscape dynamics. Phys Geogr 28:378–396

McDowell NG, Bond BJ, Hill LT, Ryan MG, Whitehead D (2011) Relationships between tree height and carbon isotope discrimination. In: Meinzer FC, Lachenbruch B, Dawson TE (eds) Size- and age-related changes in tree structure and function. Springer, Dordrecht

Meinzer FC, James SA, Goldstein G (2004) Dynamics of transpiration, sap flow and use of stored water in tropical forest trees. Tree Physiol 24:901–909

Meinzer FC, Brooks JR, Domec J-C, Gartner BL, Warren JM, Woodruff DR, Bible K, Shaw DC (2006) Dynamics of water transport and storage in conifers studied with deuterium and heat tracing techniques. Plant Cell Environ 29:105–114

Nemani RR, Keeling CD, Hashimoto H, Jolly WM, Piper SC, Tucker CJ, Myeni RB, Running SW (2003) Climate-driven increases in global terrestrial net primary production from 1982 to 1999. Science 300:1560–1563

Oliver CD, Larson BC (1996) Forest stand dynamics: update edition. Wiley, New York, p 520 p

Oren R, Sperry JS, Katul GG, Pataki DE, Ewers BD, Phillips N, Schäfer KVR (1999) Survey and synthesis of intra- and interspecific variation in stomatal sensitivity to vapour pressure deficit. Plant Cell Environ 22:1515–1526

Paulsen J, Weber UM, Körner C (2000) Tree growth near treeline: abrupt or gradual reduction with altitude? Arct Antarct Alp Res 32:14–20

Paw U KT, Gao W (1988) Applications of solutions to non-linear energy budget equations. Agric For Meteorol 43:121–145

Phillips NG, Ryan MG, Bond BJ, McDowell NG, Hinckley TM, Cermak J (2003) Reliance on stored water increases with tree size in three speciues in the Pacific Northwest. Tree Physiol 23:237–245

Phillips OL, Lewis SL, Baker TR, Chao K-J, Higuchi N (2008) The changing Amazon forest. Philos T Roy Soc B 363:1819–1827

Qian H, Joseph R, Zeng N (2009) Enhanced terrestrial carbon uptake in the northern high latitudes in the 21st century from the coupled carbon cycle climate model intercomparison project model projections. Glob Change Biol 16:641–656

Rossi S, Deslauriers A, Anfodillo T, Carrer M (2008a) Age-dependent xylogenesis in timberline conifers. New Phytol 177:199–208

Rossi S, Deslauriers A, Gri ar J, Seo J-W, Rathberger CBK, Anfodillo T, Morin H, Levanic T, Oven P, Jalkanen R (2008b) Critical temperatures for xylogenesis in conifers of cold climates. Glob Ecol Biogeog 17:696–707

Ryan MG, Phillips N, Bond BJ (2006) The hydraulic limitation hypothesis revisited. Plant Cell Environ 29:367–381

Sankaran M, Ratnam J, Hana NP (2004) Tree-grass coexistence in savannas revisited – insights from an examination of assumptions and mechanisms invoked in existing models. Ecol Lett 7:480–490

Sankaran M, Hanan NP, Scholes RJ, Ratnam J, Augustine DJ, Cade BS, Gignoux J, Higgins SI, Le Roux X, Ludwig F, Ardo J, Banyikwa F, Bronn A, Bucini G, Caylor KK, Coughenour MB, Diouf A, Ekaya W, Feral CJ, February EC, Frost PGH, Hiernaux P, Hrabar H, Metzger KL, Prins HHT, Ringrose S, Sea W, Tews J, Worden J, Zambatis N (2005) Determinants of woody cover in African savannas. Nature 438:846–849

Scheffer M, Holmgren M, Brovkin V, Claussen M (2005) Synergy between small- and large-scale feedbacks of vegetation on the water cycle. Glob Change Biol 11:1003–1012

Schimel DS, House JI, Hibbard KA, Bousquet P, Ciais P, Peylin P, Braswell BH, Apps MJ, Baker D, Bondeau A, Canadell J, Churkina G, Cramer W, Denning AS, Field CB, Friedlingstein P, Goodale C, Heimann M, Houghton RA, Melillo JM, Moore B III, Murdiyarso D, Noble I,

Pacala SW, Prentice IC, Raupach MR, Rayner PJ, Scholes RJ, Steffen WL, Wirth C (2001) Recent patterns and mechanisms of carbon exchange by terrestrial ecosystems. Nature 414:169–172

Schuur EAG (2003) Productivity and global climate revisited: the sensitivity of tropical forest growth to precipitation. Ecology 84:1165–1170

Susiloto S, Hilavuori E, Berninger F (2010) Testing the growth limitation hypothesis for subarctic Scots pine. J Ecol 98:1186–1195

Sveinbjörnsson B, Hofgaard LA (2002) Natural causes of the tundra-taiga boundary. Ambio 12:23–29

Sveinbjörnsson B, Smith M, Traustason T, Ruess RW, Sullivan PF (2010) Variation in carbohydrate source-sink relations of forest and treeline white spruce in southern, interior and northern Alaska. Oecologia 163:833–843

Thomas RQ, Canham CD, Weathers KC, Goodale CL (2010) Increased tree carbon storage in response to nitrogen deposition in the US. Nat Geosci 3:13–17

Voelker SL, Muzika R, Guyette RP, Stambaugh MC (2006) Historical CO_2 growth enhancement declines with age in Quercus and Pinus. Ecol Monogr 76:549–564

Voelker SL, Muzika R, Guyette RP (2008) Individual tree and stand level influences on the growth, vigor and decline of red oaks in the Ozarks. For Sci 54:8–20

Woodruff DR, Meinzer FC, Lachenbruch B, Johnson DM (2009) Coordination of leaf structure and gas exchange along a height gradient in a tall conifer. Tree Physiol 29:261–272

Woodruff DR, Meinzer FC, McCulloh KA (2010) Height-related trends in stomatal leaf-to-air vapor pressure deficit in a tall conifer. J Exp Bot 61:203–210

Wyckoff PH, Clark JS (2002) The relationship between growth and mortality for seven co-occurring tree species in the southern Appalachian Mountains. J Ecol 90:604–615

Zhang Y-J, Meinzer FC, Hao G-Y, Scholz FG, Bucci SJ, Takahashi FSC, Villalobos-Vega R, Giraldo JP, Cao K-K, Hoffman WA, Goldstein G (2009) Size-dependent mortality in a Neotropical savanna tree: the role of height-related adjustments in hydraulic architecture and carbon allocation. Plant Cell Environ 32:1456–1466

Zhao M, Running SW (2010) Drought-induced reduction in global terrestrial net primary production from 2000 through 2009. Science 329:940–943

Chapter 18
Consequences of Stand Age and Species' Functional Trait Changes on Ecosystem Water Use of Forests

Brent E. Ewers, Ben Bond-Lamberty, and D. Scott Mackay

Abstract Determining how trees move water from the soil to the atmosphere has been central to the study of forest productivity and hydrology for over a century. Improved predictive understanding of stand age and water yield will become even more important as humans modify forest stand age structure through forest management and/or climate change. Successional changes in plant species and/or plant functional traits increase uncertainty in estimates of stand transpiration as forest stands age because of the lack of theory on how tree species influence transpiration. This chapter investigates how changes in stand age can impact stand transpiration. We focus on reanalyzing published datasets from two of our own chronosequences of stand-scale transpiration. Our goal is to determine the minimal model complexity needed to maximally capture this variation in stand transpiration across chronosequences. We seek to answer the following question: "If we chose a reference stomatal conductance from one stand age of the entire chronosequence to put into a model, would modeled transpiration for the other ages be biased?" We then explore the relative roles of stomatal conductance, leaf and sapwood area in explaining these patterns in our study and in other published studies of forest chronosequences.

B.E. Ewers (✉)
Department of Botany, Program in Ecology, University of Wyoming,
1000 E. University Ave., 3165, Laramie, WY 82071, USA
e-mail: beewers@uwyo.edu

B. Bond-Lamberty
Joint Global Change Research Institute, 8400 Baltimore Avenue, Suite 201,
College Park, MD 20740, USA

D.S. Mackay
Department of Geography, State University of New York at Buffalo,
105 Wilkeson Quadrangle, Buffalo, NY 14261, USA

F.C. Meinzer et al. (eds.), *Size- and Age-Related Changes in Tree Structure and Function*, Tree Physiology 4, DOI 10.1007/978-94-007-1242-3_18,

1 Introduction

Determining how trees move water from the soil to the atmosphere has been central to the study of forest productivity and hydrology for over a century. The first plausible explanation for the movement of water through very tall trees was by Dixon and Joly (1895); the Cohesion-Tension theory of that publication has since withstood over a century of various alternatives, all now soundly rejected (Angeles et al. 2004; Pockman et al. 1995; Holbrook et al. 1995). The hydraulic limits of tree water transport have emerged as an important part of understanding not only limitations to tree productivity (Gower et al. 1996; Yoder et al. 1997; Ryan et al. 2006; Steppe et al. 2011), but also how hydrology changes in aging forests (Bond et al. 2007). Improved predictive understanding of stand age and water yield will become even more important as humans modify forest stand age structure through forest management (Jackson et al. 2005) and/or climate change (Gillett et al. 2004).

Early studies of forest water yield showed that yield is proportional to forest cover (Dunford and Fletcher 1947; Douglass 1983). More recent work has shown that important mechanisms controlling forest water yield include plant physiology controls over transpiration, frozen and liquid precipitation interception, and erosion changes. These mechanistic controls have been recently synthesized (Bond et al. 2007), and so we focus on stand age and transpiration here. Early studies relied on forest cover and later leaf area index (*LAI*) as predictive parameters of stand transpiration, coupled with stomatal physiology for modeling purposes (Aber and Federer 1992). Most models operating at the stand or higher spatial scales use a stomatal physiology model that is a combination of empirical parameterizations and basic physiological responses to light, vapor pressure deficit and soil moisture because a full mechanistic understanding of stomatal physiology is still lacking (Franks et al. 2007).

Successional changes in plant species and/or plant functional traits increase uncertainty in estimates of stand transpiration as forest stands age because of the lack of theory on how tree species influence transpiration. One hypothesis recently posed is that as plant species diversity increases, dry growing season canopy evapotranspiration (E_T) decreases; however, tests of this hypothesis with meta-analysis have been confounded by differences in edaphic, stand age and climatic conditions (Baldocchi 2005). Even the direction of change in E_T as forests change in dominant species from gymnosperm to angiosperm with succession can be difficult to generalize, especially if there are strong differences in the non-growing season (Ford et al. 2007; Ford and Vose 2007). In Boreal forests, an increasing proportion of conifers causes lower transpiration per unit leaf area with increasing stand age (Ewers et al. 2005); however, the opposite trend is observed in temperate forests (Stoy et al. 2006), likely due to widely varying leaf-life span in conifers (~18–80 months in these two studies) and associated leaf economic traits (Wright et al. 2004) that change with tree height (Koch et al. 2004).

Species differences in stand transpiration with succession could also be associated with changes in wood anatomy (i.e. tracheids only in gymnosperms, and ring or diffuse porous vessel arrangements in angiosperms, Brodribb and Feild 2000).

Baldocchi (2005) suggested that such variations in tree properties lead to systematic changes in sapwood area per ground area which would affect transpiration rates. In the last few decades sap flux per unit xylem area measurements have allowed species-specific transpiration responses to light, vapor pressure deficit (*VPD*) and soil moisture to be estimated at high temporal resolution. These responses can then be used to estimate canopy stomatal conductance when combined with leaf area and sapwood area scalars by inverting the Penman-Monteith (PM) equation (Monteith and Unsworth 1990; Phillips and Oren 1998). Such calculations are tractable, and have been validated against leaf-level gas exchange (Ewers et al. 2007a).

While the absolute magnitudes of maximum or reference stomatal conductance vary significantly with species (Körner 1994), plant hydraulics theory predicts that these species differences are minimized when expressed as the parameter (m), the ratio between the rate of reduction in stomatal conductance with the rate of increasing vapor pressure deficit (i.e., stomatal sensitivity to *VPD*) and a reference conductance (G_{Sref}; Oren et al. 1999). Using this approach can simplify models while maintaining physiologic rigor (Mackay et al. 2003). Future models should thus be able to capture the potential changes in stand transpiration with age that occur from a combination of tree species substitution and tree size impacts on G_{Sref}.

This chapter investigates how changes in stand age can impact stand transpiration. We focus on reanalyzing published datasets from two of our own chronosequences of stand-scale transpiration. Other chapters in this book already focus on the physiological changes occurring as individual trees increase in size and age. Thus, our goal is to determine the minimal model complexity needed to maximally capture this variation in stand transpiration across chronosequences. We seek to answer the following question: "If we chose a reference stomatal conductance from one stand age of the entire chronosequence to put into a model, would modeled transpiration for the other ages be biased?" We then explore the relative roles of stomatal conductance, leaf and sapwood area in explaining these patterns in our study and in other published studies of forest chronosequences.

2 Methodological Considerations

2.1 Succession and Chronosequences

Plant succession occurs over decades or longer; thus observational research on forest succession is challenging for human researchers, who have relatively short life spans and even shorter funding cycles. Researchers often meet this challenge by turning to the chronosequence approach–a space for time substitution concept, borrowed from geology and soil science, in which it is assumed that all drivers of the system are the same except for time (Jenny 1941). If correct, each site repeats the succession sequence of every other site in the sequence. This approach has been widely used in the study of successional dynamics in ecology from its earliest days to the present (e.g., Cowles 1899; Letcher and Chazdon 2009). Numerous studies have questioned the use of untested chronosequences (e.g., Johnson and Miyahshi 2008; Chazdon et al. 2007; Yanai et al. 2000; Fastie 1995), but parameters such as leaf area index and plant cover clearly work well with some chronosequences (Ewers and Pendall 2008). Other chronosequences have mixed results, with basal area (amount of tree stem area) but not species richness working in one (Letcher and Chazdon 2009) and species richness but not diversity in another (Foster and Tilman 2000). The main message from this is that the validity of a chronosequence may depend on the parameter being investigated, as well as the site selection of the researcher. Johnson and Miyahshi (2008) suggested that chronosequence selection should be based on mechanisms that likely change over the chronosequence and invalidate the assumption of a uniform trajectory. Combining the mixed results of the chronosequences tests with the call to mechanisms suggests that a biophysical approach that takes advantage of functional traits may work best in transpiration studies of chronosequences. Because of the tight link between basal area, sapwood area and thus transpiration, chronosequences may work well for forest transpiration studies. However, many forest types experience species change during succession, which does not work as well when tested in replicated chronosequences (Johnson and Miyahsi 2008). Further, species effects on forest transpiration are still not predictable from first principles (Baldocchi 2005).

2.2 *Tree Species' Functional Traits Effects on Transpiration*

While species have been the biological unit of analysis for community ecology, functional traits are emerging as a superior alternative (McGill et al. 2006; Kearney and Porter 2006), suggesting biophysical traits could work well in succession and transpiration studies. Baldocchi (2005) summarized the arguments at both the tree and stand scale for tree species effects on transpiration. At the tree scale, sap flux per active xylem area varied less than sapwood area per ground area which has been confirmed in tree to watershed-level scaling (Ford et al. 2007); other variations could be explained by differences in ring- vs. diffuse-porous trees, and by within-species variation in genotypes, tree age, height, and edaphic factors. Most of these factors can then be related to stomatal conductance through plant hydraulics (Ewers et al. 2005; Franks et al. 2007) providing a direct connection to transpiration.

A key part of understanding transpiration across tree stands and ages is determining what scalars are actually needed for both temporal and spatial scaling. Traditional bottom-up approaches start with leaf-level mechanisms and a few simplifying assumptions: big-leaf approaches assume that all leaves are equal, implying a simple averaging of leaf fluxes, while the multi-layer approach divides the canopy into some arbitrary number of layers to determine the profiles in leaf fluxes and their drivers (Raupach and Finnigan 1988). These approaches can be refined by modeling the amount of sun and shade leaves using light extinction coefficients (dePury and Farquhar 1997). The questions being asked and computational resources available determine which method is chosen. Sap flux estimates of canopy transpiration often represent a superior means of parameterizing and testing both of these models (Williams et al. 1996; Samanta et al. 2008) because sap flux methods can sample a much larger fraction of the transpirational flux than do leaf chamber-based methods, and they do not include other sources of water loss like eddy covariance or watershed-based methods.

The now-standard approach to scale sap flux from individual sensors to trees and stands is to first quantify radial and circumferential trends in the active xylem sapwood (Oren et al. 1998), usually near breast height. The resulting trends are then used to calculate an average sap flux based on methods varying from simple weighting schemes (Phillips et al. 1996; Clearwater et al. 1999; Ewers and Oren 2000) to dynamic functions calibrated against a point measurement (Caylor and Dragoni 2009). Recent work suggests that there may be mechanistic ways to explain these trends using tree parameters such as wood density (Barbour and Whitehead 2003) or the proportion of the active leaf area (Fiora and Cescatti 2008). Regardless of the method used, the resulting tree sap flux is multiplied by the sapwood area to obtain either tree-based or stand transpiration per unit ground area.

This type of sapwood-based scaling has now been verified in several studies. Early approaches such as those by Hatton et al. (1990) and Cermak et al. (1992) showed the importance of sapwood area and radial trends in sap flux when describing whole tree transpiration. More recent studies have demonstrated that sapwood area is more important than sap flux per xylem area in obtaining transpiration at the stand (Wullshleger et al. 2001; Ewers et al. 2002; Kagawa et al. 2009) and at the watershed (Ford et al. 2007) scales. This conclusion has been reinforced by spatially explicit transpiration studies using tens to hundreds of sap flux sensors within stands to show that sapwood area is the key spatial scalar while sap flux per unit xylem area provides the appropriate temporal response to environmental drivers (Adelman et al. 2008; Loranty et al. 2008; Traver et al. 2010; Kume et al. 2010). These results have now been extended to sampling schemes which show that only a few trees need to be sampled to obtain the temporal response to environmental drivers such as light, vapor pressure deficit and soil moisture, while sapwood area estimates are needed more extensively for stand transpiration (Mackay et al. 2010).

The fact that sap flux gives the temporal response to environmental drivers while sapwood area provides the magnitude permits these measurements to be used to explain tree physiological responses to the environment. While there is some controversy as to how well sap flux at breast height explains capacitance in larger trees (Burgess and Dawson 2008; Phillips et al. 2009), studies have shown that sap flux can be used to scale directly to leaf-chamber based stomatal conductance (Ewers et al. 2007a). Once sap flux is scaled to estimate stomatal conductance, it can then be used to explain how forest transpiration may change with successional time based on species or functional trait based approaches. Such approaches have been used to test hypotheses about how tree hydraulics explicitly regulate stomatal conductance as trees increase in size and age (McDowell et al. 2002; Ryan et al. 2000; Phillips et al. 2003). Because of the evidence showing that most trees regulate leaf water potential to prevent runaway embolism (Oren et al. 1999; Ewers et al. 2007b;

Franks et al. 2007), sap flux measurements provide a relatively simple method for testing both large- and small-scale models of tree and stand transpiration.

E_T from vegetated land surfaces can be predicted from environmental drivers using the Penman-Monteith (PM) combination equation:

$$E_{TPM} = \frac{s \cdot (R_n - G) + \rho_a c_p G_a D}{\rho_w \lambda [s + \gamma \cdot (1 + G_a / G_v)]} \tag{18.1}$$

where s is the rate of change of saturation vapor pressure with temperature [kPa °C^{-1}], R_n is net absorbed radiation [W m^{-2}], G is ground heat flux [W m^{-2}], ρ_a is air density [kg m^{-3}], c_p is the specific heat capacity of air [J kg^{-1} °C^{-1}], G_a [m s^{-1}] is aerodynamic conductance, *VPD* is vapor pressure deficit [kPa], ρ_w is density of water [kg m^{-3}], λ is latent heat of vaporization [J kg^{-1}], γ is the psychrometric constant [kPa °C^{-1}], and G_v is a combination of leaf boundary-layer (G_b) and canopy stomatal conductance (G_s). G_s is dynamic with environmental drivers at daily to seasonal time scales and can potentially change with tree size and age which can be incorporated as changes in G_{Sref} (Ewers et al. 2008).

3 Example Chronosequences of Stand Transpiration

To assess how forest aging affects canopy transpiration, we reevaluated two North American forest chronosequences of stand transpiration. The first example is from the northern hardwoods forest of the Western Great Lakes and uses the transpiration data and modeling tests from Ewers et al. (2002, 2007b, 2008) and Samanta et al. (2007, 2008). In this example we look at multiple pathways of succession from forest management culminating in an old-growth forest that is now rare in this region. The second example comes from northern North American boreal forests in which fire drives the disturbance regime and uses transpiration data and modeling results from Ewers et al. (2005), Angstmann (2009) and Bond-Lamberty et al. (2009). We take two different approaches to modeling and scaling these results. In the first example, we use a detailed stand transpiration model operating at half-hourly time scales assessed against stand transpiration with no additional scaling beyond the stand level. In the boreal example, we use a stand model that directly incorporates succession operating at regional scales to assess stand transpiration over 150 years. In both cases we ask the question "If we chose a reference stomatal conductance from one stand age of the entire chronosequence to put into a model, would modeled transpiration for the other ages be biased?"

3.1 *Northern Hardwoods Forest Chronosequence*

The stands in our northern hardwoods chronosequence are a subset of forest sites that are represented by 11 flux towers in the northern Wisconsin/upper Michigan region and which comprise the Chequamegon Ecosystem Atmosphere Study

(Davis et al. 2003, ChEAS, http://cheas.psu.edu). Sites in the ChEAS network allow for an examination of how dynamics in stand age, stand type, substrates and other spatial variables affect regional estimates of land– surface atmosphere interactions. The climate of ChEAS is northern continental, characterized by short growing seasons and cold winters. Soil moisture deficits impact transpiration and stomatal conductance rarely (~ 1 in 10 years) and only in drought years, so it is not needed in modeling the data sets used here (Ewers et al. 2002; Mackay et al. 2002, 2007; Cook et al. 2004; Desai et al. 2005; Samanta et al. 2008).

The three chronosequence stands are regenerating aspen (*Populus tremuloides*) with an understory of balsam fir (*Abies balsamea*; hereafter called RA), mature sugar maple (hereafter called MM), and old-growth sugar maple (*Acer saccharum*) and eastern hemlock (*Tsuga canadensis;* hereafter called OM). This is not a traditional chronosequence because the aspen and maple stands represent two potential pathways of succession that will likely result in old growth. The old-growth site is located 100 m north of the boundary to the Sylvania Wilderness and Recreation area, Ottawa National Forest, Michigan, USA (46° 14′ 31″ N, 89° 20′ 52″ W). The 8,500 ha Sylvania Wilderness in the Upper Peninsula of Michigan is one of few large tracts of old-growth forest in the Midwest (Frelich 1995). There is no history of logging disturbance at this site. Tree ages range 0–350 years, and dominant species are sugar maple and eastern hemlock. The forest occurs within a glacial outwash and moraine landscape (Ferrari 1999), which creates an irregular and hummocky landscape with an average slope of 10% over short distances (Davis et al. 1996; Pastor and Broschart 1990); average elevation is 517–567 m. Dominant upland soils are moderately well drained, coarse or sandy loam Spodosols (Pastor and Broschart 1990). Our study site is in an *A. saccharum*-dominated old-growth stand with *A. saccharum* comprising 71% of trees in addition to *T. canadensis* (14%), yellow birch (*Betula allenghaniensis* Britton (7%)), American basswood (*Tilia americana* L.) and American hophornbeam (*Ostrya virginiana* (Mill.) K. Koch) (8% each). Additional details and description of this site are found in Desai et al. (2005).

The MM forest is located in the Chequamegon- Nicolett National Forest, WI, USA (45° 48′ 21″ N, 90° 04′ 47″ W) and is approximately 50 km from Sylvania. Dominant species at this site are *A. saccharum* (68%), *T. americana* (24%), and green ash (*Fraxinus pennsylvanica* Marsh.) (8%). The stand is about 70 years old with a sandy loam soil. A detailed site description can be found in Cook et al. (2004, cf. Willow Creek site).

The RA stand (tree age ~ 20 years) is a result of cutting and is solely dominated by trembling aspen in the canopy with balsam fir forming an understory in sandy loam soil. The site is located in the Hay Creek Wildlife Management Area. More details of the site can be found in Ewers et al. (2007b).

3.1.1 Northern Hardwoods Ecosystem Model Description

To test how our stand scale parameters would affect ecosystem modeling we ran the Terrestrial Regional Ecosystem Exchange Simulator (TREES) (Ewers et al. 2008; Mackay et al. 2003; Samanta et al. 2007; Samanta et al. 2008) on each plot. Most of the details of this canopy model are described in these previous publications, and so here we present a few modifications that have been made to deal with more heterogeneous canopies along chronosequences.

We employed a simple whole-plant hydraulic function (Oren et al. 1999) as a constraint on C_3 photosynthesis (dePury and Farquhar 1997; Farquhar et al. 1980) in sun and shade canopy elements. The hydraulically and biochemically limited photosynthesis was then used to calculate stomatal conductance to water vapor and then transpiration for both elements in parallel. We summed the sun and shade transpiration values to obtain canopy transpiration (E_{Csim}), which was then evaluated against sap flux scaled E_C. Here we describe just the parts of the model needed to interpret the results from the current study.

Sunlit leaf area of the canopy (LAI_{sun}) used the equations from Campbell and Norman (1998) with leaf angle distribution and clumping factors appropriate for broadleaf and needle-leaf. *LAI* is canopy total half-hemispherical leaf area index and $LAI{*}P_{CC}^{-1}$ is clumped leaf area index taking into account percent canopy closure (P_{CC}) at the plot scale. Shaded leaf area (LAI_{shd}) was calculated by subtracting L_{sun} from clumped *LAI* using values from Table 18.1. OM and MM stands were dominated by overstory with mixed species and no understory species, while the RA stand had overstory aspen and understory balsam fir. For this two-level canopy we first ran TREES on the aspen to determine the total short-wave radiation absorbed by this overstory species, and then deducted this from total above-canopy short-wave radiation, with this remaining radiation serving as the input for the balsam fir canopy. P_{CC}, the Bayesian posterior prediction interval for canopy closure, was a parameter input to the model which is described in the next section.

Canopy average stomatal conductance without photosynthetic limitation was calculated as (Oren et al. 1999):

$$G_s = G_{Sref} - m \ln D \tag{18.2}$$

where G_{Sref} is reference canopy stomatal conductance (mol m^{-2} s^{-1}) at $VPD = 1$ kPa in absence of photosynthetic light limitation, and $m = -\partial G_S / \partial \ln D$ is the sensitivity of G_S to *VPD*. In mesic environments, for species that regulate leaf water potential to just prevent runaway cavitation it has been shown that $m = 0.6G_{Sref}$ (Addington et al. 2004; Ewers et al. 2000; Ewers et al. 2005; Ewers et al. 2007b; Oren et al. 1999; Wullschleger et al. 2002). For the understory balsam fir in the RA stand we calculated a reduced canopy *VPD* based on a ratio between upper canopy and below canopy *VPD* from a nearby, similarly structured aspen-fir stand (Loranty et al. 2008). Both G_{Sref} and m were derived through Bayesian parameterization.

The CO_2 conductance of sunlit and shaded canopy elements (hereafter denoted by subscript k) was calculated from stomatal and aerodynamic conductances in series. Leaf-level stomatal conductance to vapor as modified by photosynthesis was determined from Fick's First Law of Diffusion using $A_{n,k}$ as net photosynthesis (μmol m^{-2} s^{-1}), and c_a and $c_{i,k}$ as ambient and intercellular CO_2 concentrations, respectively. $A_{n,k}$ was calculated using the Farquhar photosynthesis model (dePury and Farquhar 1997;

Table 18.1 Predicted parameter distributions from Bayesian analysis, showing mean, median, and 95% posterior prediction interval for canopy closure (P_{CC}), canopy average reference stomatal conductance (G_{Sref}), canopy average sensitivity of canopy conductance to vapor pressure deficit (m), initial slope of the photosynthetic light response curve (quantum yield, ϕ), and maximum carboxylation rate (V_{cmax})

Parameter	Stand[a]	Species[b]	Mean	Median	95% Posterior interval		
P_{CC} ($m^2\ m^{-2}$)	OM	Hemlock	0.102	0.101	0.100	–	0.106
	OM	Sugar maple	0.630	0.627	0.582	–	0.693
	MM	Sugar maple	0.984	0.987	0.953	–	0.999
	RA	Trembling aspen	0.971	0.975	0.919	–	0.998
	RA	Balsam fir	0.121	0.118	0.102	–	0.154
G_{sref} (mmol $m^{-2}\ s^{-1}$)	OM	Hemlock	21.8	21.8	21.1	–	22.5
	OM	Sugar maple	30.8	30.9	29.2	–	32.5
	MM	Sugar maple	90.3	90.2	87.7	–	93.3
	RA	Trembling aspen	92.7	92.7	89.9	–	95.9
	RA	Balsam fir	73.4	73.4	69.1	–	77.7
m (mmol $m^{-2}\ s^{-1}$)	OM	Hemlock	–2.5	–2.5	–1.3	–	–3.9
	OM	Sugar maple	–14.9	–14.8	–13.4	–	–16.3
	MM	Sugar maple	–58.8	–58.8	–53.6	–	–64.3
	RA	Trembling aspen	–45.2	–45.1	–40.9	–	–49.4
	RA	Balsam fir	–37.5	–37.4	–28.4	–	–47.0
ϕ (mol CO_2 mol^{-1} photons)	OM	Hemlock	0.095	0.097	0.056	–	0.123
	OM	Sugar maple	0.097	0.104	0.037	–	0.124
	MM	Sugar maple	0.087	0.091	0.033	–	0.123
	RA	Trembling aspen	0.091	0.093	0.055	–	0.115
	RA	Balsam fir	0.106	0.108	0.073	–	0.124
V_{cmax} (μmol $m^{-2}\ s^{-1}$)	OM	Hemlock	11.0	10.6	7.1	–	18.1
	OM	Sugar maple	45.0	46.9	16.9	–	67.1
	MM	Sugar maple	31.2	29.9	9.5	–	59.2
	RA	Trembling aspen	235.0	244.2	143.8	–	273.4
	fir	Balsam fir	9.0	8.4	6.9	–	13.9

[a]*OM* = Old-growth Maple; *MM* = Mature Maple; *RA* = Regenerating Aspen
[b]hemlock = *T. canadensis*; sugar maple- *A. saccharum*; trembling aspen = *P. tremuloides*; and balsam fir = *A. balsamea*

Farquhar et al. 1980) and $c_{i,k}$ was calculated following Katul et al. (2003), both constrained by stomatal conductance to CO_2 ($g_{CO,k}$). Most parameters for the photosynthesis sub-model were taken from literature (dePury and Farquhar 1997; Medlyn et al. 2002), but quantum yield (ϕ; mol CO_2 mol^{-1} photons), which controls the light-limited response, and maximum carboxylation rate (V_{cmax}; μmol $m^{-2}\ s^{-1}$), which controls the light-saturated response, were estimated using Bayesian parameterization. Simulated canopy transpiration (E_{Csim}) was based on the PM equation, weighted by the relative amounts of sun and shade element transpiration.

3.1.2 Bayesian Model Parameterization

To estimate P_{CC}, G_{Sref}, m, φ, and V_{cmax} and propagate the uncertainty in these parameters to E_{Csim} we employed Bayesian analysis, which has been discussed in detail in Samanta et al. (2007, 2008). The posterior distribution was sampled using Markov Chain Monte Carlo (MCMC) simulation (Metropolis et al. 1953; Hastings 1970). For each study site we initialized three Markov chains with randomly generated initial parameter values. Each chain was run through 200,000 iterations, with the first half of each chain used for chain convergence and the second half used to generate sample distributions for each of the parameters. Chain convergence was verified both visually and quantitatively by estimating the factor by which the scale of the current distribution for each parameter might be reduced if the chains were continued in the limit as the number of iterations approaches infinity (Gelman 2004). From the second half of each chain we sampled every 50th iteration to obtain a total of 6,000 samples of each parameter at each site. We then ran TREES over the sampled posterior distribution of the parameters to obtain mean E_{Csim} and 95% confidence intervals.

3.1.3 Northern Hardwoods Forest Chronosequence Model Results and Discussion

Predictions of diurnal mean transpiration per unit leaf area (E_L) for RA, MM, and OM forests closely fit the observed fluxes ($P<0.001$ for the regression between measured and model E_L for all species) and the 95% posterior intervals bounded roughly 95% of the observations (Fig. 18.1). The predicted mean fluxes slightly underestimated observed fluxes for *T. canadensis* towards the end of the simulation period, but there were no other trends among the other species. Moreover, residuals plotted against observed or modeled fluxes showed no trends (not shown here) with any of the species. For a more complete appraisal of the TREES model performance on the sugar maple trees see Ewers et al. (2008), all of the original data for the diurnal means are from that publication and Ewers et al. (2002, 2007b).

Physiological parameters differed significantly among stand ages (Table 18.1). G_{Sref} and m were substantially lower, by a factor of 2–3 times, for the old-growth species than for the others. For all age classes in the chronosequence, G_{Sref} was slightly higher for broadleaf species in comparison to needle-leaf species. The relationship between m and G_{Sref} across the entire chronosequence was linear with a slope near 0.6; i.e. the changes in plant hydraulics with stand age and species just moved along the line. This result and its correlation with changes in sapwood-to-leaf area ratio were shown with data only for all of these species in Ewers et al. (2007b). Values for ϕ showed no substantial differences within or among age-classes, while V_{cmax} was substantially lower for needle-leaf species compared to broadleaf species. In general, posterior distributions were normally distributed and typically symmetric about the mean. P_{CC} was correlated with *LAI* in all species, as expected, with narrow 95% posterior prediction intervals.

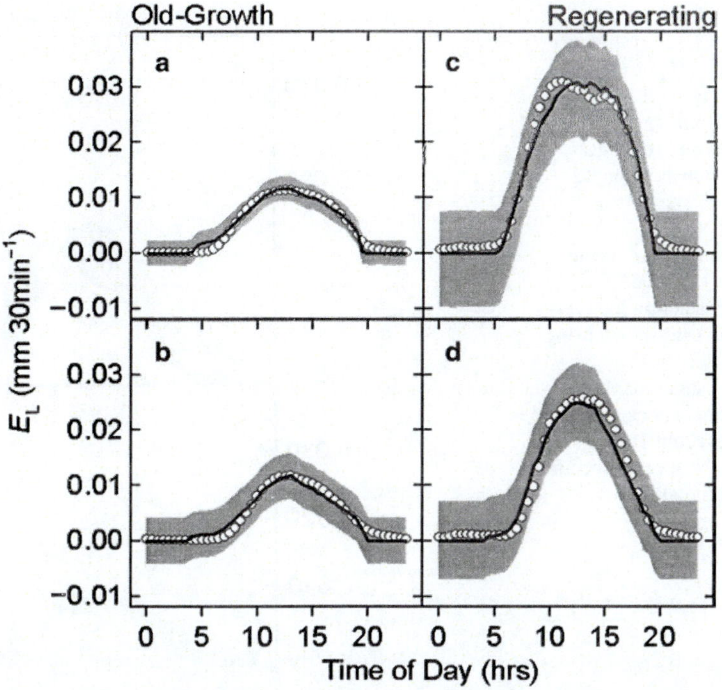

Fig. 18.1 Diurnal means of measured (*open circles*), posterior simulation means (*solid lines*), and 95% posterior prediction intervals (*grey bands*) for diurnal canopy transpiration per unit leaf area (E_L). (**a**) old-growth sugar maple (OM), (**b**) old-growth eastern hemlock (OM), (**c**) second-growth trembling aspen (RA), and (**d**) second-growth sugar maple (MM)

We show significant differences in tree hydraulic properties, as represented by lower G_{Sref} and m (Table 18.1), in the oldest site compared to the younger, regenerating sites; thus the answer to our driving question-- "If we chose a reference stomatal conductance from one stand age of the entire chronosequence to put into a model, would modeled transpiration for the other ages be biased?"-- is yes. The TREES simulation results clearly show the modeled age effect (Fig. 18.2). This is consistent with declines in transpiration per unit leaf area with stand age (Cornish and Vertessy 2001; Delzon and Loustau 2005; Domec and Gartner 2002); although, specific mechanisms for this are due more to tree height impacts on plant hydraulics than age (Delzon et al. 2004; Bond et al. 2007; Novick et al. 2009). It should be noted that hydraulic parameters for these species have previously been shown to have inter-annual variability by up to a factor of two (Ewers et al. 2007b, 2008), but not the threefold difference observed here from MM and OM forests. The photosynthetic parameters covaried with the plant hydraulic changes as expected from theory (Katul et al. 2003).

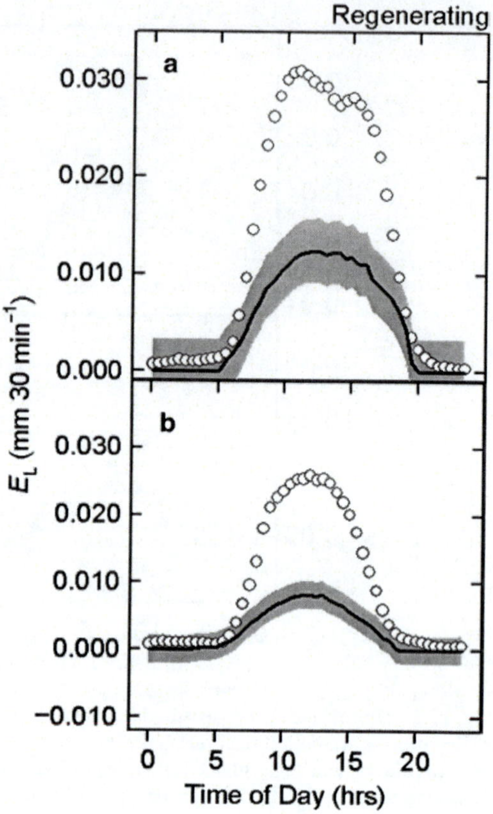

Fig. 18.2 Mean diurnal response of transpiration per unit leaf area (E_L) from sap flux as in Fig. 18.1 in (**a**) aspen from regenerating aspen (RA) and (**b**) sugar maple from mature maple (MM) (open symbols in each). In both panels, posterior simulation means (*solid lines*), and 95% posterior prediction intervals (*grey bands*) are shown using the parameters from old-growth sugar maple and eastern hemlock (both from OM) (From Table 18.1 instead of the species-specific functional traits)

3.2 *Boreal Forest Chronosequence*

After the conclusion of the Boreal Ecosystem-Atmosphere Study (BOREAS, Sellers et al. 1997; Hall 1999) project in the 1990s, a follow-on study was established to study the effects of stand age on ecosystem carbon and water cycling. The BOREAS study design focused almost exclusively on mature stands, in spite of the fact that North American boreal forests are subject to frequent fires and thus successional dynamics (Fortin et al. 1999). Thus a series of follow-on studies aimed to examine the effects of stand age and succession-driven species shifts on forest biogeochemical cycling. From a water cycling perspective, boreal ecosystems are unusual and interesting because of (i) the importance of bryophytes—nonvascular, poikilohydric plants such as *Sphagnum* and *Pleurozium*—with differing and unusual water acquisition and transport mechanisms compared to forest trees; (ii) the overwhelming dominance of the phenotypically plastic conifer black spruce, *Picea mariana*, in the poorly-drained landscape, and (iii) the interaction between water cycling and the occurrence of wildfires (Amiro et al. 2006) that shape the larger landscape into a mosaic of differently-aged stands.

The follow-on study was conducted in black spruce (*Picea mariana* Mill. BSP)-dominated chronosequences west of Thompson, Manitoba, Canada, near the BOREAS Northern Study Area (55° 53′ N, 98° 20′ W). The full chronosequence consisted of seven different-aged black spruce forests, all of which originated from stand-killing wildfire in mature forests; the youngest stands burned ~2 years previous to the study initiation, while the oldest chronosequence stands (~157 years in 2007) surrounded the BOREAS NSA tower site (Goulden et al. 1998; Dunn et al. 2007). The stands were dominated by three tree species: trembling aspen (*Populus tremuloides* Michx), black spruce, and jack pine (*Pinus banksiana* Lamb). Early-successional deciduous tree species are replaced by black spruce in the older stands; the black spruce canopy closure, at 50–60 years, is associated with drastic thinning of the understory and growth of thick feather mosses. Mean annual temperature and precipitation were ~0 °C and ~450 mm, respectively.

The carbon and water dynamics of these sites were reported in a series of papers with the more notable conclusions that (1) both well- and poorly-drained areas transitioned from carbon loss to carbon gain at 15–20 years (Litvak et al. 2003; Bond-Lamberty et al. 2004); (2) species changes over succession significantly affected tree canopy transpiration (Ewers et al. 2005) as well as the significant bryophyte contribution to E_T (Bond-Lamberty et al. 2010); and (3) soil respiration was relatively invariant except in the first few years after fire (Wang et al. 2002).

3.2.1 Biome-BGC-MV Canopy Transpiration Model Description

The boreal chronosequence simulations discussed here were described in Bond-Lamberty et al. (2007a). The model used, Biome-BGC-MV (Running and Coughlan 1988; Running and Gower 1991; Kimball et al. 1997; Bond-Lamberty et al. 2005), runs on a daily time step, using daily meteorological data, ecophysiological parameters, and general stand soil information to simulate energy, C, water, and nitrogen cycling (Running and Hunt 1993; White et al. 2000). Ecophysiological parameters define vegetation types—descriptions based on leaf habit, photosynthesis pathway, plant type, etc.—that are intended to be broadly significant. These parameters change with stand age using height based-competition within and between conifers (black spruce and jack pine) and hardwoods (trembling aspen) to adjust a maximum G_S (Bond-Lamberty et al. 2009) which can be directly related to G_{Sref} and thus linked back to plant hydraulic mechanisms (Mackay et al. 2003).

There are two ways in which water can enter a Biome-BGC simulation, via soil inflow and precipitation, and six ways in which it can exit: soil outflow, soil evaporation, snow sublimation, canopy evaporation, canopy transpiration, and volatilization in fire. A fraction of precipitation is intercepted by the plant canopy and subject to evaporation before transpiration occurs. Soil evaporation and canopy E_T are determined using the Penman-Monteith equation (Monteith 1965; Campbell and Norman 1998) and thus driven primarily by radiation and vapor pressure deficit. Leaf and canopy conductances are determined by stomatal, cuticular and

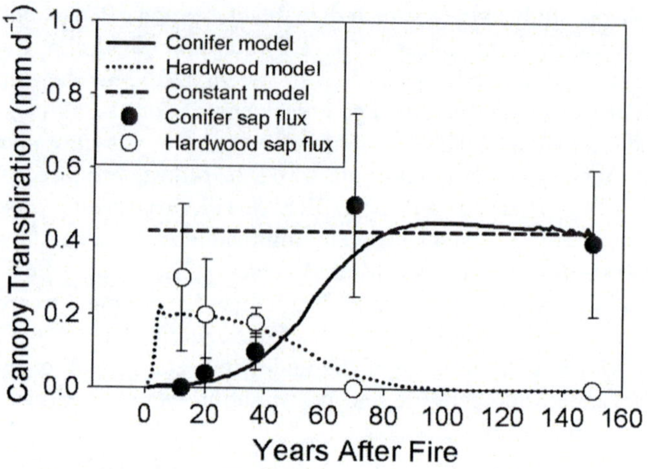

Fig. 18.3 Biome-BGC canopy transpiration simulations across the chronosequence. The model simulations are based on a conifer model simulating black spruce and jack pine and a hardwood model simulating trembling aspen. The constant model uses the potential vegetation from Northern Old Black Spruce (NOBS) site for the stomatal conductance across the entire chronosequence. Sapflux scaled estimates of canopy transpiration include standard errors taken (From Ewers et al. 2005)

boundary layer conductances (the first two in parallel, the last in series), defined on a vegetation-specific basis, which varies with successional status based on competition and is modulated by plant water stress, temperature, vapor pressure deficit and soil anoxia (Thornton 1998; Bond-Lamberty et al. 2007a, b).

3.2.2 Boreal Forest Model Results and Discussion

There is a stark difference in Biome-BGC-MV simulated canopy transpiration using a constant vs. dynamic G_{Sref} with stand age (Fig. 18.3). While this result may seem obvious, prior models of boreal evapotranspiration components do not incorporate multiple vegetation types and thus cannot account for much of the dynamics shown here (Bond-Lamberty et al. 2009). The model captures the impact of tree functional types (and thus species) on canopy transpiration because the simulations are always within the mean plus standard error of the sap-fluxed based canopy transpiration. Further, even the dynamic in the means of sap-flux based canopy transpiration is captured by the simulations (Fig. 18.3).

Because of the theory behind Eq. 18.2, the dynamics in G_{Sref} predicted by Biome-BGC-MV can be directly related to the impact of tree species and size on plant hydraulics. Trembling aspen and jack pine show the typical response of increasing sapwood-to-leaf area ratio with increasing height to compensate for increased resistance with path length while black spruce shows an opposite response (Ewers et al. 2005). The model successfully captures these differences as shown by the rise and fall in canopy transpiration in the hardwood before 40 years after fire and the drop in canopy transpiration of the conifers between 70 and 150 years (Fig. 18.3). Thus, the answer to our question "If we chose a reference stomatal conductance from one stand age of the entire chronosequence to put into a model, would modeled transpiration for the other ages be biased?"-- is yes.

4 Synthesis of North American Northern Hardwood and Boreal Forest Canopy Transpiration Chronosequences

Both of our boreal and northern hardwood chronosequences are challenged by not satisfying all the assumptions of a true chronosequence. In the case of the northern hardwoods, the first two stands represent alternative states of early succession, which violate the assumption that all stands follow the same successional trajectory. These forests may start as either aspen or maple dominated while both will end up as maple and hemlock dominated if undisturbed for centuries (Davis et al. 1994). The boreal chronosequence discussed here was not formally replicated, but the stands were shown to be consistent (in structure and net primary production fluxes) with an additional 28 stands in northern Manitoba (Bond-Lamberty et al. 2004). Although Goulden et al. (2006) used Landsat TM and ETM + data to conclude that this chronosequence comprised a valid space-for-time substitution, Bond-Lamberty and Gower (2008) noted that some parameters do not fit a chronosequence. Despite not satisfying all of the chronosequence assumptions explicitly, stand age still plays a crucial role in explaining the ecosystem fluxes from both of these forest types (Desai et al. 2005; Bond-Lamberty et al. 2004). In both cases we used a biophysical approach through modeling stomatal conductance response to environmental drivers with the models operating at very different temporal and spatial scales. This biophysical approach models not the actual species but rather their functional traits (specifically G_{Sref} and m) that provide the species and tree age/size distinction for transpiration. Because these types of traits are valid for predicting transpiration regardless of how well we get the species composition or even proportion of species right with succession (Letcher and Chazdon 2009), our approach to modeling provides a mechanism for explaining the relevant variation in tree transpiration as forest stands age.

The answer to our question "If we chose a reference stomatal conductance from one stand age of the entire chronosequence to put into a model, would modeled transpiration for the other ages be biased?" is clearly yes for the boreal and northern hardwoods chronosequences. Thus, even if our chronosequences are wildly deviant from conforming to the assumptions of true chronosequences, the large change in transpiration with stand age strongly suggests that this must be quantified and considered for regional type modeling and calculating the water cost of increasing carbon sequestration by vegetation (Jackson et al. 2005). Further, our functional trait approach can be used in larger scale models that investigate mass and energy exchanges (Bond-Lamberty et al. 2009), which would improve the ability of the models to determine the interaction between land surface processes and climate change. This argument is buttressed by the E_{T} data from FLUXNET analyzed by Baldocchi (2005) showing that differences in functional traits reflected varying amounts of species' change in E_T.

5 Synthesis of Literature Examples of Canopy Transpiration from Forest Chronosequences

To further test how stand age impacts tree transpiration, we synthesize studies that combine sap flux and chronosequence approaches in forests. We selected as examples only those studies from the literature that reported both or one of sapwood and leaf area per unit ground area and explicitly investigate the impact of stand age. We thus excluded studies that looked at the impact tree age/size within stands. From this synthesis, it is clear that stand transpiration varies with tree and stand age, across dramatically different forest types (Table 18.2 and Fig. 18.4a).

Sapwood area and leaf area are more important determinants of stand transpiration at a given set of environmental conditions than sap flux per unit xylem area (Fig. 18.4b–c). The minimum difference in transpiration across a chronosequence was 0.4 mm d^{-1} in the North Americal Boreal Mixed Poorly-Drained boreal forest type while the maximum difference was 2.4 mm d^{-1} in the North American Mixed Conifer forest type; even 0.4 mm d^{-1} is a substantial difference in water loss just from canopy trees. The variability in canopy transpiration is constrained the older the forest (stands ages close to 200 years, Fig. 18.4a); however, this may be due to lack of studies in these ages rather than a real phenomenon. Across these very different chronosequences, it is clear that sapwood area is driving transpiration when it is less than 12 m^2 ha^{-1} while the same is true of *LAI* below 5 m^2 m^{-2}. At these low to moderate values of sapwood and leaf area, the trees are limited by the ability to exchange gas with the atmosphere, and leaf area is then likely correlated to the sapwood area as suggested by the pipe-model theory (Shinozaki et al. 1964). These data point to the use of fairly simple models for areas dominated by younger stands while older stands will need more complex models to capture the climatic, edaphic and biological differences driving transpiration reflected in the higher sapwood and leaf areas.

This synthesis reinforces our premise from earlier in the chapter (Sect. 2.1) that tree (Hatton et al. 1990), stand (Traver et al. 2010) and watershed (Ford et al. 2007) approaches agree that sapwood area is crucial to determine transpiration expressed as water lost by trees per ground area. From these studies and this synthesis, we suggest that more effort be spent on quantifying sapwood areas appropriately throughout stands while the sap flux per unit xylem area measurements are used to determine the temporal response to environmental drivers (Mackay et al. 2010). Such an approach will maximize the functional trait information gleaned from the data and ultimately better inform the models using such traits. A functional trait based approach to modeling the impact of forest stand age on ecosystem water loss will require models that can capture the dynamics in stomatal conductance while simultaneously coping with other dominant factors that are changing with stand age such as soils and microclimate. These results contradict models that use a potential

Table 18.2 Literature examples of stand transpiration chronosequence data from sap flux studies

Forest type	Stand ages (years)[a]	Species[b]	Leaf area index ($m^2\ m^{-2}$)	Sapwood area ($m^2\ ha^{-1}$)	Transpiration ($mm\ d^{-1}$)	Transpiration timing	Source
European Spruce	40, 70, 110, 140	*P.a.*	6.3, 7.9, 7.6, 6.5	23, 20, 19, 14	1.6, 1.4, 1.5, 0.9	Peak month	Alsheimer et al. (1998)
European Pine	10, 32, 54, 91	*P.p.*	2.9, 2.3, 1.8, 1.8	16, 21, 20, 20	0.9, 0.5, 0.3, 0.3	Annual mean	Delzon and Loustau (2005)
Australian Old Eucalyptus	50, 90, 150, 230	*E.r.*	–	7, 6, 4, 4	1.9, –, –, 0.8	Peak month	Dunn and Conner (1993)
North American Boreal Mixed Well Drained	19, 44, 78, 158	*P.m., P.b., P.t.*	0.4, 2.0, 7.5, 6.1	5, 5, 8, 2	0.3, 0.3 ,0.9 ,0.2	Mean growing season	Ewers et al. (2005), Angstmann (2009)
North American Boreal Mixed Poorly Drained	19, 44, 78, 158	*P.m., P.b., P.t.*	0.3, 0.5, 0.6, 1.0	3, 5, 1, 4	0.6, 0.6, 0.1, 0.2	Mean growing season	Angstmann (2009)
North American Northern Hardwoods	20, 70, 350	*P.t., A.b., A.s.,T.c.*	3.5, 5.4, 7.1	18, 24, 20	2.1, 3.9, 2.0	Mean growing season	Ewers et al. (2007b, 2008)
Australian Young Eucalyptus	2, 4, 5, 6, 7, 8	*E.g.*	1.0, 2.7, 2.8, 3.7, 3.5, 2.7	1, 6, 7, 11, 11, 9	0.4, 1.3, 1.6, 1.4, 1.9, 1.1	Annual mean	Forrester et al. (2009)
North American Pine	14, 250	*P.p.*	1.0, 2.1	–, –	0.8,1.0	Peak annual	Irvine et al. (2002)
North American Mixed Conifer	40, 450	*A.r.,Ps.m.; T.h.*	10.6, 12.1	22, 16	3.6,1.2	Peak growing season	Moore et al. (2004)
Asian Boreal	28, 56, 138, 203, 383, 243	*P.s.*	0.6, 1.6, 1.3, 1.5, 1.6, 0.6	5, 11, 6, 8, 7, 4	0.6, 1.3, 0.3, 0.7, 0.8, 0.4	Mean growing season	Zimmerman et al. (2000)

Forest type is based on the single common name or a general forest type if multiple species measured; stand ages are estimated from tree ages or time of disturbance depending on study; species list includes all that were measured for entire chronosequence where some species may be absent in specific stands; sapwood area is the sum of all species in each; transpiration is the sum of canopy transpiration of all species in the stand; transpiration timing indicates the method by which the daily average transpiration was estimated and source is the single or multiple studies contributing to the data

[a]If replications were conducted at the same stand age, all of the replications were averaged

[b]*P.a.–Picea abies, P.p–Pinus pinaster, E.r.–Eucalyptus regnans, E.g.–Eucalyptus globulus, P.p.–Pinus ponderosa, P.s.–Pinus sylvestris, A.r. Alnus rubra, Ps.m.–Pseudotsuga menziesii, T.h. Tsuga heterophylla, P.s.–Pinus sylvestris, P.m.–Picea mariana, P.b., Pinus banksiana, P.t.–Populus tremuloides, A.b.–Abies balsamea, A.s.–Acer saccharum, T.s.–Tusga canadensis*

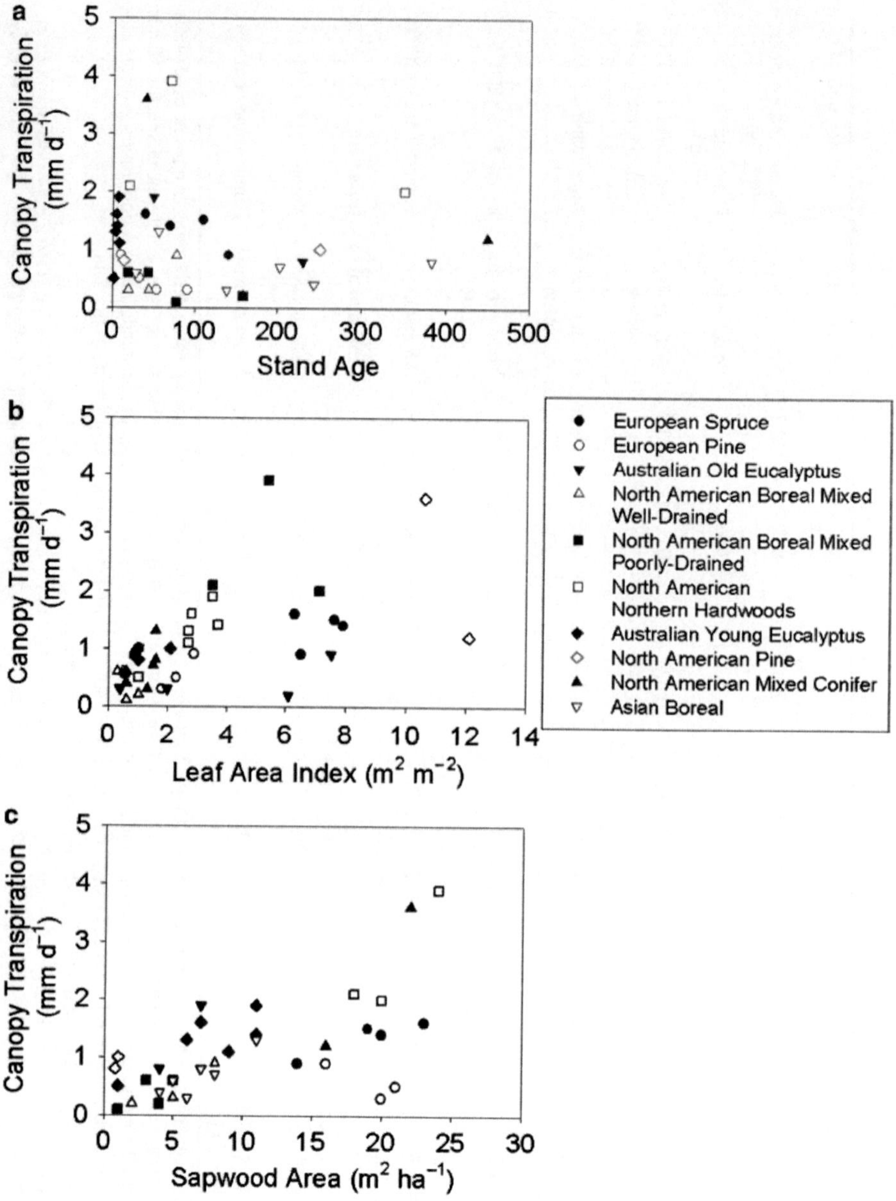

Fig. 18.4 Summary of sap flux estimates of canopy transpiration from chronosequences reviewed in Table 18.2 using the columns of stand age (**a**), leaf area index (**b**) and sapwood area (**c**) as explanatory variables

vegetation approach (i.e., end of a chronosequence--constant model Fig. 18.3) such as IBIS (Foley et al. 1996; Kuckarik et al. 2006) to model regional vegetation feedback loops on global climate change. Such models are unfortunately reinforced by global FLUXNET datasets that look at mature but rarely regenerating stands (e.g., Teuling et al. 2009). One approach to improve such models is to incorporate height driven ecosystem demography into ecosystems, which are then constrained by LiDAR estimates (Hurtt et al. 2004). Our approach here strongly suggests that the stomatal conductance portion of revised or new models of canopy transpiration should be based on emerging plant hydraulic knowledge that provides realistic but relatively simple representation of plant functional traits.

6 Conclusion

We tested whether using stomatal conductance could capture the dynamic in transpiration with forest age. To do this we answered the question "If we chose a reference stomatal conductance from one stand age of the entire chronosequence to put into a model, would modeled transpiration for the other ages be biased?" with a resounding yes. We found that obtaining the right stomatal conductance was crucial for accurate models in two different chronosequences. This strongly suggests that stomatal conductance can be an integrator of inter- and intra-species change in tree transpiration with forest age. If we had tried to use a single reference canopy stomatal conductance, it would not have been able to capture the variability in transpiration with stand age despite the suggestion that hydraulic limitation was consistently acting on the trees. Because we used a biophysical approach, even if our and other researchers' chronosequences do not fit the assumptions, the results are still useful. Further, our synthesis of sap flux based estimates of tree transpiration shows a large successional dynamic, and suggests that our approach to modeling is crucial in the face of anthropogenic changes to forest age, structure and species composition. We have now provided the framework for a mechanistically rigorous, yet simple approach based on tree hydraulics to measuring and modeling stand transpiration with changing forest age and/or species composition.

References

Aber JD, Federer CA (1992) A generalized, lumped-parameter model of photosynthesis, evapotranspiration and net primary production in temperate and boreal forest ecosystems. Oecologia 92:463–474

Addington RN, Mitchell RJ, Oren R et al (2004) Stomatal sensitivity to vapor pressure deficit and its relationship to hydraulic conductance in *Pinus palustris*. Tree Physiol 24:561–569

Adelman JD, Ewers BE, Mackay DS (2008) Use of temporal patterns in vapor pressure deficit to explain spatial autocorrelation dynamics in tree transpiration. Tree Physiol 28:647–658

Alsheimer M, Kostner B, Falge E et al (1998) Temporal and spatial variation in transpiration of Norway spruce stands within a forested catchment of the Fichtelgebirge, Germany. Ann Sci For 55:103–123

Amiro BD, Barr AG, Black TA et al (2006) Carbon, energy and water fluxes at mature and disturbed forest sites, Saskatchewan, Canada. Agric For Meteorol 136:237–251

Angeles G, Bond B, Boyer JS et al (2004) The cohesion-tension theory. New Phytol 163:457–459

Angstmann JL (2009) Spatial and temporal variability of tree transpiration and its drivers along a soil drainage gradient in the boreal black spruce forest. University of Wyoming, Laramie

Baldocchi DD (2005) The role of biodiversity on the evaporation of forests. In: Scherer-Lorenzen M, Korner C, Schulze E-D (eds) Ecological studies 176 forest diversity and function: temperate and boreal systems. Springer, Heidelberg

Barbour MM, Whitehead D (2003) A demonstration of the theoretical prediction that sap velocity is related to wood density in the conifer Dacrydium cupressinum. New Phytol 158:477–488

Bond BJ, Meinzer FC, Brooks JR (2007) How trees influence the hydrological cycle in forest ecosystems. In: Wood PJ, Hannah DM, Sadler JP (eds) Hydroecology and ecohydrology: past, present and future. Wiley, Hoboken

Bond-Lamberty B, Gower ST (2008) Decay and fragmentation of coarse woody debris: revisiting a boreal black spruce chronosequence. Ecosystems 11(6):831–840

Bond-Lamberty B, Wang C, Gower ST (2004) Net primary production and net ecosystem production of a boreal black spruce fire chronosequence. Glob Change Biol 10:473–487

Bond-Lamberty B, Gower ST, Ahl DE et al (2005) Reimplementation of the BIOME-BGC model to simulate successional change. Tree Physiol 25:413–424

Bond-Lamberty B, Gower ST, Ahl DE (2007a) Improved simulation of poorly drained forests using Biome-BGC. Tree Physiol 27:703–715

Bond-Lamberty B, Peckham SD, Ahl DE et al (2007b) The dominance of fire in determining carbon balance of the central Canadian boreal forest. Nature 450:89–92

Bond-Lamberty B, Peckham SD, Gower ST et al (2009) Effects of fire on regional evapotranspiration in the central Canadian boreal forest. Glob Change Biol 15:1242–1254

Bond-Lamberty B, Gower ST, Amiro BD, Ewers BE (2010) Measurement and modeling of bryophyte evapotranspiration in a boreal forest chronosequence. Ecohydrology. doi:DOI: 10.1002/eco.118

Brodribb TJ, Feild TS (2000) Stem hydraulic supply is linked to leaf photosynthetic capacity: evidence from New Caledonian and Tasmanian rainforests. Plant Cell Environ 23:1381–1388

Burgess SSO, Dawson TE (2008) Using branch and basal trunk sap flow measurements to estimate whole-plant water capacitance: a caution. Plant Soil 305:5–13

Campbell GS, Norman JM (1998) An introduction to environmental biophysics. Springer, New York

Caylor KK, Dragoni D (2009) Decoupling structural and environmental determinants of sap velocity: part I. Methodological development. Agric For Meteorol 149:559–569

Cermak J, Ulehla J, Kucera J et al (1992) Radial velocity profiles of water flow in stems of spruce and oak and response of spruce tree to severing. Tree Physiol 10:367–380

Chazdon RL, Letcher SG, Van Breugel M et al (2007) Rates of change in tree communities of secondary Neotropical forests following major disturbances. Phil Trans R Soc B 362:273–289

Clearwater MJ, Meinzer FC, Andrade JL et al (1999) Potential errors in measurement of nonuniform sap flow using heat dissipation probes. Tree Physiol 19:681–687

Cook BD, Davis KJ, Wang W et al (2004) Carbon exchange and venting anomalies in an upland deciduous forest in northern Wisconsin, USA. Agric For Meteorol 126:271–295

Cornish PM, Vertessy RA (2001) Forest age-induced changes in evapotranspiration and water yield in a eucalypt forest. J Hydrol 242:43–63

Cowles HS (1899) The ecological relations of vegetation on the sand dunes of Lake Michigan. Bot Gaz 27:97–117

Davis MB, Parshall TE, Ferrari JB (1994) Historical development of alternate communities in a hemlock-hardwood forest in northern Michigan, USA. In: May R, Webb N, Edwards P (eds) Large-scale ecology and conservation biology. Blackwell, Oxford

Davis MB, Sugita S, Calcote RR et al (1996) Landscape heterogeneity of hemlock-hardwood forest in northern Michigan. In: Davis MB (ed) Northeastern old-growth forest. Island Press, Washington, DC

Davis KJ, Bakwin PS, Yi C et al (2003) The annual cycles of CO2 and H2O exchange over a northern mixed forests as observed from a very tall tower. Glob Change Biol 9:1278–1293

Delzon S, Loustau D (2005) Age-related decline in stand water use: sap flow and transpiration in a pine forest chronosequence. Agric For Meteorol 129:105–119

Delzon S, Sartore M, Granier A et al (2004) Radial profiles of sap flow with increasing tree size in maritime pine. Tree Physiol 24:1285–1293

dePury DGG, Farquhar GD (1997) Simple scaling of photosynthesis from leaves to canopies without the errors of big-leaf models. Plant Cell Environ 20:537–557

Desai AR, Bolstad PV, Cook BD et al (2005) Comparing net ecosystem exchange of carbon dioxide between an old-growth and mature forest in the upper Midwest, USA. Agric For Meteorol 128:33–55

Dixon HH, Joly J (1895) On the ascent of sap. Phil Trans R Soc Lond B 186:563–576

Domec J-C, Gartner BL (2002) Age- and position-related changes in hydraulic versus mechanical dysfunction of xylem: inferring the design criteria for Douglas-fir wood structure. Tree Physiol 22:91–104

Douglass JE (1983) The potential for water yield augmentation from forest management in the Eastern United States. Water Res Bull 19:351–358

Dunford EG, Fletcher PW (1947) Effect of removal of stream-bank vegetation upon water yield. Trans Am Geophys Union 28:105–110

Dunn GM, Connor DJ (1993) An analysis of sap flow in mountain ash (Eucalyptus regnans) forests of different age. Tree Physiol 13:321–336

Dunn AL, Barford CC, Wofsy SC et al (2007) A long-term record of carbon exchange in a boreal black spruce forest: means, responses to interannual variability, and decadal trends. Glob Change Biol 13:577–590

Ewers BE, Oren R (2000) Analyses of assumptions and errors in the calculation of stomatal conductance from sap flux measurements. Tree Physiol 20:579–589

Ewers BE, Pendall E (2008) Spatial patterns in leaf area and plant functional type cover across chronosequences of sagebrush ecosystems. Plant Ecol 194:67–83

Ewers BE, Oren R, Sperry JS (2000) Influence of nutrient versus water supply on hydraulic architecture and water balance in *Pinus taeda*. Plant Cell Environ 23:1055–1066

Ewers BE, Mackay DS, Gower ST et al (2002) Tree species effects on stand transpiration in northern Wisconsin. Water Res Res. doi:10.1029/2001WR000830

Ewers BE, Gower ST, Bond-Lamberty B et al (2005) Effects of stand age and tree species composition on transpiration and canopy conductance of boreal forest stands. Plant Cell Environ 28:660–678

Ewers BE, Oren R, Kim H-S et al (2007a) Effects of hydraulic architecture and spatial variation in light on mean stomatal conductance of tree branches and crowns. Plant Cell Environ 30:483–496

Ewers BE, Mackay DS, Samanta S (2007b) Interannual consistency in canopy stomatal conductance control of leaf water potential across seven tree species. Tree Physiol 27:11–24

Ewers BE, Mackay DS, Tang J et al (2008) Intercomparison of Sugar Maple (*Acer saccharum* Marsh.) stand transpiration responses to environmental conditions from the Western Great Lakes Region of the United States. Agric For Meteorol 148:231–246

Farquhar GD, Caemmerer SV, Berry JA (1980) A biochemical-model of photosynthetic co2 assimilation in leaves of c-3 species. Planta 149:78–90

Fastie CL (1995) Causes and ecosystem consequences of multiple pathways of primary succession at Glacier Bay, Alaska. Ecology 76:1899–1916

Ferrari JB (1999) Fine-scale patterns of leaf litterfall and nitrogen cycling in an old-growth forest. Can J For Res 29(3):291–302

Fiora A, Cescatti A (2008) Vertical foliage distribution determines the radial pattern of sap flux density in *Picea abies*. Tree Physiol 28:1317–1323

Foley JA, Prentice C, Ramankutty N et al (1996) An integrated biosphere model of land surface processes, terrestrial carbon balance, and vegetation dynamics. Glob Biogeochem Cy 10:603–628

Ford CR, Vose JM (2007) *Tsuga canadensis* (L.) Carr mortality will impact hydrologic processes in southern appalachian forest ecosystems. Ecol Appl 17:1156–1167

Ford CR, Hubbard RM, Kloeppel BD et al (2007) A comparison of sap flux-based evapotranspiration estimates with catchment-scale water balance. Agric For Meteorol 145:176–185

Forrester DI, Collopy JJ, Morris JD (2009) Transpiration along an age series of *Eucalyptus globulus* plantations in southeastern Australia. For Ecol Manage 259:1754–1760

Fortin M-J, Payette S, Marineau K (1999) Spatial vegetation diversity index along a postfire successional gradient in the northern boreal forest. Ecoscience 6:204–213

Foster BL, Tilman D (2000) Dynamic and static views of succession: testing the descriptive power of the chronosequence approach. Plant Ecol 146:1–10

Franks PJ, Drake PL, Froend RH (2007) Anisohydric but isohydrodynamic: seasonally constant plant water potential gradient explained by stomatal control mechanism incorporating variable plant hydraulic conductance. Plant Cell Environ 30:19–30

Frelich LE (1995) Old forest in the lake states today and before European settlement. Nat Areas J 15:157–167

Gelman A (2004) Parameterization and Bayesian modeling. J Am Stat Assoc 99:537–545

Gillett NP, Weaver AJ, Zwiers FW et al (2004) Detecting the effect of climate change on Canadian forest fires. Geophys Res Let. doi:10.1029/2004GL020876

Goulden ML, Wofsy SC, Harden JW et al (1998) Sensitivity of boreal forest carbon balance to soil thaw. Science 279:214–217

Goulden ML, Winston GC, McMillan A et al (2006) An eddy covariance mesonet to measure the effect of forest age on land–atmosphere exchange. Glob Change Biol 12:1–17

Gower ST, McMurtrie RE, Murty D (1996) Aboveground net primary production decline with stand age: potential causes. Trends Eco Evol 11:378–382

Hall FG (1999) BOREAS in 1999: experiment and science overview. J Geophys Res Atmos 104:27627–27639

Hastings WK (1970) Monte Carlo sampling methods using Markov chains and their applications. Biometrika 57:97–109

Hatton TJ, Catchpole EA, Vertessy RA (1990) Integration of sap flow velocity to estimate plant water use. Tree Physiol 6:201–209

Holbrook NM, Burns MJ, Field CB (1995) Negative xylem pressures in plants: a test of the balancing pressure technique. Science 270:1193–1194

Hurtt GC, Dubayah R, Drake J et al (2004) Beyond potential vegetation: combining LiDAR data and a height-structured model for carbon studies. Ecol Appl 14:873–883

Irvine J, Law BE, Anthoni PM, Meinzer FC (2002) Water limitations to carbon exchange in old-growth and young ponderosa pine stands. Tree Physiol 22:189–196

Jackson RB, Jobbagy EG, Avissar R et al (2005) Trading water for carbon with biological carbon sequestration. Science 310:1944–1947

Jenny H (1941) Factors of soil formation: a system of quantitative pedology. McGraw-Hill, New York

Johnson EA, Miyanishi K (2008) Testing the assumptions of chronosequences in succession. Ecol Lett 11:419–431

Kagawa A, Sack L, Duarte K et al (2009) Hawaiian native forest conserves water relative to timber plantation: species and stand traits influence water use. Ecol Appl 19:1429–1443

Katul G, Leuning R, Oren R (2003) Relationship between plant hydraulic and biochemical properties derived from a steady-state coupled water and carbon transport model. Plant Cell Environ 26:339–350

Kearney M, Porter WP (2006) Ecologists have already started rebuilding community ecology from functional traits. Trends Ecol Evol 21:481–482

Kimball JS, White MA, Running SW (1997) BIOME-BGC simulations of stand hydrological processes for BOREAS. J Geophys Res Atmos 102(D24):29043–29051

Koch GW, Sillett SC, Jennings GM et al (2004) The limits to tree height. Nature 428:851–854

Korner C (1994) Leaf diffusive conductances in the major vegetation types of the globe. In: Schulze E-D, Caldwell MM (eds) Ecophysiology of photosynthesis. Springer, Heidelberg

Kucharik CJ, Barford CC, Maayar ME et al (2006) A multiyear evaluation of a dynamic global vegetation model at three Ameriflux forest sites: vegetation structure, phenology, soil temperature, and CO2 and H2O vapor exchange. Eco Model 196:1–31

Kume T, Tsuruta K, Komatsu H et al (2010) Effects of sample size on sap flux-based stand-scale transpiration estimates. Tree Physiol 30:129–138

Letcher SG, Chazdon RL (2009) Rapid recovery of biomass, species richness, and species composition in a forest chronosequence in northeastern Costa Rica. Biotropica 41:608–617

Litvak M, Miller S, Wofsy SC et al (2003) Effect of stand age on whole ecosystem CO_2 exchange in the Canadian boreal forest. J Geophys Res Atmos. doi:10.1029/2001JD000854

Loranty MM, Mackay DS, Ewers BE et al (2008) Environmental drivers of spatial variation in whole-tree transpiration in an aspen-dominated upland-to-wetland forest gradient. Water Res Res 44:1–15

Mackay DS, Ewers BE, Cook BD et al (2002) Effects of aggregated classifications of forest composition on estimates of evapotranspiration in a northern Wisconsin forest. Glob Change Biol 8:1253–1265

Mackay DS, Ahl DE, Ewers BE et al (2003) Physiological tradeoffs in the parameterization of a model of canopy transpiration. Adv Water Resour 26:179–194

Mackay DS, Ewers BE, Cook BD et al (2007) Environmental drivers of evapotranspiration in a shrub wetland and an upland forest in Northern Wisconsin. Water Res Res. doi:10.1029/2006WR005149

Mackay DS, Ewers BE, Loranty MM et al (2010) On the representativeness of plot size and location for scaling transpiration from trees to a stand. J Geophy Res. doi:10.1029/2009JG001092

McDowell NG, Phillips N, Lunch CK et al (2002) Hydraulic limitation and compensation in large old Douglas-fir trees. Tree Physiol 22:763–774

McGill BJ, Enquist BJ, Weiher E et al (2006) Rebuilding community ecology from functional traits. Trends Ecol Evol 21:178–185

Medlyn BE, Dreyer E, Ellsworth D et al (2002) Temperature response of parameters of a biochemically based model of photosynthesis. II. A review of experimental data. Plant Cell Environ 25:1167–1179

Metropolis N, Rosenbluth AW, Rosenbluth MN et al (1953) Equations of state calculations by fast computing machines. J Chem Phys 113:121–144

Monteith JL (1965) Evaporation and environment. Paper presented at 19th symposium of the society for experimental biology: the state and movement of water in living organisms, Cambridge University Press, Cambridge

Monteith JL, Unsworth MH (1990) Principles of environmental physics. Edward Arnold, London

Moore GW, Bond BJ, Jones JA, Phillips N, Meinzer FC (2004) Structural and compositional controls on transpiration in a 40- and 450-year-old riparian forests in western Oregon, USA. Tree Physiol 24:481–491

Novick K, Oren R, Stoy P et al (2009) The relationship between reference canopy conductance and simplified hydraulic architecture. Adv Water Res 32:809–819

Oren R, Phillips N, Katul G et al (1998) Scaling xylem sap flux and soil water balance and calculating variance: a method for partitioning water flux in forests. Ann For Sci 55:191–216

Oren R, Sperry JS, Katul GG et al (1999) Survey and synthesis of intra- and interspecific variation in stomatal sensitivity to vapour pressure deficit. Plant Cell Environ 22:1515–1526

Pastor J, Broschart M (1990) The spatial pattern of a northern conifer-hardwood landscape. Landscape Ecol 4:55–68

Phillips N, Oren R (1998) A comparison of daily representations of canopy conductance based on two conditional time-averaging methods and the dependence of daily conductance on environmental factors. Ann For Sci 55:191–216

Phillips N, Oren R, Zimmerman R (1996) Radial patterns of xylem flux in non-, diffuse-, and ring-porous tree species. Plant Cell Environ 19:983–990

Phillips N, Ryan MG, Bond BJ et al (2003) Reliance on stored water increases with tree size in three species in the Pacific Northwest. Tree Physiol 23:237–245

Phillips N, Scholz FG, Bucci SJ et al (2009) Using branch and basal trank sap flow measurements to estimate whole-plant water capacitance comment on Burgess and Dawson (2008). Plant Soil 315:315–324

Pockman WT, Sperry JS, O'Leary JW (1995) Sustained and significant negative water pressure in xylem. Nature 378:715–716

Raupach MR, Finnigan JJ (1988) Single layer models of evaporation from plant communities are incorrect, but useful, whereas multi-layer models are correct but useless: discussion. Aust J Plant Physiol 15:705–716

Running SW, Coughlin JC (1988) A general model of forest ecosystem processes for regional applications. I. Hydrological balance, canopy gas exchange and primary production processes. Eco Model 42:125–154

Running SW, Gower ST (1991) FOREST-BGC, a general model of forest ecosystem processes for regional applications. II. Dynamic carbon allocation and nitrogen budgets. Tree Physiol 9:147–160

Running SW, Hunt RE (1993) Generalization of a forest ecosystem process model for other biomes, BIOME-BGC, and an application for global-scale models. In: Ehleringer JR, Field CB (eds) Scaling physiologic processes: leaf to globe. Academic, San Diego

Ryan MG, Bond BJ, Law BE et al (2000) Transpiration and whole-tree conductance in ponderosa pine trees of different heights. Oecologia 124:553–560

Ryan MG, Phillips N, Bond BJ (2006) The hydraulic limitation hypothesis revisited. Plant Cell Environ 29:368–381

Samanta S, Mackay DS, Clayton MK et al (2007) Bayesian analysis for uncertainty estimation of a canopy transpiration model. Water Res Res. doi:04410.01029/02006WR005028

Samanta S, Clayton MK, Mackay DS et al (2008) Quantitative comparison of canopy conductance models using a Bayesian approach. Water Res Res. doi:10.1029/2007WR006761

Sellers PJ, Hall FG, Kelly D et al (1997) BOREAS in 1997: experiment overview, scientific results, and future directions. J Geophys Res 102:28731–28769

Shinozaki K, Yoda K, Hozumi K et al (1964) A quantitative analysis of plant form, the pipe model theory. II. Further evidence of the theory and its application in forest ecology. Jap J Ecol 14:133–139

Steppe K, Niinemets U, Teskey RO (2011) Tree size- and age-related changes in leaf physiology and their influence on carbon gain. In: Meinzer FC, Lachenbruch B, Dawson T (eds) Size- and age-related changes in tree structure and function. Springer Tree Physiology Series, Dordrecht

Stoy PC, Katul GG, Siqueira MBS et al (2006) Separating the effects of climate and vegetation on evapotranspiration along a successional chronosequence in the southeastern US. Glob Change Biol 12:2115–2135

Teuling AJ, Hirschi M, Ohmura A et al (2009) A regional perspective on trends in continental evaporation. Geo Res Let. doi:10.1029/2008GL036584

Thornton PE (1998) Regional ecosystem simulation: combining surface- and satellite-based observation to study linkages between terrestrial energy and mass budgets. PhD thesis, University of Montana, Missoula

Traver E, Ewers BE, Mackay DS et al (2010) Tree transpiration varies spatially in response to atmospheric but not edaphic conditions. Funct Ecol 24:273–282

Wang C, Bond-Lamberty B, Gower ST (2002) Soil surface CO_2 flux in a boreal black spruce fire chronosequence. J Geophys Res Atmos. doi:10.1029/2001JD000861

White MA, Thornton PE, Running SW et al (2000) Parameterization and sensitivity analysis of the BIOME_BGC terrestrial ecosystem model: net primary production controls. Earth Interact 4:1–85

Williams M, Rastetter EB, Fernandes DN et al (1996) Modeling the soil-plant-atmosphere continuum in a Quercus-Acer stand at Harvard Forest: the regulation of stomatal conductance by light, nitrogen and soil/plant hydraulic properties. Plant Cell Environ 19:911–927

Wright IJ, Reich PB, Westoby M et al (2004) The worldwide leaf economics spectrum. Nature 428:821–827

Wullschleger SD, Hanson PJ, Todd DE (2001) Transpiration from a multi-species deciduous forest as estimated by xylem sap flow techniques. For Ecol Manage 143:205–213

Wullschleger SD, Gunderson CA, Hanson PJ (2002) Sensitivity of stomatal and canopy conductance to elevated CO2 concentration - interacting variables and perspectives of scale. New Phytol 153:485–496

Yanai RD, Arthur MA, Siccama TG et al (2000) Challenges of measuring forest floor organic matter dynamics: repeated measures from a chronosequence. For Ecol Manage 138:273–283

Yoder BJ, Ryan MG, Waring RH et al (1997) Evidence of reduced photosynthetic rates in old trees. For Sci 40:513–527

Zimmerman R, Schulze E-D, Wirth C et al (2000) Canopy transpiration in a chronosequence of central Siberian pine forests. Glob Change Biol 6:25–37

Index

F.C. Meinzer et al. (eds.), *Size- and Age-Related Changes in Tree Structure and Function*, Tree Physiology 4, DOI 10.1007/978-94-007-1242-3,

K

L

M

N

O

P

R

S